In the frontiers of Computational Science

Lecture Series on Computer and Computational Sciences
Editor-in-Chief and Founder: Theodore E. Simos

Volume 3

In the frontiers of Computational Science

Lectures presented in the International Conference
of Computational Methods in Science and Engineering
(ICCMSE 2005)

Recognised Conference by the European Society of
Computational Methods in Sciences and Engineering
(ESCMSE)

Editors:

George Maroulis and Theodore Simos

///VSP///

ISBN: 90 6764 442 0

© Copyright 2005 by Koninklijke Brill NV, Leiden, The Netherlands.
Koninklijke Brill NV incorporates the imprints Brill Academic Publishers, Martinus Nijhoff Publishers and VSP

A C.I.P. record for this book is available from the Library of Congress

PRINTED IN THE NETHERLANDS BY RIDDERPRINT BV, RIDDERKERK
COVER DESIGN: ALEXANDER SILBERSTEIN

Brill Academic Publishers
P.O. Box 9000, 2300 PA Leiden
The Netherlands

Lecture Series on Computer
and Computational Sciences
Volume 3, 2005, pp. i-ii

In the frontiers of Computational Science

Lectures presented in the
International Conference of Computational Methods in Science and Engineering
(**ICCMSE 2005**)
Corinth, Greece
21-26 October 2005

Preface

This volume brings together lectures presented in the International Conference of Computational methods in Science and Engineering (ICCMSE 2005), held in Corinth, Greece. These are the lectures of Invited Speakers and Symposium Organizers. The collection inaugurates a tradition of excellence for the ICCMSE. Invited Speakers and Symposium Organizers are expected to present important work and the text of their talk is included in this volume for the benefit of a wider audience of specialists. We have especially in mind the informative character of these papers for young scientists, so our speakers have to present their results with the widest possible purview in mind.

The content of the papers bears upon new developments of Computational Science pertinent to Physics, Chemistry, Biology, Medicine, Mathematics and Engineering. It is important at this point to emphasize the profound unity of Natural Science brought upon by the harmony of mathematical reasoning. This bold synthesis appears early in Pythagorean philosophy. **Archytas** of Taras, in the preface of his work *On harmonics* , had this to say:

Mathematicians seem to me to have excellent discernment, and it is not at all strange that they should think correctly about the particulars that are; for inasmuch as they can discern excellently about the physics of the universe, they are also likely to have excellent perspective on the particulars that are. Indeed, they have transmitted to us a keen discernment about the velocities of the stars and their risings and settings, and about geometry, arithmetic, astronomy, and, not least of all, music. These seem to be sister sciences, for they concern themselves with the first two related forms of being [number and magnitude]. [1]

Early scientific thought paved the way and this bold vision has never been lost sight of. Thus, the growth of Computational Science represents a major element of progress for modern science. The ambition of the ICCMSE is to create a fertile ground where advancement in distinct fields is accomplished through interaction and exchange. The vehicle of this bold move forward is Computational Science.

Molecular Science is a privileged ground for the application and evaluation of new mathematical tools and computational methods. In recent years, novelty and progress with greatest conceivable speed is common experience. This flavor of research findings carrying many consequences for distant fields is easily evidenced in the lectures collected in this volume.

[1] K Freeman, *Ancilla to the Pre-Socratic Philosophers* (Oxford, 1971).

Professor **George Maroulis**
Department of Chemistry
Faculty of Science
University of Patras
GR-26500 Patras
GREECE

Professor **Theodore Simos**
Department of Computer Science and Technology
Faculty of Sciences and Technology,
University of Peloponnese,
GR-221 00 Tripolis
GREECE

Brill Academic Publishers
P.O. Box 9000, 2300 PA Leiden,
The Netherlands

Lecture Series on Computer
and Computational Sciences
Volume 3, 2005, pp. iii-iv

European Society of Computational Methods in Sciences and Engineering (ESCMSE)

Aims and Scope

The *European Society of Computational Methods in Sciences and Engineering (ESCMSE)* is a non-profit organization. The URL address is: http://www.uop.gr/escmse/

The aims and scopes of *ESCMSE* is the construction, development and analysis of computational, numerical and mathematical methods and their application in the sciences and engineering.

In order to achieve this, the *ESCMSE* pursues the following activities:

•Research cooperation between scie ntists in the above subject.
•Foundation, development and organization of national and international conferences, workshops, seminars, schools, symposiums.
•Special issues of scientific journals.
•Dissemination of the research results.
•Participation and possible representation of Greece and the European Union at the events and activities of international scientific organizations on the same or similar subject.
•Collection of reference material relative to the aims and scope of *ESCMSE.*

Based on the above activities, *ESCMSE* has already developed an international scientific journal called **Applied Numerical Analysis and Computational Mathematics (ANACM).** This is in cooperation with the international leading publisher, **Wiley-VCH.**

ANACM is the official journal of *ESCMSE.* As such, each member of *ESCMSE* will receive the volumes of **ANACM** free of charge.

Categories of Membership

European Society of Computational Methods in Sciences and Engineering (ESCMSE)

Initially the categories of membership will be:

• **Full Member (MESCMSE):** PhD graduates (or equivalent) in computational or numerical or mathematical methods with applications in sciences and engineering, or others who have contributed to the advancement of computational or numerical or mathematical methods with applications in sciences and engineering through research or education. Full Members may use the title MESCMSE.

• **Associate Member (AMESCMSE):** Educators, or others, such as distinguished amateur scientists, who have demonstrated dedication to the advancement of computational or numerical or mathematical methods with applications in sciences and engineering may be elected as Associate Members. Associate Members may use the title AMESCMSE.

• **Student Member (SMESCMSE):** Undergraduate or graduate students working towards a degree in computational or numerical or mathematical methods with applications in sciences and engineering or a related subject may be elected as Student Members as long as they remain students. The Student Members may use the title SMESCMSE

• **Corporate Member:** Any registered company, institution, association or other organization may apply to become a Corporate Member of the Society.

Remarks:

1. After three years of full membership of the European Society of Computational Methods in Sciences and Engineering, members can request promotion to Fellow of the European Society of Computational Methods in Sciences and Engineering. The election is based on international peer-review. After the election of the initial Fellows of the European Society of Computational Methods in Sciences and Engineering, another requirement for the election to the Category of Fellow will be the nomination of the applicant by at least two (2) Fellows of the European Society of Computational Methods in Sciences and Engineering.

2. All grades of members other than Students are entitled to vote in Society ballots.

3. All grades of membership other than Student Members receive the official journal of the ESCMSE Applied Numerical Analysis and Computational Mathematics (ANACM) as part of their membership. Student Members may purchase a subscription to ANACM at a reduced rate.

We invite you to become part of this exciting new international project and participate in the promotion and exchange of ideas in your field.

VSP International
Science Publishers
P.O. Box 346, 3700 AH Zist
The Netherlands

*Lecture Series on Computer
and Computational Sciences*
Volume 3, 2005, pp. v-vi

Table of Contents

Brill Academic Publishers
P.O. Box 9000, 2300 PA Leiden
The Netherlands

*Lecture Series on Computer
and Computational Sciences*
Volume 3, 2005, pp. 1-9

Recent improvements in the treatment of large molecular systems: modified parameters for hybrid Quantum Mechanics-Molecular Mechanics calculations.

Y. Moreau and X. Assfeld[1]

Equipe de Chimie et de Biochimie Théoriques, UMR UHP-CNRS 7565,
Faculté des Sciences et Techniques,
Université Henri Poincaré,
54506 Vandoeuvre-lès-Nancy Cedex, France.

Received 10 August, 2005; accepted 12 August 2005

Abstract: An adapted QM:MM potential to describe the first solvation shell (or part of it) around the QM (or QM/MM) solute with the bulk effects being treated by means of a continuum model, is defined. A well-chosen set of test systems, containing functional groups of biological interest, indicate that the QM:MM interaction is larger than the QM:QM one. Since the TIP3P potential, used in this study, was developed to reproduce condensed phase systems this overestimation was expected. It is shown that a simple modification of the existing TIP3P parameters can greatly improve the accuracy of the solute-solvent interaction. The new model, named NP, proposes to increase the σ parameter and to decrease the ε parameter of the Lennard-Jones potential of the oxygen atom of the water molecule only. The classical parameters of the solute (used to compute non-bonded interaction) are kept unchanged. In addition, the atomic point charges are scaled by a factor of 0.8302 reducing the dipole moment of a single molecule to a value close to the experimental gas phase measurement. The transferability and the additivity properties of the model are evaluated on the ethanol-water complexes. Furthermore, the NP parameters are used in the study of the water dimer and of the Claisen rearrangement.

Keywords: QM/MM; QM:MM; solvent effects; water parameters

1. Introduction

Everyday chemistry involves now more and more complex systems to achieve a better control on the products formed. The reagents are optimally designed to selectively react together, and the medium in which the reaction takes place is precisely chosen. As a consequence, the size of the molecules under study has considerably increased and the surrounding conditions are more draconian. During the XXe century, quantum chemistry was mainly devoted to reproduce experimental data on small systems. It is now one of its challenge to predict, and to explain of course, how these complex molecular systems behave.

Unfortunately, performing quantum chemical calculations on quite large systems (thousands of atoms) is still out of reach at an acceptable level of accuracy. One could wait for computer performances, either the computer speed and/or the storage capacity, to increase enough to run such calculations but, even with the optimistic estimate of Moore's law[1], one shall wait more than thirty years! Hence, it is immediately seen that one must seek for alternative approaches.

During the last ten years, our community has developed a wide range of hybrid method allying Quantum Mechanics (QM) and Molecular Mechanics (MM) to tackle this problem[2-5]. A small part of the system is treated with QM and the rest with MM. In this paper we will distinguish between two types of technique depending on the separation between the two sub-systems. If part of a molecule is treated with QM and the remaining of the very same molecule is handled with MM then we will use the so-called QM/MM acronym. However, if a complete molecule is modeled by QM and all the other molecules are described by MM, then we define the QM:MM symbol. Both techniques are based on the

---

[1] Corresponding author. E-mail: Xavier.Assfeld@uhp-nancy.fr

same philosophy recognizing that chemical phenomena are localized in space. Hence, the center of interest is a small part of the whole system and is described by means of quantum mechanical tools to render properly the bond breaking or forming process that could occurr. The left over atoms are treated by means of molecular mechanics force fields to account for constraining steric interactions and/or specific solute-solvent interactions that could hardly be taken into account by mean of continuous averaged representations.

We have recently proposed a three-layer method that combines both a hybrid QM/MM method to describe the solute and a continuum solvation model[6] to take into account solvent effects[7] and we named it LSCF/MM/SCRF for Local Self-Consistent Field/Molecular Mechanics/Self-Consistent Reaction Field. For the first time both solvent influence and steric hindrance were studied simultaneously on the asymmetric Diels-Alder reaction between cyclopentadiene and (–)-menthyl acrylate. Both effects have been previously examined separately[8, 9]. Solvent effects have been previously investigated on the model methyl acrylate molecule[8], hence lacking the encumbering (–)-menthyl group influence which was taken into account five years later[9] but without the solvent contribution. Our results[7] show that taken in conjunction, both interactions lead to conclusions that couldn't have been guessed otherwise. For example, the s-trans isomer stabilization induced by the electrostatic solvent effect occurs only when steric hindrance is negligible.

Even if the three-layer model corresponds to a nice improvement, the continuum representation of the solvent suffers from several limitations. For example, it has been shown that the explicit consideration of hydrogen bonds is crucial[10] to estimate solvent effect on rate constant. Unfortunately, the directionality of specific solute-solvent interactions like hydrogen bonds can not be represented by means of continuum model. Moreover, the dynamical aspect of the interaction is completely neglected. It would then be desirable to incorporate explicit solvent molecules in the modeling of solvated large chemical systems, as has already been discussed earlier[11, 12].

If the solvent molecule participate actively to the reaction coordinate then it needs to be described by QM. However, if it only has a solvation role, then MM can be used to save computer time. In this paper our aim is to use as few as possible molecules of solvent, described with MM, in the vicinity of the reactive center (treated at the QM level) so that they can influence the reactivity of the system. We are thus not directly concerned by the interaction between the MM part of the solute and the classical molecules of solvent. Since water is the most widely used solvent, and almost the only one in biology, we will focus our attention on this specific molecule. We will show how well hydrogen bond interactions are reproduced by QM:MM methods for selected model solute molecules. Since most MM force fields for H_2O are developed to reproduce liquid water properties they are not adapted for calculations on small solute-solvent dimers, and we expect that some modifications will be needed[13-15]. In the first section the comparison between QM, MM, and QM:MM hydrogen bonded 1:1 complexes is presented. The modifications of the QM:MM interaction is explained in the second section and are applied on illustrative systems in the third section.

2. Comparison between QM, MM, and QM:MM.

In order to test the QM:MM interaction between a classical water molecule and a quantal solute molecule, a set of model molecules is constructed in such a way that the most common functional groups that can undergo hydrogen bonding are represented. Moreover, since macromolecules are often biomolecules, the functional groups are those present in most of the side chain of amino acids. The model molecules are monosubstituted ethane molecules with the general formula C_2H_5G where G is taken form the following list: COOH, COO^-, NH_2, NH_3^+, OH, O^-, SH, and S^-. From these eight molecules, ten hydrogen bonded complexes are formed with one water molecule, since the OH and the SH groups can be hydrogen atom acceptor or donor. The QM level of theory is MP2/6-311G(2d,2p). The MP2 method is required to account for the electronic correlation that is quite important for intermolecular interactions. The basis set is of triple-ζ quality with a double set of polarization functions. Although some anions are present in the set, we decide to use no diffuse function to avoid the "overpolarization" effect[16-18] but instead to add a second set of polarization functions. The Amber[19] force field is chosen for the MM calculations. The atomic point charges are determined with the RESP[20] algorithm based on MP2/6-31G* results. The modified[21] TIP3P[22] potential was used for the water molecule for its simplicity and to be coherent with previous studies[11, 13]. For the QM:MM calculations, one has to note that the QM part is of course polarized by the MM classical atomic charges and that the Amber parameters are used for the van-der-Waals QM:MM interactions. The complexation energies (corrected from BSSE for the QM calculations) are gathered in Table 1.

Table 1. Complexation energies (in kcal/mol) for the ten hydrogen bonded complexes treated either at the MP2/6-311G(2d,2p) (QM) level, or with the Amber and TIP3P force fields (MM), or at the combined MP2/6-311G(2d,2p):{Amber+TIP3P} (QM/MM) level. The complexes corresponding to the general formula $C_2H_5G...H_2O$ are referenced with the details of the G group. For OH and SH complexes, d means that the solute molecule is hydrogen donor and a that it is an acceptor. Differences with respect to the QM results are collected in the Δ columns.

G	QM	MM	Δ_{MM}	QM:MM	$\Delta_{QM:MM}$
COOH	−8.44	−10.53	2.09	−13.80	5.36
COO⁻	−16.86	−22.41	5.55	−22.51	5.65
NH₂	−5.71	−6.12	0.41	−9.15	3.44
NH₃⁺	−17.42	−18.29	0.87	−20.13	2.71
OH (d)	−4.09	−6.19	2.10	−8.65	4.56
OH (a)	−5.54	−5.80	0.26	−7.30	1.76
O⁻	−18.78	−20.05	1.27	−20.23	1.45
SH (d)	−1.65	−2.99	1.34	−5.77	4.12
SH (a)	−3.72	−3.46	−0.26	−5.76	2.04
S⁻	−12.05	−18.33	6.28	−17.78	5.73

One can readily see that the MM results slightly overestimated the interaction energy compared to the QM calculations. This is quite understandable since the TIP3P potential and the RESP atomic charges determined at the MP2/6-31G* level are designed to model condensed phase systems where the polarization is greater than in gas phase complexes. Nevertheless, the difference between the two methods is less than 2 kcal for the neutral and the cationic systems and one can conclude that the agreement is fair. For the anionic complexes however the difference is larger, up to 6.28 kcal/mol. This could certainly be attributed to the lack of diffuse functions in the QM computations. Looking at the QM:MM results, it is evident that the complexation energy is largely overestimated compared either to the QM reference or to the MM values. Although not shown here, the hydrogen bond distances are also too small at the QM:MM level compared to the QM one. Several effects could explain this behavior. Among other we can cite that the MM molecule is not polarizable and, as already noted, that the TIP3P potential give results compatible with condensed phase and not to gas phase. It is clear that the MM parameters used here are not adapted for QM:MM calculations. However, if one plots the QM:MM values versus the QM ones, then one can find the regression line equation $E_{QM:MM} = 0.98\ E_{QM} - 3.70$ (in kcal/mol)—the regression coefficient is 0.99. The slope of 0.98 shows that the error is quite constant and close to 3.70 kcal/mol. Hence, we could expect that with slight modifications the QM:MM results could be improved significantly.

3. Modifications of the QM:MM interaction.

The QM:MM interaction is composed of two terms corresponding to the non-bonded interactions in classical molecular mechanics. The first term characterizes the van der Waals interaction and is generally treated by means of a Lennard-Jones potential. Hence, the parameters used depend both on the solvent's atom type and on the solute's atom type. The second term corresponds to the electrostatic interaction between the atomic point charges of the MM atoms and the nuclear charges and the electrons of the QM solute. In our approach, this interaction is explicitly taken into account in the Fockian allowing the polarization of the electronic wave function. One can see that the only parameters one can play with in the electrostatic contribution are those of the MM part.

Some other groups have already proposed modifications of the QM:MM interaction. Gao and coworkers[13] have chosen to modify the Lennard-Jones parameters of the QM atoms and to keep the standard TIP3P unaltered. On the other hand, Tu and Laaksonen[14] have modified the solvent parameters. They consider first the modification of the van der Waals interaction, and in a second time the adjustment of the electrostatic interaction. Finally, Ruiz and coworkers[15] have shown that a better accuracy can be reached by using a Buckimgham type repulsion potential or a pseudo-potential on the classical atom.

Since our aim is to define an adapted water model to be used in our multi-layer approach, we decide to adjust the water parameters only. In that way, the model won't be solute dependent. In addition, only the QM:MM interaction will be modified, since the MM:MM interaction seems to be correctly estimated. The TIP3P model contains only three parameters, the Two Lennard-Jones parameters of the oxygen atom and its point charge (the hydrogen point charges deriving from it by a multiplication by −1/2), that we decide to vary. The new set of parameters is obtained from a simplex procedure[23]. The set of data against which the parameters are optimized are the 6 neutral complexes, each at three different geometries. The first geometry is the QM equilibrium geometry, and the second (third) one corresponds to the QM optimized geometry with the H-bond stretched (squeezed) by 0.1 Å from the equilibrium distance. In order to fully take advantage of the optimization procedure, the QM calculations are performed at a higher level of theory than in the previous section, MP2/6-311++G(2d,2p). However, the QM:MM computation uses the MP2/6-31G* level only to save computer time without loosing accuracy thanks to the optimization of the parameters. The score function S to minimize is:

$$S(\sigma,\varepsilon,Q)=\sum_{i=1}^{18}\left(E_i^{QM:MM}-E_i^{QM}\right)^2 \tag{1}$$

where σ and ε are the Lennard-Jones parameters and Q the scale factor for the atomic charges. E_i correspond to the complexation energy. The optimization process stops when the following three conditions are fulfilled:

$$\sum_{i=1}^{4}\left|\sigma_i-\frac{1}{4}\sum_{j=1}^{4}\sigma_j\right|<\delta_\sigma \tag{2}$$

$$\sum_{i=1}^{4}\left|\varepsilon_i-\frac{1}{4}\sum_{j=1}^{4}\varepsilon_j\right|<\delta_\varepsilon \tag{3}$$

$$\sum_{i=1}^{4}\left|Q_i-\frac{1}{4}\sum_{j=1}^{4}Q_j\right|<\delta_Q \tag{4}$$

The three threshold δ are set equal to 10^{-4} in the respective units.

Table 2. Lennard-Jones parameters (σ in Å, ε in kcal/mol) of the oxygen atom of the water molecule and the charge scaling factor (Q) for the TIP3P potential, for the adapted version of Tu and Laaksonen (TL), and from this work (NP).

	TIP3P	TL	NP
σ	1.7683	2.17	2.0627
ε	0.1520	0.041	0.0245
Q	1.0	0.84	0.8302

The final parameters are given in Table 2, together with the original TIP3P[22] parameters and the adapted parameters of Tu and Laaksonen[14] (hereafter noted TL). It is noteworthy that our parameters (hereafter denoted NP[24]) are very close those obtained by Tu and Laaksonen, although obtained from a very different way. Tu and Laaksonen use the water dimer to determine their parameters, and vary either the Lennard-Jones parameters or the point charges but not both simultaneously. The variation from the TIP3P parameters are easily understood. In the previous section we show that the QM:MM interaction was too strong, leading to overrated interaction energy and to shortened hydrogen bond distance. The decrease of ε means that the well depth will be smaller and accordingly that the interaction energy will be reduced. The increase of σ will enlarge the hydrogen bond length. In addition the scaling factor of 0.8302 gives a dipole moment to the classical water molecule of 1.95 Debye, very close to the dipole moment of an isolated molecule. Hence, the NP parameters seems to perfectly correspond to what we were seeking, and they are tested against various systems in the next section.

4. Application of the new parameter set.

1:1 complexes.

The first, for obvious reasons, system on which the new parameters are tested is of course the ten hydrogen bonded complexes studied above. The geometry of the complexes is fully optimized. The results are given in Table 3. Concerning the six neutral systems that were used to determine the parameters, one can see that the maximum error is close to one kcal/mol only. The average error is 0.48 kcal/mol for these 6 complexes. This clearly shows that our optimization procedure—using three points to describe the shape of the hypersurface of potential—is robust enough since all interaction energies are reproduced with an acceptable accuracy. It is also noteworthy that the error is either positive or negative showing that no systematic discrepancy remains. When considering the four charged complexes, one can remark that, except for O^- for which the difference is larger than 7 kcal/mol, the error is largely reduced although it remains close to two kcal/mol. This indicates that for charged system it is necessary to improve further the description of the QM:MM interaction. This could be done by either using a polarizable force field or by treating the water molecule with quantum mechanics.

Table 3. Complexation energies (in kcal/mol) for the ten hydrogen bonded complexes treated either at the MP2/6-311++G(2d,2p) (QM) level, or at the combined MP2/6-31G*:{Amber+NP} (QM/MM) level. The complexes corresponding to the general formula $C_2H_5G...H_2O$ are referenced with the details of the G group. For OH and SH complexes, d means that the solute molecule is hydrogen donor and a that it is an acceptor. Differences with respect to the QM results are collected in the Δ columns.

G	QM	QM:MM	$\Delta_{QM:MM}$
COOH	−8.77	−8.56	−0.21
COO⁻	−18.07	−15.98	−2.08
NH₂	−6.94	−6.69	−0.25
NH₃⁺	−16.65	−14.17	−2.48
OH (d)	−4.32	−5.46	1.14
OH (a)	−5.32	−4.86	−0.46
O⁻	−21.83	−14.70	−7.13
SH (d)	−2.02	−2.74	0.72
SH (a)	−4.16	−4.26	0.10
S⁻	−14.08	−14.29	0.21

Transferability and cooperative effects.

The second property one can look at is the transferability of the new parameters when different levels of theory are used. Three methods are considered: Hartree-Fock, the B3LYP[25] functional, and MP2[26]. These three methods are used in conjunction to four basis set: 3-21G, 6-31G*, 6-311G**, and 6-311++G(2d,2p). The resulting twelve levels of theory are applied on the complex formed by the ethanol molecule and one or two water molecules. In addition to the transferability, the cooperative effect will be tested. The standard TIP3P potential is used between the two water molecules. The interaction energies are gathered in Table 4 and 5. The reference values are obtained at the MP2/6-311++G(2d,2p) (corrected from BSSE) and are equal to −5.32 and −14.42 kcal/mol for the complexes with one and two water molecules respectively. For the complex with two water molecules the interaction energy is defined as the difference between the complex energy and the energy of the ethanol molecule and two times the energy of a water molecule.

Table 4. Interaction energies (in kcal/mol) of the complex formed by the ethanol molecule and one water molecule for various levels of theory. The full QM reference is −5.32 kcal/mol.

	NP			TIP3P		
	RHF	B3LYP	MP2	RHF	B3LYP	MP2
3-21G	−5.56	−4.65	−4.84	−8.36	−7.08	−7.34
6-31G*	−5.07	−4.54	−4.86	−7.62	−6.92	−7.35
6-311G**	−4.96	−4.52	−4.53	−7.44	−6.84	−6.85
6-311++G(2d,2p)	−4.99	−4.69	−4.79	−7.45	−7.00	−7.13

Table 5. Interaction energies (in kcal/mol) of the complex formed by the ethanol molecule and two water molecules for various levels of theory. The full QM reference is −14.42 kcal/mol.

	NP			TIP3P		
	RHF	B3LYP	MP2	RHF	B3LYP	MP2
3-21G	−14.24	−13.37	−13.60	−19.79	−18.64	−18.95
6-31G*	−14.03	−13.44	−14.00	−19.57	−18.80	−19.62
6-311G**	−13.84	−13.45	−13.48	−19.30	−18.83	−18.86
6-311++G(2d,2p)	−14.18	−13.99	−14.14	−19.86	−19.65	−19.87

From the data it is clear that both the TIP3P or the NP parameters present the same type of transferability property. This means that either TIP3P or NP could be used with any level of theory and give about the same results. However, it is also easily seen that the results obtained with the NP parameters are closer to the full QM calculation than the ones obtained with TIP3P. The largest difference between the full QM reference and the QM:NP results is 0.80 kcal/mol (to be compared to 3.04 for QM:TIP3P) for the complex with one water molecule. For the complex with two water molecules the error is 1.05 kcal/mol for the QM:NP calculation and 5.45 kcal/mol for the QM:TIP3P method. Hence, it is clear that nonetheless the NP parameters give accurate interaction energy, but they are also suitable for any level of theory, with the same accuracy. Furthermore the accuracy of the model is conserved when several molecules are used.

Water dimer.

The third example we would like to present here is the water dimer. We choose this system because it is the one on which Tu et al.[14a] fit their parameters, and a comparison could be made. The level of theory is the same as the one they choose, HF/6-311G**. In the water dimer, one molecule is an hydrogen atom donor and the other one is an acceptor. Hence, two complexes can be envisaged depending on which molecule is treated by quantum mechanics. Values are reported in Table 6. One can see that our model gives slightly smaller values for the interaction energy than the TL one (approximately 0.2 kcal/mol). The geometrical parameters are also is relatively good agreement (the largest discrepancy being 0.14 Å for the OO distance). It is then remarkable that our model performs almost as well as the TL model which was specifically devoted to the water dimer (and is therefore certainly the most accurate one for this system).

Table 6. Interaction energies (in kcal/mol), oxygen–oxygen distance (in Å), and the O…H–O angle (in °) of the water dimer complex for both the Tu and Laaksonen[14] (TL) and the NP parameters, at the HF/6-311G** level of theory. In the QM:MM (MM:QM) calculations, the hydrogen atom donor(acceptor) is treated with QM.

	E_{int}(kcal/mol)		d_{OO} (Å)		a_{OHO} (°)	
	TL	NP	TL	NP	TL	NP
QM:MM	−5.32	−5.19	2.93	2.79	177.3	178.5
MM:QM	−5.13	−4.94	2.95	2.82	175.6	177.0

Claisen rearrangement.

Finally, the last system on which our parameters are tested against is a chemical reaction, since it is the final goal of our work. The aliphatic Claisen rearrangement is a [3,3]-sigmatropic rearrangement in which an allyl vinyl ether is converted thermally to an unsaturated carbonyl compound. Many theoretical studies have been performed and solvent effects are well known[27-30]. We focus our attention on two major features. The first one is the differential solvation of the carbonyl with respect to ether, leading to an enhancement of the exothermicity of the reaction and to a decrease of the barrier height. It has been shown that hydrogen bonds[27, 28] are responsible for the better solvation of the carbonyl compound. The action of this effect is along the reaction coordinate. The second point is the displacement along a direction more or less orthogonal to the reaction coordinate. This direction correspond to the separation of the system into two charged species (see Figure 1).

Table 7. Reaction energies (ΔE in kcal/mol), activation energies (E_a in kcal/mol), distances of the two forming bonds (d in Å), sum of the Mulliken atomic charges of the fragment not containing the oxygen atom (δq in e) for the Claisen rearrangement (see figure 1) of allyl vinyl ether, treated as an isolated molecule (Isolated), or in a dielectric continuum (Continuum), or with a classical water molecule H bonded (H bonded complex), or as the H bonded cluster solvated by the continuum (Solvated complex).

	ΔE	E_a	$d\,C...O$	$d\,C...C$	δq
Isolated molecule	−17.86	27.69	1.914	2.322	0.215
Continuum	−18.40	27.26	1.921	2.333	0.227
H bonded complex	−18.79	25.93	1.935	2.359	0.248
Solvated complex	−18.99	25.58	1.937	2.361	0.254

To test these two effects the complex between the allyl vinyl ether and a water molecule is studied. The water is described by means of MM with the NP parameters, whereas the solute is treated with QM. In addition, the long range bulk effects are taken into account by means of a continuum model[6]. The B3LYP/6-31G** level of theory was used as in reference 30. The Amber[19] force field parameters were chosen for the non bonded QM:MM interaction. The relative dielectric constant of water (78.4) is used for the polarizable continuum and a molecular shape cavity is used. The multipolar expansion is performed until the sixth order. Only the "chair" conformation transition state is considered since it is the most stable[31].

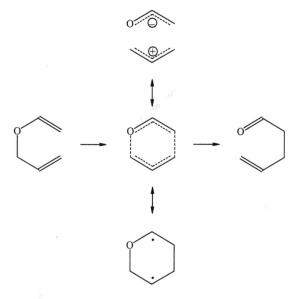

Figure 1. Claisen rearrangement of allyl vinyl ether. Schematic representation of limit mesomer forms for the transition state.

The energies, the relevant geometrical parameters and the intramolecular charge transfer (see Figure 1) are collected in Table 7. One can see that the both the reaction energy and the activation barrier are lowered by the presence of the solvent treated either implicitly as a continuum or explicitly as a discrete molecule. When used together, the two solvent models don't give simple additive contributions. One needs to note however that since we use only one explicit water molecule and because the cavity size is slightly overestimated[32], the stabilization induced by the solvent is underestimated in our calculations[28]. However, all trends are correctly reproduced. From the energy values one can also remark that the transition state is more solvated than the product, *i.e.* the carbonyl compound, since the barrier height decreases by 2.11 kcal/mol and the reaction energy by 1.13 only. This feature shows that a significant displacement along the coordinate orthogonal to the reaction coordinate is present in the solvated transition state. This displacement can be viewed from the two forming (C–C) and breaking (C–O) bonds which both increase under the action of the solvent. Again, although both solvent model influence the reaction in the same direction, their contributions are not simply additive. Our model correctly predict the higher polarity of the transition state as illustrated by the sum of the Mulliken atomic charges (δq) of the CH_2–CH–CH_2 fragment (Figure 1).

5. Conclusion.

The aim of this paper was to define an adapted QM:MM potential to describe the first solvation shell (or part of it) around the QM (or QM/MM) solute, the bulk effects being treated by means of a continuum model. Series of test systems, containing functional groups of biological interest, indicate that the QM:MM interaction is larger than the QM:QM one. Since the TIP3P potential, used in this study, was developed to reproduce condensed phase systems it was expected that when used in small QM:MM cluster, the interaction energy would be overestimated. In this paper we have shown that a simple modification of the existing TIP3P parameters can greatly improve the accuracy of the solute-solvent interaction. The new model, named NP, proposes to increase the σ parameter and to decrease the ε parameter of the Lennard-Jones potential of water only. The classical parameters of the solute (used to compute non-bonded interaction) are kept unchanged. In addition, the atomic point charges are scaled by a factor of 0.8302 reducing the dipole moment of a single molecule close to the experimental gas phase value. The parametrization procedure was tested against the geometry optimization of the molecules forming the data set, and was shown to be robust enough. The transferability and the additivity of the model are evaluated on the ethanol-water complexes. Nonetheless the NP potential possess a slightly better transferability property than the TIP3P model when used with various levels of theory, but it also correctly reproduces the cooperative effect of water when several water molecules are present. Furthermore, the NP parameters are used in the study of the water dimer and compared to the TL model developed by Tu and Laaksonen. Although the TL potential was specifically created for the description of the water dimer, both models gives very similar results. It is noteworthy that the two parametrization procedure, TL and NP, give very close parameters, albeit obtained in very different ways. Finally solvent effects on the Claisen rearrangement are studied. Our model correctly predict the effect of water along the reaction coordinate but also along an orthogonal coordinate. This proves that both energetic and geometric aspects, or equivalently, both the electronic and the nuclear polarization are properly handled by the hybrid QM:MM/continuum model. Although the accuracy is improved for neutral system it is clear that this type of model is not adequate in case of very strong interaction with anionic solute. In that case, using explicit water molecules described with quantum mechanics is recommended. In forthcoming paper the NP parameter will be used in our QM/MM:MM/Continuum model.

References

[1] In 1965, Intel co-founder Gordon Moore states that the number of transistors on a chip doubles about every two years. Hence, the computer performance doubles accordingly.
[2] X. Assfeld, and J.-L. Rivail, *Chem. Phys. Lett.* **263** 100 (1996).
[3] N. Ferré, X. Assfeld, J.-L. Rivail, *J. Comput. Chem.* **23** 610 (2002).
[4] N. Ferré, and X. Assfeld, *J. Chem. Phys.* **117** 4119 (2002).
[5] N. Ferré, and X. Assfeld, *J. Mol. Struct.* (THEOCHEM) **632** 83 (2003).
[6] (a) D. Rinaldi, A. Bouchy, J.-L. Rivail, and V. Dillet, *J. Chem. Phys.* **120** 2343- (2004). (b) D. Rinaldi, and R. R. Pappalardo, SCRFPAC; Quantum Chemistry Program Exchange, Indiana Univeristy: Bloomington, IN, 1992; program no. 622.
[7] Y. Moreau, P.-F. Loos, and X. Assfeld, *Theor. Chem. Acc.* **112** 228 (2004).
[8] M. F. Ruiz-Lopez, X. Assfeld, J. I. Garcia, J. A. Mayoral, and L. Salvatella, *J. Am. Chem. Soc.* **115** 8780 (1993).
[9] L. Salvatella, A. Mokrane, A. Cartier, and M. F. Ruiz-López, *J. Org. Chem.* **63** 4664 (1998).
[10] J. Chandrasekhar, S. sharillskul, and W. L. Jorgensen, *J. Phys. Chem. B* **106** 8078 (2002).
[11] Q. Cui, *J. Chem. Phys.* **117** 4720 (2002).
[12] S. Chalmet, D. Rinaldi, and M. F. Ruiz-López, *Int. J. Quant. Chem.* **84** 559 (2001).
[13] (a) J. Gao, *Acc. Chem. Res.* **29** 227 (1996). (b) M. Freindhorf, and J. Gao, *J. Comp. Chem.* **17** 386 (1995).
[14] Y. Tu, and A. Laaksonen, *J. Chem. Phys.* **111** 7519 (1999). (b) Y. Tu, and A. Laaksonen, *J. Chem. Phys.* **113** 11264 (2000).
[15] S. Chalmet, and M. F. Ruiz-López, *Chem. Phys. Lett.* **329** 154 (2000).
[16] P. Surján, and J. Ángyán, *Chem. Phys. Lett.* **225** 258 (1994).
[17] D. Wei, and D. R. Salahub, *Chem. Phys. Lett.* **224** 291 (1994).
[18] M. Ben-Nun, and T. J. Martínez, *Chem. Phys. Lett.* **290** 289 (1998).
[19] W. Cornell, P. Cieplak, C. Bayly, I. Gould, K. Merz Jr., D. Fergusson, D. Spellmeyer, T. Fox, J. Cladwell, and P. Kollman, *J. Am. Chem. Soc.* **117** 5179 (1995).

[20] C. Bayly, P. Cieplak, W. Cornell, and P. A. Kollman, *J. Phys. Chem.* **97** 10269 (1993).

[21] M. J. Field, P. A. Bash, and M. Karplus, *J. Comp. Chem.* **11** 700 (1999).

[22] W. L. Jorgensen, J. Chandreskhar, J. D. Madura, R. W. Impey, and M. L. Klein, *J. Chem. Phys.* **79** 926 (1982).

[23] Numerical Recipes in FORTRAN: The Art of Scientific Computing, 2nd ed. Cambridge, England: Cambridge University Press, pp. 402-406 and 423-436, 1992.

[24] NP stands for "Nos Paramètres", the French translation of "our parameters".

[25] (a) Becke, A. D. *J. Chem. Phys.* **98** 5648 (1993). (b) Lee, C.; Yang, W.; Parr, R. G. *Phys. Rev. B* **37** 785 (1988).

[26] C. Møller et M. S. Plesset, *Phys. Rev.* **46** 618 (1934).

[27] J. J. Gajewski, *Acc. Chem. Res.* **30** 219 (1997).

[28] (a) M. M. Davidson, I. H. Hillier, R. J. Hall, and N. A. Burton, *J. Am. Chem. Soc.* **116** 9294 (1994). (b) R. J. Hall, M. M. Davidson, N. A. Burton, and I. H. Hillier, *J. Phys. Chem.* **99** 921 (1995).

[29] D. L. Severance, and W. L. Jorgensen, *J. Am. Chem. Soc.* **114** 10966 (1992).

[30] H. Hu, M. N. Kobrak, C. Xu, and S. Hammes-Schiffer, *J. Phys. Chem. A* **104** 8058 (2000).

[31] R. L. Vance, N. G. Rondan, K. N. Houk, F. Jensen, W. Thatcher Borden, A. Komornicki, and E. Wimmer, *J. Am. Chem. Soc.* **110** 2314 (1988).

[32] To ensure a rapid convergence of the multipolar expansion a scaling factor of two is applied on the van der Waals radii.

Brill Academic Publishers
P.O. Box 9000, 2300 PA Leiden
The Netherlands

Lecture Series on Computer
and Computational Sciences
Volume 3, 2005, pp. 10-17

Relativistic Effects in the Chemistry of very Heavy and super heavy Molecules

K. Balasubramanian[1]

Chemistry & Material Science Directorate
Lawrence Livermore National Laboratory, PO Box 808, L-268
Livermore, CA 94550
And
California State University, East bay,
Hayward CA 94542

Received 15 July, 2005; accepted 12 August 2005

Abstract: Relativistic effects are very significant for molecules and clusters containing very heavy and super heavy elements. We demonstrate this further with our recent results of relativistic computations that included complete active space multi-configuration interaction (CAS-MCSCF) followed by multi-reference configuration interaction (MRSDCI) computations with up to 50 million configurations of transition metal and main group clusters. We shall also be presenting our recent works on substituted fullerenes and actinide complexes of environmental concern. My talk will emphasize these unusual features and trends concerning structure and spectroscopic properties of these very heavy species. We compare the properties of not only the ground electronic states, but also several excited electronic states. We have also carried out extensive computations on very heavy clusters such as gold clusters; ruthenium clusters and assignment of the observed spectra have been suggested. It is shown that the gold clusters exhibit anomalous trends compared to copper or silver clusters. For example, Jahn-Teller distortion is quenched in the case of Au_3 by spin-orbit coupling, and for the first time the spin-orbit component of the Au_3 ground state has been observed experimentally. We have also carried out relativistic computations for the electronic states of the newly discovered super heavy elements and yet to be discovered elements such as 113 (eka-thallium) 114 (eka-lead) and 114$^+$. Many unusual periodic trends in the energy separations of the electronic states of the elements 114, 113, 114$^+$, (113)H, 114H, and Lawrencium and Nobelium compounds. We will be presenting our results on uranyl, plutonyl and meputunyl complexes in aqeous solution using a combined quantum chemical and PCM models for solvation. We have employed coupled cluster levels of theory to obtain the frequencies and equilibrium geometries of these complexes.

Keywords: Relativistic Effects, Very heavy Clusters, super heavy clusters

This is the year of centennial celebration of publication of Einstein's famous 1905 paper on special theory of relativity. Indeed it is an exciting and very fitting that the celebrated work of Einstein penetrated many branches of science and changed our thinking in many ways. In atomic theory and more recently in chemical context, it is now well recognized that Einstein's special theory of relativity becomes important particularly in the properties of very heavy and superheavy elements and their molecules, as inner electrons in these species travel with speeds comparable to the speed of light. This is a consequence of a large nuclear charge, which is the driving force of the inner electrons, which speed up in order to keep balance with the increased electrostatic attraction. For example, the speed of the 1s electron of gold, which has 79 protons is estimated to be ~60% of the speed of light. Thus to treat the properties of such species one needs to embrace both relativity and quantum mechanics, the latter was viewed with skepticism by Einstein. Nevertheless the combination of two theories provides the founding blocks of modern relativistic quantum mechanics, which is most relevant in dealing with the properties of molecules containing very heavy and superheavy elements.

Relativistic effects are defined as differences in the observable properties of electrons as consequence of the correct speed of light, as opposed to infinite speed. Many relativistic quantum

---

[1]. E-mail: balu@llnl.gov

approaches start with the relativistic analog of the Schrödinger equation well known as the Dirac equation. Then electron correlation effects are introduced in some combination with relativistic techniques since relativistic and electron correlation effects can be coupled in such species with very heavy elements. Although relativistic effects are more important for the core electrons, the valence electrons too experience such relativistic effects as to cause substantial differences in the chemical and spectroscopic properties of atoms and molecules containing heavy atoms such as third row transition metal species, sixth-row main group elements, actinides and superheavy elements.

Relativistic effects can cause substantial changes to te properties of molecules containing very heavy atoms[1,2]. For example, the contraction of the 6s valence orbital of the gold atom leads to shorter bonds in gold clusters and gold compounds the ionization potentials of the copper, silver and gold atoms. As one goes down a given group in the periodic table, one expects a monotonic decrease in the ionization energy, since the outermost electron is farther from the nucleus. However, gold is an anomaly in that its ionization potential is not only higher than silver but also copper. This anomaly of the gold atom is due to the relativistic stabilization of the outermost 6s electron of the atom, thus, making it strongly bound, leading to higher ionization energy. Yellow color of gold is attributed to relativity.

The inertness of the $6p^2$ shell of the lead atom and even more so the 7p2 shell of element 114 are both due to both mass-velocity and spin-orbit relativistic effects. The relativistic stabilization of the 6s orbital of mercury leads to a larger 6s-6p promotion energy in Hg compared to Cd and Zn. The increase for mercury is a consequence of relativistic stabilization of the $6s^2$ shell, which leads to a larger ^1S-^1P separation for Hg compared to Cd. The $6s^2$ shell cannot, thus, form a very strong bond, unless promotion into 6p is achieved. This is the reason for the fact that Hg_2, in its ground state, forms only a van der Waals complex, while in its excited state, it is bound. Mercury is thus a liquid at room temperature due to the formation of weak clusters of mercury atoms, which undergo metal-nonmetal transition by hybridization with 6p as the cluster size increases.

Spin-orbit coupling is another relativistic effect, which can alter the reactivity and spectroscopic properties of very heavy and superheavy species. The spin-orbit coupling can increase or decrease the bond lengths depending on, which states mix. Likewise it can destabilize the chemical bonding as in the case of Bi_2, which forms a less stable bonding than a triple bond. On the other hand, the PtH molecule is considerably more stable compared to PdH. The ground state of the platinum atom is a triplet arising from $5d^9 6s^1$, while it is singlet for Pd arising from the $4d^{10}$ configuration. This is primarily due to the relativistic stabilization of the 6s orbital of Pt, which overcomes the enhanced stability attributed to the closed-shell d^{10} configuration. Consequently, the differences in the chemistry of Pt and Pd containing systems arise from this and the larger spin-orbit splitting in Pt compared to Pd. The entire third row transition metal atoms react more compared to the second row atoms due to relativity. We will demonstrate analogy between mercury and element 114.

In mathematical terms relativity results in a "four-vector" formalism of angular momentum that couples both the spin and spatial angular momenta via spin-orbit coupling and thus a relativistic electron has memory of only total angular momentum symmetry. This results in novel double group symmetry due to the spin-orbit coupling **L.S** operator. Since this operator changes sign upon rotation by 360°, the periodicity of the identity operation is broken and is thus no longer the identity operation of the group, as exemplified in Figure 1 with a Mobius strip, as one completes a 360° rotation along the Mobius surface there is a sign change since one goes from inside of the surface to the outside. This requires the introduction of a new operation R in the normal point group of a molecule that corresponds to the rotation by 360° which is not equal to E, the identity operation. Hence we have the double group and double-valued representations in relativistic quantum chemistry of molecules with heavy atoms.

The author[3] has recently formulated the symmetry double groups of non-rigid molecules recently. The double group character tables of such species have been derived using wreath product groups and their double groups. These character tables grow astronomically in sizes due to large number of operations in the double group. It may also be noted that the double group is not a direct product of the normal single group and the two-valued operational groups. It is this aspect that introduces even-dimensional two-valued representations into the double group. Thus the relativistic spinor representations of molecules containing heavy atoms do not conform to normal spin and spatial symmetries. Two electronic states that have the same symmetry or transform as the same double group representation can mix regardless of their spin multiplicities and spatial symmetries.

Figure 1. A Mobius strip depiction for the double group symmetry of a relativistic hamiltonian. The introduction of spin-orbit coupling into the relativistic Hamiltonian changes the periodicity of the normal point group symmetry into a double group symmetry, as rotation by 360 deg is not the identity operation.

We provide here two different set of species to demonstrate the importance of relativity in the chemistry and bonding of very heavy and superheavy species. The first set of species is from the compounds of late actinides namely, compounds of Nobelium and Lawrencium[4]. The compounds of Nobelium and lawrencium surprisingly exhibit unusual non-actinide properties in that the chemistry of these species is principally determined by the 7s and 7p orbitals rather than the 5f or 6d shells. Since hydrides are the simplest of all species, we have considered high-level relativistic computations for the lawrencium and nobelium dihydrides. The ground and first excited states of lawrencium and nobelium arise from the 7s and 7p shells, and thus the potential energy surfaces of these species are unusual in having 7p characteristics. Both molecules form stable bent ground states reminiscent of a sp^2 hybridization with equilibrium bond angles near 120°. The lawrencium compounds exhibit unusual characteristics due to avoided crossings of the potential energy surfaces. As a result of spin-orbit coupling, the $^2B_2$ state of LrH_2 undergoes avoided crossing with the $^2A_1$ state in the spin double group, which reduces the barrier for insertion of Lr into H_2. The Nobelium compounds are considerably less stable compared to the lawrencium compounds due to the relativistic stabilization of the 7s shell of the nobelium atom. The barrier for the insertion of Lr into H_2 is lowered by relativity (spin-orbit coupling) while No has to surpass a larger barrier due to the relativistic stabilization of the $7s^2$ shell. Lawrencium is the only element in the actinide series with unusually low ionization potential, and NoH_2 has an unusually large dipole moment of 5.9 Debye. We have found that the lawrencium and nobelium compounds have periodic similarities to the thallium and radium compounds, respectively.

The ground state of Lr is not $7s^2 6d$ as listed in many websites, for example, the Los Alamos periodic table, but $7s^2 7p^1$. One could have expected Lr to be similar to Ac or La but in fact the change of this configuration makes it different. The $^2D_{3/2}$ state of Lr is however close to the $^2P_{1/2}$ state. That is, the $^2D_{3/2}$-$^2P_{1/2}$ splitting seems quite sensitive to the level of theory. In any case we find that the ground state of Lr is the $^2P_{1/2}$ state arising from the 7s7p configuration and the spin-orbit splitting of 7p as measured by the $^2P_{3/2}$-$^2P_{1/2}$ energy separation is 7790 cm^{-1}. The ionization potential of Lr is 4.87 eV, substantially smaller than No or any other actinide. Thus unusually low ionization energy of Lr would result in the ionic character of the Lr compounds. The large drop in the IP from No to Lr is a consequence of relativistic stabilization of the 7s orbital of Lr which leads to very stable Lr$^+$ with $7s^2$ closed-shell configuration, while in the case of No the ground state of the neutral No is stabilized due to the $7s^2$ configuration of No relative to No$^+$. Thus the IP of No is substantially larger than Lr. Our computations on the energy levels and atomic states of Lr and Lr$^+$ are consistent with the previous relativistic Fock space coupled cluster study of Eliav et al[5]

Figure 2 shows our computed potential energy surfaces of the electronic states of LrH_2 in the

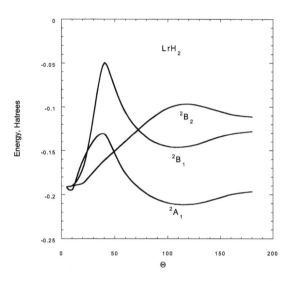

Figure 2 . The potential energy surfaces of LrH_2 in the absence of spin-orbit coupling.

absence of spin-orbit coupling. All three lowest double electronic states considered here correlate into the $Lr(^2P) + H_2$ dissociation limit in the absence of spin-orbit coupling. Among these $^2A_1$ state is the lowest, which forms a stable ground state near a bond angle of $120°$ and this minimum is more stable than the $Lr(^2P) + H_2$ dissociation limit. The ground state of LrH_2 is 12 kcal/mole more stable than $Lr(^2P) + H_2$ at the CASSCF level and 16 cal/mole more stable at the second-order CI level. Figure 3 shows the computed potential energy surfaces of LrH_2 including spin-orbit effects. Comparison of thse potential energy surfaces reveals dramatic differences. These differences arise from avoided crossings introduced by spin-orbit coupling. As seen from Fig.2 the $^2B_2$ curve crosses the $^2A_1$ curve. Both states correlate into the same E representation in the double group and thus the $^2A_1$ state with open-shell spin α can strongly couple with the $^2B_2$ state with spin β in the region of curve crossing. The near proximity of the $^2B_1$ state near the dissociation limit could also introduce strong mixing due to spin-orbit coupling. Thus the $^2B_2$ state and $^2A_1$ state undergo avoided crossing, which leads to the lowering of the barrier for insertion of Lr into H_2.

The potential energy surfaces dissociate quite differently when spin-orbit effects are included, as can be seen from Figure 2. The ground state of LrH_2 including spin-orbit effects dissociates into $Lr(^2P_{1/2}) + H_2$. The first excited state arises from the $Lr(^2D_{3/2}) + H_2$ dissociation, and the potential energy surface looks very different from the corresponding $^2B_1$ state in the absence of spin-orbit effects due to avoided crossings. The curve has a substantially smaller barrier followed by a shallow minimum. The potential energy curves arising from $Lr(^2D_{5/2})$ and $Lr(^2P_{3/2})$ exhibit substantially larger barriers, and all of these states form obtuse minima. The substantial differences and the shapes of the potential energy surfaces can be rationalized by consideration of the composition of the electronic states including spin-orbit effects as a function of the H-Lr-H bond angle. At $\theta = 20°$, the E(I) state is

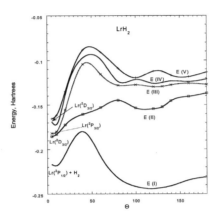

Figure 3 The computed potential energy surfaces of LrH$_2$ including spin-orbit effects.

composed of 61% $^2B_2$, 6.5% $^2B_1$, 9%$^4B_2$ + $^2B_2$ ($1a_1^2 2a_1 3a_1 1b_2$). At $\theta = 40°$, the E(I) state becomes 80% $^2B_2$, 4% $^4A_1$, and 3% $^2B_2$ (II), but at $\theta = 60°$ this state is 92% $^2A_1$ 0.8% $^2B_2$, and 0.5% $^2B_1$. Thus the $^2A_1$ and $^2B_2$ states undergo avoided crossing near 50° regions. Near the minimum bond angle of 118°, the E(I) state becomes 94% $^2A_1$ ($1a_1^2 2a_1 1b_2^2$), 0.5% $^2B_1$ The dominant contribution remains the same at longer bond distances but the second important contributor changes to $^4A_2$ near linear geometries. It is also interesting to note that at the linear limit the state attains substantial Lr (6d) character as the σ_g orbital arises from the interaction of the Lr(6dσ) with the hydrogen 1s orbitals all with same signs. This contrasts with the bent geometry for which the Lr (7p) makes a substantial contribution. The E(II) state is 59% $^2B_1$, 10% $^2B_2$ (II), 6% $^2B_2$ (III), 5% $^2A_1$ at $\theta = 20°$ but at $\theta = 40°$ this state becomes 77% $^2A_1$, 9% $^2B_2$, and 3% $^2B_2$. At $\theta = 60°$ this state becomes 80% $^2B_2$, 3% $^2B_2$ (II), 1% $^2A_1$. Near the minimum ($\theta = 120°$) this state becomes 34% $^2B_1$, 32% $^2B_1$ (II), 7% $^2A_1$(6a$_1$), an d3% $^2A_1$(5a$_1$). Near the linear geometry this state becomes 43% $^2B_1$, 37% $^2A_1$ (II) so that the state would correlate into a ω state of 3/2. The analysis of the selected states in the double group reveals the complexity of the electronic states and how the states vary as a function of the bond angle and geometry. The number of avoided crossings exhibited by these electronic states including spin-orbit effects leads to the shapes of the potential energy curves in Figure 2.

Figure 4 shows our computed potential energy surfaces of NoH$_2$ in the absence of spin-orbit coupling while Fig 5 shows the corresponding curves with spin-orbit effects. Since the No atom has a closed-shell $5f^{14} 7s^2$ configuration in contrast to the open-shell configuration of Lr, we expect No to be less reactive with H$_2$. The No($^1S_0$) state does not insert into H$_2$ as it has to surpass a large barrier in the absence of spin-orbit coupling. On the other hand, the $^3B_2$ state crosses the $^1A_1$ state before the barrier is reached, and thus spin-orbit effect may have substantial impact in reducing the barrier height. The $^3A_1$ and $^3B_1$ states are considerably higher in energy and they arise from the excited No(^3P) + H$_2$ species. Among these the $^3A_1$ state forms a shallow minimum in the obtuse bond angle. Comparison of Figs4 and 5 exemplfy the differences due to spin-orbit coupling in these electronic states.

We have been interested in relativistic computations of the properties of molecules containing superheavy elements such as 113[6] and 114. We have computed the electronic states of (113)H, the eka-thallium hydride and 114H$_2$ the eka-lead dihydride. We have demonstrated that the 6d-electron correlation-spin-orbit effects are so large for (113)H that they lead to significant shortening of the (113)-H bond and stabilization of the bond. It is shown that the periodic trends of (113)H are such that (113)H has a knight's move relation to AuH in exhibiting unusual stability and d correlation-spin-orbit

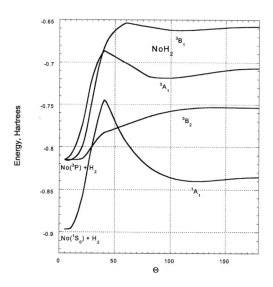

Figure 4 The potential energy surfaces of NoH$_2$ in the absence of spin-orbit coupling.

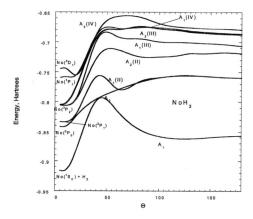

Figure 5 The potential energy surfaces of NoH$_2$ in the absence of spin-orbit coupling.

effects. We have also shown that (114)H$_2$ exhibits similarity to HgH$_2$ and also considerable spin-orbit coupling that mixes the singlet ($^1A_1$) state of (114)H$_2$ with the triplet state ($^3B_1$). The computed potential energy curves of (113)H are shown below on Fig. 6

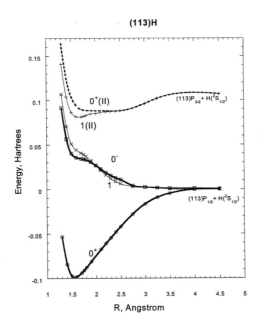

Figure 6. The potential energy surfaces (113)H with of spin-orbit coupling.

An unusual feature of the ground state of (113)H is dramatic shortening of the bond length by valence spin-orbit coupling and core-valence spin-orbit coupling. That is, the r_e value of the 0^+ ground state is 1.782 Å when no excitations from the 6d shells of 113 are allowed. The corresponding dissociation energy is 1.45 eV. However when excitations are allowed from the 6d orbitals from a multi-reference set of configurations, there is significant bond contraction and stabilization. The origin of the contraction can be understood through analysis of the composition of the 0^+ state. The $^3\Pi$ (0^+) state originating from the excitation of the 6d (σ) shell to 7p (π) orbital is the most important contributor to the contraction of the (113)-H bond. This leads to 4 real configurations, viz., 6d$(\sigma)\alpha$ $1\pi_x\alpha$, 6d$(\sigma)\beta$ $1\pi_x\beta$, 6d$(\sigma)\alpha$ $1\pi_y\alpha$, and 6d$(\sigma)\beta$ $1\pi_y\beta$ in addition to other reference configurations discussed already. These reference configurations couple with other valence multi-configurations causing the contraction of the (113)-H bond. Consequently this is a core-valence multi-reference spin-orbit and correlation effect. We have also considered the importance of relativity in bonding and spectroscopic properties of actinide complexes of relevance to the environment.[7-10]. Since these topics have been considered in depth elsewhere, we refer to these works for further details concerning such actinide complexes.

Acknowledgments

This research was performed under the auspices of the US department of Energy by the University of California, LLNL under contract number W-7405-Eng-48 while the work at California State university was supported by National Science Foundation and basic energy sciences division of US department of energy.

References

[1] K.Balasubramanian, "Relativistic Double Group Spinor Representations of Non-rigid Molecules", *Journal of Chemical Physics*, **120**, 5524-5535(2004)

[2] K.Balasubramanian, *Relativistic Effects in Chemistry, Part A: Theory and Techniques*, Wiley-interscience, New York, NY p301 1997.

[3] K.Balasubramanian, *Relativistic Effects in Chemistry, Part B: Applications*, Wiley-interscience, New York, NY p527 1997.

[4] K.Balasubramanian, *J. Chem. Phys.*, **116**, 3568 (2002).

[5] E.Eliav, U. Kaldor and Y. Ishiwata, *Phys., Rev. A.*, **52**, 291 (1995).

[6] K.Balasubramanian, *Chem. Phys. Lett.*, **361**, 297 (2002).

[7] Z.Cao and K. Balasubramanian, "Theoretical Studies of hydrated complexes of uranyl, neptunyl and plutonyl in aqueous solution: $UO_2^{2+}(H_2O)_n$, $NpO_2^{2+}(H_2O)_n$, and $PuO_2^{2+}(H_2O)_n$", J. Chem. Phys. *In Press* (2005)

[8] D.Chaudhuri and K.Balasubramanian, "Electronic Structure and Spectra of Plutonyl Complexes and their hydrated forms: PuO_2CO_3 and $PuO_2CO_3.nH2O$ (n=1,2)", *Chemical Physics Letters*, **399**, 67 (2004)

[9] D.Majumdar and K.Balasubramanian, "Theoretical studies on the nature of uranyl–silicate, uranyl–phosphate and uranyl–arsenate interactions in the model $H_2UO_2SiO4 \cdot 3H_2O$, $HUO_2PO_4 \cdot 3H_2O$, and $HUO_2AsO_4 \cdot 3H_2O$ molecules", *Chemical Physics Letters*, **397**, 26 (2004).

[10] K Balasubramanian, "Relativity and the Periodic Table", (D. Rouvary & R. B. King, editors), Periodic Table into the 21st Century, 2004

Brill Academic Publishers
P.O. Box 9000, 2300 PA Leiden,
The Netherlands

*Lecture Series on Computer
and Computational Sciences*
Volume 3, 2005, pp. 18-26

Generalized Hopf algebra fundamental formula for non-orthogonal group functions

Patrick Cassam-Chenaï[1]

Laboratoire J. A. Dieudonné,
CNRS- Université de Nice Sophia-Antipolis,
Faculté des Sciences, Parc Valrose,
06108 Nice cedex 2, France.

Received 23 June, 2005; accepted 1 August, 2005

Abstract: The fundamental formula defining a Hopf algebra structure is generalized to the case of iterated exterior products and coproducts in view of calculating matrix elements between non-orthogonal group functions.

Keywords: Hopf algebra, non-orthogonal group functions, molecular electronic calculation

Mathematics Subject Classification: 81R50, 15A75

PACS: 31.25.Qm , 31.15.Ar

1 Introduction

Recently, we have proposed a new, general method for solving the molecular vibrational Schrödinger equation called the "vibrational mean-field configuration interaction" (VMFCI) method [1, 2, 3]. The effectiveness of the VMFCI method has led us to transpose its principle in the context of electronic calculations. So, following the reverse way that has led to the development of the vibrational self-consistent field (VSCF), vibrational configuration interaction (VCI), vibrational multiconfiguration self-consistent field (VMCSCF), vibrational complete active space self-consistent field (VCASSCF), ... by analogy with pre-existing electronic methods, we have developed an electronic version of the MFCI called the "electronic mean-field configuration interaction" (EMFCI) method [4].

Briefly speaking, the EMFCI is a variational method which consists in contracting together groups of electronic degrees of freedom in the mean-field of the other degrees. If the same partition of electronic degrees of freedom is iterated, an SCF method is obtained. Making coarser partitions (i.e. including more degrees in the same groups) and discarding the high energy states, the Full-CI limit can be approached. At each calculation step, an approximation of the ground state and of many excited states is obtained.

More precisely, the EMFCI optimises group functions [5, 6, 7, 8, 9, 10, 11, 12, 13, 14, 15, 16, 17], but in contrast with the usual group function theory, the too restrictive [18, 19] "strong-orthogonality" condition is not enforced. So, at a given SCF step, it finds the "best" (with respect to the variational principle) n-electron wave functions of the form

$$\Psi = \Psi_1 \wedge \cdots \wedge \Psi_r , \tag{1}$$

---

[1]Corresponding author. E-mail: cassam@unice.fr

where $\wedge$ denotes Grassmann's exterior product (see Appendix), and Ψ_i an n_i-electron function, with $\sum_i n_i = n$. No orthogonality constraint is imposed on the Ψ_i, in particular, if for all i, $n_i = 2$, AGP functions such as, $\Psi_1 \wedge \cdots \wedge \Psi_1$, are considered by the variational process. Such a general form for the wave functions makes the computation of Hamiltonian and overlap matrix elements particularly difficult.

We have already drawn attention to the tools provided by the Hopf algebra for the quantum theory of Fermionic systems [20, 21]. In this contribution, we show, after generalizing the fundamental formula defining a Hopf algebra structure to the case of iterated exterior products and coproducts, that the Hopf algebra formalism enables one to derive matrix element expressions which respect the structure of group functions. As a result, one can avoid repeating a large number of operations for each element when calculating the Hamiltonian and overlap matrices.

2 Useful Hopf algebra formulas

2.1 The Hopf algebra formalism

The Fermionic Fock space, $\wedge\mathcal{H}$, associated to a one-particle Hilbert space, $\mathcal{H}$, provided with the exterior product, $\mathcal{X}$, and the exterior coproduct, $\mathcal{Y}$, has the structure of an Hopf algebra, (see Ref. [20] or Appendix). Here, the exterior product is regarded as a map from the tensor product $\wedge\mathcal{H} \otimes \wedge\mathcal{H}$ to $\wedge\mathcal{H}$,

$$
\begin{aligned}
\mathcal{X} : \wedge\mathcal{H} \otimes \wedge\mathcal{H} &\longmapsto \wedge\mathcal{H} \\
\Phi \otimes \Psi &\longrightarrow \mathcal{X}(\Phi \otimes \Psi) = \Phi \wedge \Psi \ ,
\end{aligned}
\tag{2}
$$

where $\wedge$ denotes the usual Grassmann product. The exterior coproduct or simply "coproduct" acts on single configuration wave functions in the following way,

$$
\begin{aligned}
\mathcal{Y} : \wedge\mathcal{H} &\longmapsto \wedge\mathcal{H} \otimes \wedge\mathcal{H} \\
\psi_1 \wedge \cdots \wedge \psi_n &\longrightarrow \mathcal{Y}(\psi_1 \wedge \cdots \wedge \psi_n) = \sum_{(I,\bar{I}) \in \mathcal{P}_n} \rho_{I,\bar{I}} \psi_{i_1} \wedge \cdots \wedge \psi_{i_{Card(I)}} \otimes \psi_{\bar{i}_1} \wedge \cdots \wedge \psi_{\bar{i}_{n-Card(I)}} \ ,
\end{aligned}
$$

where $\mathcal{P}_n$ is the set of all possible partitions of $\{1, \ldots, n\}$ into two ordered sequences, $I := (i_1 < \cdots < i_{Card(I)})$ and $\bar{I} := (\bar{i}_1 < \cdots < \bar{i}_{n-Card(I)})$, $\rho_{I,\bar{I}}$ is the sign of the permutation reordering the concatenated sequence $I//\bar{I}$ in increasing order and $Card(I)$ is the number of elements in I. The action of the coproduct on a general wave function,

$$
\Psi = \sum_l c_l \, \psi_{l,1} \wedge \cdots \wedge \psi_{l,n} \ ,
\tag{3}
$$

is obtained from Eq.(3) by linearity.

A convenient notation which encapsulates the expansion of Ψ as well as the sign factor $\rho_{I,\bar{I}}$ is,

$$
\Psi_I \otimes \Psi_{\bar{I}} := \rho_{I,\bar{I}} \sum_l c_l \, \psi_{l,i_1} \wedge \cdots \wedge \psi_{l,i_{Card(I)}} \otimes \psi_{l,\bar{i}_1} \wedge \cdots \wedge \psi_{l,\bar{i}_{n-Card(I)}} \ ,
\tag{4}
$$

so that the coproduct writes,

$$
\mathcal{Y}(\Psi) = \sum_{(I,\bar{I}) \in \mathcal{P}_n} \Psi_I \otimes \Psi_{\bar{I}} \ .
\tag{5}
$$

It has been found advantageous, in particular in view of computational application, to introduce iterated products, $\mathcal{X}^{[k]}$, and coproducts $\mathcal{Y}^{[k]}$ recursively as follows. The kth iterated product

$$
\mathcal{X}^{[k]} := \mathcal{X} \circ (\mathcal{X}^{[k-1]} \otimes Id) \ ,
\tag{6}
$$

$(k > 0, \mathcal{X}^{[0]} := Id)$, acts on $(k+1)$-component tensor products to form their $(k+1)$-component exterior product counterparts,

$$\mathcal{X}^{[k]}(\Phi_0 \otimes \cdots \otimes \Phi_k) = \Phi_0 \wedge \cdots \wedge \Phi_k . \tag{7}$$

The kth iterated coproduct

$$\mathcal{Y}^{[k]} := \underbrace{(Id \otimes \cdots \otimes Id}_{k-1 factors} \otimes \mathcal{Y}) \circ \mathcal{Y}^{[k-1]} , \tag{8}$$

$(k > 0, \mathcal{Y}^{[0]} := Id)$, decomposes a Fermionic wave function into a $(k+1)$-component tensor product. More explicitly, denoting by $\mathcal{P}_n^{[k]}$ the set of all possible partitions of $\{1, \ldots, n\}$ into $(k+1)$ ordered sequences, $I_0 := (i_{0_1} < \cdots < i_{0_{Card(I_0)}}), \ldots, I_k := (i_{k_1} < \cdots < i_{k_{Card(I_k)}}), (\mathcal{P}_n^{[1]} \equiv \mathcal{P}_n)$, and by $\rho_{I_0, \ldots, I_k}$ the sign of the permutation reordering the concatenated sequence $I_0 // \cdots // I_k$ in increasing order, the action of the kth iterated coproduct on a general n-Fermion wave function, (Eq. (3)), is,

$$\mathcal{Y}(\Psi) = \sum_{(I_0, \ldots, I_k) \in \mathcal{P}_n^{[k]}} \rho_{I_0, \ldots, I_k} \sum_l c_l \, \psi_{l, i_{0_1}} \wedge \cdots \wedge \psi_{l, i_{0_{Card(I_0)}}} \otimes \cdots \otimes \psi_{l, i_{k_1}} \wedge \cdots \wedge \psi_{l, i_{k_{Card(I_k)}}} , \tag{9}$$

or, more simply, generalizing the notation of Eq. (4),

$$\mathcal{Y}(\Psi) = \sum_{(I_0, \ldots, I_k) \in \mathcal{P}_n^{[k]}} \Psi_{I_0} \otimes \cdots \otimes \Psi_{I_k} . \tag{10}$$

We specify further by $\mathcal{Y}_{i_0, \ldots, i_k}^{[k]}$ the component of the iterated coproduct corresponding to the decomposition of an n-Fermion wave function into the tensor product of $(k+1)$ wave functions of $i_0, \ldots, i_k$-particles respectively, with $\sum_{l=0}^{k} i_l = n$. Denoting by, $\mathcal{P}_{n, i_1, \ldots, i_k}^{[k]}$, the subset of $\mathcal{P}_n^{[k]}$ consisting of partitions, $(I_0, \ldots, I_k)$, such that $Card(I_1) = i_1, \ldots, Card(I_k) = i_k$, (hence also such that $Card(I_0) = n - \sum_{l=1}^{k} i_l = i_0$), the expression of $\mathcal{Y}_{i_0, \ldots, i_k}^{[k]}(\Psi)$ is obtained from Eq.(10) by limiting the summation on $\mathcal{P}_n^{[k]}$ to $\mathcal{P}_{n, i_1, \ldots, i_k}^{[k]}$,

$$\mathcal{Y}_{i_0, \ldots, i_k}^{[k]}(\Psi) = \sum_{(I_0, \ldots, I_k) \in \mathcal{P}_{n, i_1, \ldots, i_k}^{[k]}} \Psi_{I_0} \otimes \cdots \otimes \Psi_{I_k} . \tag{11}$$

2.2 Generalized twist operator

Let us define a generalized twist operator $T^{(p,q)}$, (p and $q > 1$) on the tensor product of $p.q$, Fermionic Hilbert spaces, $\wedge^{n_i^j} \mathcal{H}$, with fixed number of particles, n_i^j, $i \in 1, \cdots, p$ and $j \in 1, \cdots, q$. Let $\Phi := \phi_1^1 \otimes \cdots \otimes \phi_p^1 \otimes \cdots \otimes \phi_1^q \otimes \cdots \otimes \phi_p^q$ be an element of $\wedge^{n_1^1} \mathcal{H} \otimes \cdots \otimes \wedge^{n_p^1} \mathcal{H} \otimes \cdots \otimes \wedge^{n_1^q} \mathcal{H} \otimes \cdots \otimes \wedge^{n_p^q} \mathcal{H}$, then by definition

$$T^{(p,q)}[\Phi] = \rho_{n_1^1, \ldots, n_p^1, \ldots, n_1^q, \ldots, n_p^q} \, \phi_1^1 \otimes \cdots \otimes \phi_p^q \otimes \cdots \otimes \phi_1^q \otimes \cdots \otimes \phi_p^q \tag{12}$$

where $\rho_{n_1^1, \ldots, n_p^1, \ldots, n_1^q, \ldots, n_p^q}$ is the sign of the permutation which would reorder the tensorial components ϕ_i^j in their initial order if we had exterior products in place of tensor products. More explicitly,

$$\rho_{n_1^1,\ldots,n_p^1,\ldots,n_1^q,\ldots,n_p^q} = (-1)^{\sum\limits_{i=1}^{p-1} \sum\limits_{j=2}^{q} \sum\limits_{k=j+1}^{p} \sum\limits_{l=1}^{i-1} n_i^j \cdot n_k^l} \tag{13}$$

The definition of $T^{(p,q)}$ is extented by multilinearity to the $p.q^{th}$ tensorial power, $\bigotimes^{p.q} \wedge\mathcal{H} := \underbrace{\wedge\mathcal{H} \otimes \cdots \otimes \wedge\mathcal{H}}_{p.q \ factors}$, of the Fermionic Fock space (i. e. the exterior algebra), $\wedge\mathcal{H}$.

In the following, for any sequence of integers, $l_0 := 0 < l_1 < l_2 < \cdots < l_s < l_{s+1} := p.q$, we will make use of the notation

$$T^{(p,q)}_{l_0+1,\ldots,l_1}[\Psi] \otimes T^{(p,q)}_{l_1+1,\ldots,l_2}[\Psi] \otimes \cdots \otimes T^{(p,q)}_{l_s+1,\ldots,l_{s+1}}[\Psi] := T^{(p,q)}[\Psi] \tag{14}$$

where $T^{(p,q)}_{l_r+1,\ldots,l_{r+1}}[\Psi]$, refers to the tensor product components $l_r + 1,\ldots,l_{r+1}$, extracted from $T^{(p,q)}[\Psi]$. This allows one to encapsulate in the notation, (in exactly the same way as was done for the coproduct in Eq.(5)), the signs $\rho_{n_1^1,\ldots,n_q^p,\ldots,n_1^1,\ldots,n_q^p}$ and the expansion of Ψ, when $T^{(p,q)}$ is applied to a general wave function.

Note that, $T^{(2,2)} = Id \otimes T \otimes Id$ where T is the twist operator defined in [20].

2.3   Generalized Hopf algebra fundamental relations

The generalization of the Hopf algebra fundamental relation (see Appendix, Eq.(30)) to the case of iterated product, $\mathcal{X}^{[p-1]}$, $(p > 1)$, and coproduct , $\mathcal{Y}^{[q-1]}$, $(q > 1)$, is,

$$\mathcal{Y}^{[q-1]} \circ \mathcal{X}^{[p-1]} = \underbrace{(\mathcal{X}^{[p-1]} \otimes \cdots \otimes \mathcal{X}^{[p-1]})}_{q factors} \circ T^{(q,p)} \circ \underbrace{(\mathcal{Y}^{[q-1]} \otimes \cdots \otimes \mathcal{Y}^{[q-1]})}_{p factors}, \tag{15}$$

or for a particular coproduct component,

$$\mathcal{Y}^{[q-1]}_{n_1,\ldots,n_q} \circ \mathcal{X}^{[p-1]} = \sum_{n_i^j, \sum_{j=1}^p n_i^j = n_i} \underbrace{(\mathcal{X}^{[p-1]} \otimes \cdots \otimes \mathcal{X}^{[p-1]})}_{q factors} \circ T^{(q,p)} \circ (\mathcal{Y}^{[q-1]}_{n_1^1,\ldots,n_1^p} \otimes \cdots \otimes \mathcal{Y}^{[q-1]}_{n_q^1,\ldots,n_q^p}). \tag{16}$$

Applied to an element $\Phi_1 \otimes \cdots \otimes \Phi_p$ of $\wedge^{n^1}\mathcal{H} \otimes \cdots \otimes \wedge^{n^p}\mathcal{H}$, we obtain

$$\mathcal{Y}^{[q-1]}_{n_1,\ldots,n_q}(\Phi_1 \wedge \cdots \wedge \Phi_p) = \sum_{n_i^j, \sum_{j=1}^p n_i^j = n_i, \sum_{i=1}^q n_i^j = n^j} \underbrace{(\mathcal{X}^{[p-1]} \otimes \cdots \otimes \mathcal{X}^{[p-1]})}_{q factors} \qquad \circ \quad T^{(q,p)}$$

$$\circ \quad \left(\mathcal{Y}^{[q-1]}_{n_1^1,\ldots,n_q^1}(\Phi_1) \otimes \cdots \otimes \mathcal{Y}^{[q-1]}_{n_1^p,\ldots,n_q^p}(\Phi_p)\right). \tag{17}$$

The proof is easy. It is a direct consequence of the fact that to extract in all possible manners q subsequences out of p concatenated sequences (left hand side) amounts to first extract in all possible manners q subsequences out of each of the p sequences, then to rearrange the p first subsequences so-obtained together, the p second subsequences together, and so on till the p q^{th} subsequences, and finally to concatenate each set of p rearranged subsequences (right hand side). The signs arising from the Pauli principle are taken care of by the definitions of $\mathcal{Y}$ and $T^{(q,p)}$.

A Formula, similar to Eq.(15), has already appeared in the Hopf algebra literature [22].

3   Application to the calculation of the integrals appearing in the EMFCI method

3.1   General integral formula between group functions

Let h_p be a p-particle operator and $h_p * Id := \mathcal{X} \circ (h_p \otimes Id) \circ \mathcal{Y}$ the operator it induces on the Fock space (see Ref. [20] or Appendix, Eq. (33)). Let $\Phi_1 \wedge \cdots \wedge \Phi_q \in (\wedge^{n_1} \mathcal{H}) \wedge \cdots \wedge (\wedge^{n_q} \mathcal{H})$ and $\Psi_1 \wedge \cdots \wedge \Psi_r \in \left(\wedge^{m^1} \mathcal{H} \right) \wedge \cdots \wedge \left(\wedge^{m^r} \mathcal{H} \right)$ be two, so-called group functions. Since the operator $h_p * Id$ preserves the number of particle we can assume that the two group functions have the same number of particles, say n, so that $n_1 + \cdots + n_q = m^1 + \cdots + m^r = n$. The matrix element of $h_p * Id$ between the two functions is,

$$\langle \Phi_1 \wedge \cdots \wedge \Phi_q | (h_p * Id)[\Psi_1 \wedge \cdots \wedge \Psi_r] \rangle =$$
$$\langle \Phi_1 \otimes \cdots \otimes \Phi_q | \mathcal{Y}^{[q-1]}_{n_1,\ldots,n_q} \circ \mathcal{X} \circ (h_p \otimes Id) \circ \mathcal{Y}_{p,n-p} \circ \mathcal{X}^{[r-1]}[\Psi_1 \otimes \cdots \otimes \Psi_r] \rangle \quad , \qquad (18)$$

where we have used the Laplace formula (see Ref. [20] or Appendix, Eq. (32)), the fact that h_p is a p-particle operator and definitions already introduced. Recall that $\mathcal{X} \equiv \mathcal{X}^{[1]}$ and $\mathcal{Y} \equiv \mathcal{Y}^{[1]}$, we notice that there are two places in Eq.(18) where Eq.(17) can be applied. Denoting by $\Sigma'_{n^l_k, m^j_i}$ the summations constrained by the relations appearing in Eq.(17) and using the special notation of Eqs.(10) and (14) (and obvious generalisations of it allowing us to perform algebraic manipulations on the tensorial components), the matrix element becomes ,

$$\langle \Phi_1 \wedge \cdots \wedge \Phi_q | (h_p * Id)[\Psi_1 \wedge \cdots \wedge \Psi_r] \rangle =$$
$$\Sigma'_{n^l_k, m^j_i} \prod_{s=1}^{q} \langle \Phi_s | \mathcal{X} \circ T^{(q,2)}_{2s-1,2s} \left[\mathcal{Y}^{[q-1]}_{n^1_1,\ldots,n^1_q} \circ h_p \circ \mathcal{X}^{[r-1]} \circ T^{(2,r)}_{1,\ldots,r} \left[\mathcal{Y}_{m^1_1,m^1_2}[\Psi_1] \otimes \cdots \otimes \mathcal{Y}_{m^r_1,m^r_2}[\Psi_r] \right] \otimes \right.$$
$$\left. \mathcal{Y}^{[q-1]}_{n^2_1,\ldots,n^2_q} \circ \mathcal{X}^{[r-1]} \circ T^{(2,r)}_{r+1,\ldots,2r} \left[\mathcal{Y}_{m^1_1,m^1_2}[\Psi_1] \otimes \cdots \otimes \mathcal{Y}_{m^r_1,m^r_2}[\Psi_r] \right] \right] \rangle =$$
$$\Sigma'_{n^l_k, m^j_i} \sum_{(I_j, \bar{I}_j) \in \mathcal{P}_{m^j, m^j_2}} \rho_{m^1_1, m^1_2,\ldots,m^r_1, m^r_2} \prod_{s=1}^{q} \langle \Phi_s | \mathcal{X} \circ T^{(q,2)}_{2s-1,2s} \left[\mathcal{Y}^{[q-1]}_{n^1_1,\ldots,n^1_q} \circ h_p [(\Psi_1)_{I^1} \wedge \cdots \wedge (\Psi_r)_{I^r}] \otimes \right.$$
$$\left. \mathcal{Y}^{[q-1]}_{n^2_1,\ldots,n^2_q} \circ \mathcal{X}^{[r-1]} \circ T^{(2,r)}_{r+1,\ldots,2r} [(\Psi_1)_{\bar{I}^1} \wedge \cdots \wedge (\Psi_r)_{\bar{I}^r}] \right] \rangle =$$
$$\Sigma'_{n^l_k, m^j_i} \rho_{n^1_1,\ldots,n^1_q, n^2_1,\ldots,n^2_q} \rho_{m^1_1, m^1_2,\ldots,m^r_1, m^r_2} \times$$
$$\sum_{\substack{(K^1_1,\ldots,K^l_q) \in \mathcal{P}^{[q-1]}_{p,n^1_2,\ldots,n^1_q} \\ (I_j, \bar{I}_j) \in \mathcal{P}_{m^j, m^j_2}}} \prod_{s=1}^{q} \langle \Phi_s | (h_p [(\Psi_1)_{I^1} \wedge \cdots \wedge (\Psi_r)_{I^r}])_{K^1_s} \wedge ((\Psi_1)_{\bar{I}^1} \wedge \cdots \wedge (\Psi_r)_{\bar{I}^r})_{K^2_s} \rangle. \qquad (19)$$

The initial n-Fermion integral has therefore been reduced explicitly to products of n_k-Fermion integrals ($k \in \{1,\ldots,q\}$) involving only a single component of the wave function on the left hand side. However, care must be taken that the different n_k-Fermion integrals are connected by the elements of the partitions $(K^l_1,\ldots,K^l_q)$. A symmetrical formula can be obtained by exchanging the part played by bras and kets.

3.2   Particular integrals occuring in the EMFCI method

In the EMFCI method [4] applied to the usual, Coulombic, non relativistic, molecular Hamiltonian, the integrals to be calculated correspond to the matrix elements calculated in the previous section with $p = 2$, $q = r$ and all but one contraction, say contraction 1, have possibly different wave functions, that is to say $\Phi_2 = \Psi_2 = \Gamma_2, \ldots, \Phi_q = \Psi_q = \Gamma_2$, where Γ_s is the "ground state" of contraction s.

Eq. (19) has been used to derive explicit, recursive formulas, for such matrix elements. Since this equation respects, as far as allowed by the Fermionic symmetry, the structure of the group functions, a large part of the calculations involving the Γ_s which is common to all matrix elements, does not need to be repeated as Φ_1 and Ψ_1 vary. The whole Hamiltonian and overlap matrices can thus be computed at limited cost.

The method has been implemented in a local version of the quantum chemistry code tonto (http://www.theochem.uwa.edu.au/tonto) for the simplest case of EMFCI which does not break the spin symmetry in the case of a closed-shell system, that is to say, the non-orthogonal geminal SCF method (NOGSCF). Results [23] show that the NOGSCF method dissociates correctly and that it can recover all of the correlation energy (to a few tenth of milliHartree) in cases where the antisymmetrized product of strongly orthogonal geminals (APSG) ansatz recovers only about 1/3 of the CCSD(T) correlation energy [18]. This shows the importance of the non-orthogonality in obtaining an accurate geminal wave function.

Acknowledgment

The author wishes to thank Dr. B. Sutcliffe for his careful reading of the manuscript and his fruitful comments and suggestions.

Appendix

In this appendix, we recall some basic properties of the Grassmann product and of the exterior Hopf algebra. In particular, we introduce the Laplace Formula and the convolution to make the article more self-contained.

Grassmann product and Fermionic exterior algebra

Denoting by $\mathcal{H}$ the Hilbert space of spinorbitals, the n-fermion Hilbert space induced by $\mathcal{H}$ is the n-th exterior power of $\mathcal{H}$, denoted by $\wedge^n \mathcal{H}$. An element Ψ of $\wedge^n \mathcal{H}$ is an n-fermion wave function. It is more specifically termed a "single configuration" function if it can be cast in the form of a simple (Grassmann's) exterior product of n spinorbitals $\psi_1, \ldots, \psi_n$,

$$\Psi = \psi_1 \wedge \psi_2 \wedge \cdots \wedge \psi_n , \tag{20}$$

which corresponds in the second quantization of a fermionic system formalism to the quantum state

$$\Psi = a_1^\dagger \cdots a_n^\dagger |0\rangle , \tag{21}$$

where $a_i^\dagger$ is the creation operator of a fermion in the spinorbital ψ_i.

The fermionic Fock space of the second quantization $\wedge \mathcal{H}$ is the direct sum of all the $\wedge^n \mathcal{H}$,

$$\wedge \mathcal{H} := \bigoplus_{n \geq 0} \wedge^n \mathcal{H}, \tag{22}$$

where $\mathcal{H}$ is identified to $\wedge^1 \mathcal{H}$ and the field of complex or real numbers, denoted by $\mathbb{K}$, is identified to $\wedge^0 \mathcal{H}$ through the "unit map", $\mathcal{U}$,

$$\begin{aligned} \mathcal{U} : \mathbb{K} &\longmapsto \wedge \mathcal{H} \\ \lambda &\longrightarrow \mathcal{U}(\lambda) = \lambda \in \wedge^0 \mathcal{H}. \end{aligned} \tag{23}$$

Provided with Grassmann's exterior product, $\wedge$, it acquires the mathematical structure of an "exterior algebra". Note that when one of its factors is a scalar λ, the exterior product reduces to the multiplication by this scalar,

$$\lambda \wedge \Psi = \Psi \wedge \lambda = \lambda\Psi. \tag{24}$$

The fermionic symmetry which traduces the Pauli exclusion principle, is built-in in this exterior algebra because of the following antisymmetry relation between exterior products of 1-particle functions:

$$\mathcal{X}(\phi \otimes \psi) := \phi \wedge \psi = -\psi \wedge \phi =: -\mathcal{X}(\psi \otimes \phi), \tag{25}$$

which entirely determines the behavior of an n-fermion wave function under the symmetric group $\mathcal{S}_n$. That is to say, for an n-fermion single configuration and a permutation, σ, with signature $|\sigma|$:

$$\psi_1 \wedge \cdots \wedge \psi_n = (-1)^{|\sigma|}\psi_{\sigma(1)} \wedge \cdots \wedge \psi_{\sigma(n)}. \tag{26}$$

The general case is obtained from Equation (26) by linearity.

Exterior coproduct and Fermionic Hopf algebra

The coproduct is defined in Section 2.1. The definition consists in inverting the arrows in the definition of the exterior product Eq. (2). The idea behind the coproduct is to split an n-fermion single configuration function into a p and an $(n-p)$-fermion single configuration functions in all possible ways, where p ranges from 0 to n, the exterior product of the two parts so-obtained giving back the initial function, the sign of the reordering permutation being taken care of. The definition extends by linearity to general wave functions.

As an example, let us write down the formula for $\mathcal{Y}$ acting on a 3-fermion single configuration,

$$
\begin{aligned}
\mathcal{Y}(\psi_a \wedge \psi_b \wedge \psi_c) &= \psi_a \wedge \psi_b \wedge \psi_c \otimes 1 + \psi_a \wedge \psi_b \otimes \psi_c - \psi_a \wedge \psi_c \otimes \psi_b \\
&\quad + \psi_b \wedge \psi_c \otimes \psi_a + \psi_a \otimes \psi_b \wedge \psi_c - \psi_b \otimes \psi_a \wedge \psi_c \\
&\quad + \psi_c \otimes \psi_a \wedge \psi_b + 1 \otimes \psi_a \wedge \psi_b \wedge \psi_c.
\end{aligned} \tag{27}
$$

Another important mapping is the counit map,

$$
\begin{aligned}
\mathcal{V} : \wedge\mathcal{H} &\longmapsto \mathbb{K} \\
\Psi &\longrightarrow \mathcal{V}(\Psi) = 0 \quad if \quad \Psi \in \wedge^n\mathcal{H}, \qquad n > 0 \\
\lambda &\longrightarrow \mathcal{V}(\lambda) = \lambda \quad if \quad \lambda \in \wedge^0\mathcal{H}.
\end{aligned} \tag{28}
$$

The counit is compatible with the coproduct in the sense that,

$$\forall\Psi \in \wedge\mathcal{H}, \ (\mathcal{V} \otimes Id) \circ \mathcal{Y}(\Psi) = (Id \otimes \mathcal{V}) \circ \mathcal{Y}(\Psi) = \Psi. \tag{29}$$

When the Fock space is endowed with the coproduct and the counit it becomes a coalgebra. Furthermore, the Fock space has the structure of a Hopf algebra since the algebra and the coalgebra structures are compatible in the sense that the following relationship, generalized in Section 2.3, (and others of minor interest to us, not presented here) is fulfilled:

$$\mathcal{Y} \circ \mathcal{X} = (\mathcal{X} \otimes \mathcal{X}) \circ (Id \otimes T \otimes Id) \circ (\mathcal{Y} \otimes \mathcal{Y}), \tag{30}$$

where T is the twisting map (generalized in Section 2.2):

$$
\begin{aligned}
\forall\Phi \in \wedge^p\mathcal{H} \quad &, \quad \forall\Psi \in \wedge^q\mathcal{H} \\
T(\Psi \otimes \Phi) &= (-1)^{pq}\Phi \otimes \Psi.
\end{aligned} \tag{31}
$$

The definition of T extends to more general elements of the Fock space by linearity.

The fundamental Hopf algebra relation, Eq.(30), expresses the fact that the same decomposition of the product of two single configurations into a tensor product of 2 subconfigurations is obtained by applying the coproduct to the exterior product of the two single configurations, or alternatively (second member of Eq.(30)), by first splitting each single configuration separately ($\mathcal{Y} \otimes \mathcal{Y}$), then grouping the first tensorial components of each decomposition together and the second components together ($Id \otimes T \otimes Id$), and finally by taking the exterior product of the first components on the one hand and the exterior product of the second components on the other hand ($\mathcal{X} \otimes \mathcal{X}$).

A straightforward consequence of this relation is the famous Laplace formula used to expand a determinant. A somewhat generalized version of the Laplace formula amounts to the following, duality-like relationship between the exterior product and the coproduct: $\forall \Phi, \Psi, \Theta \in \wedge \mathcal{H}$ we have:

$$\langle \mathcal{X}(\Theta, \Phi) | \Psi \rangle \equiv \langle \Theta \wedge \Phi | \Psi \rangle = \langle \Theta \otimes \Phi | \mathcal{Y}(\Psi) \rangle. \tag{32}$$

Finally, The Hopf algebra structure induces an associative algebra structure on the space of linear endomorphisms of $\wedge \mathcal{H}$. The product is called the "convolution product", is denoted by, $*$, and is distinct from the composition product, $\circ$. If O is a p-particle operator (a map from $\wedge^p \mathcal{H}$ to $\wedge \mathcal{H}$), it induces an operator $\hat{O}$ on the whole of $\wedge \mathcal{H}$ defined by:

$$\hat{O} := O * Id = \mathcal{X} \circ (O \otimes Id) \circ \mathcal{Y}, \tag{33}$$

where Id is the identity map on the Fermionic algebra. The restriction of the operator $\hat{O}$ to $\wedge^q \mathcal{H}$ is equal to 0 if $q < p$.

References

[1] P. Cassam-Chenaï and J. Liévin, Int. J. Quantum Chem. **93**, 245-264 (2003).

[2] P. Cassam-Chenaï, J. Quant. Spectrosc. Radiat. Transfer **82**, 251-277 (2003).

[3] P. Cassam-Chenaï and J. Liévin, J. Comp. Chem. , (In Press).

[4] P. Cassam-Chenaï, to be published , (2005).

[5] E. Kapuy, Acta Physica. Hung. **9**, 237 (1958).

[6] E. Kapuy, Acta Physica. Hung. **10**, 125 (1959).

[7] E. Kapuy, Acta Physica. Hung. **11**, 409 (1959).

[8] E. Kapuy, Acta Physica. Hung. **12**, 185 (1960).

[9] E. Kapuy, Acta Physica. Hung. **12**, 351 (1960).

[10] E. Kapuy, Acta Physica. Hung. **13**, 345 (1961).

[11] E. Kapuy, Acta Physica. Hung. **13**, 461 (1961).

[12] E. Kapuy, Acta Physica. Hung. **15**, 177 (1962).

[13] R. McWeeny, Proc. Roy. Soc. (London) **A253**, 242 (1959).

[14] R. McWeeny, Rev. Mod. Phys. **32**, 335 (1960).

[15] E. Kröner, Z. Naturforsch. **15a**, 260 (1960).

[16] M. Klessinger and R. McWeeny, J. Chem. Phys **42**, 3343-3354 (1965).

[17] M. Klessinger, J. Chem. Phys **53**, 225-232 (1970).

[18] E. Rosta and P.R. Surjàn, Int. J. Quantum Chem. **80**, 96-104 (2000).

[19] E. Rosta and P.R. Surjàn, J. Chem. Phys **116**, 878 (2002).

[20] P. Cassam-Chenaï, F. Patras, J. Math. Phys. **44**, 4884-4906 (2003).

[21] P. Cassam-Chenaï, F. Patras, Phys. Let. A **326**, 297-306 (2004).

[22] F. Patras, J. of Algebra **170**, 547-566 (1994).

[23] P. Cassam-Chenaï and D. Jayatilaka, to be published , (2005).

[24] P.R. Surjàn, "an introduction to the theory of orbitals" in *Correlation and Localization*, Topics in Current Chemistry **203**, (P.R. Surjàn ed., Springer, Berlin, 1999)

Brill Academic Publishers
P.O. Box 9000, 2300 PA Leiden
The Netherlands

*Lecture Series on Computer
and Computational Sciences*
Volume 3, 2005, pp. 27-34

Theoretical investigation of the nonlinear optical properties
of chiral species

Benoît CHAMPAGNE [1]

Laboratoire de Chimie Théorique Appliquée,
Facultés Universitaires Notre-Dame de la Paix
Rue de Bruxelles, 61,
B-5000 Namur, Belgium

Received 10 August, 2005; accepted 12 August 2005

Abstract: This paper discusses our recent investigations on the theoretical design of chiral NLO compounds. It focuses on two types of NLO responses, sum-frequency generation of isotropic media, which presents a pure electric contribution, and mixed electric-magnetic second harmonic generation. Helicenes and related spiral π-conjugated compounds are shown to be suitable candidates to achieve large chirality-based NLO responses.

Keywords: nonlinear optical properties, chirality, *ab initio* and semi-empirical quantum chemistry methods, helicenes, sum frequency generation, second harmonic generation.

PACS: 33.15.Kr, 42.65.An

1. Introduction

Chirality, a term introduced in 1884 by Lord Kelvin for describing the property of any object which cannot be brought to coincide with its mirror image, is of central importance in many phenomena ranging from enzyme-substrate recognition to materials science [1]. In particular, the life molecules, the amino acids and sugars, which are components of proteins and DNA, are chiral. Similarly, many natural products are chiral and for many drugs only one enantiomer (one of the two mirror-image forms) presents activity. The complete determination of molecular stereochemistry remains nowadays a major challenge for science as well as the origin of homochirality of many natural compounds. Chiral molecules and assemblies present specific properties which can be used to unravel the configurations of the many stereogenic centers of biomolecules. Pasteur already showed that enantiomers rotate the polarization plane of linearly polarized light in opposite direction. This phenomenon, called *optical rotation* (OR), has subsequently been widely used as a structural indicator in the determinations of stereogenic centers in molecules, as evidenced, for instance, by the first paper of the Journal of the American Chemical Society [2]. *Circular dichroism* (CD), the difference of absorption by a sample of left *versus* right circularly polarized light, is also a well-known phenomenon employed to demonstrate the presence of chiral species and to deduce their configurations. Like OR, it results from interferences between induced oscillating electric and magnetic moments. The vibrational analog of CD (VCD) evidenced for the first time in 1974 [3] is now a well-recognized method in the pharmaceutical industry [4]. Similarly,vibrational Raman optical activity (VROA), a differential Raman scattering of circularly polarized light, of which the first spectra were reported more than 30 years ago [5], is now undergoing a rapid development towards structure determination [6]. Recently, using theoretical tools, we showed that the VROA signal is sensitive to the helical pitch of polymer [7].

Besides these linear phenomena, in the presence of intense electromagnetic fields, chirality induces many nonlinear effects, making chiral molecules and supramolecular assemblies appropriate species for specific applications in optics and electro-optics. Chirality is at the origin of second

---

[1] benoit.champagne@fundp.ac.be

harmonic generation (SHG) [8] and sum-frequency generation (SFG) [9] phenomena, two types of second-order NLO processes. In the case of SHG, chirality provides a mixed electric-magnetic response (or electric dipole-electric quadrupole response since these nonlocal effects are generally difficult to distinguish) that can be used to probe molecular organization at interfaces [10]. SFG is associated with a pure electric response but (electronic and/or vibrational) resonance enhancement was so far necessary to make the SFG signal observable [11-12]. Other second-order NLO effects related to chirality encompass two-photon CD [13], two-photon OR [14], magnetochiral birefringence [15], and inverse magnetochiral birefringence [16]. Chirality-based higher-order phenomena have also been predicted and, in some cases, demonstrated. Among these, third-order nonlinear CD was evidenced recently [17], electric-field-induced SFG was recently observed in solutions of binaphtol [18], while BioCARS was predicted to probe the vibrational modes of chiral molecules, which are both Raman and hyper-Raman active [19]. More details as well as a generalization of the NLO phenomena description can be found in Ref. 20.

Rational design of NLO-phores and materials is a multidisciplinary field that includes, in addition to their synthesis and characterization, the prediction and interpretation of these properties from theoretical simulations. As reviewed recently [21], our approach consists in developing and applying theoretical tools. Although much has been done in this area to account for the effects of electron correlation and of the surroundings as well as for the vibrational contributions, especially for designing organic π-conjugated systems presenting large second-order NLO responses, little has been achieved for chirality-based properties [22]. The purpose of this paper is to discuss our recent works on the theoretical design of chiral NLO compounds. Part of these investigations will concentrate on a special type of chirality: helicity. Indeed, helical structures are attracting widespread interest in materials and life [23] sciences: DNA forms a double helix while Langmuir-Blodgett films of substituted helicenes can exhibit substantial second-order nonlinear optical responses [10].

2. Sum-Frequency Generation of chiral liquids

In isotropic media, within the electric dipole approximation, three-wave mixing (TWM) in second-order NLO processes is only symmetry-allowed for sum- and difference-frequency generations when the constituents are chiral and present an enantiomeric excess [9]. The TWM in chiral liquids is related to the completely antisymmetric isotropic component of the first hyperpolarizability (β) tensor, the pseudoscalar $\overline{\beta}$ quantity:

$$\overline{\beta}\left(-\omega_\sigma;\omega_1,\omega_2\right)=\frac{1}{6}\varepsilon_{\chi\zeta\eta}\beta_{\chi\zeta\eta}\left(-\omega_\sigma;\omega_1,\omega_2\right)$$
$$=\frac{1}{6}\left(\beta_{xyz}-\beta_{xzy}+\beta_{zxy}-\beta_{zyx}+\beta_{yzx}-\beta_{zyx}\right)$$

(1)

where χ, ζ, and η are Cartesian indices and the Einstein notation is assumed and $\omega_\sigma=\omega_1+\omega_2$ $\overline{\beta}$ is zero when the two incident frequencies are identical or one is zero. In our case, $\omega_1=2\omega=2\omega_2$. Although $\overline{\beta}$ changes sign with the molecular handedness, sum-frequency generation (SFG) cannot distinguish between enantiomers because the scattering power is proportional to the square of the polarization, and therefore to the square of $\overline{\beta}$. In off-resonance conditions, the chirality-allowed SFG phenomenon is weak. In addition to experimental works, where $\overline{\beta}$ was resonantly enhanced in order to enable demonstration of the phenomenon [11-12], theoretical works have investigated design strategies to tune the chiral SFG response. The first evaluation employs a phenomenological model based on a simple single-centre molecular orbital approach where the transition moments and energies are fixed to arbitrary – although realistic – values [9]. This model was used to estimate the magnitude of the SFG response as well as the importance of its frequency dispersion. Then, *ab initio* approaches have been used on model chiral compounds, R-(+)-propylene oxide and R-monofluoro-oxirane [9, 24-25]. Both Configuration Interaction Singles (CIS) and Time-Dependent Hartree-Fock (TDHF) methods were employed to evaluate the first hyperpolarizability tensor components. At the CIS level, the latter are expressed as summations over states (SOS), which enables to introduce a phenomenological damping factor associated to the population decay rate of the corresponding excited state. Numerous investigations have adopted the SOS/CIS scheme but it was combined with semi-empirical Hamiltonians. *Ab initio* SOS/CIS investigations are less frequent because the excitation energies are overestimated due to the neglect of electron correlation effects. Although different recipes exist to correct for these limitations [26], the so-called "scissor operator" has been used in the SOS/CIS SFG determinations. It consists in downshifting all excitation energies by the same amount, which can be determined using experimental data. Moreover, both semi-empirical and *ab initio* SOS/CIS approaches can suffer from the truncation of the summations imposed by the limited computational resources. An illustration of the evolution of the SFG first hyperpolarizability with the number of excited states is given in Fig. 1 for R-(+)-propylene oxide. Contrary to the vector projection of β on the dipole moment – the experimental quantity in electric-field-induced SHG – for push-pull π-conjugated systems, which is dominated by a few excited states [27], many excited states contribute to $\overline{\beta}(-3\omega;2\omega,\omega)$ of R-(+)-propylene oxide, a compound without chromophore. A similar conclusion has been drawn for R-monofluoro-oxirane [24]. Moreover, as discussed in Ref. 24, except in the vicinity of resonances, introducing a finite damping factor has a negligible effect on the SFG first hyperpolarizabilities. At the TDHF level, the successive electric field responses of the density matrix are obtained iteratively [28] to provide frequency-dependent first hyperpolarizabilities. The extension of the original schemes to the SFG response has recently been achieved within the *2n + 1* rule [25]. Fig. 1 compares the *ab initio* SOS/CIS and TDHF dispersion curves for R-(+)-propylene oxide. Provided a similar scissor operator is used in both, the two methods provide very similar $\overline{\beta}(-3\omega;2\omega,\omega)$ values. Further *ab initio* investigations will account for the effects of electron correlation.

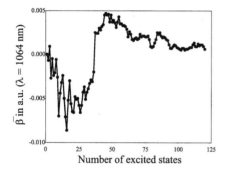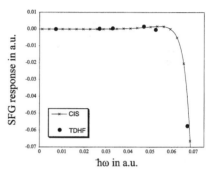

Figure 1: (left) CIS/6-311++G** $\overline{\beta}(-3\omega;2\omega,\omega)$ of R-(+)-propylene oxide as a function of the number of excited states included in the SOS expression (λ = 1064 nm, scissor operator = 2.675 eV); (right) dispersion of $\overline{\beta}(-3\omega;2\omega,\omega)$ of R-(+)-propylene oxide determined at the CIS and TDHF levels of approximation using the 6-311++G** basis set.

Since large first hyperpolarizabilities are generally associated with π-conjugation as well as specific substitutions by donor/acceptor groups, helicenes appear as natural candidates to exhibit large $\overline{\beta}(-3\omega;2\omega,\omega)$ values. Indeed, helicenes (Fig. 2) are spiral molecules made of ortho-fused benzene rings where the π-conjugated segment is the helix itself [29]. These structure/property investigations [30-31] were carried out adopting the TDHF/AM1 scheme. On the one hand, this semi-empirical scheme enables to study – in a reasonable amount of computational time – systems containing more than 100 atoms whereas it provides sufficient accuracy. Several times, it was shown that for evaluating the first hyperpolarizability, the TDHF/AM1 scheme performs better than the *ab initio* Hartree-Fock scheme in comparison with post-Hartree-Fock methods. In particular, for the hyper-Rayleigh scattering (HRS) response of tetrathia-[7]-helicene, its 2,13,diamino-substituted and 2,amino-13,nitro-substituted analogs, the MP2/6-31G β_{HRS} data are in the ratio 0.18:0.74:1.00 whereas the static TDHF/AM1 values are in the ratio 0.24:0.68:1.00 [31]. Moreover, for α,ω-nitro,amino-polyenes with 3 double bonds, the TDHF/AM1, TDHF/6-31G, and CCSD/6-31G dc-Pockels (λ = 1907 nm) longitudinal first hyperpolarizability values are in the ratio 0.83 : 0.48 : 1.00 whereas for SHG, 0.93 : 0.48 : 1.00, respectively [32].

Fig. 2 shows that, with the exception of the smallest helicenes which are planar or approximately planar, $\overline{\beta}(-3\omega;2\omega,\omega)$ increases almost linearly with the size of the helicene [30]. A similar behavior has been found for heliphenes, another class of spiral compounds where fused benzene rings alternate with cyclobutadiene rings [33], showing therefore that the amplitude of the chiral SFG response depends more on the size of the helix than on its nature.

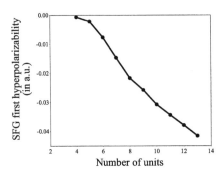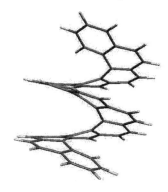

Figure 2 : (left) SFG first hyperpolarizability of increasingly large helicenes determined at the TDHF/AM1 level of approximation for $\omega_1 = 2\ \omega_2 = 1.0$ a.u. (right) representation of [13]-H.

Substitution effects investigated for tetrathia-[7]-helicene [31] demonstrate that appropriately-chosen positions of the donor/acceptor pair can enhance $\bar{\beta}(-3\omega; 2\omega, \omega)$. In particular, strong acceptor groups lead to larger SFG responses when placed on the central sites (M, M') than at the ends of the helix (N, N'). This favorable situation leads to a $\bar{\beta}(-3\omega; 2\omega, \omega)$ values one order of magnitude larger than for the non substituted species. When donor groups are used as substituents, the effect of their position (centre, end) is reduced, the most favorable case being to substitute end positions. Then, combining these two effects (NO_2 groups in M and M' positions while NH_2 groups in N and N' positions) leads to a $\bar{\beta}(-3\omega; 2\omega, \omega)$ value 22 times larger than for the non substituted species. These relative effects of donors and acceptors can also explain the variations of $\bar{\beta}(-3\omega; 2\omega, \omega)$ when substituting tetrathia-[7]-helicene by one D/A pair. Nevertheless, much work remains to be done in order to systematically unravel which functional groups maximize the SFG pseudoscalar and what is the effect of the nature of the helix. Indeed, similar investigations carried out for hexahelicene have shown little correlation between substituent positions and SFG response [30]. Another direction of investigation to enhance the chirality-based second-order NLO properties, consists in oxidizing or reducing the helicene, as discussed in Ref. 34 for the HRS response. Indeed, adding or removing an electron creates a geometrical defect, which can enhance the electron delocalization and therefore $\bar{\beta}(-3\omega; 2\omega, \omega)$.

Table 1. TDHF/AM1 $\bar{\beta}(-3\omega; 2\omega, \omega)$ (in a.u.; $\lambda = 1907$ nm) of substituted tetrathia-[7]-helicenes as a function of the number and position of the D/A NH_2/NO_2 substituents. Representation of tetrathia-[7]-helicene and numbering of the substituent positions.

M	M'	N	N'	$\bar{\beta}(-3\omega; 2\omega, \omega)$
H	H	H	H	-0.38
H	H	NO_2	NH_2	-0.05
NH_2	H	NO_2	H	0.52
NH_2	H	H	NO_2	-0.27
NO_2	H	H	NH_2	-1.97
NO_2	H	NH_2	H	-2.36
NO_2	NH_2	H	H	-0.58
NO_2	NO_2	H	H	-3.51
NH_2	NH_2	H	H	-0.87
H	H	NO_2	NO_2	0.67
H	H	NH_2	NH_2	-1.30
NO_2	NO_2	NH_2	NH_2	-8.43
NH_2	NH_2	NO_2	NO_2	1.35

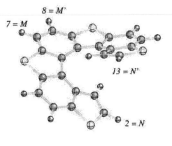

3.   Mixed electric-magnetic NLO responses of oriented films of helicenes

Stimulated by experimental investigations due to Persoons and co-workers [10, 35], the mixed electric-magnetic SHG first hyperpolarizabilities have been evaluated for helicenes with and without substituents [36]. They have been computed using the random phase approximation (RPA) [37] and the 6-31G* basis set as quadratic response functions:

$$\beta_{ijk}^{eem}(-2\omega;\omega,\omega) = \frac{1}{2}\langle\langle r_i; r_j; L_k \rangle\rangle_{\omega,\omega} \tag{2}$$

$$\beta_{ijk}^{mee}(-2\omega;\omega,\omega) = \frac{1}{2}\langle\langle L_i; r_j; r_k \rangle\rangle_{\omega,\omega} \tag{3}$$

where r and L stand for the electric dipole and angular momentum operators, respectively. The superscripts refer to the electric-dipole (e) and magnetic-dipole (m) interactions. Together with the pure electric-dipole contributions, these microscopic quantities have been used to evaluate the macroscopic second-order NLO responses of films of helicenes, oriented along the z-direction and presenting C_∞ symmetry [38]. In the simulation, the helicene axis defines a θ angle with respect to the macroscopic z-axis. The macroscopic responses are obtained by averaging the miscroscopic responses over the various free orientations of the helix. For the special case where $\theta = 0°$,

$$\chi_{zzz}^{eee} = N\beta_{z'z'z'}^{eee} \tag{4}$$

$$\chi_{yzy}^{eem} = \frac{N}{2}\beta_{x'z'x'}^{eem} \tag{5}$$

$$\chi_{zzz}^{mee} = N\beta_{z'z'z'}^{mee} \tag{6}$$

where N is the number density of the helicenes and x', y', and z' the Cartesian coordinates in the molecular frame. The effective SHG polarization for this experimental configuration is:

$$\vec{P}_{eff}(2\omega) = \chi_{zzz}^{eee}E_z(\omega)E_z(\omega)\vec{k} + \left(\chi_{yzy}^{eem} + \chi_{zzz}^{mee}\right)E_z(\omega)E_z(\omega)\vec{j} \tag{7}$$

where $\vec{k}$ and $\vec{j}$ are the unit vectors along the z and y laboratory axes, respectively (the laser beam propagates along the x-axis) while E_z is the electric field amplitude. Subsequently, the ratio between the mixed electric-magnetic terms and the pure electric contributions is given by:

$$R = \frac{\chi_{yzy}^{eem} + \chi_{zzz}^{mee}}{\chi_{zzz}^{eee}} = \frac{\beta_{x'z'x'}^{eem} + 2 \beta_{z'z'z'}^{mee}}{2 \beta_{z'z'z'}^{eee}} \tag{8}$$

Since in these RPA quadratic response calculations the gauge invariance is not ensured, the mixed electric-magnetic responses are origin-dependent. The center of mass was chosen as the origin of the system of axes. Then, by comparing the 6-31G* results to these obtained using the 6-31G, 6-31G**, and 6-311G* basis sets, it was found that basis sets effects are of similar magnitude for both the mixed electric-magnetic responses and the pure electric contribution [36]. As already mentioned in the Introduction, there also exists a electric-dipole electric-quadrupole contribution. However, since helicenes are known to present enhanced magnetic-dipole effects (optical rotation and circular dichroism [39]), the electric-quadrupole contribution was assumed to be negligible with respect to the former. When applying this theoretical scheme to tetrathia-[7]-helicene and bistermethylsilyl tetrathia-[7]-helicene (Fig. 3), structure-property relationships have been evidenced. For tetrathia-[7]-helicene, the dominant contribution comes from χ_{zzz}^{mee} and R = -1.45 whereas for bis-termethylsilyl tetrathia-[7]-helicene, the pure electric contribution is dominant and R = 0.28. However, both χ_{zzz}^{mee} and χ_{yzy}^{eem} are larger in bis-termethylsilyl tetrathia-[7]-helicene than in tetrathia-[7]-helicene. Subsequent TDDFT/B3LYP/6-31G* calculations have been performed on these helicenes in order to further address the relationships between molecular structures and NLO properties [40]. It was found that upon substitution by two SiMe$_3$ groups, the first important CD peak shifts from 3.26 eV to 3.17 eV whereas its intensity increases by about 40%.

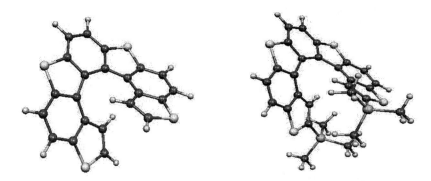

Figure 3: Representation of (left) tetrathia-[7]-helicene and (right) bis-termethylsilyl tetrathia-[7]-helicene.

4. Conclusions and outlook

Chirality-based NLO responses constitute another challenge for the design of new materials and the related theoretical and experimental investigations. In this paper, using quantum chemistry tools, several of these aspects are addressed for the purely-electric sum-frequency generation of isotropic media as well as the mixed electric-magnetic second harmonic generation responses. In particular, helicenes of which the π-electron delocalizable network is associated with the helix itself, are shown to be suitable candidates to achieve large NLO responses.

Acknowledgments

The author thanks Dr. Edith BOTEK, Prof. Koen CLAYS, Prof. Peer FISCHER, Vincent LIÉGEOIS, Prof. André PERSOONS, Dr. Olivier QUINET, and Dr. Milena SPASSOVA for fruitful discussions. He also thanks the Belgian National Fund for Scientific Research for his Research Director position.

References

[1] J.M. Hicks (Ed.) Chirality: Physical Chemistry, ACS Symposium Series 810 (ACS, Washington, 2002).

[2] P. de P. Ricketts, J. Am. Chem. Soc. **1**, 2 (1879).

[3] G. Holzwarth, E.C. Hsu, H.S. Mosher, T.R. Faulkner et A. Moscowitz, J. Am. Chem. Soc. **96**, 251 (1974).

[4] L.A. Nafie, Annu. Rev. Phys. Chem. **48**, 357 (1997).

[5] L.D. Barron, M.P. Bogaard et A.D. Buckingham, J. Am. Chem. Soc. **95**, 603 (1973).

[6] I.H. McColl, E.W. Blanch, L. Hecht, N.R. Kallenbach, and L.D. Barron, J. Am. Chem. Soc. **126**, 5076 (2004); G. Zuber and W. Hug, Helv. Chim. Acta **87**, 2208 (2004).

[7] V. Liégeois, O. Quinet, and B. Champagne, J. Chem. Phys. **122**, 214304 (2005).

[8] T. Petralli-Mallow, T.M. Wong, J.D. Byers, H.I. Yee, and J.M. Hicks, J. Phys. Chem. **97**, 1383 (1993); M. Kauranen, T. Verbiest, E.W. Meijer, E.E. Havinga, M.N. Teerenstra, A.J. Schouten, R.J.M. Nolte, and A. Persoons, Adv. Mater. **7**, 641 (1995).

[9] J.A. Giordmaine, Phys. Rev. **138**, A1559 (1965); P. Fischer, D.S. Wiersma, R. Righini, B. Champagne, and A.D. Buckingham, Phys. Rev. Lett. **85**, 4253 (2000).

[10] T. Verbiest, S. van Elshocht, M. Kauranen, L. Hellemans, J. Snauwaert, C. Nuckolls, T.J. Katz, and A. Persoons, Science **282**, 913 (1998); F. Hache, H. Mesnil, and M.C. Schanne-Klein, J. Chem. Phys. **115**, 6707 (2001).

[11] M.A. Belkin, T.A. Kulakov, K.-H. Ernst, L. Yan, and Y.R. Shen, Phys. Rev. Lett. **85**, 4474 (2000); M.A. Belkin, S.H. Han, X. Wei, and Y.R. Shen, Phys. Rev. Lett. **87**, 113001 (2001); M.A. Belkin and Y.R. Shen, Phys. Rev. Lett. **91**, 213907 (2003).

[12] P. Fischer, K. Beckwitt, F.W. Wise, and A.C. Albrecht, Chem. Phys. Lett. **352**, 463 (2002); P. Fischer, F.W. Wise, and A.C. Albrecht, J. Phys. Chem. A **107**, 8232 (2003).

[13] I. Tinoco, J. Chem. Phys. **62**, 1006 (1975); T. Verbiest, M. Kauranen, A. Persoons, M. Ikonen, J. Kurkela, and H. Lemmetyinen, J. A. Chem. Soc. **116**, 9203 (1994); M.C. Schanne-Klein, F. Hache, A. Roy, C. Flytzanis, and C. Payrastre, J. Chem. Phys. **108**, 9436 (1998); T. Petralli-Mallow, A.L. Plant, M.L. Lewis, and J.M. Hicks, Langmuir **16**, 5960 (2000).

[14] J.D. Byers, H.I. Yee, and J.M. Hicks, J. Chem. Phys. **101**, 6233 (1994); R. Cameron and G.C. Tabisz, Mol. Phys. **90**, 159 (1997).

[15] G.J.L.A. Rikken and E. Raupach, Nature (London) **390** 493 (1997).

[16] G. Wagnière, Phys. Rev. A **40**, 2437 (1989).

[17] F. Hache, H. Mesnil, and M.C. Schanne-Klein, Phys. Rev. B **60**, 6405 (1999); H. Mesnil and F. Hache, Phys. Rev. Lett. **85**, 4257 (2000).

[18] P. Fischer, A.D. Buckingham, K. Beckwitt, D.S. Wiersma, and F.W. Wise, Phys. Rev. Lett. **91**, 173901 (2003).

[19] N.I. Koroteev, Biospectrosc. **1**, 341 (1995).

[20] G.H. Wagnière, *Linear and Nonlinear Optical Properties of Molecules* (Verlag Hel.v. Chim. Acta, Basel, 1993).

[21] B. Champagne and B. Kirtman, in *Handbook of Advanced Electronic and Photonic Materials and Devices*, edited by H.S. Nalwa, Vol. 9, *Nonlinear Optical Materials*, (Academic Press, San Diego, 2001), Chap. 2, p. 63.

[22] D. Jonsson, Y. Luo, K. Ruud, P. Norman, and H. Ågren, Chem. Phys. Lett. **288**, 371 (1998); S. Coriani, M. Pecul, A. Rizzo, P. Jørgensen, and M. Jaszunski, J. Chem. Phys. **117**, 6417 (2002).

[23] H. Jiang, C. Dolain, J.M. Léger, H. Gornitzka, and I. Huc, J. Am. Chem. Soc. **126**, 1034 (2004).

[24] B. Champagne, P. Fischer, and A.D. Buckingham, Chem. Phys. Lett. **331**, 83 (2000).

[25] O. Quinet and B. Champagne, Int. J. Quantum Chem. **85**, 463 (2001).

[26] See for instance, A. Kohn and C. Hättig, J. Chem. Phys. **119**, 5021 (2003); S. Grimme and E.I. Izgorodina, Chem. Phys. **305**, 223 (2004).

[27] For a general survey, see D.R. Kanis, M.A. Ratner, and T.J. Marks, Chem. Rev. **94**, 195 (1994); for an *ab initio* illustration, see M. Spassova, V. Monev, I. Kanev, B. Champagne, D.H. Mosley, and J.M. André, in *Quantum Systems in Chemistry and Physics*, Vol. 1: *Basic Problems and Model Systems*, edited by A. Hernandez-Laguna et al. (Kluwer, Dordrecht, 2000) 101.

[28] H. Sekino and R.J. Bartlett, J. Chem. Phys. **85**, 976 (1986); S.P. Karna and M. Dupuis, J. Comp. Chem. **12**, 487 (1991).

[29] R.H. Martin, Angew. Chem. (Int. Ed.) **13**, 649 (1974).

[30] E. Botek, B. Champagne, M. Turki, and J.M. André, J. Chem. Phys. **120**, 2042 (2004).

[31] B. Champagne, J.M. André, E. Botek, E. Licandro, S. Maiorana, A. Bossi, K. Clays, and A. Persoons, ChemPhysChem **5**, 1438 (2004).

[32] D. Jacquemin, B. Champagne, and C. Hättig, Chem. Phys. Lett. **319**, 327 (2000); E. Botek and B. Champagne, Appl. Phys. B **74**, 627 (2002).

[33] S. Han, A.D. Bond, R.L. Disch, D. Holmes, J.M. Schulman, S.J. Teat, K.P.C. Vollhardt, and G.D. Whitener, Angew. Chem. Int. Ed. **41**, 3227 (2002).

[34] E. Botek, M. Spassova, B. Champagne, I. Asselberghs, A. Persoons, and K. Clays, Chem. Phys. Lett., in press.

[35] Th. Verbiest, S. Sioncke, A. Persoons, L. Vyklicky, and T. Katz, Angew. Chem. Int. Ed. **41**, 3882 (2002).

[36] E. Botek, J.M. André, B. Champagne, Th. Verbiest, and A. Persoons, J. Chem. Phys. **122**, 234713 (2005).

[37] J. Linderberg and Y. Öhrn, *Propagators in Quantum Chemistry* (Wiley Interscience, New York, 2004); H. Hettema, H.J.Aa. Jensen, P. Jørgensen, and J. Olsen, J. Chem. Phys. **97**, 1174 (1992).

[38] S. Sioncke, Th. Verbiest, and A. Persoons, Opt. Mater. **21**, 7 (2002).

[39] F. Furche, R. Ahlrichs, C. Wacksmann, E. Weber, A. Sobanski, F. Vögtle, and S. Grimme, J. Am. Chem. Soc. **122**, 1717 (2000); S. Grimme, Chem. Phys. Lett. **339**, 380 (2001).

[40] M; Spassova *et al.*, in preparation.

Brill Academic Publishers
P.O. Box 9000, 2300 PA Leiden,
The Netherlands

*Lecture Series on Computer
and Computational Sciences*
Volume 3, 2005, pp. 35-50

Grid Enabled Molecular Dynamics: classical and quantum algorithms

S. C. Farantos[a,b,1], **S. Stamatiadis**[a,b], **L. Lathouwers**[c] and **R. Guantes**[d]

[a] Institute of Electronic Structure and Laser, Foundation for Research and Technology-Hellas,
Iraklion 711 10, Crete, Greece.
[b]Department of Chemistry, University of Crete, Iraklion 711 10, Crete, Greece.
[c]Department of Wiskunde-Informatica, University of Antwerp, Groenenborgerlaan 171, B2020,
Antwerp, Belgium.
[d]Instituto de Matematicas y Fisica Fundamental, Consejo Superior de Investigaciones Cientificas,
Serrano 123, 28006 Madrid, Spain.

Received 5 August, 2005; accepted in revised form 12 August, 2005

Abstract: Molecular simulations have become a powerful tool in investigating the microscopic behavior of matter as well as in calculating macroscopic observable quantities. The predictive power and the accuracy of the methods used in molecular calculations are closely related to the current computer technology. Thus, the rapid advancement of Grid computing, i.e. the utilization of geographically distributed computers connected by relatively high latency networks, is expected to influence extensively the progress of computational sciences, provided algorithms which can utilize the hundreds and even thousands of the available computers in the Grid exist. In this lecture we review some of our computational methods, classical and quantum, used in small molecules, which seem promising for studying large scale in time and molecular size problems. In particular, in classical molecular simulations we are searching for specific trajectories connecting two regions of phase space (rare events) by solving two-point boundary value problems with multiple shooting techniques. In quantum dynamics we argue that using variable order finite difference methods for solving the Schrödinger equation in a cartesian coordinate system result in sparse Hamiltonian matrices which can make large scale problem solving feasible.

Keywords: Quantum molecular dynamics, classical molecular dynamics, finite differences, multiple shooting methods, Grid computing.

Mathematics Subject Classification: 0.2.60-x

PACS: 0.2.60.Lj, 0.2.70.Bf, 02.70.Ns, 34.30.+h, 82.20.Ln

1 Introduction

Computational Chemistry is a topic whose progress is closely related to the current technology of computers. Molecular simulations are important in designing new materials, pharmaceuticals, in Biological Chemistry and Physics. In general, Molecular Computational Sciences face two challenges in the twenty first century. The first has to do with the size of the systems; studies which cover molecules from two atoms to nanostructures and the macroscopic states of matter are required. The second challenge is related to time and the dynamics of the systems; phenomena

---

[1]Corresponding author. E-mail: farantos@iesl.forth.gr

which last from femtoseconds to seconds are important in material and biological sciences. Grid (distributed) computing could be the solution in bridging the gaps in time and space. A Grid of computers is envisioned as a seamless, integrated, computational and collaborative environment embracing different categories of distributed systems. Following the classification introduced by Foster and Kesselman ("The Grid: the blueprint for a new computing infrastructure") [1], computational Grids are categorized into five major classes of applications. In summary, these classes identify a specific context of applications such as supercomputing applications, high-throughput computing, on-demand computing, data-intensive computing and collaborative computing. This new paradigm in scientific computing is rapidly developing. Several Grid testbeds have been deployed around the world among which the European Grid. The ENACTS project [2] with its well planned and in depth studies has shown that Grids of computers are well developed not only at the national level but also as a European multi-national integration.

In our days, the most successful computational Grid model is the one based on internet by accessing thousands of PCs via running the programs as screensavers. The first such application was SETI@home to analyze the data from radio telescopes looking for signs of extraterrestrial life. Taking its inspiration from SETI, a protein folding project called Folding@Home (http://www.stanford.edu/group/pandegroup/folding) has been in operation at Stanford University for several years. Another active academic project is Predictor@home (http://predictor. scripps.edu/) which is aiming to structures prediction. This is a pilot of the Berkeley Open Infrastructure for Network Computing (BOINC), a software platform for distributed computing using volunteer computer resources. Another action on the human protein folding problem has been taken by the World Community Grid (WCG) (http://www.worldcommunitygrid.org/) which is supported by United Devices (UD). UD also runs the life-science research hub (http://www.grid.org/) to search for drug candidates to treat the smallpox virus.

In order to use successfully a worldwide distributed computing environment of hundreds or even thousands of heterogeneous processors such as the Grid, communications among these processors should be minimum. There are not many molecular simulations using the Grid environment such as to allow us to point out the strategy one should adopt in writing codes for molecular applications. A practical rule is to allow each processor to work independently, even though calculations are repeated, and only if something important happens to one of them, then they communicate. The common ingredient of the above successful applications in Internet Grid Computing is the computational algorithm employed, which guarantees minimum and asynchronous communications. Obviously, there are few problems which can be solved with such algorithms. Most of the interesting problems of Computational Chemistry require solutions of a large number of linear equations that can not be solved without significant communication among the computer nodes. On the other hand, parallelized codes written for large parallel machines may be of no good use when thousands of computers should be exploited connected by high latency networks. Thus, new algorithms and new programming paradigms suitable for distributed computing should be investigated in order to exploit Computational Grids [3].

In this lecture we review some of our methods used in classical and quantum dynamics of highly excited small molecules and we examine their potential to be utilized in a Grid computational environment. In quantum dynamics we shall investigate the combination of time evolution of wave packets in cartesian coordinate representations with angular momentum projections and the realization of this technique via generalizable time evolution schemes and variable order finite differencing methods for computing the derivatives. With respect to classical dynamics we shall explore the idea of searching specific trajectories which connect regions of phase space that correspond to different conformations of the molecule by formulating the problem as a two-point boundary value and using multiple shooting techniques.

2   Computational Methods

Our understanding of basic molecular phenomena is strongly based on our ability to successfully solve the Schrödinger equation for realistic molecular systems. This is the aim of Quantum Molecular Dynamics (QMD) and it is the way to obtain reliable quantitative predictions. However, very useful physical insight can be gained from the use of a causal time dependent picture where, for instance, the path of a molecular encounter or a vibrational motion can be traced in configuration space. Therefore, a lot of efforts have been directed towards the development of methods which combine classical description with quantum corrections. Accurate wave packet propagation, together with knowledge of the proper classical orbits of the system, is one of the most complete views that we can have for a molecular process because exact numerical results are supplemented by a very clear physical picture. Such a program is accomplished by locating families of periodic orbits which then can be used to determine initial configurations for the wave packets [4, 5].

In the next two subsections we describe our approaches to QMD and Classical Molecular Dynamics (CMD) which are based on discretizing the configuration space and time respectively. The grid representation of space and time make the algorithms suitable for using them in a Grid of computers.

2.1   Quantum Molecular Dynamics

Grid methods for the solution of the Schrödinger equation are nowadays one of the most powerful and exploited tools both for the time independent (known as Discrete Variable Representation (DVR)) and time dependent pictures. Concerning the last one, it was the introduction of the Fourier Pseudospectral (PS) method which provided the necessary accuracy and computational efficiency to compete with the traditional variational techniques and to make feasible the task (for an excellent review see Ref.[6]). In the time dependent picture or wave packet propagation, the wave packet is advanced in time by an evolution operator which, if the Hamiltonian $\hat{H}$ is time independent, is an exponential function $\hat{U}(t; \hat{H}) = e^{-i\hat{H}t}$ (we put $\hbar = 1$). This can be approximated by, for instance, a polynomial expansion [7]. The basic operation is therefore reduced to the evaluation of the action of the Hamiltonian operator onto the wave packet, $\hat{H}\Psi$. In long time propagation this operation has to be repeated many times, and thus, its efficiency will generally determine the computational cost of the problem. In practice, the wave function is often discretized in a grid using a collocation method and the Hamiltonian operator is represented by a matrix. If we considered only local operators this matrix would be diagonal, but the Hamiltonian includes the kinetic energy operator which is non local in the coordinate domain. If the number of collocation points in each coordinate α is n_α, the Hamiltonian matrix will contain of the order of $N \times n$ nonzero elements, where N is the total number of collocation points $N = \prod_\alpha n_\alpha$ and $n = \sum_\alpha n_\alpha$.

In the time independent picture one is concerned with the diagonalization of the Hamiltonian matrix constructed in the same way to obtain the eigenvalues and eigenvectors. Usually, an iterative procedure such as the Lanczos method and variants [8] is used for efficient diagonalization, which again involves products of the Hamiltonian matrix with vectors as the basic step. It is clear that important computational savings are possible when the Hamiltonian matrix is sparse or its action on the wave function can be calculated efficiently.

A way to increase significantly the sparsity of the Hamiltonian matrix is by the use of *local* methods. Here, local means that the action of the kinetic energy operator (or the Laplacian) on the wave function is approximated by using only local information or neighboring grid points. Although the derivative of a function is a local property, the wave function is defined on the whole configuration space and a piecewise representation of a function by a local polynomial approximation, generally, converges more slowly than a spectral representation.

In this work we review some recent advances in Finite Difference algorithms which may result in sparse Hamiltonian matrices. However, developing algorithms for QMD suitable to use the Grid computing technologies we must first decide the coordinate system to express the molecular Hamiltonian.

2.1.1   The angular momentum projection method

In their 1996 review of Quantum Molecular Dynamics [9] R.E. Wyatt and J.Z.H. Zang state "it is very important to develop and apply methods that can be extended to moderately sized polyatomic molecules (5 to 12 atoms)". In spite of the diversity of theoretical approaches and the ever growing computational facilities it is safe to say that the calculation of vibration-rotation spectra or state to state cross sections for five or more atomic systems is all but routine. We will first examine the reasons for this state of affairs and then propose a scheme that can satisfy the above ambitions.

The standard way of describing a polyatomic system in the absence of external fields is to introduce the center of mass, three Euler angles and $3N-6$ internal coordinates. The Hamiltonian in mass weighted cartesian coordinates x_{km} is written

$$\hat{H} = -\frac{1}{2}\sum_{k,m}\frac{\partial^2}{\partial x_{km}^2} + V(R), \tag{1}$$

where, $k = 1, 2, 3$ denotes the three cartesian coordinates, $m = 1, \ldots, N$ the atoms and $V(R)$ the potential energy as a function of the internal coordinates R.

The transformation of the original cartesian Hamiltonian to the above general curvilinear coordinates yields the generic form

$$\hat{H} = \hat{H}_0\left(q, \frac{\partial}{\partial q}\right) + \sum_k A_k\left(q, \frac{\partial}{\partial q}\right)J_k + \sum_{k,l}\frac{1}{2}I_{kl}^{-1}(q)J_kJ_l. \tag{2}$$

J_k denote the components of angular momentum and I_{kl} are the components of the moment of inertial tensor.

In the above equation we recognize the vibrational, rotational and the vibrational-rotational coupling parts of the Hamiltonian. We will refer to the derivation of Eq. (2) as *frame transformation* (FT). The problem with Eq. (2) is twofold. First, there is the analytical derivation of the Hamiltonian parts in curvilinear coordinates $q_1, \ldots, q_{3N-6}$, and Euler angles ϕ, θ, γ or angular momentum operators. Not only this is a cumbersome task for more than three atoms but this procedure has to be repeated for every choice of curvilinear coordinates and angles. Secondly, the general expression for the vibrational Hamiltonian reads

$$\hat{H}_0\left(q, \frac{\partial}{\partial q}\right) = -\frac{1}{2}\sum_{i,j}\frac{\partial}{\partial q_i}\left(\sqrt{g}g^{ij}\frac{\partial}{\partial q_j}\right) + V(q), \tag{3}$$

where g^{ij} is the metric tensor of the curvilinear coordinates. The appearance of cross derivatives implies that the number of partial differential operators necessarily exceeds $(3N-6)(3N-5)/2$. Therefore, the computational effort scales as N^2.

The question then arises whether an alternative theoretical approach can be devised that avoids the inherent problems of frame transformation and that can readily be extended to five or more atoms. Our basic starting point is to not separate rotations at the operator level but to use projection techniques to conserve total angular momentum. An alternative version of Eq. (2) is to introduce N Jacobi vectors the last one of which is the center of mass. The Hamiltonian can then be written as

$$\hat{H} = -\frac{1}{2}\sum_{k,m}\frac{\partial^2}{\partial u_{km}^2} + V(R(u)), \tag{4}$$

which contains the $3N - 3$ kinetic energies of the Jacobi coordinates u_{km}, $k = 1, 2, 3$ and $m = 1, \dots, N - 1$.

We first observe that all molecular properties such as spectra, cross sections, etc., can be obtained by Fourier transforming the appropriate time-correlation function. For specified total angular momentum, J, the vibration-rotation spectra require the calculation of the autocorrelation function

$$C_J(t) = < \Psi(0)| \exp(-i\hat{H}_J t)|\Psi(0) >, \tag{5}$$

while for other properties the appropriate operator has to be inserted in the matrix element. Consider the rotations about the center of mass which constitute the rotation group of the system. The corresponding operators on the system wave functions are

$$R(\Omega) = \exp(-i\phi J_z) \exp(-i\theta J_y) \exp(-i\gamma J_z), \tag{6}$$

while, the associated irreducible representations are the Wigner functions

$$D^J_{MK}(\Omega) = \exp(-iK\phi)d^J_{MK}(\theta) \exp(-iL\gamma). \tag{7}$$

From the group elements and the irreducible representations one can construct the group projection operators

$$P^J_{MK} = \frac{2J + 1}{8\pi^2} \int d\Omega D^J_{MK}(\Omega)R(\Omega). \tag{8}$$

Their properties have been listed and proven elsewhere [10]. The ones we need in the present context are

$$\sum_J P_J = 1, \text{ with } P_J = \sum_K P^J_{KK}, \tag{9}$$

i.e., one can construct the projector onto the space of total angular momentum J as the K sum of the operators P^J_{KK}. The operators P_J, being true projection operators, are idempotent and hermitian and, because the Hamiltonian is rotationally invariant, they commute with $\hat{H}$

$$P_J^2 = P_J, \quad P_J^\dagger = P_J, \quad [\hat{H}, P_J] = 0. \tag{10}$$

Since, the operator $\hat{H}_J$ appearing in Eq.(5) is really $P_J \hat{H} P_J$, one can rewrite the formula for the autocorrelation function as

$$C_J(t) = < \Psi(0)|P_J \exp(-i\hat{H}t)P_J|\Psi(0) > . \tag{11}$$

This expression is the key to the method. Indeed, it shows that evaluating $C_J(t)$ can be regarded as a three stage process: the projection of the initial state onto the subspace J, the time evolution of the initial state under the total Hamiltonian and the calculation of the overlap of the projected and time evolved states.

The projection of the initial state is a one time numerical integration and is computationally irrelevant versus the time evolutions. The time evolution of the $3N - 6$ variables involves cross derivatives in a $2J + 1$ dimensional matrix. The alternative propagation under Eq. (4) uses $3N - 3$ strictly separated variables. One can make a rough estimate of the ratio of the computational effort of calculating the action of Eq. (2) versus Eq. (4). Assuming that all variables are mapped on a grid of P points we find $(2J + 1)(3N - 6)P^{-3}$.

However there is an extra savings device for the evolution under Eq. (4). Indeed, the wave function will not extend over the entire Jacobi hypercube but rather it is confined to the corresponding hypercube.

2.1.2 Variable order finite difference method

Now we proceed to compute the action of the Hamiltonian operator onto the wave function. In previous papers [11, 12], we reported results from the application of a variable order Finite Difference (FD) method to approximate the action of a Hamiltonian operator on the wave function in the time-dependent Schrödinger equation or the Hamiltonian matrix elements in the time-independent picture. One, two and three dimensional model potentials in radial and angular coordinates were used to investigate the accuracy and the stability of these methods, whereas in a companion paper [13], the time-dependent Schrödinger equation was solved for the van der Waals system Ar_3. The impetus for this project was given by recent advances in high order Finite Difference approximations. We mainly refer to the limit of infinite order Finite Difference formulas with respect to global Pseudospectral methods (PS) investigated by Fornberg [14], and Boyd's work [15] which views Finite Difference methods as a certain sum acceleration of Pseudospectral techniques.

Finite Difference approximations of the derivatives of a function $F(x)$ can be extracted by interpolating $F(x)$ with Lagrange polynomials, $P(x)$. This allows one to calculate the derivatives analytically at arbitrarily chosen grid points and with a variable order of approximation. The Lagrange fundamental polynomials of order $N - 1$ are defined by

$$L_k(x) = \prod_{j=1}^{N}{}' (x - x_j) / \prod_{j=1}^{N}{}' (x_k - x_j), \quad k = 1, 2, ..., N, \tag{12}$$

where the prime means that the term $j = k$ is not included in the products. The values of $L_k(x_j)$ are zero for $j \neq k$ and one for $j = k$ by construction. The function can then be approximated as

$$F(x) \approx P_N(x) = \sum_{k=1}^{N} F(x_k) L_k(x). \tag{13}$$

P_N is a polynomial of order $N - 1$. In Ref.[11] we discussed how FD is related to the Sinc-DVR method by taking the limit in the two above mentioned senses:

i) An infinite order limit of centered FD formulas on an equi-spaced grid yields the Discrete Variable Representation result when we use as a basis set the Sinc functions ($\text{Sinc}(x) \equiv \sin(\pi x)/\pi x$) [14, 17]. Although, this limit is defined formally as N, the number of grid points used in the approximation, tends to infinity, some theoretical considerations [14] as well as numerical results [11] lead us to expect that the accuracy of the FD approximation is the same to that of the DVR method as we approach the full grid to calculate the FD coefficients.

ii) FD can also be viewed as a sum acceleration method which improves the convergence of the Pseudospectral approximation [15]. The rate of convergence is, however, non-uniform in the wave number, giving very high accuracy for low wave numbers and poor accuracy for wave numbers near the aliasing limit [18]. However, this does not cause a severe practical limitation, since, by increasing the number of grid points in the appropriate region we can have an accurate enough representation of the true spectrum in the range of interest. This is one property which makes FD useful as an alternative to the common DVR [19] and other PS methods such as Fast Fourier Transform techniques (FFT) [6, 7].

The current interest in Finite Difference methods is fully justified when solutions of the Schrödinger equation are required for multidimensional systems such as polyatomic molecules. The present most popular methods employed in quantum molecular dynamics are the Fast Fourier Transform and the Discrete Variable Representation techniques. FFT generally uses hypercubic grid domains which result in wasted configuration space sampling. A large number of the selected configuration points correspond to high potential energy values, which do not contribute to the eigenstates that we are seeking. Global DVR methods allow us to choose easily the configuration points which are relevant to the states we want to calculate, but still, we must employ in each dimension all grid points. Local methods such as FD have the advantages of DVR but also produce sparse matrices which make them more economical in computer memory and time provided the PS accuracy is achieved at lower order than the high order limit.

There are some other benefits for FD with respect to global Pseudospectral methods. Convergence can be examined not only by increasing the number of grid points but also by varying in a systematic way the order of approximation of the derivatives. Finite Difference methods may incorporate several boundary conditions and choose the grid points without necessarily relying on specific basis functions. The topography of the multidimensional molecular potential functions is usually complex. The ability of using non equi-spaced grids is as important as keeping the grid points in accordance to the chosen energy interval. The computer codes for a FD representation of the Hamiltonian can be efficiently parallelized, since the basic operation is the multiplication of a vector by a sparse matrix. The ongoing, active research in the relevant field produced various approaches to the efficient distribution of a sparse matrix, especially a "regular" one, i.e., one formed through the discretization of a differential equation on a grid, and the subsequent multiplication with an appropriately distributed vector. The developed methods tackle, in various degrees, the requirements of load balancing and minimal communication cost across a processor grid. The overview presented in [20] is a starting point for the available techniques.

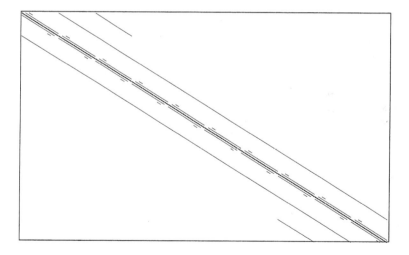

Figure 1: Pattern of a Hamiltonian matrix for a system of three degrees of freedom with the kinetic energy operator evaluated by finite difference approximation using three point stencils.

In Figure 1 we show schematically the Hamiltonian matrix when we employ a 3-point difference method to calculate the second derivatives in a three dimensional system. The high sparsity of the matrix is inherent, and this is the best for distributed computing.

As an example we will study a two dimensional system employed in several investigations in the past, mainly in the connection between classical and quantum dynamics. One of us [21] carried out extensive studies of the periodic orbit structure of this system and its relation to quantum mechanics. The system is described by the Hamiltonian:

$$H = \frac{1}{2}(p_x^2 + p_y^2) + \frac{1}{2}(\omega_x^2 x^2 + \omega_y^2 y^2) - \epsilon x^2 y, \tag{14}$$

where the parameters are $\omega_x^2 = 0.9$, $\omega_y^2 = 1.6$ and $\epsilon = 0.08$. The time dependent Schrödinger equation was solved numerically using a Chebychev expansion [7] for the propagation in time, while the action of the Hamiltonian operator on the wave function was evaluated with the FFT method and with matrix-vector multiplication using FD in order to study the convergence and compare the computation times. The resulting vibrational spectrum was obtained as usual with the Fourier transform of the autocorrelation function of the wave packet

$$I(E) = \frac{1}{2\pi} \int_{-\infty}^{\infty} \exp(iEt) < \Psi(\vec{x},0)|\Psi(\vec{x},t) > dt. \tag{15}$$

We propagated a Gaussian wave packet initially localized on a 1:2 resonance periodic orbit at a high energy ($E = 23$ a.u., see Fig. 12b in Ref.[21]). It is interesting to plot the wave packet in configuration space and compare the solution of the Schrödinger equation using the two methods. In both cases we used a rectangular grid with 64 points in each dimension, and $\Delta x = 0.3175$. We wanted to obtain a high resolution for the spectrum ($\Delta E \simeq 0.056$), and thus, we propagated the wave packet for 1024 time steps. At $t = 1/4$ of the total time we took a snapshot of the wave function. In Fig. 2a we show the wave function obtained with the FFT method (the potential energy contour is superimposed on the same plot at the mean energy of the wave packet). In Figs 2b, c and d we show the same wave packet obtained with a second, fourth and seventh order Finite Difference approximation, respectively. The order of approximation (M) yields stencils with $2M+1$ points. From Fig. 2 we can see that convergence of the wave function is approached with the FD method, even for a wave packet quite spread in configuration space. Finer details reproduced by even higher order FD approximations, however, will not affect very much the spectrum, since, this is an average of the propagated wave function over the configuration space.

The resulting spectra are shown in Fig. 3 (here we omit the second order FD for the sake of clarity, although the comparison is very poor as expected). Even with order $M = 7$, the only appreciable difference is in the intensities at high energies.

We recall that, a FD approximation is equivalent to a Taylor expansion series of the kinetic energy spectrum, therefore, in the limit of $\Delta x \to 0$ we recover the exact spectrum irrespectively of the order of the approximation. It is worth to investigate then how the FD converges to the Fourier method as we increase the order as well as we decrease the grid spacing. We examine the differences of the central eigenvalue ($E = 22.18$) from that obtained with the FFT method with 64 points in each dimension. When this difference is less than the resolution in the power spectrum the results are considered identical. Of course, for higher eigenvalues we should increase further the order or decrease the grid spacing to converge to the desired resolution, but the general behavior is seen in Fig. 4 where we used 64, 80, 100 and 120 points ($\Delta x = 0.3175, 0.2532, 0.2020$ and 0.1681, respectively). It is seen that convergence in both directions is quite fast, although it is computationally cheaper on a single computer to increase the order of the approximation and take less grid points than to increase the number of grid points (by increasing the order by one we should add two more grid points to evaluate the Laplacian, but we should increase N by about 20 points at low order to get the same reduction in the error).

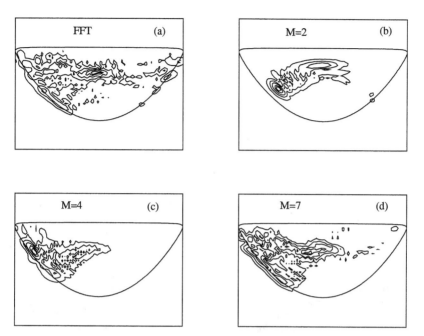

Figure 2: Snapshots of the wave packet at $t = 28$ a.u. (1/4 of total propagation time) with a rectangular grid and 64 grid points in each dimension. (a) FFT method; (b) FD with $M = 2$ (second order); (c) FD 4th order; (d) FD 7th order. The stencils are with $2M + 1$ points.

2.2 Classical Molecular Dynamics

The term Molecular Dynamics usually implies a Monte Carlo method which involves the integration of classical trajectories in phase space and a random selection for their initial conditions. Dealing with large molecules convergence is not an easy task and it usually requires proper sampling of phase space as well as the integration of the equations of motion for very long times or the sampling of a large number of trajectories. Therefore, it is useful if we know the type of trajectories we want to sample, for example those which visit the regions in configuration space of reactants and products in chemical reactions or trajectories which originate and end to particular phase space structures like stable and unstable manifolds. In those cases where the initial and final states of a trajectory satisfy a known relation, it is preferable to solve a two-point boundary value (2PBV) problem than randomly sampling initial conditions for integrating the equations of motion and accepting or rejecting trajectories according to the final state of the trajectory. A typical 2PBV problem is the location of periodic orbits (PO).

2.2.1 The 2PBV problem

A 2PBV problem is formulated in the following way. The phase space of a molecule is described with the generalized coordinates $q_i(t)$, $i = 1, \ldots, N$ and their conjugate momenta $p_i(t)$, $i = 1, \ldots, N$ which are functions of time t. For the simplification of the mathematical equations we define the

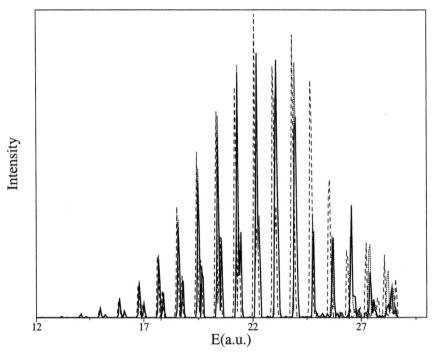

Figure 3: Power spectra obtained from the correlation function of the wave packet shown in Fig. 2. FFT(solid line), FD 4th order (dashed line) and FD 7th order (dotted line). The stencils are with $2M + 1$ points.

column vector

$$\vec{x} = (\vec{q}, \vec{p})^T. \tag{16}$$

Using $\vec{x}$ we can write Hamilton equations in the form

$$\frac{d\vec{x}(t)}{dt} = \dot{\vec{x}}(t) = J\nabla H[\vec{x}(t)] \quad (0 \leq t \leq T), \tag{17}$$

where H is the Hamiltonian function, and J is the symplectic matrix

$$J = \begin{pmatrix} 0_N & I_N \\ -I_N & 0_N \end{pmatrix}. \tag{18}$$

0_N and I_N are the zero and unit $N \times N$ matrices respectively. $J\nabla H(\vec{x})$ is a vector field, and J satisfies the relations,

$$J^{-1} = -J \quad \text{and} \quad J^2 = -I_{2N}. \tag{19}$$

We want to find trajectories whose the initial, $\vec{x}(0)$, and final point, $\vec{x}(T)$, satisfy the equation

$$\vec{B}[\vec{x}(0), \vec{x}(T); T] = 0. \tag{20}$$

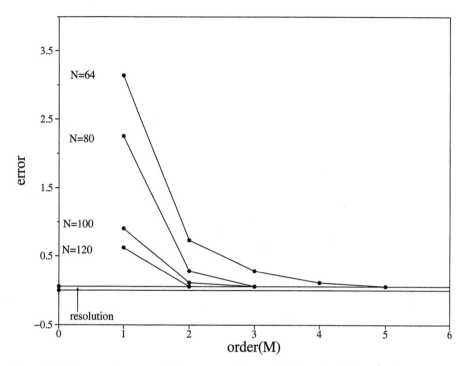

Figure 4: Differences in energy with respect to the central eigenvalue ($E = 22.18$) in the FFT spectrum of Fig. 3 as a function of the order of approximation. Rectangular grids with 64, 80, 100 and 120 points were used.

Treating the time T as a parameter we are searching for families of trajectories which are solutions of the Hamilton equations of motion and simultaneously satisfy the two-point boundary conditions, Eq. (20). For example, trajectories with known distance of the initial and final points

$$\vec{x}(T) - \vec{x}(0) = \vec{\xi},\tag{21}$$

the boundary conditions are

$$\vec{B}[\vec{x}(0), \vec{x}(T)] = \vec{x}(T) - \vec{x}(0) - \vec{\xi} = 0.\tag{22}$$

$\xi = 0$ yields the boundary conditions for periodic orbits.

The common procedure to solve a 2PBV problem is to cast it into an *initial value problem* through an iterative method (Ref.[5] and references therein). We consider the initial values of the coordinates and momenta, $\vec{s}$, of an approximate solution

$$\vec{s} = \vec{x}(0),\tag{23}$$

as independent variables of the nonlinear functions

$$\vec{B}(\vec{s}; T) = \vec{B}[\vec{s}, \vec{x}(T; \vec{s})].\tag{24}$$

$\vec{B}$ parametrically depends on the period T. We denote the roots of boundary equations as $\vec{s}_*$, i.e.,

$$\vec{B}(\vec{s}_*;T) = 0. \tag{25}$$

Hence, if $\vec{s}$ is a nearby value to the solution $\vec{s}_*$ we can compute the functions $\vec{B}(\vec{s})$ by integrating Hamilton equations for the time interval $[0,T]$. By appropriately modifying the initial values $\vec{s}$ we hope to converge to the solution, that is, $\vec{s} \to \vec{s}_*$ and $B \to 0$.

A common procedure to find the roots of Eqs (25) is the Newton-Raphson method. This is an iterative scheme and at each iteration, k, we update the initial conditions of the orbit

$$\vec{s}_{k+1} = \vec{s}_k + \Delta \vec{s}_k. \tag{26}$$

The corrections $\Delta \vec{s}_k$ are obtained by expanding Eqs (25) in a Taylor series up to the first order

$$\vec{B}(\vec{s}_{k+1};T) = \vec{B}(\vec{s}_k + \Delta \vec{s}_k; T) \approx \vec{B}(\vec{s}_k;T) + \frac{\partial \vec{B}}{\partial \vec{s}_k} \Delta \vec{s}_k = 0, \tag{27}$$

or

$$\vec{B}(\vec{s}_k;T) + \left[\frac{\partial \vec{B}}{\partial \vec{s}_k} Z_k(0) + \frac{\partial \vec{B}}{\partial \vec{x}_k(T)} Z_k(T) \right] \Delta \vec{s}_k = 0. \tag{28}$$

The matrix (Jacobian)

$$Z_k(t) = \frac{\partial \vec{x}_k(t;\vec{s}_k)}{\partial \vec{s}_k}, \tag{29}$$

is the *Fundamental Matrix*, which is evaluated by integrating the *variational equations*

$$\dot{\vec{\zeta}}(t) = A(t)\vec{\zeta}(t) \quad (0 \le t \le T), \tag{30}$$

where the second derivatives of the Hamiltonian with respect to coordinates and momenta are needed

$$A(t) = J\partial^2 H[\vec{x}(t)]. \tag{31}$$

These equations calculate the linearized part of the difference of two initially neighboring trajectories in time, $\vec{\zeta} = \vec{x}' - \vec{x}$. The Fundamental Matrix is also a solution of the variational equations as can be seen by taking the time derivatives of Eq. (29)

$$\dot{Z}(t) = A(t)Z(t). \tag{32}$$

Thus, to perform the kth iteration in the Newton-Raphson method we first integrate for time $[0,T]$ the differential equations

$$\dot{\vec{x}}_k(t) = J\nabla H[\vec{x}_k(t)]$$
$$\dot{Z}_k(t) = A_k(t)Z_k(t),$$

with initial conditions

$$\vec{x}_k(0) = \vec{s}_k$$
$$Z_k(0) = I_{2N}.$$

Then, we solve the linear algebraic equations

$$\left[\frac{\partial \vec{B}}{\partial \vec{s}_k} Z_k(0) + \frac{\partial \vec{B}}{\partial \vec{x}_k(T)} Z_k(T) \right] \Delta \vec{s}_k = -\vec{B}(\vec{s}_k;T), \tag{35}$$

in order to find the initial conditions for the $(k+1)^{th}$ iteration (Eq. (26)).

For a periodic orbit the boundary conditions become

$$[Z_k(T) - I_{2N}]\Delta\vec{s}_k = -\vec{B}(\vec{s}_k; T). \tag{36}$$

The Fundamental Matrix at period T is called *Monodromy Matrix*, $M = Z(T)$, and its eigenvalues are used for the stability analysis of the periodic orbit. Details can be found in Ref.[22].

Once we have located one member of the family of the trajectories that we are looking for, we can apply continuation techniques [23] to find trajectories at different times T. This is done by using T as the control parameter. Usually, for small increments of the time T linear extrapolation methods are sufficient [5].

2.2.2   The multiple node 2PBV algorithm

Let us assume that we divide the time T in $(m-1)$ time intervals by choosing m nodes. The idea of the multiple shooting method is to integrate $m-1$ trajectories -one for each time sub-interval- and to check if the final values of each trajectory coincide with the initial conditions of the next. To formulate this idea we introduce the scaled time, $\tau = t/T, \quad (0 \le \tau \le 1)$,

$$0 = \tau_1 < \tau_2 < \cdots < \tau_{m-1} < \tau_m = 1. \tag{37}$$

Thus, for the simple shooting method $m = 2$.

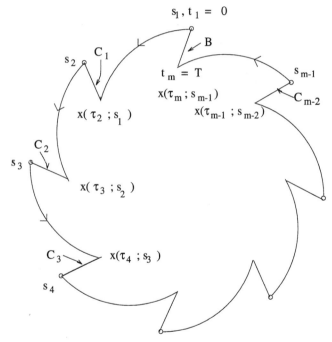

Figure 5: A schematical representation of the multiple shooting procedure applied to periodic orbits.

In this section we drop the index for the iterations k, and we use the index j to denote the nodes in time. If the initial conditions of the trajectory at each node j is $\vec{s}_j$ at time τ_j, and the final value of the trajectory at time τ_{j+1} is denoted by $\vec{x}(\tau_{j+1}; \vec{s}_j)$, then $(m-2)$ *continuity conditions* should be satisfied (for a graphical representation see Fig. 5)

$$\vec{C}_j(\vec{s}_j, \vec{s}_{j+1}) = \vec{x}(\tau_{j+1}; \vec{s}_j) - \vec{s}_{j+1} = 0, \quad j = 1, 2, \cdots, m-2, \tag{38}$$

together with the boundary conditions

$$\vec{B}(\vec{s}_{1*}, \vec{s}_{m-1*}) = 0. \tag{39}$$

Now, we have to solve $(m-1)$ initial value problems, and for that we adopt the Newton-Raphson method

$$\vec{C}_j(\vec{s}_j + \Delta\vec{s}_j, \vec{s}_{j+1} + \Delta\vec{s}_{j+1}) \approx \vec{C}_j(\vec{s}_j, \vec{s}_{j+1}) + \frac{\partial\vec{C}}{\partial\vec{s}_j}\Delta\vec{s}_j + \frac{\partial\vec{C}}{\partial\vec{s}_{j+1}}\Delta\vec{s}_{j+1} = 0. \tag{40}$$

These equations become

$$\vec{C}_j(\vec{s}_j, \vec{s}_{j+1}) + Z_j(\tau_{j+1})\Delta\vec{s}_j - \Delta\vec{s}_{j+1} = 0, \quad 1 \leq j \leq m-2, \tag{41}$$

where,

$$Z_j(\tau_{j+1}) = \frac{\partial\vec{x}(\tau_{j+1}; \vec{s}_j)}{\partial\vec{s}_j}. \tag{42}$$

Linearizing the boundary conditions we get

$$\vec{B}(\vec{s}_1 + \Delta\vec{s}_1, \vec{s}_{m-1} + \Delta\vec{s}_{m-1}) = \vec{B}(\vec{s}_1, \vec{s}_{m-1}) + \frac{\partial\vec{B}}{\partial\vec{s}_1}Z_1(0)\Delta\vec{s}_1 + \frac{\partial\vec{B}}{\partial\vec{s}_{m-1}}Z_{m-1}(1)\Delta\vec{s}_{m-1} = 0, \tag{43}$$

Hence, Eqs (41,43) provide a linear system with $2N(m-1)$ unknown variables. For periodic orbit boundary conditions these equations in matrix form are written as

$$\begin{bmatrix} Z_1 & -I_{2N} & 0 & \cdots & 0 & 0 \\ 0 & Z_2 & -I_{2N} & \cdots & 0 & 0 \\ \cdots & \cdots & \cdots & \cdots & \cdots & \cdots \\ 0 & 0 & 0 & \cdots & Z_{m-2} & -I_{2N} \\ -I_{2N} & 0 & 0 & \cdots & 0 & Z_{m-1} \end{bmatrix} \begin{bmatrix} \Delta\vec{s}_1 \\ \Delta\vec{s}_2 \\ \cdots \\ \Delta\vec{s}_{m-2} \\ \Delta\vec{s}_{m-1} \end{bmatrix} = - \begin{bmatrix} \vec{C}_1 \\ \vec{C}_2 \\ \cdots \\ \vec{C}_{m-2} \\ \vec{B} \end{bmatrix} \tag{44}$$

Although each block Z_i is not sparse the total matrix is sparse and it can be solved with special routines for distributed computing [24, 25, 26].

3 Conclusions

The algorithms described in the previous sections were used extensively in the past mainly with triatomic molecules [27, 28, 29]. Currently, we are developing the computer codes to apply the classical and quantum algorithms to molecules with biological interest.

Acknowledgment

Support from the Greek Ministry of Education and European Union through the postgraduate program EPEAEK, "Applied Molecular Spectroscopy", is gratefully acknowledged. LL thanks the Institute of Electronic Structure and Laser-FORTH for the hospitality during his visits to Crete.

References

[1] I. Foster and C. Kesselman, (Editors.). *The Grid: Blueprint for a new Computing Infrastructure*, Morgan Kaufmann (1999).

[2] *European Network for Advanced Computing Technology for Science* (ENACTS), http://www.epcc.ed.ac.uk/enacts/.

[3] http://tccc.iesl.forth.gr/general/intro/pdf/108.pdf

[4] S. C. Farantos, Exploring Molecular Vibrations with Periodic Orbits, *Int. Rev. Phys. Chem.* **15**, 345–374 (1996).

[5] S. C. Farantos, POMULT: A Program for Computing Periodic Orbits in Hamiltonian Systems Based on Multiple Shooting Algorithms, *Comp. Phys. Comm.* **108**, 240–258 (1998).

[6] R. Kosloff, Quantum Molecular Dynamics on Grids, *Dynamics of Molecules and Chemical Reactions*, Editors: R. E. Wyatt and J. Z. H. Zhang, Marcel Dekker Inc., N.Y., 185–230 (1996).

[7] R. Kosloff, Propagation Methods for Quantum Molecular Dynamics, *Ann. Rev. Phys. Chem.* **45**, 145–178 (1994).

[8] (a) C. Lanczos, *J. Res. Nat. Bur. Stand.* **45**, 58 (1950); (b) C. Iung and C. Leforestier, Direct calculation of overtones: Application to the CD_3H molecule, *J. Chem. Phys.* **102**, 8453–8461 (1995); (c) R. E. Wyatt, Computation of high-energy vibrational eigenstates: Application to C_6H_5D, *J. Chem. Phys.* **103**, 8433–8443 (1995); (d) R. B. Lehoucq, D. C. Sorensen and C. Yang, *Solution of Large Scale Eigenvalue Problems with Implicitly Restarted Arnoldi Methods*, (SIAM, Philadelphia (1998)).

[9] *Dynamics of Molecules and Chemical Reactions*, Editors: R. E. Wyatt and J. Z. H. Zhang, Marcel Dekker, Inc., N.Y. (1996).

[10] J. Broeckhove and L. Lathouwers, Quantum Dynamics and Angular Momentum Projection, *Numerical Grid Methods and their Applications to Schrödinger Equation*, Editor: C. Cerjan, Kluwer Academic Publishers, 49–56 (1993).

[11] R. Guantes and S. C. Farantos, High Order Finite Difference Algorithms for Solving the Schrödinger Equation in Molecular Dynamics, *J. Chem. Phys.* **111**, 10827–10835 (1999).

[12] R. Guantes and S. C. Farantos, High Order Finite Difference Algorithms for Solving the Schrödinger Equation in Molecular Dynamics. II. Periodic variables, *J. Chem. Phys.* **113**, 10429–10437 (2000).

[13] R. Guantes, A. Nezis and S. C. Farantos, Periodic Orbit - Quantum Mechanical Investigation of the Inversion Mechanism of Ar_3, *J. Chem. Phys.* **111**, 10836–10842 (1999).

[14] (a) B. Fornberg, *A Practical Guide to Pseudospectral Methods*, Cambridge Monographs on Applied and Computational Mathematics (Cambridge Univ. Press, **1** (1998)); (b) B. Fornberg and D. M. Sloan, *Acta Numerica*, 203-267 (1994).

[15] J. P. Boyd, A fast algorithm for Chebyshev, Fourier, and sinc interpolation onto an irregular grid, *J. Comp. Phys.* **103**, 243–257 (1992).

[16] J. P. Boyd, Sum-accelerated pseudospectral methods: Finite differences and sech-weighted differences, *Comp. Methods Appl. Mech. Engrg.* **116**, 1–11 (1994).

[17] D. T. Colbert and W. H. Miller, A novel discrete variable representation for quantum mechanical reactive scattering via the S-matrix Kohn method, *J. Chem. Phys.* **96**, 1982–1991 (1992).

[18] W. H. Press, B. P. Flannery, S. A. Teukolsky and W. T. Vetterling, *Numerical Recipes*, Cambridge University Press (1986).

[19] (a) J. V. Lill, G. A. Parker and J. C. Light, Discrete variable representations and sudden models in quantum scattering theory, *Chem. Phys. Lett.* **89**, 483–489 (1982); (b) J. C. Light, I. P. Hamilton and J. V. Lill, Generalized discrete variable approximation in quantum mechanics, *J. Chem. Phys.* **82**, 1400–1409 (1985); (c) S. E. Choi and J. C. Light, Determination of the bound and quasibound states of ArHCl van der Waals complex: Discrete variable representation method, *ibid.* **92**, 2129–2145 (1990).

[20] Brendan Vastenhouw and Rob H. Bisseling, A Two-Dimensional Data Distribution Method for Parallel Sparse Matrix-Vector Multiplication, *SIAM Review* **47**, 67–95 (2005) .

[21] M. Founargiotakis, S. C. Farantos, G. Contopoulos and C. Polymilis, Periodic Orbits, Bifurcations and Quantum Mechanical Eigenfunctions and Spectra, *J. Chem.Phys.* **91**, 1389–1402 (1989).

[22] R. Seydel, *From Equilibrium to Chaos: Practical bifurcation and stability analysis*, Elsevier (1988).

[23] E. L. Allgower and K. Georg, *Numerical Continuation Methods*, Springer series in computational mathematics, **13**, Berlin:Springer-Verlag (1993).

[24] Satish Balay, Kris Buschelman, William D. Gropp, Dinesh Kaushik, Matthew G. Knepley, Lois Curfman McInnes, Barry F. Smith and Hong Zhang, PETSc, http://www.mcs.anl.gov/petsc (2001).

[25] Satish Balay, Kris Buschelman, Victor Eijkhout, William D. Gropp, Dinesh Kaushik, Matthew G. Knepley, Lois Curfman McInnes, Barry F. Smith and Hong Zhang, PETSc Users Manual, ANL-95/11 - Revision 2.1.5, Argonne National Laboratory (2004).

[26] Satish Balay, Victor Eijkhout, William D. Gropp, Lois Curfman McInnes, Barry F. Smith, Efficient Management of Parallelism in Object Oriented Numerical Software Libraries, in *Modern Software Tools in Scientific Computing*, Editors: E. Arge, A. M. Bruaset and H. P. Langtangen, 163–202, Birkhäuser Press (1997).

[27] H. Ishikawa, R. W. Field, S. C. Farantos, M. Joyeux, J. Koput, C. Beck and R. Schinke, HCP - CPH Isomerization: Caught in the Act, *Annual Review of Physical Chemistry* **50**, 443-484 (1999).

[28] M. Joyeux, S. C. Farantos and R. Schinke, Highly Excited Motion in Molecules: Saddle-Node Bifurcations and their Fingerprints in Vibrational Spectra, *J. Phys. Chem.* **106**, (feature article) 5407–5421 (2002).

[29] M. Joyeux, S. Yu. Grebenshchikov, J. Bredenbeck, R. Schinke and S. C. Farantos, Intramolecular Dynamics Along Isomerization and Dissociation Pathways, *"Geometrical Structures of Phase Space in Multi-Dimensional Chaos"*, Editors: I. Prigogine, Stuart Rice, Advances in Chemical Physics **130**, 267–303 John Wiley & Sons,(2005).

Brill Academic Publishers
P.O. Box 9000, 2300 PA Leiden,
The Netherlands

*Lecture Series on Computer
and Computational Sciences*
Volume 3, 2005, pp. 51-58

Model Energy Functions for Metallic Clusters and Surfaces

René Fournier[1], Min Zhang[2], Yasaman Soudagar[3], and Mark Hopkinson

Department of Chemistry, York University
4700 Keele Street, Toronto, Ontario, M3J 1P3 Canada

Received 24 June, 2005; accepted in revised form 1 August, 2005

Abstract: A function we call *Scaled Morse Potential* (SMP) is proposed for large scale atomistic simulations of metals. It is a sum of Morse potentials, each of which depends on the coordination of the two interacting atoms. The SMP fits the diatomic and bulk precisely and interpolates well when the coordination is intermediate between these two limits. It is qualitatively correct for clusters and surfaces, but a detailed comparison to experiments and Density Functional Theory (DFT) shows some shortcomings. The accuracy and applicability of the SMP is improved by adding a purely ionic term calculated from atomic charges. These charges depend on the geometry of the system and are calculated within the electronegativity equalization (EE) model.

Keywords: Clusters, Surfaces, Empirical Potentials, Electronegativity

PACS: 36.40.-c, 34.20.Cf, 68.35.Bs, 68.47.De

1   Scaled Morse Potential

The Morse potential[1] is a simple function that describes the energy curve of covalent diatomics:

$$D(r; D_e, R_e, a_e) = D_e[e^{-2a_e(r-R_e)} - 2e^{-a_e(r-R_e)}] \tag{1}$$

Its parameters R_e, D_e, and a_e are readily obtained from spectroscopic constants. The Morse function also describes the cohesive energy curve of bulk solids near the minimum[2]. Of course, the diatomic and bulk parameters differ greatly. The coordination of atoms at metal surfaces with defects and adatoms vary anywhere between 3 and 12, and as a result, surfaces (and clusters) are very poorly described by simple energy functions. It turns out that, for s-valence metals, the cohesive energy E_c is nearly equal to $\sqrt{12} \times (D_e/2)$[3]. If we express the energy as a sum of pairwise contributions, this implies $D_{12} \approx D_e \div \sqrt{12}$, or more generally, $D_n \approx D_e \div \sqrt{n}$, where D_n is the well depth for the interaction between atoms having coordination n. This assumption gives surface energies per atom of $0.13\, E_c$ for fcc(111) and $0.18\, E_c$ for fcc(100), in excellent agreement with the empirical relation $\gamma = 0.16\, \Delta H_{sub}$ between surface tension and heat of sublimation of metals[4]. So we propose to describe the energy of N interacting metal atoms by a sum of pairwise Morse functions where D_e, R_e, and a_e are functions of the atomic coordinations n_j, and α, β, and γ are

---

[1]Corresponding author. E-mail: renef@yorku.ca
[2]Department of Physics, York University
[3]Département de Génie Physique, Université de Montréal

parameters used to interpolate between the values appropriate for the diatomic and bulk.

$$U_{SMP} = \sum_{i>j} U_{ij} = \sum_{i>j} D(r_{ij}; R_{ij}, D_{ij}, a_{ij}) \tag{2}$$

$$D_{ij} = (D_i + D_j)/2 \quad ; \quad R_{ij} = (R_i + R_j)/2 \quad ; \quad a_{ij} = (a_i + a_j)/2$$
$$D_j = D_e n_j^\alpha \quad ; \quad R_j = R_e n_j^\beta \quad ; \quad a_j = a_e n_j^\gamma \tag{3}$$

Our working definition of n_j is based on a smooth cut-off function $f(r)$ of interatomic distance,

$$n_j = \sum_{i \neq j} f(r_{ij}) \tag{4}$$

$$
\begin{aligned}
f(r) &= 1 & r \leq C_1 \\
&= 0 & r \geq C_2 \\
&= \frac{1}{2} - \frac{1}{2} sin[\frac{\pi(r - R_{mid})}{\Delta}] & C_1 < r < C_2
\end{aligned} \tag{5}
$$

where C_1 and C_2 are cut-off limits, $\Delta = (C_2 - C_1)$, and $R_{mid} = (C_1 + C_2)/2$. For simplicity, we take $C_1 = R_{NN}$ and $C_2 = \rho R_{NN}$, where R_{NN} is the bulk nearest-neighbour distance and $\rho = 1.4$ for all elements. This defines a cut-off region starting at the first, and extending until the second, fcc coordination shell. Long-range interactions are problematic. Keeping the long-range part of the function unchanged in a pairwise potential makes the connection to coordination fuzzy, and more importantly, it produces an unphysical interlayer expansion near surfaces[7]. In first-principles methods, long-range interactions are screened by other atoms. We chose to "screen" U_{ij} of Eq. 2 by multiplying it by the cut-off function in Eq. 5. This introduces a big error in the dissociation limit (large r) for any system. But this flaw has little or no consequence for properties that depend on structures near energy minima, and it can be fixed by replacing R_{NN} by $Max\{R_{NN}, d_{min}^i\}$ in the equations for C_1 and C_2, where d_{min}^i is the distance between atom i and its nearest neighbor. Assuming $n_j = 12$ for both fcc and bcc, we get the six parameters for each element directly from six experimental data, the diatomic D_e, R_e and ω_e (harmonic frequency)[5], and the bulk E_c, R_{NN}, and B (bulk modulus)[6] (Table 1). There is a satisfying regularity in these parameters: α is always close to $-\frac{1}{2}$ (especially so for Cu, Ag, Au), β is always within $[0.057, 0.079]$ except for the open-shell metals V (0.164), Nb (0.133), and Mo (0.141), and γ is nearly zero for all elements except Cs (−0.117), V (−0.094), Nb (−0.079), and Mo (−0.056). Not surprisingly, the elements for which the SMP parameters are anomalous are the ones that exhibit multiple bonding or whose electronic configuration changes significantly between diatomic and bulk. With SMP, the bcc structure is favored for the alkali and fcc is favored for all others. So the structure of bulk V, Nb, and Mo are incorrect in SMP, presumably because multiple bonding is not accounted for in this simple model.

2   SMP Results and Discussion

Surface energies and relaxation. — Calculations were done for Ni, Cu, and Ag clusters of 279, 239, and 271 atoms defining seven-layer slabs of fcc(111), fcc(100), and fcc(110) respectively. Geometries were relaxed by energy minimization. The SMP percent relaxation in the first (Δ_{12}) and second (Δ_{23}) interlayer spacings agree qualitatively with experimental data[4] (Table 2) and the best effective medium methods[7]. Note that Δ_{ij} values from different experiments vary by as much as 50%, and additive potentials without cut-off are quite wrong: they predict a large expansion of the first few interlayer spacings. The SMP also yields surface energies similar to the best effective

Table 1: Parameters of the scaled Morse potential

	D_e (eV)	R_e (Å)	a_e (Å^{-1})	α	β	γ
Li	1.068	2.673	0.859	-0.560	0.062	-0.012
Na	0.730	3.079	0.856	-0.556	0.079	-0.001
K	0.520	3.905	0.765	-0.492	0.069	-0.028
Cs	0.397	4.470	0.737	-0.463	0.079	-0.117
V	2.786	1.760	2.193	-0.505	0.164	-0.094
Ni	2.059	2.155	1.433	-0.412	0.058	0.002
Cu	2.046	2.220	1.413	-0.506	0.057	0.000
Nb	4.826	2.078	1.787	-0.535	0.133	-0.079
Mo	4.150	1.930	2.199	-0.506	0.141	-0.056
Ag	1.662	2.480	1.486	-0.490	0.062	-0.024
Pt	3.153	2.400	1.629	-0.473	0.058	-0.002
Au	2.302	2.470	1.694	-0.518	0.062	-0.020

medium methods[7]. Experimental surface energies are measured for multi-faceted solids near their melting point. To allow meaningful comparisons, we studied quasi-spherical cluster fragments of the fcc crystal ranging from $n = 1961$ to $n = 15683$ atoms. We did partial geometry optimizations that typically recover 85% of the relaxation energy, and extrapolated results to infinite cluster size. The numerical uncertainty on our extrapolated values is roughly 4%. The experimental data we use are 0 K linear extrapolations of surface tensions from Ref. [7]. The SMP and experimental surface energies (in Jm^{-2}) are: 2.75 (SMP) and 2.49 (expt) for Ni; 1.81 (SMP) and 2.04 (expt) for Cu; 1.23 (SMP) and 1.38 (expt) for Ag. This level of agreement is satisfying, it is comparable to that of more complicated potentials that involve fits to experimental or first-principles data.

Clusters. — We found the lowest energy isomer at each cluster size in Table 3 by doing

Table 2: Relaxed surface energies σ (in J/m^2) and percent change in interlayer spacings Δ. Values in parentheses are averages of experimental values listed in Table 2.3a of Ref. [4].

		σ	Δ_{12}	Δ_{23}
fcc(111)	Ni	2.24	−0.90 (−0.6)	−0.11
	Cu	1.47	−0.84 (−0.7)	−0.09
	Ag	1.00	−0.90 (0)	−0.09
fcc(100)	Ni	2.67	−1.88 (−3.9)	+0.23
	Cu	1.76	−1.89 (−1.1)	+0.28 (+1.7)
	Ag	1.20	−1.91	+0.28
fcc(110)	Ni	2.83	−4.58 (−7.4)	+0.85 (+3.2)
	Cu	1.87	−4.48 (−7.7)	+1.01 (+2.6)
	Ag	1.27	−4.63 (−7.3)	+1.01 (+4.2)

hundreds of searches that start with a random geometry, followed by simulated annealing, and then conjugate gradient optimization. The SMP minima are high symmetry compact structures with the well-known pentagonal bipyramid (PBP) growth pattern. Their structure differ from Lennard-Jones clusters only in details. But *first-principles* calculations give very different structures: small

Table 3: Cluster dissociation energies (ΔE) in eV from SMP, collision induced dissociation experiments (in parentheses, [11]), and DFT calculations (in brackets, [9]).

n	V	Ni	Ag	Nb
3	2.60 (1.42)	2.21 (0.82)	1.59 [0.92]	4.30 (4.10)
4	3.32 (3.67)	2.93 (1.51)	2.04 [2.29]	5.42 (6.26)
5	3.17 (3.08)	2.87 (2.65)	1.99 [1.93]	5.30 (5.46)
6	3.22 (4.03)	2.83 (3.25)	1.97 [2.67]	5.26 (5.59)
7	3.75 (3.73)	3.42 (2.15)	2.32 [2.07]	6.07 (6.51)
8	3.20 (4.11)	2.79 (2.60)	1.95 [2.70]	5.21 (6.07)
9	3.66 (3.51)	3.39 (2.63)	2.30 [1.65]	5.98 (5.25)
10	3.48 (3.93)	3.32 (2.64)	2.24 [2.61]	5.76 (6.45)
11	3.39 (3.78)	3.30 (2.73)	2.22 [2.11]	5.66 (5.40)
12	3.35 (4.09)	3.75 (3.25)	2.45 [2.65]	5.82
13	4.20 (4.63)	4.36 (3.45)	2.85	7.07
MAD	0.50	0.76	0.40	0.50

($n \leq 6$) clusters of Li, Na, and Ag[8, 9] have planar minima, while V and Nb clusters adopt compact, often low-symmetry, structures[10]. This problem is common to all empirical potentials because they ignore details of electronic structure such as spin subshells, Jahn-Teller distortions, orbital symmetry and hybridization. This should have less serious consequences as one goes to very large clusters, surfaces and bulk. The SMP and DFT energies of Ag_n[9] agree in an average sense but not in detail (Table 3): the mean absolute difference (MAD) between the two is 0.40 eV which amounts to 30% of $(E_c - D_e)$. This is partly due to even-odd alternations caused by spin subshells: SMP overestimates all odd n ΔE and and underestimates all even n ΔE of Ag_n. If we average the ΔE's of consecutive pairs, the MAD decreases from 0.40 to 0.19 eV. The MAD between SMP and DFT for 41 bond lengths in Ag_n (n =3–9) is 0.07 Å which is 13% of the range in the DFT bond lengths. For Nb_n clusters the MAD is 0.13 Å, or 19% of the range in DFT. The MAD in harmonic frequencies calculated by SMP and DFT is 16 cm^{-1} (7% of the range) in Ag_n clusters, and it is 41 cm^{-1} (11% of the range) in Nb_n clusters. The largest errors occur when DFT predicts Jahn-Teller distortion.

3   Electronegativity Equalization (EE)

The idea behind EE[12] is that elements have different tendencies to gain and lose electrons. As atoms come together to form a N-atom system, we assume the existence of atoms-in-molecules (AIMs) that acquire net charges. The Coulombic interactions between charged AIMs, and the energy change in charging them, is a convenient model for calculating electrostatic energies in N-atom systems. The tendency of an atom (or any system) to gain or lose electrons can be expressed by the power series expansion of energy as a function of number of electrons (N), or charge $Q = Z - N$:

$$U(Q) = U(0) + \chi^0 Q + \eta^0 Q^2 + \frac{1}{6}\left(\partial^3 U/\partial Q^3\right) Q^3 + \frac{1}{24}\left(\partial^4 U/\partial Q^4\right) Q^4 + \dots \qquad (6)$$

where $\chi^0 = \partial U/\partial Q$ is the electronegativity, and $\eta^0 = \left(\frac{1}{2}\right)\partial^2 U/\partial Q^2$ is the absolute hardness[13]. One can estimate the electronegativity and absolute hardness of atoms by finite difference using experimental atomic ionization energies (IEs) $(U(1) - U(0)$ and electron affinities $(U(0) - U(-1)$ or directly by calculating the energy derivatives in DFT. But the electronegativity and hardness of an

AIM is not quite the same as for a free atom. We get the electronegativity and hardness of AIMs by calculating the $U(Q)$ curve in high-symmetry reference systems (for example, 10-atom rings) by DFT, subtracting out the Coulombic interaction term, and fitting the remainder to a polynomial in Q. The fit expresses only the part of the energy associated with charging the AIMs in the reference system, which is what we need. The electronegativity of a N-atom system, $\chi = \partial U/\partial Q$, is the same everywhere. So, to be consistent, we must assume that the electronegativity of the AIMs are all equal, $\chi_i = \chi_1$, $i = 2, N$. This, and charge conservation ($\sum_i Q_i = Q_{tot}$), gives us a system of N equations with solutions that are the N AIM charges Q_i. In the original implementation by Rappe and Goddard[12], the power series expansion of $U(Q)$ was truncated after the quadratic term, but we keep all terms up to Q^4 for better accuracy and to avoid unphysical charges, for example, greater than +1 or even +3 for a Li AIM. Such unphysical charges can occur in systems with atoms having large electronegativity differences. Keeping all terms up to Q^4 is formally equivalent to having a charge dependent hardness $\eta'(Q) = \eta^0 + c_3 Q + c_4 Q^2$. The coefficients in the system of equations depend on the hardness of AIMs, so in our scheme they depend on the Q_i's and we have to solve the equations iteratively to self-consistency. This is not a problem and we typically reach self-consistent charges in fewer than 20 iterations. Once we have Q_i's, we calculate an approximate electrostatic energy as a sum of AIM charging energies and pairwise screened Coulombic interactions:

$$U_{EE} = \sum_i \Delta U(Q_i) + \sum_i \sum_{j>i} J_{ij} Q_i Q_j \tag{7}$$

The screening function that we use was proposed by Louwen and Vogt[14]:

$$J_{ij} = \left(R_{ij}^3 + \gamma_{ij}^{-3} \right)^{-1/3} \tag{8}$$

$$\gamma_{ij} = 2(\eta_i^0 \eta_j^0)^{1/2} \tag{9}$$

It correctly reduces to $1/R_{ij}$ in the limit of large distances and to twice the average of the two atoms' hardnesses in the limit of very small distances. Since the SMP is based on a purely covalent model of bonding, we simply add U_{SMP} and U_{EE} to create the SMP-EE model.

4 SMP-EE Results and Discussion

Heteronuclear Dimers — We calculated the bond dissociation energy and bond length for several heteronuclear metal dimers by SMP, SMP-EE, and by DFT with several different types of exchange-correlation. After comparing to experiment we find that the DFT method that best agrees with experiment is the gradient corrected exchange-correlation functional often denoted "BP86"[15]. It gives a root-mean square deviation (RMSD) from experiment of 0.16 eV on bond energies and 0.14 Å on bond lengths for the molecules in Table 4 for which experimental values are available. The RMSD for SMP-EE are 0.12 eV on bond energies and 0.18 Å on bond lengths. We also report SMP and local spin density (SVWN) results in Table 4 as an example of a relatively inaccurate first-principles method. Putting the four sets of calculations in increasing order of RMSD for dissociation energies, we have SMP-EE < BP86 < SVWN < SMP; for bond lengths, the order is BP86 < SVWN < SMP-EE < SMP. SMP-EE is clearly better than SMP and, for the type of elements in Table 4, it gives an accuracy comparable to a good DFT method (BP86). *Larger*

Systems — We are currently testing SMP-EE in larger systems and have only preliminary results. In one test we calculated the atomization energies and IEs of five substitutional isomers of the Ag_5Li_5 cluster and compared to DFT results. The mean absolute deviation between atomization energies calculated by DFT and SMP is 2.8 eV (or 16% of the mean DFT energy, 17.05 eV). With SMP-EE, the mean absolute deviation decreases to 0.9 eV (5%). However, relative isomer

Table 4: Dissociation energies and bond lengths of heteronuclear metal dimers.

Dissociation energy (eV)

	SMP	SMP-EE	SVWN	BP86	expt
LiNa	0.883	0.886	0.90	0.74	0.86505[17, 18]
LiK	0.745	0.796	0.78	0.62	0.811[19, 20]
LiCu	1.478	2.029	2.17	1.86	1.9451[21]
LiAg	1.332	1.775	2.13	1.76	2.06[22, 26], 1.89[21]
NaK	0.616	0.642	0.76	0.59	0.6455[23, 20]
NaAg	1.101	1.599	1.96	1.58	1.59[24]
KCu	1.031	1.912	1.90	1.63	
KAg	0.929	1.668	1.87	1.53	
CuAg	1.844	1.847	2.43	1.87	1.74±0.10[25]

Equilibrium bond length (Å)

	SMP	SMP-EE	SVWN	BP86	expt
LiNa	2.87	2.87	2.884	2.935	3.20[5]
LiK	3.23	3.21	3.314	3.377	
LiCu	2.44	2.37	2.359	2.396	2.2618[26]
LiAg	2.57	2.52	2.419	2.459	2.41[21]
NaK	3.47	3.45	3.401	3.501	3.59[5]
NaAg	2.76	2.69	2.599	2.681	
KCu	2.94	2.81	2.957	3.044	
KAg	3.11	3.00	3.016	3.093	
CuAg	2.35	2.35	2.342	2.409	2.3735[25]

energies do not improve significantly. The IEs for the same five isomers of Ag_5Li_5 are (EE/DFT): 6.90/5.99, 5.99/5.46, 5.85/5.35, 5.94/5.32, 5.07/4.69. The trend in EE follows closely DFT. This is encouraging, and represents a big advantage of SMP-EE over SMP, because the IE is often the first experimental property known about a cluster and it often correlates with chemical reactivity[16]. As another test, we calculated the work function (in eV) of bulk Li and Ag by extrapolating cluster IEs to infinite atom number. The agreement with experiment is good: 3.02 vs 2.93 (expt) for Li, and 5.10 vs 4.64 (expt) for Ag. The SMP-EE method is computationally very efficient. We used it to search the lowest energy substitutional isomers of 55-atom bimetallic clusters with the Mackay icosahedron geometry. This is a complex problem, as there are roughly $55!/((55-n)!n!)$ substitutional isomers of $A_nB_{(55-n)}$. With SMP-EE, we could afford to do a combined simulated annealing and conjugate gradient optimization where we typically evaluate the energy on the order of 10^9 times in runs that take only a few days on a PC.

5   Conclusions

We devised a simple energy model for metals (SMP-EE) which gives qualitatively correct structures and IEs, and gives correct trends for interaction energies as a function of cluster size, atomic coordinations, and substitutional isomerism in bimetallic systems. It can not reproduce energy variations caused by orbital symmetry and electronic shell closing, therefore it is not a reliable tool for small clusters. But it is computationally very efficient, so it can be used to study large clusters that are interesting in their own right (for example, 55-atom metal clusters with a Mackay icosahedron base structure) or as models of metal catalysts. Obviously an empirical model like SMP-EE can not be as accurate and reliable as a first-principles theory like DFT. In systems where it can be applied, either directly or through the use of statistics as in the cluster expansion technique[27], DFT is surely referable to SMP-EE. But it is very difficult and impractical to carry any kind of DFT calculation on large, finite, metal clusters that lack symmetry or a crystal-like structure. SMP-EE can provide a useful level of accuracy and computational efficiency for this kind of system. Since EE provides atomic charges and IEs it can be used to characterize, in a crude way, the electronic structure of metallic particles and surfaces. Along with structure predictions, the SMP-EE atomic charges and IEs could help to better understand reactivity at the surface of metal catalysts. More work remains to be done for validating SMP-EE by comparing to more DFT calculations, especially in bigger systems. It is also desirable to make SMP-EE more accurate by fixing some of the flaws in the physics of the model. We are currently looking at the possibility of making AIM electronegativities depend on their coordination, so that we can describe charges at the surface of elemental solids. Another possibility is to use a simple electronic model, such as a s-only tight-binding hamiltonian, to estimate the HOMO-LUMO gap (ie, the hardness of the N-atom system) and derive from it a small energy term that would mimick shell closings and Jahn-Teller distortions in small clusters.

Acknowledgment

The authors wish to thank the National Science and Engineering Research Council (NSERC) of Canada for financial support.

References

[1] P. Morse, Phys. Rev. 34, 57 (1929); M. Dagher, H. Kobeissi, M. Kobressi, J. D'Incan, and C. Effantin, J. Comp. Chem. 14, 1320 (1993);

[2] J.H. Rose, J. Ferrante, and J.R. Smith, Phys. Rev. Lett. 47, 675 (1981).

[3] M. Methfessel, D. Hennig, and M. Scheffler, Appl. Phys. A 55, 442 (1992).

[4] G.A. Somorjai, *Introduction to Surface Chemistry and Catalysis*, (Wiley, New York, 1994).

[5] K. P. Huber, and G. Herzberg, *Molecular Spectra and Molecular Structure, Vol. IV., Constants of Diatomic Molecules*, (Van Nostrand, Toronto, 1979).

[6] C. Kittel, *Introduction to Solid State Physics*, 7th edition (Wiley, New York, 1996).

[7] S.B. Sinnott, M.S. Stave, T.J. Raeker, and A.E. DePristo, Phys. Rev. B 44, 8927 (1991).

[8] G. Gardet, F. Rogemond and H. Chermette, J. Chem. Phys. 105, 9933 (1996); U. Röthlisberger and W. Andreoni, J. Chem. Phys. 94, 8129 (1991); V. Bonačić-Koutecký *et al.*, P. Fantucci, and J. Koutecký, Phys. Rev. B 37, 4369 (1988).

[9] R. Fournier, J. Chem. Phys. 115, 2165 (2001).

[10] X. Wu and A.K. Ray, J. Chem. Phys. 110, 2437 (1999); R. Fournier, T. Pang, and C. Chen, Phys. Rev. A 57, 3683 (1998);

[11] D.A. Hales, L. Lian, and P.B. Armentrout, Int. J. Mass Spectrom. and Ion Processes, underline{102}, 269 (1990); L. Lian, C.-X. Su, and P.B. Armentrout, J. Chem. Phys. $\underline{96}$, 7542 (1992); C.-X. Su, D.A. Hales, and P.B. Armentrout, J. Chem. Phys. $\underline{99}$, 6613 (1993).

[12] A. K. Rappe and W. A. Goddard III, J. Phys. Chem. $\underline{95}$, 3358 (1991).

[13] R.G. Parr and W. Yang, *Density Functional Theory of Atoms and Molecules* (1989, Oxford University Press).

[14] N. Jaap and E.T.C. Vogt, J. Mol. Catalysis $\underline{A\ 134}$, 63 (1998).

[15] A. D. Becke, Phys.Rev. A $\underline{38}$, 3098 (1988); J. P. Perdew, Phys.Rev. B $\underline{33}$, 8822 (1986).

[16] A. Bérces, P.A. Hackett, L. Lian, S.A. Mitchell, and D.M. Rayner, J. Chem. Phys. $\underline{108}$, 5476 (1998), and references therein.

[17] F. Engelke, G. Hennen, and K. H. Meiwes, Chem. Phys. $\underline{66}$, 391 (1982).

[18] C. E. Fellows, J. Chem. Phys. $\underline{94}$, 5855 (1991).

[19] F. Engelke, H. Hage, and U. Sprick, Chem. Phys. $\underline{88}$, 443 (1984).

[20] K. F. Zmbov, C. H. Wu, and H. R. Ihle, J. Chem. Phys. $\underline{67}$, 4603 (1977).

[21] L. R. Brock, A. M. Knight, J. E. Reddic, J. S. Pilgrim, and M. A. Duncan, J. Chem. Phys. $\underline{106}$, 6268 (1997).

[22] J.S. Pilgrim, M.A. Duncan, Chem. Phys. Lett. $\underline{232}$, 335 (1995); L.R. Brock, A.M. Knight, J.E. Reddic, J.S. Pilgrim, M.A. Duncan, J. Chem. Phys. $\underline{106}$, 6268 (1997).

[23] E. J. Breford and F. Engelke, J. Chem. Phys. $\underline{71}$, 1994 (1979).

[24] A. Stangassinger, A.M. Knight, M. A. Duncan, Chem. Phys. Lett. $\underline{266}$, 189 (1997).

[25] G.A. Bishea, N. Marak, and M.D. Morse, J. Chem. Phys. $\underline{95}$, 5618 (1991).

[26] L. M. Russon, G. K. Rothschopf, and M. D. Morse, J. Chem. Phys. $\underline{107}$, 1079 (1997).

[27] J.M. Sanchez, F. Ducastelle, and D. Gratias, Physica A $\underline{128}$, 334 (1984); S. Müller, J. Phys.: Condens. Matter $\underline{15}$, R1429 (2003), and references therein.

Brill Academic Publishers
P.O. Box 9000, 2300 PA Leiden
The Netherlands

*Lecture Series on Computer
and Computational Sciences*
Volume 3, 2005, pp. 59-74

Computational Electron Spectroscopy - Application to Magnesium Clusters

J. Jellinek[*] and P.H. Acioli[†]

Chemistry Division, Argonne National Laboratory, Argonne, Illinois 60439, USA

Received 26 June, 2005; accepted 12 August, 2005

Abstract: A brief overview of the methodology and applications of computational electron spectroscopy based on density functional theory and recently formulated robust and accurate scheme for conversion of the Kohn-Sham eigenenergies into electron binding energies is presented. The applications are given for magnesium clusters. They illustrate the power of the approach in studies of size, structure, and charge state specific properties of finite systems.

Keywords: Clusters, isomers, spectra of electron binding energies, size-dependence of properties, size-induced transition to metallicity

PACS: 36.40.-c, 31.15.Ew, 33.60.-q, 71.30.+h

1. Introduction

The electronic features of physical systems - be they atoms, molecules, clusters, or extended systems - are among their most fundamental attributes. Accurate knowledge of the electronic energy levels and electronic density of states is central as they define many other physical properties and characteristics. The remarkable recent progress in experimental photoelectron spectroscopies [1] converted electrons into messengers, which, when appropriately interrogated, disclose a wealth of information on a broad variety of system properties. This makes electron spectroscopy a particularly valuable tool in studies of atomic and molecular clusters.

The hallmark of cluster systems is that their characteristics change with size (i.e., the number of atoms and/or molecules in them). This change may be, and in the range of small sizes indeed is, nonmonotonic and seemingly unpredictable. The finiteness of size also introduces couplings between the different properties, which may not have bulk analogs. An example is the role of the charge. In many macroscopic systems addition or withdrawal of electrons has no effect on their properties. Changing the charge state of a finite system, however, may result not only in a shift of its discrete electronic energy levels, but also in changes in its energetically preferred geometric structure and its optical, magnetic, and chemical features. Consequently, the variety and variability of cluster properties is much broader than those of either individual atoms and molecules, or macroscopic systems.

The same is true for different physical phenomena. For example, the finite-size analog of melting – a first-order phase transition in crystalline bulk materials – is a complex multistage transformation that takes place over a finite range of energy or temperature [2]. The peculiarities of this transformation depend on both the cluster size and material [2,3]. Another example is the fact that atomic clusters of nominally metallic elements may not possess metallic attributes up to a certain size or size-range [4].

How does the size-induced transition to metallicity take place? How do the optical, magnetic, and chemical properties of finite systems evolve with their size and why do they evolve the way(s) they do? Electron spectroscopy is one of the most efficient ways to address the questions raised above. In most cases, however, the measured spectra alone cannot give complete answers. This is especially true with regard to the fundamental nature and mechanisms of the size dependence. A comprehensive

---

[*] Corresponding author. E-mail: jellinek@anl.gov

[†] Permanent address: Instituto de Física, Universidade de Brasília, Brasília, DF, 70919-970, Brazil.

understanding of these cannot be achieved without complementary theoretical investigations. A prerequisite for connecting experimental electron spectroscopy with theory is the ability of the latter to compute accurately and efficiently spectra of electron energy levels (or, equivalently, electron binding energies) for systems of interest.

These spectra can, in principle, be obtained employing techniques of quantum chemistry. However, because of their computational complexity, these techniques, especially the higher level ones, are practical only for systems with a relatively small number of electrons. An alternative is to use the density functional theory (DFT). One has to remember, though, that the single-particle eigenenergies of the Kohn-Sham (KS) realization of DFT correspond to auxiliary quasiparticles, rather than real electrons, and they require corrections that convert them into eigenenergies of electrons.

In this paper we briefly outline our recently formulated scheme for conversion of the KS eigenenergies into electron binding energies [5] and discuss the results of its application to magnesium clusters [6-8]. In particular, we compare the spectra of electron binding energies computed for these clusters with the ones that have been measured for them and use the comparison to analyze the cluster properties and phenomena embedded and reflected in these spectra. The conversion scheme is outlined in the next section. In Section 3 we give some details on the computational methodology employed. The results and their analyses are presented in Section 4. A brief summary is given in Section 5.

2. From Kohn-Sham Eigenenergies to Electron Binding Energies

As mentioned above, in order to use the computationally efficient DFT in studies of single-electron properties of systems, one has to have a way to convert the single-particle KS eigenenergies into eigenenergies (or, alternatively, binding energies) of electrons. A brief critical review of the conversion schemes proposed in the past is given in Ref. 5. Here we recap the main elements of a new general and accurate protocol we have formulated recently [5].

The main idea behind this protocol is that the binding energy of an arbitrary M-th electron of a nominally N-electron system, $1 \leq M \leq N$, can be computed rigorously within any version of DFT when this electron is the most external one; any electron can be made the most external by removing all the electrons "above" it. The binding energy $BE_{HOMO}(M) \equiv BE_M(M)$ of the "top" ("HOMO") M-th electron can be obtained as

$$BE_{HOMO}(M) = E(M-1) - E(M), \tag{1}$$

where $E(M)$ is the DFT-defined ground-state total energy of the system with M electrons. The correction $\Delta_{HOMO}(M) \equiv \Delta_M(M)$ that converts the negative of the top M-th (HOMO) KS eigenenergy $\varepsilon_{HOMO}(M) \equiv \varepsilon_M(M)$ of the M-electron system into the binding energy of the M-th electron is

$$\Delta_{HOMO}(M) = BE_{HOMO}(M) - (-\varepsilon_{HOMO}(M)). \tag{2}$$

Our ultimate objective is the correction term $\Delta_M(N)$ that converts the negative of an arbitrary M-th KS eigenenergy $\varepsilon_M(N)$ of an N-electron system into the binding energy of the M-th electron in the N-electron system. The recipe for computing $\Delta_M(N)$ should take into account the shift in the value of the M-th KS eigenenergy from $\varepsilon_M(M)$, for which the correction term is defined by Eq. (2), to $\varepsilon_M(N)$ as the total charge of the system is increased from M to N.

Assume that the values of $\Delta_K(K) = \Delta_{HOMO}(K)$ have been obtained as defined by Eqs. (2) and (1) for all $K = M, M+1,..., N$. Our prescription for computation of the correction $\Delta_M(M+1)$ that converts the negative of $\varepsilon_M(M+1)$ into the binding energy of the M-th electron in the $M+1$-electron system,

$$BE_M(M+1) = -\varepsilon_M(M+1) + \Delta_M(M+1), \tag{3}$$

is

$$\Delta_M(M+1) = \Delta_M(M) + [\Delta_{M+1}(M+1) - \Delta_M(M)]\alpha_M(M+1), \tag{4}$$

where

$$\alpha_M(M+1) = \frac{\varepsilon_M(M+1) - \varepsilon_M(M)}{\varepsilon_{M+1}(M+1) - \varepsilon_M(M)} \,. \tag{5}$$

The meaning of Eqs. (4) and (5) is transparent. If an increase of the total charge of the system from M to $M+1$ does not shift the value of $\varepsilon_M(M)$, i.e., $\varepsilon_M(M+1) = \varepsilon_M(M)$, then $\Delta_M(M+1) = \Delta_M(M)$. If, on the other hand, $\varepsilon_M(M+1) = \varepsilon_{M+1}(M+1)$, i.e., the M-th and the $M+1$-th KS eigenenergies of the $M+1$-electron system are degenerate, then $\Delta_M(M+1) = \Delta_{M+1}(M+1)$. Typically, the value of $\varepsilon_M(M+1)$ is between those of $\varepsilon_M(M)$ and $\varepsilon_{M+1}(M+1)$. In this case, the correction $\Delta_M(M+1)$ is defined by Eq. (4) as a linear in $\alpha_M(M+1)$, Eq. (5), interpolation between $\Delta_M(M)$ and $\Delta_{M+1}(M+1)$.

As one adds one more electron to the system, the additive correction $\Delta_{M+1}(M+2)$ that converts the negative of the KS eigenenergy $\varepsilon_{M+1}(M+2)$ into the binding energy of the $M+1$-th electron in the $M+2$-electron system is obtained by applying again Eqs. (4) and (5) with the values of all the subscripts and the arguments in the parentheses incremented by 1. The knowledge of $\Delta_M(M+1)$ and $\Delta_{M+1}(M+2)$ allows for computation of the correction $\Delta_M(M+2)$ that is to be added to the negative of the KS eigenenergy $\varepsilon_M(M+2)$ in order to convert it into the binding energy of the M-th electron in the $M+2$-electron system. The correction $\Delta_M(M+2)$ is defined as

$$\Delta_M(M+2) = \Delta_M(M+1) + [\Delta_{M+1}(M+2) - \Delta_M(M+1)]\alpha_M(M+2), \tag{6}$$

where

$$\alpha_M(M+2) = \frac{\varepsilon_M(M+2) - \varepsilon_M(M+1)}{\varepsilon_{M+1}(M+2) - \varepsilon_M(M+1)} \,. \tag{7}$$

Adding successively one electron at a time and repeating each time the procedure described above, one eventually arrives at the sought value of $\Delta_M(N)$.

The general prescription for computation of $\Delta_M(N)$ can be formulated as

$$\Delta_M(N) = \Delta_M(N-1) + [\Delta_{M+1}(N) - \Delta_M(N-1)]\alpha_M(N), \tag{8}$$

where

$$\alpha_M(N) = \frac{\varepsilon_M(N) - \varepsilon_M(N-1)}{\varepsilon_{M+1}(N) - \varepsilon_M(N-1)} \,. \tag{9}$$

The corrections $\Delta_M(N-1)$ and $\Delta_{M+1}(N)$ in the r.h.s. of Eq. (8) are themselves obtained through recursive application of Eqs. (8) and (9), until they are reduced to $\Delta_{HOMO}(K)$ with $K = M, M+1, ..., N$. The corrections $\Delta_{HOMO}(K)$ are computed using Eq. (2). The KS eigenenergies $\varepsilon_{HOMO}(K)$ are to be used in Eq. (9) in conjunction with the corrections $\Delta_{HOMO}(K)$ in the r.h.s. of Eq. (8). As a byproduct, the procedure also yields the values of $\Delta_{M+1}(N)$, $\Delta_{M+2}(N)$, ..., $\Delta_{N-1}(N)$. The schematic for the computation of the corrections $\Delta_K(N)$, $K = M, M+1, ..., N$, is shown in Fig. 1.

Further technical details on this conversion scheme as well as its accuracy can be found in Ref. 5 (cf. also Ref. 9). Here mention only that: 1) It is applicable to any implementation of DFT; 2) It uses only ground state properties rigorously defined within DFT; 3) It yields orbital-specific corrections to the KS eigenenergies; and 4) It furnishes highly accurate electron binding energies, provided the total electronic energies in different charge states of the system are reproduced by the chosen version of DFT sufficiently accurately.

3. Computational Details

The computations were performed within the gradient-corrected DFT. Restricted and unrestricted wavefunctions were utilized for closed-shell (neutral Mg_n) and open-shell (anionic Mg_n^-) systems, respectively. The choice of the exchange-correlation functional and the pseudopotential/basis set was governed by tests that considered a variety of alternatives; the details can be found in Ref. 7. For

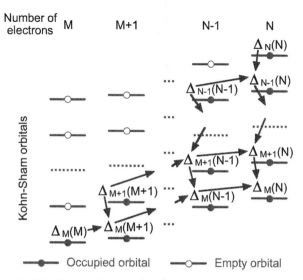

Figure 1: Schematic of the interpolation procedure for computing the corrections $\Delta_K(N)$ (see the text for details).

magnesium, the combination of the Becke exchange [10] and Perdew correlation [11] functionals (BP86) used in conjunction with the Wadt-Hay pseudopotential [12] representation of the ionic cores of the atoms emerged as the best. The remaining two external 2s electrons of each Mg atom and the extra electron of anionic clusters were described by a (21| 21) contracted Gaussian basis set.

The search for structural forms was performed for both neutral and anionic clusters; as a rule, photoelectron spectroscopy (PES) measurements are performed on the latter. The equilibrium structures of the clusters were obtained using gradient-based techniques. A variety of different initial guess configurations was considered for each cluster size and charge, and the optimizations were performed over all the degrees of freedom. Normal mode analysis was used to separate stable structures (isomers) from those that correspond to saddle points of the corresponding potential energy surfaces.

4. Results and Discussion

A. Structures

The most stable isomers of the Mg_n and Mg_n^-, n=2-22, clusters identified in our computations are shown in Figs. 2 and 3, respectively. For the most part, the lowest energy structures of the neutral Mg_n agree with, or are close to, those found in earlier studies [13-16]. Mg_{15} and Mg_{16} are exceptions: their most stable forms shown in Fig. 2 appear to be new. The lowest energy structures of these clusters as identified in Ref. 15 emerge in our computations as well, but as the second isomer. With the exception of Mg_{18}, the preferred spin-multiplicity state of Mg_n, n=2-22, is singlet. For Mg_{18} it is triplet.

Only small anionic Mg_n^-, n=1-7, clusters were considered in earlier theoretical studies [17] (cf. also Ref. 16). In agreement with earlier computations [17] and experimental evidence [18], we find that Mg^- is not a stable species. Our most stable forms of Mg_6^- and Mg_7^- shown in Fig. 2 are different from those found in Ref. 17 – an octahedron for Mg_6^- and a distorted pentagonal bipyramid for Mg_7^-. These emerge in our computations as well, but as the second isomer. With the exception of Mg_{18}^-, the preferred spin-multiplicity state of Mg_n^-, n=2-22, is doublet. For Mg_{18}^- it is quartet.

Inspection of Figs. 2 and 3 shows that the most stable forms of magnesium clusters may depend on their charge. As defined by our computations, the lowest energy isomers of Mg_n differ from those of Mg_n^- for n=6, 7, 8, 11, 12, 18, and 21. As discussed below, the dependence of the preferred structure of a cluster on its charge state has an important implication regarding the interpretation of the PES measured on anionic cluster systems.

A characteristic feature of clusters is that they form many locally stable isomeric forms. For a given material and cluster size, the energy ordering of and the energy gaps between the isomers may depend

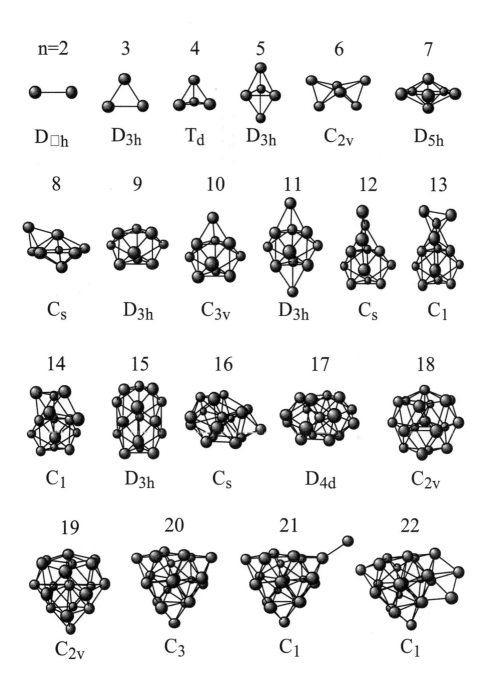

Figure 2: The most stable structures and their symmetries for the neutral Mg_n, n=2-22, clusters.

on the cluster charge. Figure 4 shows the first three fully relaxed isomers of the neutral and anionic Mg_{16}. One notices that addition of an electron to the neutral cluster changes the energy ordering of its second (C_{3v}) and third (C_s) isomers, and the energy gaps between the anionic equilibrium structures are larger than those between their neutral counterparts. Figure 5 displays the first three fully relaxed isomers of the neutral and anionic Mg_{18}. Here, the anionic isomers are very close in energy, whereas

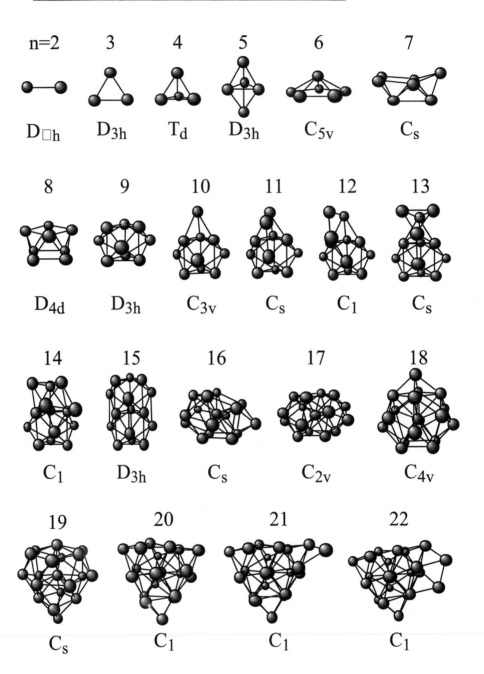

Figure 3: The most stable structures and their symmetries for the anionic Mg_n^-, n=2-22, clusters.

their neutral counterparts are separated by larger energy gaps, and the third (C_{4v}) isomer of Mg_{18} becomes the most stable structural form of Mg_{18}^-. The energy changes accompanying the structural relaxations caused by addition of an electron to different isomers of the neutral clusters can be assessed from Figs. 6 and 7; these show the differences in the configurational energies of the anionic Mg_{16}^- and Mg_{18}^-, respectively, computed in the native structures of both the anionic and neutral isomers. For

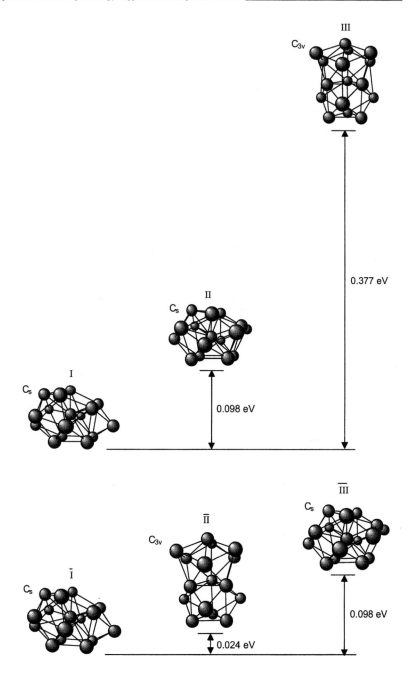

Figure 4: The first three isomers of the anionic Mg_{16}^{-} (Roman numerals) and neutral Mg_{16} (Roman numerals with bars) clusters.

Mg_{16}^{-} the relaxation energies are 0.032 eV (relaxation of structure $\bar{I}$ into structure I), 0.111 eV ($\overline{III}$ into II), and 0.101 eV ($\overline{II}$ into III). For Mg_{18}^{-} they are 0.078 eV ($\overline{III}$ into I), 0.056 eV ($\bar{I}$ into II), and 0.039 eV ($\overline{II}$ into III).

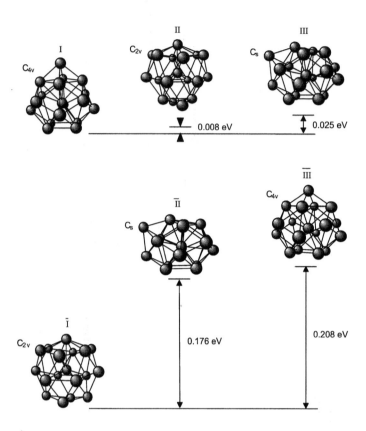

Figure 5: The first three isomers of the anionic Mg_{18}^- (Roman numerals) and neutral Mg_{18} (Roman numerals with bars) clusters.

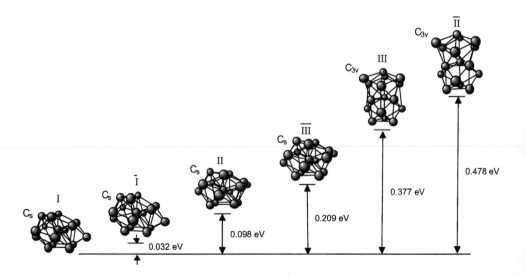

Figure 6: Energy ordering of Mg_{16}^- in different structures shown in Fig. 4.

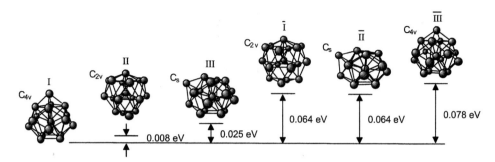

Figure 7: Energy ordering of Mg_{18}^- in different structures shown in Fig. 5.

B. Electronic Properties

We turn now to electronic properties of clusters. Figure 8 shows the spectra of electron binding energies computed [7,8] and measured [19] for Mg_4^-. The computations predict that this cluster forms a single, tetrahedral isomer. Barring accidental coincidence, the remarkable agreement between the theoretical and measured data serves as an experimental corroboration of the tetrahedral form of Mg_4^-.

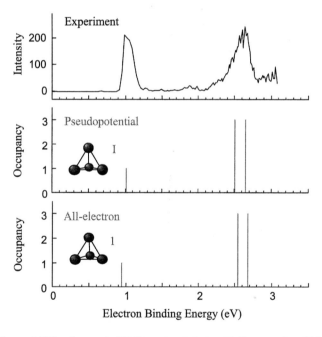

Figure 8: Measured [19] and computed [7,8] spectra of electron binding energies of Mg_4^-. The quality of the pseudopotential based computations can be gauged through comparison with the results of an all-electron treatment.

The spectra computed [7,8] for different structures of Mg_{16}^- are shown together with the measured [19] spectrum of this cluster in Fig. 9. Inspection of this figure leads to the following observations. The pattern of three groups of binding energies in the computed spectrum of isomer I fits very well the shape of the measured spectrum. The pattern of the computed spectrum of isomer II is different from that of isomer I, but it falls entirely under the measured spectrum. The relatively broad nature of the latter is an indication that the experiment probes both isomers I and II. However, that is not the case for isomer III; it can be excluded as a contributor to the measured spectrum because the binding energy of its most external electron is not represented in this spectrum. Comparison of spectra of Mg_{16}^- computed

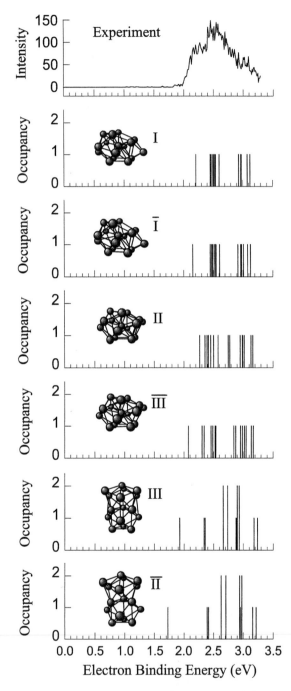

Figure 9: Computed electron binding energies and the occupancy of the corresponding orbitals (vertical bars) in different structures of Mg_{16}^- defined in Fig. 4. The measured spectrum [19] is also shown.

in structures $\bar{I}$, $\overline{III}$, and $\bar{II}$ with those computed in structures I, II, and III, respectively, (cf. Fig. 4) allows one to assess the changes in the electron binding energies accompanying the relaxations from equilibrium structures native for the neutral isomers to those native for the anionic isomers. Relaxation

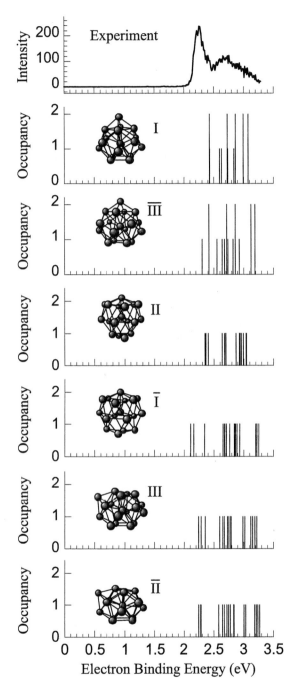

Figure 10: Computed electron binding energies and the occupancy of the corresponding orbitals (vertical bars) in different structures of Mg_{18}^- defined in Fig. 5. The measured spectrum [19] is also shown.

of structure $\bar{I}$ into structure I causes only minor changes in the electron binding energies. Relaxation of $\overline{III}$ into II and of $\overline{II}$ into III has a more noticeable effect.

Data similar to those in Fig. 9, but for Mg_{18}^-, are shown in Fig. 10. The computed spectra for all three isomers of this cluster (I, II, and III) fall under the measured spectrum, and most probably all three contribute to it. The shifts in binding energies accompanying the relaxations from equilibrium structures native for the neutral isomers to those native for the anionic isomers (cf. Fig. 5) are moderate, but distinct.

The results presented and discussed above illustrate how high accuracy computations of spectra of electron binding energies, when contrasted with data obtained in high accuracy PES measurements, allow for identification of the structures of clusters actually generated and interrogated experimentally. Size-specific features in the spectra can be used to analyze size-driven phenomena. One of the most intriguing among these is the mentioned above size-induced transition to metallicity [4,6,7,13,15,16,19-28]. Is this transition monotonic or not? At which cluster size, or size range, does it take place? What is (are) the indicator(s) of finite-size metallicity? The answers to these questions are largely unknown.

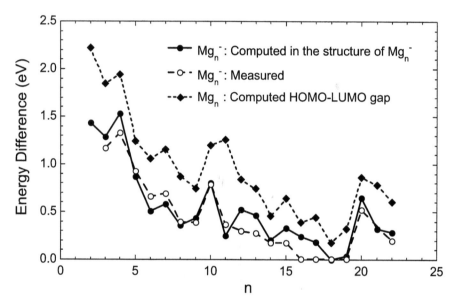

Figure 11: Computed [6,7] and measured [19] difference in the binding energies of the two most external electrons of anionic Mg_n^- clusters as a function of cluster size. The computed [6,7] HOMO-LUMO gaps of the neutral Mg_n are also shown. The computed results are for the lowest energy structures of the clusters.

Since direct measurements of electron transport in small and medium size clusters are a challenge yet to be overcome, the approach used most often to investigate the size-induced nonmetal-to-metal transition is based on the following considerations. Bulk metals are systems with zero gap between the conduction and valence bands. In clusters energy bands are replaced by discrete energy levels, and the difference between the highest occupied molecular orbital (HOMO) and the lowest unoccupied molecular orbital (LUMO) plays the role of the band gap. Thus, one can watch how the HOMO-LUMO gap of a cluster evolves with its size and anticipate a closer of this gap in metal clusters as their size increases.

The reasoning that underlies the experimental studies proceeds as follows. In an anion of a cluster the extra electron occupies the LUMO of the neutral cluster. Thus through PES measurements of the binding energies of the two most external electrons in anionic clusters one can, neglecting the relaxation of the energy levels, determine the HOMO-LUMO gap.

Figure 11 shows the graphs of the difference between the binding energies of the two most external electrons in anionic Mg_n^- clusters as computed [6,7] in their most stable isomeric forms and extracted from the measured [19] PES, together with the graph of the HOMO-LUMO gap computed [6,7] in the lowest energy structures of the neutral Mg_n, all considered as a function of n. Although globally the trend in the n-dependence of the HOMO-LUMO gap follows that in the computed and measured differences in the binding energies of two top electrons in Mg_n^-, the two quantities are quite different. Not only are the HOMO-LUMO values larger, but they also display local behavior that is peculiar to them. For example, in going from n=10 to n=11 the HOMO-LUMO graph turns upward, whereas the other two graphs turn downward. On the other hand, the data computed and measured for Mg_n^- are in excellent agreement with each other. The inescapable message of Fig. 11 is that, contrary to the above line of reasoning, PES measurements on anionic clusters (more generally, finite size anionic systems) do not, in general, probe the HOMO-LUMO gap of the corresponding neutral clusters (systems). The main cause for that is not the neglect of the shifts in the energy levels as a consequence of addition of an extra electron; as is clear from Fig. 12, which shows the HOMO-LUMO gap and the difference in the binding energies of the two top electrons in Mg_n^- computed assuming that the anions are frozen in the most stable structures of their neutral counterparts, this shift, except for the smallest clusters (n<5), is indeed only minor.

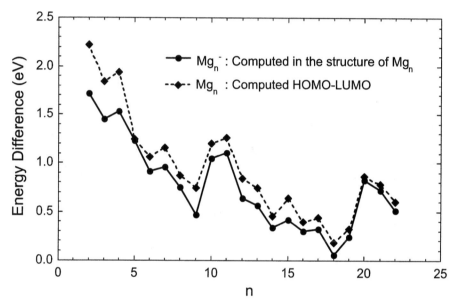

Figure 12: Computed difference in the binding energies of the two most external electrons of anionic Mg_n^- clusters frozen in the most stable structures of their neutral counterparts. The HOMO-LUMO gaps of the latter are also shown.

The main reason for the possible difference between the HOMO-LUMO gap of neutral clusters and the measured gap between the binding energies of the two most external electrons of their anionic counterparts (defined as the difference between the vertical detachment energies of these electrons) is the fact that addition of an electron to a cluster (especially a small one) may change its preferred geometric form, which in turn changes the spectrum of the electron energy levels. This effect of the extra charge is embedded in the measured PES and should be taken into account in their interpretation.

Does the preceding discussion mean that, because PES measurements on anions do not necessarily probe the HOMO-LUMO gap, they are inappropriate for studying the size-induced transition to metallicity? The negative answer to this question follows from the recognition that the difference between the binding energies of the two most external electrons in anionic finite systems is as a legitimate finite size analog of the bulk band gap as the HOMO-LUMO gap of neutral systems. The two energy differences approach each other as the system grows in size and both converge to the bulk band gap in the limit of large systems.

The role of charge in the size-induced transition to metallicity is another finite size effect. The transition can manifest itself differently in different charge states of a finite system. A comprehensive understanding and description of the transition should incorporate these various manifestations. The fact that, as shown in Fig. 11, both the computed and the measured difference in the binding energies of the two top electrons in Mg_n^- clusters decreases, even if not monotonically, as n increases and vanishes at n=18 is consistent with evolution towards a metallic state. It is also clear, however, that what is presented in Fig. 11 is only a stage in this evolution. In fact, the data in the figure indicate that the gap between the binding energies reopens for n>18 (see also Ref. 19). The studies will have to be extended to clusters of larger sizes. For a quantitative account of the contribution of different isomers into a PES measured at different temperatures (such measurements are yet to become routine) one will have to include the entropic effect. This is a nontrivial task, as it requires evaluation of anharmonic vibrational densities of states. Most importantly, we will have to develop a better fundamental understanding of what constitutes a finite-size analog (or analogs) of bulk metallicity and how it (they) evolves with the size of systems.

5. Summary

This paper discusses computational electron spectroscopy as an efficient tool to study properties of atomic clusters or, more generally, finite systems. It outlines a correction scheme, which allows for conversion of the Kohn-Sham eigenenergies of the density functional theory into eigenenergies (or, equivalently, binding energies) of real electrons; thereby the scheme extends the applicability of the traditional (time-independent) DFT to single-electron properties. It can be combined with any implementation of DFT, and it furnishes highly accurate results.

The illustrations are given through applications to magnesium clusters. We present and analyze results on the structures of neutral and anionic Mg_n and use them to emphasize that the energy ordering of and the energy gaps between the different isomeric forms may depend on the cluster charge. This has important implications regarding the interpretation of the experimental photoelectron spectra, which are normally measured on anionic species.

We show how a comparison of computed isomer-specific spectra of electron binding energies with measured PES allows for identification of the structural form(s) of clusters actually generated and probed experimentally. We point out what is needed for a quantitative evaluation of the contribution of different isomeric forms to a measured PES.

Finally, we give a brief discussion of the extremely interesting and technologically important phenomenon of size-induced nonmetal-to-metal transition. The transition to metallicity in finite systems is more multifaceted and complex than in bulk materials. As an illustration, we demonstrate the role of the charge state. But the transition may also be a function of the structural forms, which, in their turn, may also depend on the charge. The fundamental understanding of finite-size metallicity and its evolution to the bulk metallic state is yet to be developed.

Acknowledgments

We thank Kit Bowen for providing us the experimental PES data on Mg_n^- clusters. This work was supported by the Office of Basic Energy Sciences, Division of Chemical Sciences, Geosciences, and Biosciences, U.S. Department of Energy under Contract number W-31-109-Eng-38.

References

[1] C.-Y. Ng (Ed.): *Photoionization and Photodetachment*, Parts I and II, World Scientific, Singapore, 2000; and references therein.

[2] R. S. Berry, Phases and phase changes of small systems, *Theory of Atomic and Molecular Clusters* (Editor: J. Jellinek), Springer, Heidelberg (1999) 1-26; and references therein.

[3] J. Jellinek and E.B. Krissinel, Alloy clusters: structural classes, mixing, and phase changes, *Theory of Atomic and Molecular Clusters* (Editor: J. Jellinek), Springer, Heidelberg (1999) 277-308; and references therein.

[4] B. von Issendorff and O. Cheshnovsky, Metal to insulator transition in clusters, *Annual Review of Physical Chemistry* **56**, 549-580 (2005); and references therein.

[5] J. Jellinek and P.H. Acioli, Converting Kohn-Sham eigenenergies into electron binding energies, *The Journal of Chemical Physics* **118**, 7783-7796 (2003)

[6] P.H. Acioli and J. Jellinek, Electron binding energies of anionic magnesium clusters and the nonmetal-to-metal transition, *Physical Review Letters* **89**, 213402 (2002).

[7] J. Jellinek and P.H. Acioli, Magnesium clusters: structural and electronic properties and the size-induced nonmetal-to-metal transition, *The Journal of Physical Chemistry A* **106**, 10919-10925 (2002); **107** 1610 (2003).

[8] P.H. Acioli and J. Jellinek, Theoretical determination of electron binding energy spectra of anionic magnesium clusters, *The European Physical Journal D* **24**, 27–32 (2003).

[9] J.M. Mercero, J.M. Matxain, X. Lopez, D.M. York, A. Largo, L.A. Erikson, and J.M. Ugalde, Theoretical methods that help understanding the structure and reactivity of gas phase ions, *International Journal of Mass Spectrometry* **240**, 37-99 (2005).

[10] A.D. Becke, Density-functional exchange-energy approximation with correct asymptotic behavior, *Physycal Review A* **38**, 3098-3100 (1988).

[11] J.P. Perdew, Density-functional approximation for the correlation energy of the inhomogeneous electron gas, *Physical Review B* **33**, 8822-8824 (1986).

[12] W.R. Wadt and P.J. Hay, *Ab initio* effective core potentials for molecular calculations. Potentials for main group elements Na to Bi, *The Journal of Chemical Physics* **82**, 284-298 (1985)

[13] V. Kumar and R. Car, Structure, growth, and bonding nature of Mg clusters, *Physical Review B* **44**, 8243-8255 (1991).

[14] U. Rothlisberger, W. Andreoni, and P. Giannozzi, Thirteen-atom clusters: equilibrium geometries, structural transformations, and trends in Na, Mg, Al, and Si, *The Journal of Chemical Physics* **96**, 1248-1256 (1992).

[15] A. Köhn, F. Weigend, and R. Alrichs, Theoretical study on clusters of magnesium, *Physical Chemistry Chemical Physics* **3**, 711-719 (2001).

[16] J. Akola, K. Rytkonen, and M. Manninen, Metallic evolution of small magnesium clusters, *The European Physical Journal D* **16**, 21-24 (2001).

[17] F. Reuse, S.N. Khanna, V. de Coulon, and J. Buttet, Pseudopotential local-spin-density studies of neutral and charged Mg_n ($n \leq 7$) clusters, *Physical Review B* **41**, 11743-11759 (1990).

[18] H. Hotop and W.C. Lineberger, Binding energies in atomic negative ions: II, *Journal of Physical and Chemical Reference Data* **14**, 731-750 (1985).

[19] O.C. Thomas, W. Zheng, S. Xu, and K.H. Bowen, Jr., Onset of metallic behavior in magnesium clusters, *Physical Review Letters* **89**, 213403 (2002).

[20] T. Diederich, T. Döppner, J. Braune, J. Tiggesbaunker, and K.-H. Meiwes-Broer, Electron delocalization in magnesium clusters grown in supercold helium droplets, *Physical Review Letters* **86**, 4807-4810 (2001).

[21] C. Brechignac, M. Broyer, Ph. Cahuzac, G. Delacretaz, P. Labastie, and L. Wöste, Size dependence of inner-shell autoionization lines in mercury clusters, *Chemical Physics Letters* **120**, 559-563 (1985).

[22] K. Rademann, B. Kaiser, U. Even, and F. Hensel, Size dependence of the gradual transition to metallic properties in isolated mercury clusters, *Physical Review Letters* **59**, 2319-2321 (1987).

[23] C. Brechignac, M. Broyer, Ph. Cahuzac, G. Delacretaz, P. Labastie, J.P. Wolf, and L. Wöste, Probing the transition from van der Waals to metallic mercury clusters, *Physical Review Letters* **60**, 275–278 (1988).

[24] H. Haberland, B. von Issendorf, J. Yufeng, and T. Kolar, Transition to plasmonlike absorption in small Hg clusters, *Physical Review Letters* **69**, 3212-3215 (1992).

[25] R. Busani, M. Folkers, and O. Cheshnovsky, Direct observation of band-gap closure in mercury clusters, *Physical Review Letters* **81**, 3836-3839 (1998).

[26] M.E. Garcia, G.M. Pastor, and K.H. Bennemann, Theory for the change of the bond character in divalent-metal clusters, *Physical Review Letters* **67**, 1142-1145 (1991).

[27] X.G. Gong, Q.Q. Zheng, and Y.Z. He, Electronic structures of magnesium clusters, *Physics Letters A* **181**, 459-464 (1993).

[28] P. Delaly, P. Ballone, P., and J. Buttet, Metallic bonding in magnesium microclusters, *Physical Review B* **45**, 3838-3841 (1992).

Brill Academic Publishers
P.O. Box 9000, 2300 PA Leiden
The Netherlands

*Lecture Series on Computer
and Computational Sciences*
Volume 3, 2005, pp. 75-78

Crystal Orbital theory for polarization of quasilinear periodic polymers in electric fields

Bernard KIRTMAN

Department of Chemistry and Biochemistry, University of California, Santa Barbara, CA. 93106 (USA).

Received 5 August, 2005; accepted 12 August, 2005

Abstract: A crystal orbital (band structure) treatment is presented for the linear and nonlinear polarization of a quasilinear periodic polymer in a spatially uniform, time-dependent electric field. Both the translational symmetry-breaking and spatial unboundedness of the interaction potential are taken into account. Although applicable at all levels, we emphasize the Hartree-Fock and density functional theory (DFT) formulations. A comparison with the molecular Hamiltonian and physical interpretation of the various terms is given. On this basis it can be seen why model calculations show the sawtooth potential model to be inadequate for π-conjugated organic molecules. Geometry relaxation in a finite static field, which is used to determine vibrational (hyper)polarizabilities, is analyzed as a special case. It is shown that using an optimized effective potential (OEP) for exact-exchange avoids the catastrophic overshoot in DFT calculations of static nonlinear (as well as linear) polarization for long chains. The correlation effects that remain are important, but not adequately described by conventional functionals. Efforts to develop an appropriate functional and studies dealing with the exact-exchange time-dependent OEP will be described.

Keywords: electric field polarization of polymers, linear and nonlinear optical properties, crystal orbitals, exact-exchange optimized effective potential.

1. Introduction

There are two complementary general procedures for evaluating the physical properties of quasilinear periodic polymers. One of these is the finite oligomer method [1-3] whereby calculations are carried out on oligomers of increasing size and the results are extrapolated to the long, or infinite, chain limit. This approach has the major advantage that one can directly employ the extensive quantum chemical codes that have been developed for dealing with properties of ordinary molecules. The primary unresolved issue is to how to extrapolate accurately, particularly for properties that converge slowly with chain length, such as the linear and nonlinear response to an electric field by highly polarizable electronically delocalized polymers. Important progress has been made recently in using physical models to determine the form of the extrapolation function [4], in applying numerical convergence acceleration techniques [5], and in the development of linear scaling methods for alleviating size limitations [6-8]. In this presentation, however, we will focus on the other alternative, namely the crystal orbital (or band structure) approach [9,10].

In band structure calculations a perfect repeating unit cell structure is assumed. The translational symmetry allows one to avoid extrapolation by applying cyclic boundary conditions through the use of Bloch orbitals. On the other hand, it is not obvious how to incorporate the interaction with a spatially uniform, time-dependent electric field within the Bloch orbital formalism since the usual scalar potential destroys the translational symmetry and is also spatially unbounded. A solution of this theoretical problem has been developed in collaboration with Gu and Bishop [11,12]. Our KGB treatment is based on replacing the scalar potential with the translationally invariant vector potential along with a treatment of the unbounded dipole moment operator due to Blount [13]. It was initially developed in the context of time-dependent Hartree-Fock (TDHF) theory but can be generalized to other time-dependent quantum chemical methods as will be shown.

2. Time-dependent crystal orbital Hartree-Fock treatment

We begin with a review of the crystal orbital TDHF equation for the LCAO coefficients, C_{pn}^k, that determine the crystal orbitals for Bloch wavevector k and band n

$$\phi_{nk}(\mathbf{r}) = \frac{1}{\sqrt{2N+1}} \sum_{l=-N}^{+N} e^{ikla} \sum_p C_{pn}^k \chi_p^l(\mathbf{r}) \tag{1}$$

in terms of the AOs χ_p^l of unit cell l. In the presence of a time-dependent electric field, $E(t)$, these time-dependent crystal orbital coefficients are determined by :

$$\sum_p \left[F_{pq}^k C_{pn}^k + \Theta_{qp}^k C_{pn}^k - iS_{qp}^k \frac{\partial}{\partial t} \right] C_{pn}^k + E(t)[M_{qp}^k + iS_{qp}^k \frac{\partial}{\partial k}]C_{pn}^k \right] = \sum_\mu \sum_p S_{qp}^k C_{p\mu}^k \varepsilon_{\mu n}^k \tag{2}$$

In Eq.(1) N is the number of neighboring cells whose interactions with the reference cell ($l = 0$) are explicitly taken into account. The derivation of Eq.(2), following the KGB theory, assumes a non-metallic (i.e. non-zero bandgap) system. In the field-free case only the $F_{qp}^{(0)k} C_{pn}^k$ term would appear on the *lhs*. Here the zeroth-order AO Fock matrix for a particular k, i.e $F_{qp}^{(0)k}$, is given by the sum over unit cells:

$$F_{qp}^{(0)k} = \sum_{l=-N}^{+N} e^{ikla} < \chi_q^0 | \hat{F}^{(0)} | \chi_p^l > \tag{3}$$

and similarly for the overlap matrix S_{qp}^k as well as the matrices Θ_{qp}^k and M_{qp}^k defined immediately below. The matrix $\Theta_{..}$ accounts for the terms in the field-dependent Hamiltonian that arise from the usual electron-electron (coulomb and exchange) interactions. Thus, if we were to ignore the superscript k, and substitute the usual dipole moment operator for $M + i S \partial/\partial k$, Eq.(2) would be the familiar TDHF equation for a molecule in a time-dependent electric field.

The key difference between Eq.(2) and the corresponding molecular TDHF equation lies in the $M + i S \partial C/\partial k$ term. $E(t)M$ arises from the *intra*cellular sawtooth potential $E(t)$ $(z - la)$. It represents the unit cell dipole interaction with the external field. On the other hand, $iE(t) S \partial C/\partial k$ is an *inter*cellular charge flow term. The total polarization, which is just the expectation value of $M + i S \partial C/\partial k$, has often been calculated in the past by using a piecewise dipole potential (i.e $E(t)z$) and neglecting the intercellular charge flow contribution. However, recent work with Springborg and Dong [14], shows that this procedure is not adequate for typical π-conjugated organic chains. Evaluation of the intercellular charge flow term is complicated by the fact that the crystal orbital eigenvectors C_{pn}^k are, in general, complex. For each k they may be multiplied by an arbitrary phase factor. In order to have well defined derivatives a continuity condition must be imposed on the phase angle as is done in KGB. Even with that condition, however, the permanent dipole moment is only determined up to an integer multiple of the unit cell length. This means that the dipole moment calculated for a centrosymmetric polymer, such as polyacetylene, may be different from zero. Even for a polymer with an asymmetric unit cell, however, only a very crude approximate value need be known in order to compensate for this ambiguity. On the other hand, the linear and nonlinear polarizabilities determined by the perturbation theory solution of Eq.(2) are unaffected by the arbitrariness in the phase angle [12].

The non-uniqueness of the permanent dipole moment poses a potential problem for the calculation of vibrational polarizabilities and hyperpolarizabilties using the nuclear relaxation approach. In nuclear relaxation methods one must optimize the geometry in the presence of a finite (static) field. If the contribution of the permanent dipole moment to the polarization varies from one geometry to the next, then the optimization cannot be carried out. A study of this problem has recently been done in collaboration with Springborg. In the course of that investigation a new continuity condition has been developed. Furthermore, the fundamental TDHF equation has been reformulated in a way that makes it more amenable to a self-consistent-field solution in the presence of a finite field (as opposed to the perturbation theory solution given in KGB). The results of model calculations based on these developments will be presented.

3. Treatment of electron correlation

By definition the TDHF treatment does not take into account *true* electron correlation effects. Such effects have been incorporated into molecular calculations by employing a quasi-energy formulation that may also be used to derive the TDHF equation [15,16]. In principle this approach could be utilized within the framework of our crystal orbital response theory. In some ways, however, it seems that it might eventually be more fruitful to apply a density functional theory (DFT) approach. On the other hand, the catastrophic failure of conventional DFT to predict the response of long molecular chains to static electric fields is well-known [17,18]. It has been shown, however, that an accurate treatment of the self-interaction correction (SIC) by means of an exact-exchange optimized effective potential (OEP) alleviates this difficulty for static linear polarizabilities [19,20]. Based on recent work with Champagne, Yang, Toro-Labbé and Bulat [21], as well as Bonness, we will see that the same conclusion holds for static nonlinear polarizabilities. The latter calculations also reveal that the correlation contribution, which can be quite significant, is not accounted for by conventional correlation functionals. Our most recent efforts to develop appropriate functionals will be described. This is an essential ingredient if computational advantages are to be realized from DFT.

Assuming that static fields can be successfully handled there remains the problem of adapting the OEP method to treat the response to time-dependent fields. The extension of OEP to time-dependent OEP (TDOEP) has proven to be less than completely straightforward . Very recently Hirata, Ivanov, Bartlett and Grabowski [22] have implemented an exact-exchange TDOEP algorithm for dynamic polarizabilities within the adiabatic approximation and a successful application to several atoms and diatomic molecules was made. The current status of our own TDOEP studies aimed at molecular chains will be reported here.

References

[1] B. Kirtman, W.B. Nilsson and W.E. Palke, Solid State Commun. **46**, 791 (1983); B. Kirtman, Chem. Phys. Lett, **143**, 81 (1988).

[2] G.J.B. Hurst, M. Dupuis and E. Clementi, J. Chem. Phys. **89**, 385 (1988).

[3] B. Champagne, D.H. Mosley and J.-M. Andre, Int. J. Quantum Chem. Symp. **27**, 667 (1993)

[4] K.N. Kudin, R. Car and R. Resta, J. Chem. Phys. **122**, 134907 (2005).

[5] E.J. Weniger and B. Kirtman, Comput. Math. Appl., **45**, 189 (2003).

[6] X.S. Li, J.M. Millam, G.E. Scuseria, M.J. Frisch and H.B. Schlegl, J. Chem. Phys. **119**, 7651 (2003).

[7] V. Weber, A.M.N. Niklasson and M. Challacombe, Phys. Rev. Lett. **92**, 193002 (2004); **A.M.** Niklasson and M. Challacombe, Phys. Rev. Lett. **92**, 193001 (2004).

[8] F.L. Gu, Y. Aoki, A. Imamura, D.M. Bishop and B. Kirtman, Mol. Phys. **101**, 1487 (2003); J. Korchowiec, F.L. Gu, A. Imamura, B. Kirtman and Y. Aoki, Int. J. Quantum Chem. **102**, 785 (2005).

[9] J.-M. Andre, Adv. Quantum Chem. **12**, 65 (1980).

[10] J.J. Ladik, *Quantum Theory of Polymers as Solids*, Plenum Press, New York (1988).

[11] B. Kirtman, F.L. Gu and D.M. Bishop, J. Chem. Phys. **113**, 1294 (2000).

[12] D.M. Bishop, F.L. Gu and B. Kirtman, J. Chem. Phys. **114**, 7633 (2001).

[13] E.I. Blount, in *Solid State Physics*, edited by F. Seitz and D. Turnbull (Academic, New York, 1962), Vol.13, p.305.

[14] M. Springborg, B. Kirtman and Y. Dong, Chem. Phys. Lett. **396**, 404 (2004).

[15] F. Aiga, K. Sasagane and R. Itoh, J. Chem. Phys. **99**, 3779 (1993); K. Sasagane, F. Aiga and R. Itoh, J. Chem. Phys. **99**, 3738 (1993).

[16] C. Hattig and B.A. Hess, J. Chem. Phys. **105**, 9948 (1996).

[17] B. Champagne, E.A. Perpete, S.J.A. van Gisbergen, E.-J. Baerends, J.G. Snijders, C. Soubra-Ghaoui, K.A. Robins and B. Kirtman, J. Chem. Phys. **109**, 10489 (1998); *erratum* **110**, 11664 (1999).

[18] S.J.A. van Gisbergen, P.R.T. Schipper, O.V. Gritsenko, E.-J. Baerends, J.G. Snijders, B. Champagne and B. Kirtman, Phys. Rev. Lett. **83**, 694 (1999).

[19] P. Mori-Sanchez, Q. Wu and W. Yang, J. Chem. Phys, **119**, 11001 (2003).

[20] S. Kummel, L. Kronik and J.P. Perdew, Phys. Rev. Lett. **93**, 213002 (2004).

[21] F.A. Bulat, A. Toro-Labbe, B. Champagne, B. Kirtman and W. Yang, J. Chem. Phys. **123**, 014319 (2005).

[22] S. Hirata, S. Ivanov, R.J. Bartlett and I. Grabowski, Phys. Rev. A **71**, 032507 (2005); S. Hirata, S. Ivanov, I Grabowski and R.J. Bartlett, J. Chem. Phys. **116**, 6468 (2002).

Brill Academic Publishers
P.O. Box 9000, 2300 PA Leiden
The Netherlands

Lecture Series on Computer
and Computational Sciences
Volume 3, 2005, pp. 79-84

On the radius of gyration of a dendritic block copolymer of first generation

M. Kosmas

Department of chemistry, University of Ioannina, Ioannina, Greece

Received 5 August, 2005; accepted 12 August, 2005

Abstract: Our previous works on linear and star polymers are extended to dendritic block copolymer chains of an extra generation and two different species. On the ends of a star of fa branches of molecular weight Na each, which constitute the inner region of the dendrimer the new generation is built. This outer generation consits of fa new stars each one of fb branches having totally fafb branches everyone having a molecular weight Nb. Working at the critical dimensionality d=4 above which polymer chains behave ideally we determine the radius of gyration of the dendritic block copolymer in an effort to see the effects of the two different molecular weights, the two different number of branches and the rest characteristics of the chain and the solvent, on the size of the macromolecule.

Keywords: radius of gyration, dendritic block copolymers

PACS: 05.10.Cc, 61.41.+e, 82.35.Jk

1. Introduction

In order of real polymer chains in a solvent to be described by ideal random walk statistics the addition of the effects from the interactions between the chain units must be included[1]. If we consider the chain as a continuous curve R(s), we can start with a probability distribution of the structure:

$$P[R(s)] = Exp\left\{ - \int_0^n \dot{R}(s)^2 ds - u \int_0^n ds \int_0^n ds' \, \delta^d[R(s) - R(s')] \right\} \tag{1}$$

where the first term in the exponent, which includes the square of the derivative of R(s) with respect to the contour length s, ensures the connectivity of the ideal chain while the second is the energy term which expresses the interactions between all pairs of chain points when they come close in space. Macroscopic properties of the chain are taken as the averages over the probability P[R(s)] and depend on the molecular weight n of the chain which is proportional to the number of the units of the chain and the interaction parameter u which describes average repulsions between the units when positive and attractions when it is negative. By means of the probability eq.1 and scaling analysis it is easily seen that each macroscopic property like the mean end to end square distance $<R^2>$ or the mean radius of gyration $<S^2>$ of the chain obtain the form:

$$<S^2> = <S^2>_0 \, F(un^{(4-d)/2}) \quad , \qquad <S^2>_0 = n/6 \tag{2}$$

where $<S^2>_0$ is the radius of gyration of the ideal chain when the interactions vanish, u=0. The dependence of the linear dimensions of the chain on the square root of the molecular weight and its length indicates the chaotic character of the flexible conformations due to the many foldings which the chain can have. It is far from having an organized structure like that of a rod where the linear dimensions of the macromolecule would be proportional to the length of the chain. Turning on the interactions and for repulsions between the chain units the size of the chain becomes larger and the enlargement is given by the function F which depends on the specific combination of the interaction parameter u and the molecular weight n of the chain, shown in eq.2.

Another important thing to notice from eq 2 is that the nonidiality of the problem is different in different dimensionalities d. Increasing d the interaction term reduces and the effects are smaller. As a matter of fact for d>4 and large molecular weight n, the interaction parameter $un^{(4-d)/2}$ is negligible, making the chain behave ideally. We have used this observation and worked some time ago at d=4 where the nonidealities are the less possible[2]. For linear chains at d=4 first order perturbation theory pass through the evaluation of the one loop diagram $\curlyvee$ which for the mean end to end square distance obtains the value

$$\curlyvee \ = 2u \ \int_{0}^{n}\int_{j}^{n}\frac{1}{(j-i)}didj = 2u\,n\{Log[n]-(1)\}$$

and gives for the mean end to end square distance the result $<R^2>=n\{1 + 2u\,(Log[n]-1)\}$. For the corresponding value of mean radius of gyration we take that:

$$<S^2>_1 = (n/6)[1+2u(Log[n]-(13/12))] \quad . \tag{3}$$

There is a fixed point where the dependence of the properties becomes of the power law form. At this point the value of the interaction parameter can be determined from second order calculations, after the evaluation of the contribution of the second order diagrams, $\widetilde{\sigma\sigma}$, $\bigotimes_{2}$, $\underline{\sigma}$ and it is equal to $u^*=\varepsilon/16$ where $\varepsilon=4-d$. The radius of gyration obtains the form $<S^2>\sim n^{1+(\varepsilon/8)}$ and the critical exponent $1+(\varepsilon/8)$ thus determined is in accordance with the results from renormalization group theory and proper experiments. It has been shown that the structure $<S^2>_1 = (n/6)[1+2u(Logn+F)]$ of the macroscopic properties keeps the same for chains of other architectures like rings stars combs but it can also be extended to chains of two different kinds of units like block copolymers[3]. The larger the value of the function F is the harder to reach the limit of exponents where the Log[n] is dominant. Comparing for example the radius of gyration with the mean end to end square distance of the linear chain we see that the critical exponant region is reached to higher n values for the radius of gyration than the mean end to end square distance in accordance with the enumeration results from a computer.

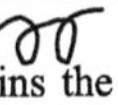

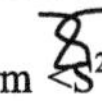

2.Dendrimers of first generation

Dendritic polymers have structures which though easy to describe it is hard to be studied. The usual homodendritic polymers made from one kind of monomer have a central core from where fa branches start. From the ends of these branches new stars emerge of fb functionality producing the first generations of fafb branches, and so on. Important properties of these polymeric structures are those correlated with their average geometric characteristics. Many voids existing in their interior make them capable to absorb and carry smaller molecules, while the many end units make them very drastic[4]. We apply our previous experience on the evaluation of the radius of gyration, to the specific block copolymer dendrimer of two different kinds of monomers shown in Fig.1.

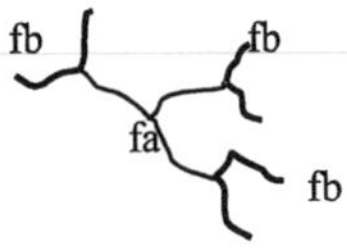

Fig. 1. A dendrimer of two generations. The 0th generation cosists of a star chain of fa branches of na length each. From the ends of the fa branches new stars start forming the 1st generation of fafa branches of nb legth each.

The intensity of interaction between two a units is equal to ua , between two units of the b kind equal to ub while that between one a and one b is equal to uab. In order to evaluate the contribution of all interactions to all segments four indices are necessary making the number of different necessary diagrams of the order of many decades. Grouping these diagrams in characteristic classes we managed to sum all contributions giving for the radius of gyration of the block copolymer dendrimer the expression

$$<S^2>_d = \{M / [fa^2(1+fbx)^3]\}.$$

$$\{[-(1/3)+(1/2)fa - fbx + (3/2) fafbx + (1/2)fafbx^2fb2x2][1+2uaLog(M)]$$

$$+ \ [fafb2x2-(1/3)fbx3+(1/2)fafb2x2][1+2ub \ Log(M)] +F_d\} \tag{4}$$

where M=na fa +nb fa fb is the total molecular weight of the chain and F_d a function which does not depend on M but on fa,fb, the interaction parameters ua,ub,uab and the ratio x=nb/na of the lengths of the two different branches. Beyond the uaLog[M] and ubLog[M] of eq.4 which give rise to the critical exponents in the limit of large molecular weights the full study of the radius of gyration can be done by means of eq.4.

Families of graphs of $<S^2>_d$ are given in the following figures in an effort to describe the dependence of the behaviour of the radius of gyration of the macromolecule on fa , fb ,x=nb/na and the interaction parameters. In Fig.2 we see that keeping the total molecular weight M of the macromolecule constant and increasing fa, a reduction of the size is observed, like in star polymers. A decrease of the

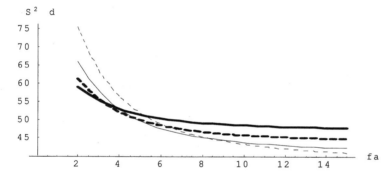

Fig.2. The radius of gyration of a dendritic block copolymer as a function of the number fa of the branches of the central star for four different numbers of branches fb[1,2,3,4] of the stars of the extra generation. The total molecular weight M=100 is kept constant , while x=1 , ua=ub=uab=0.5 are also constants.

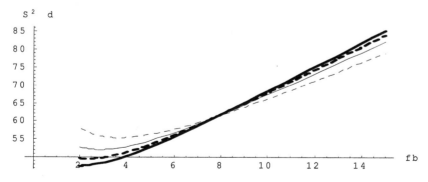

Fig.3. The radius of gyration of a dendritic block copolymer as a function of the number fb of the branches of the central star for four different numbers of branches fa[3,4,5,6] of the core star. The total molecular weight M=100 ,while x=1 , ua=ub=uab=0.5 are also constants.

length and the extension of each branch takes place in this case which means smaller occupation volumes for the macromolecule and a smaller radius of gyration. Opposite behavior is observed increasing the number fb of the branches(Fig.3) of the outer shell and this because of the accumulation of mass far from the central region of the macromolecule. The radius of gyration which beyond the size expresses the spreading of the distribution of monomers becomes larger. A combination of the two effects is observed in Fig.4 where a dependence of the radius of gyration on the ratio x=nb/na of the lengths of the two branches is given. Increasing nb when it is smaller than na and x obtains small values the accumulation of mass in the outer shell is dominant and an increase of the radius of gyration takes place. Increasing nb further na becomes negligible and the macromolecule behaves like a star of nanb branches with a decrease of the radius of gyration.

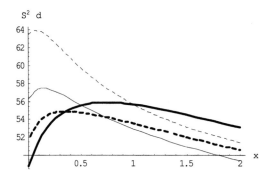

Fig.4. The radius of gyration of a dendritic block copolymer as a function of the ratio x=nb/na of the two lengths of the two different numbers of branches fa,fb[3,3(light dashed),4,4(light),5,5(heavy dashed),6,6(heavy)]. The total molecular weight M=100 and ua=ub=uab=0.5 are constants.

Many further dependences can be found quantitatively from eq.4 which can describe effects like the quality of the solvent , the temperature but also many classes of dendritic polymers having various branch molecular weights. The interaction parameters usually increase both on increasing the quality of the solvent but also on increasing the temperature. Base on this and by means of eq (4) further plots can be drawn to describe the dependence on u's and through them the dependence on the solvent quality and the temperature. From the plots of

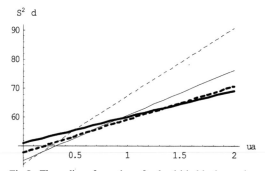

Fig.5. The radius of gyration of a dendritic block copolymer as a function of the interaction parameter ua. The ratio of the two lengths x=na/nb=1, while the two different numbers of branches go as: fa,fb[3,3(light dashed),4,4(light),5,5(heavy dashed),6,6(heavy)]. The total molecular weight M=100 and ub=uab=0.5 are constants.

Fig.5 it is seen that increasing the repulsions between the a units an increase of the radius of gyration takes place because of the expansion of the interior star. For a smaller number fa of branches the interior star is more flexible and the effects of expansion larger. Going to the dependence on the ub

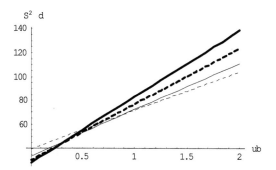

Fig.6. The radius of gyration of a dendritic block copolymer as a function of the interaction parameter ub. The ratio of the two lengths x=na/nb=1, while the two different numbers of branches go as: fa,fb[3,3(light dashed),4,4(light),5,5(heavy dashed),6,6(heavy)]. The total molecular weight M=100 and ua=uab=0.5 are constants.

interactions of the stars of the outside shell an increase again is observed (Fig.6). No restriction on the flexibility of this shell occurs which leads to larger effects for a larger number of b branches.

2. Conclusions

Two main opposite effects have been found to govern the average radius of gyration of a dendritic block copolymer of first generation. The accumulation of mass in the central region of the macromolecule which decreases the radius of gyration of the dendrimer and the accumulation of mass to the outside shell which gives rise to an increase of the radius of gyration of the macromolecule. Quantitative results and graphs have been given based on the analytical eq. 4 which has been derived by means of first order calculations at the critical dimensionality d=4. From the same equation many further dependences can be found quantitatively concerning many classes of dendritic block copolymers. We can describe for example dendritic polymers having various branch molecular weights where the total molecular weight is not kept constant. An extension of the present work to dense dendritic systems where interactions between different macromolecules are important, is also possible[5].

Acknowledgments

This research was funded by the program "Pythagoras I" of the Operational Program for Education and Initial Vocational Training of the Hellenic Ministry of Education under the 3rd Community Support Framework and the European Social Fund.

References

[1] M. Kosmas, Configurational properties of dilute polymer solutions, dimensionality and perturbation theory J. Phys. A: Math. Gen. **14** (1981) 931-943; K. Freed, Renormalization group theory of macromolecules, Wiley, New York, (**1987**).

[2] M. K. Kosmas : On the mean radius of gyration of a polymer chain J. Phys. A: Math. Gen. **14** (1981) 2779-2788; S. G. Whittington, M. K. Kosmas, D S Gaunt : The radius of gyration of a branch of a uniform star polymer, J. Phys. A: Math. Gen. **21** (1988) 4211-4216.

[3] Y. S. Sdranis and M. K. Kosmas: On the Conformational Behavior of an ABC Triblock Copolymer Molecule, *Macromolecules* **24**,(1991), 1341-1351

[4] D. J.Pochan, L.Pakstis, E.Huang, C.Hawker, R.Vestberg, J. Pople: Architectural Disparity Effects in the Morphology of Dendrimer-Linear Coil Diblock Copolymers *Macromolecules;* **35,** (2002) 9239-9242; M. A.Johnson, J.Iyer, P. T.Hammond: Microphase Segregation of PEO-PAMAM Linear-Dendritic Diblock Copolymers *Macromolecules; 37*,(2004) 2490-2501.

[5] M. Kosmas, C. Vlahos:Comparison of the stability of blends of chemically identical and different homopolymers in the bulk and in a film, J.Chem.Phys.**119**(2003)4043-51; C. Vlahos, M. Kosmas: On the miscibility of chemically identical linear homopolymers of different size, Polymer,**44**(2003) 503-7.

Brill Academic Publishers
P.O. Box 9000, 2300 PA Leiden
The Netherlands

*Lecture Series on Computer
and Computational Sciences*
Volume 3, 2005, pp. 85-92

From the Gas to Condensed Phases. Comprehensive Theoretical Investigations of the Formation of Clusters.

Jerzy Leszczynski,[a,1] Szczepan Roszak,[a,b] and Pawel Wielgus[a,b]

[a]Computational Center for Molecular Structure and Interactions, Department of Chemistry, Jackson State University, P. O. Box 17910, J. R. Lynch Street, Jackson, Mississippi 39217 USA
[b]Institute of Physical and Theoretical Chemistry, Wroclaw University of Technology, Wyb. Wyspianskiego 27, 50-370 Wroclaw Poland

Received 27 June, 2005; accepted 12 August, 2005

Abstract: Condensed phases may be considered as infinite structures of polymerized molecules. In the case of weak intermolecular interactions "bulk" properties can be accurately modeled by the corresponding properties of clusters. Especially interesting is the evolution of the structures and properties from a single molecule or bare ion to the "bulk" phase. Modern experimental and theoretical studies provide data on such an evolution. The cohesion energy of molecules in a urea crystal, the position of Cl^- in the water solution, and interactions of proton in argon environment provide examples of such studies. The present chapter discusses applications of the cluster model to the predictions of the above characteristics of the selected condense phase species.

Keywords: quantum chemistry, *ab initio*, molecular interactions, molecular and ionic clusters

Mathematics SubjectClassification: PACS

PACS: 31.15.Ar, 36.40.Wa, 31.15.-p

1. Introduction

The properties of condensed phases are to a large extent determined by the properties of molecules or ions that are building blocks of "bulk" systems and intermolecular interactions between these objects [1-3]. Intermolecular interactions are constituted mostly by two-body contributions, although exceptions from the above rule are also known [4]. The additive properties of intermolecular forces allow simplifying the description of extended systems by the use of mathematical formulas representing fragment-fragment interactions. Collections of particular formulas and parameters are known in chemical informatics under the common name "force fields" [5]. These force field parameters are widely used in chemistry, physics, biology, and material science.

The intermolecular interaction energy is determined by the properties of a single molecule such as electron density distribution (e. g., atomic charge and dipole moments) and polarizability. The interaction energy consists also terms which are of two-body character (e. g., exchange interactions) [6]. The computation of interaction energies and its decomposition into physically meaningful terms is one of key goals of quantum chemistry.

The additive character of intermolecular interactions leads to the possible path of understanding of complex systems through the observation of the evolution of properties of interest from the single molecule (or bare ion) to the bulk properties of the condensed phase. The structure of the cluster is determined due to interactions between the selected central molecule (or ion) and molecules in its close environment. The convergence process of the selected property to one in the bulk phase gives

---

[1] Corresponding author. E-mail: jerzy@ccmsi.us

interesting information concerning the properties of liquids or crystals. Modern experimental techniques, such as photoelectron, vibrational, or mass spectroscopy, allows the properties of the sequential building up of clusters to be measured.

There is a number of intrigue and not always straightforwardly explainable observations obtained by the experimental techniques. The measurement indicates [7] that the gas phase basicity of NH_2^- is higher than that of H^-, and the reverse order of basicity of amide and hydride ions is observed in the bulk ammonia solvent. The solvation of the bare anions lowers the basicity of NH_2^- and the binding of two solvent molecules is sufficient to reverse this order [8]. Regarding this property only two ammonia molecules are required to qualitatively reproduce the bulk solution. Studies of $Cl^-(H_2O)_n$ complexes have found that small clusters are formed with Cl^- present on the surface of water clusters [9]. Molecular dynamics simulations indicate that more than 10 water molecules are needed for Cl^- to drown in the liquid and properly reproduce the structure observed in the real solution [10]. Three examples of approaches are presented in this chapter to indicate the usefulness of the cluster approach in studies of bulk properties of crystal, liquid, and gas phases.

2. The crystal of urea

A comprehensive investigation of the nature of interactions in the crystal of urea involves calculations of interaction energies of the central molecule in the systematically extended molecular crystal [11]. The interaction energy of a urea molecule with its environment corresponds to the cohesion energy, which measured value amounts to 21 kcal/mol [12]. The selected consecutive clusters representing crystal include structures with 6 (Figure 1), 10, 14, 18, 22, 26, and 38 (Figure 2) ligands.

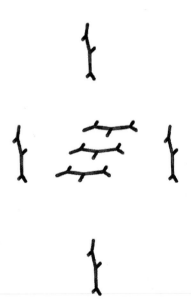

Figure 1: The molecular clusters to model a real environment of urea crystal by 6 ligands.

The results of the analysis are reported in Table 1. The total interaction energy in the smallest cluster selected amounts to -52.7 kcal/mol. This value is twice as large as the experimental result of interaction energy per molecule. The available periodic Hartree-Fock study by Dovesi et al. [13] leading to -28 kcal/mol may indicate the importance of the crystal field. Such an effect however should also be visible in cluster approach calculations of large clusters. Table 1 indicates a clear trend of the reduction of magnitude of interactions per single molecule with the growth of the cluster. Additionally, after correcting the determined energies for the restricted basis set deficiency (the estimated correction of -5 kcal/mol) and correlation energy effects (the estimated MP2 correction of -5 kcal/mol) the predicted

Table 1: The components of the interaction energy of urea molecule with its immediate crystal phase neighborhood. Calculations were performed at the HF/3-21G level of theory [11]. All values in kcal/mol.

Number of ligands	$\varepsilon_{el}^{(10)}$	ε_{ex}^{HL}	ΔE_{del}^{HF}	ΔE^{HF}
6	-75.2	41.5	-22.4	-56.1
10	-78.4	45.6	-25.3	-58.1
14	-75.5	46.0	-22.9	-52.5
18	-73.9	46.3	-22.3	-49.9
22	-69.9	46.2	-20.8	-44.5
26	-68.4	46.2	-20.4	-42.5
38	-62.6	46.1	-18.2	-34.7

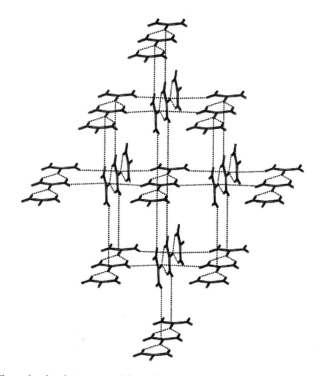

Figure 2: The molecular clusters to model a real environment of urea crystal applying 38 ligands.

interaction energy for the largest studied cluster amounts to -25 kcal/mol. The above value approaches reasonably the measured result. One should be also aware that even for the largest cluster selected here the linear ribbons (parts of the molecular crystal   Figure 2) consists of three molecules only. Based on the predicted trends one concludes that further extension of the cluster size would probably lead to even better agreement with experiment. The absolute values of all interaction energy components, except for the ε_{ex}^{HF} term, decrease with increasing cluster size. The exchange effects are local in nature and quickly converge. The largest influence on the total interaction energy is due to electrostatic forces.

3. The Cl⁻ anion in solution

The question concerning the position of the chloride anion in water clusters is important for understanding the solution phenomena in bulk solutions. Ion solvation processes are governed by interactions between ion and solvent molecules as well as interactions among the molecules of solvent. The competition between these interactions determines the arrangement of the solvent around the ion.

Table 2: Consecutive formation energies for $Cl^-(H_2O)_n$ complexes from MP2 calculations [9]. Energies in kcal/mol.

N	D_e	D_e BSSE corrected	D_e Exper.
1	15.70	14.57	14.68
2	16.79	15.60	13.01
3	18.59	15.84	11.07
4	17.18	16.40	10.61
5	13.22	-	9.48

The formation of $Cl^-(H_2O)_n$ clusters is based on the energetical interplay between Cl···H-O and O···H-O hydrogen bonds [9]. A crude estimation based on the assumption of additivity of interactions (taking into account the total dissociation energy of $Cl^-(H_2O)_n$ complexes (Table 2)) and the number of ionic and molecular hydrogen bonds, and assuming that the dissociation energy of the O···H- O bridge

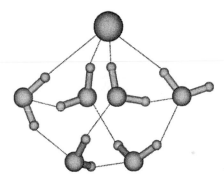

Figure 3: The molecular structure of the $Cl^-(H_2O)_5$ complex.

is equal to the experimental dissociation energy of the water dimer [14] indicates the steep decrease of the single Cl···H-O bond strength (14.75, 11.06, 7.84, and 7.14 kcal/mol for n=1-4, respectively) following the cluster growth. The inclusion of correlation effects favors the formation of structures with the second solvation shell for $Cl^-(H_2O)_5$ (Figure 3). The calculated dissociation energies (Table 2) are close to the results of the available theoretical studies and reproduce reasonably experimental data for larger clusters. An important conclusion from the structure optimization indicates that small clusters favor Cl^- in the "surface" position which is in agreement with the molecular dynamics predictions [10].

4. The solvation of proton in an argon environment

Clusters of ions with noble gas elements are of considerable interest for contemporary theoretical and experimental research [2,3]. Such clusters, although not leading to the condensed phase, significantly change the properties of pure gases which may effect a number of essential processes for

Table 3: The successive dissociation energies of Ar_nH^+ complexes calculated at the MP2 and CCSD(T) levels of theory [15]. D_e and D_o correspond to electronic and spectroscopic dissociation energies. Energies in kcal/mol.

N	D_e MP2	D_e CCSD(T)	D_o CCSD(T)
1	94.6	96.5	92.5
2	16.4	15.5	15.4
3	2.1	1.9	1.8
4	2.3	2.1	2.0
5	2.3	2.1	2.0
6	1.5		
7	2.3		

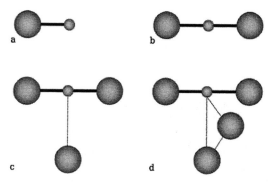

Figure 4: Minimum energy structures for Ar_nH^+ clusters: a) ArH^+ b) Ar_2H^+ c) Ar_3H^+ and d) $Ar_4H.^+$ Calculations were performed at the MP2/6-311G++(3df,3pd) level of theory.

both the science and technology. The negligible interactions between argon-argon atoms in comparison to proton-argon interactions lead to structures driven almost exclusively by properties of the central ion [15]. The ArH^+ cation possesses the significant dissociation energy (Table 3). The bonding in the complex with the second argon atom Ar_2H^+ is weaker although large enough to form a stable and well defined core cation (Figure 4). Further complexation is governed by the van der Waals interactions that are by one order of magnitude smaller from that stabilizing Ar_2H^+. The coordination of an Ar atom to Ar_2H^+ leads to the T-shape C_{2v} structure. The newly added argon is weakly bound and represents the first member of the new, second shell (B) located on the plane perpendicular to the $Ar-H^+-Ar$ axes (Figure 4). The capacity of the second shell amounts to five (Figure 5). Interestingly, the ligands

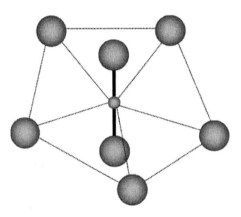

Figure 5: Minimum energy structures for Ar_5H^+ cluster. Calculations were performed at the MP2/6-311G++(3df,3pd) level of theory.

possess significant freedom within the shell and their relative positions are governed by weak Ar-Ar interactions (0.29 kcal/mol). The Ar-Ar interactions are clearly visible as distortions from the perfect D_{5h} geometry of the Ar_7H^+ complex.

5. Conclusions

The cluster approach provides an effective and computationally efficient way to study condensed phases. Selected properties of the liquid or solid phases that are of the local nature such as cohesion energy, thermodynamics of reactions, interaction energies, electron ionization and affinity phenomena are well described taking into account the closest neighborhood only. Modern experimental techniques as well as theoretical methods allow studies of clusters in the "step-wise" manner. Information regarding the evolution of clusters from bare molecule or ion to extended clusters provides information on the convergence of particular molecular properties and as a consequence, leads to the model connecting molecular and bulk properties.

The cohesion energy corresponds to intermolecular interaction energy. For the crystal of urea a rather slow convergence of this property with an increase of cluster size was observed. The detailed study indicates however, the proper trends, even for small clusters making the cluster approach a useful technique supplementing an expensive periodic Hartree-Fock approach.

The investigations of solutions are difficult because of their dynamic nature. The cluster studies provide useful information leading to the development of force-fields. The results of cluster properties are useful as a first order approximation to "bulk" phenomena. The predicted structures of $Cl^-(H_2O)_n$ indicate that the Cl^- anion prefers a position that is on top of the water cluster, and extended clusters are needed to even qualitatively reproduce the solution.

The clustering around ions leads to complexes which possess properties being a function of the size of the clusters. The formation of shells leads to characteristic phenomenon known as magic numbers. The proton in the noble gas environment forms the stable Ar_2H^+ cation, which further interacts with neighbor atoms by van der Waals forces and forms the second solvation shell. The examples presented in this work indicate the usefulness of the cluster approach as a theoretical technique supporting and complementing the experimental endeavors.

Acknowledgments

This work was supported by NSF EPSCOR Grant No. 99-01-0072-08, CREST Grant No. HRD-01-25484, a Wroclaw University of Technology Grant, and the AHPCRC under the agreement number DAAH04-95-2-00003, contract number DAAH04-95-C-0008, the contents of which do not necessarily reflect the position or policy of the government, and no official endorsement should be inferred. The Mississippi Center for Supercomputing Research and Wroclaw Center of Computing and Networking are acknowledged for a generous allotment of computer time.

References

[1] S. Roszak and J. Leszczynski, Clusters, The Intermediate State of Matter, *Computational Material Science.* (Editor: J. Leszczynski) *Theoretical and Computational Chemistry* 15 67-84 (2004).

[2] S. Roszak and J. Leszczynski, Ab Initio Studies of the Micro-Solvation of Ions, *The Journal of Physical Chemistry* A107, 949-955 (2003).

[3] S. Roszak and J. Leszczynski, Ionic Clusters with Weakly Interacting Components – Magic Numbers Rationalized by the Shell Structure, *Computational Chemistry: Reviews of Current Trends* (Editor: J. Leszczynski), World Scientific, 6 179-196 (2001).

[4] I. G. Kaplan, S. Roszak, and J. Leszczynski, Nature of Binding in the Alkaline-Earth Clusters: Be_3, Mg_3, and Ca_3, *The Journal of Chemical Physics* 113 6245-6252 (2000).

[5] A. Warshel, *Computer Modeling of Chemical Reactions in Enzymes and Solutions*, Wiley-Interscience, New York, 1991.

[6] G. Chalasinski and M. M. Szczesniak, Origins of Structure and Energetics of van der Waals Clusters from Ab Initio Calculations, *Chemical Reviews* 94 1723-1765 (1994).

[7] J. T. Snodgrass, J. V. Coe, C. B. Freidhoff, K. M. McHugh, S. T. Arnold, and K. H. Bowen, Negative Ion Photoelectron Spectroscopy of $NH_2^-(NH_3)_1$ and $NH_2^-(NH_3)_2$: Gas Phase Basicities of Partially Solvated Anions, *The Journal of Physical Chemistry* 99 9675-9680 (1995).

[8] S. Roszak, Theoretical Study of Properties of H^- and NH_2^- Complexes with Neutral Ammonia Solvent Molecules, *The Journal of Chemical Physics* 105 7569-7572 (1996).

[9] R. Gora, S. Roszak, and J. Leszczynski, Properties and Nature of Interactions in $Cl^-(H_2O)_n$ n=1,6 Clusters: A Theoretical Study, *Chemical Physics Letters* 325 7-14 (2000).

[10] L. Perera and M. L. Berkowitz, Many Body Effects in Molecular Dynamics Simulations of $Na^+(H_2O)_n$ and $Cl^-(H_2O)_n$ Clusters, *The Journal of Chemical Physics* 99 4236-4237 (1993).

[11] R. W. Gora, W. Bartkowiak, S. Roszak, and J. Leszczynski, A New Theoretical Insight into the Nature of Intermolecular Interactions in the Molecular Crystal of Urea, *The Journal of Chemical Physics* 117, 1031-1039 (2002).

[12] K. Suzuki, S. Onishi, T. Koide, and S. Seki, *Bulletin of the Chemical Society of Japan*, 29 127-129 (1956).

[13] R. Dovesi, M. Causa, R. Orlando, C. Roetti, and V. R. Saunders, Ab Initio Approach to Molecular Crystals: A Periodic Hartree-Fock Study of Crystalline Urea, *The Journal of Chemical Physics* 92, 7402-4711 (1990).

[14] L. A. Curtiss, D. J. Frurip, and M. Blander, Studies in Molecular Association in H_2O and D_2O Vapors by Measurement of Thermal Conductivity, *The Journal of Chemical Physics* 71 2703-2711 (1979).

[15] K. T. Giju, S. Roszak, J. Leszczynski, A Theoretical Study of Protonated Argon Clusters: Ar_nH^+ (n=1-7). *The Journal of Chemical Physics* 117 4803-4809 (2002).

Brill Academic Publishers
P.O. Box 9000, 2300 PA Leiden
The Netherlands

Lecture Series on Computer
and Computational Sciences
Volume 3, 2005, pp. 93-101

Interaction-induced polarizability and hyperpolarizability effects in CO_2···Rg, Rg = He, Ne, Ar, Kr and Xe

A.Haskopoulos and G.Maroulis[*]

Department of Chemistry, University of Patras, GR-26500 Patras, Greece

Received 26 June, 2005; accepted 12 August, 2005

Abstract: We present and analyze interaction dipole moments and (hyper)polarizabilities for the weakly bound systems CO_2···Rg where Rg = He, Ne, Ar, Kr and Xe. All interaction properties have been calculated from finite-field self-consistent field (SCF) and second-order Møller-Plesset perturbation theory (MP2) with large gaussian-type basis sets of near Hartree-Fock quality. The interaction properties vary monotonically with the size of the Rg atom. We observe a diverging behavior of the effect for the interaction dipole moment and mean dipole (hyper)polarizability of the T- and L-shaped CO_2···Rg complexes. Qualitatively similar results for the two configurations are obtained for the anisotropy of the dipole polarizability. The mean interaction hyperpolarizability increases significantly with the size of the rare gas atom. At the MP2 level of theory we find for the T-shaped configuration $\overline{\gamma}_{int}(CO_2 \cdots Rg)/e^4 a_0^4 E_h^{-3}$ = -11.66 (He), -25.88 (Ne), -108.2 (Ar), -207 (Kr) and -460 (Xe).

Keywords: Carbon dioxide, rare gases, interaction dipole moment, interaction (hyper)polarizability.

1. Introduction

The theory of electric moments and polarizabilities is of central importance to the theory intermolecular forces and the interpretation of relevant phenomena [1,2]. The analysis and rigorous interpretation of collision- and interaction-induced spectroscopic observations relies essentially on the possibility of accurate descriptions of the interaction electric properties [3-5]. Systems involving atoms and small molecules constitute a privileged ground for the observation of collision-induced phenomena. For obvious reasons, such systems are ideal for high-level theoretical investigations. They offer the possibility of systematic computational studies that enrich our knowledge leading to an unambiguous rapprochement between theory and experiment [6].

In previous work we explored various computational aspects of interaction dipole moment and (hyper)polarizability calculations [7-11]. In this paper we present a study of the properties of carbon dioxide interacting with rare gas atoms, CO_2···Rg, Rg = He, Ne, Ar, Kr and Xe. Model experimental studies of CO_2···Rg systems are available in the literature [12]. Our aim is to observe patterns and regularities in the evolution of the interaction properties with the size of the Rg atom.

2. Theory

The energy and the electric multipole moments of an uncharged molecule perturbed by a weak, static electric field can be expanded in terms of the permanent properties of the free molecule and the components of the field, as [13-15]

$$E \equiv E(F_\alpha, F_{\alpha\beta}, F_{\alpha\beta\gamma}, F_{\alpha\beta\gamma\delta}, ...)$$

$$= E^0 - \mu_\alpha{}^0 F_\alpha - (1/3)\Theta_{\alpha\beta}{}^0 F_{\alpha\beta} - (1/15)\Omega_{\alpha\beta\gamma}{}^0 F_{\alpha\beta\gamma} - (1/105)\Phi_{\alpha\beta\gamma\delta}{}^0 F_{\alpha\beta\gamma\delta} + ...$$

$$- (1/2)\alpha_{\alpha\beta}F_\alpha F_\beta - (1/3)A_{\alpha,\beta\gamma}F_\alpha F_{\beta\gamma} - (1/6)C_{\alpha\beta,\gamma\delta}F_{\alpha\beta}F_{\gamma\delta}$$

---

[*] Corresponding author. E-mail: maroulis@upatras.gr

$$- (1/15)E_{\alpha,\beta\gamma\delta}F_\alpha F_{\beta\gamma\delta} + ... - (1/6)\beta_{\alpha\beta\gamma}F_\alpha F_\beta F_\gamma - (1/6)B_{\alpha\beta,\gamma\delta}F_\alpha F_\beta F_{\gamma\delta} + ...$$

$$- (1/24)\gamma_{\alpha\beta\gamma\delta}F_\alpha F_\beta F_\gamma F_\delta + ... \tag{1}$$

where F_α, $F_{\alpha\beta}$, etc. are the field, field gradient, etc. at the origin. E^0, μ_α^0, $\Theta_{\alpha\beta}^0$, $\Omega_{\alpha\beta\gamma}^0$ and $\Phi_{\alpha\beta\gamma}^0$ are the energy and the dipole, quadrupole, octopole and hexadecapole moment of the free molecule. The second, third and fourth-order properties are the dipole and quadrupole polarizabilities and hyperpolarizabilities $\alpha_{\alpha\beta}$, $\beta_{\alpha\beta\gamma}$, $\gamma_{\alpha\beta\delta}$, $A_{\alpha,\beta\gamma}$, $C_{\alpha\beta,\gamma\delta}$, $E_{\alpha,\beta\delta}$ and $B_{\alpha\beta,\gamma\delta}$. The subscripts denote Cartesian components. A repeated subscript implies summation over x, y and z. The number of independent components needed to specify the non-vanishing tensors is regulated by symmetry [13].

For sufficiently weak fields the expansion of Eqs 1 converges rapidly. Thus, the finite-field method [16] offers the possibility of a direct approach to the calculation of electric moments, polarizabilities and hyperpolarizabilities. In the case of a homogeneous electric field, Eq 1 reduces to a much simpler one,

$$E = E^0 - \mu_\alpha^0 F_\alpha - (1/2)\alpha_{\alpha\beta}F_\alpha F_\beta - (1/6)\beta_{\alpha\beta\gamma}F_\alpha F_\beta F_\gamma - (1/24)\gamma_{\alpha\beta\gamma\delta}F_\alpha F_\beta F_\gamma F_\delta + ... \tag{2}$$

In Eq 2, the coefficients are the dipole properties: dipole polarizability ($\alpha_{\alpha\beta}$), first ($\beta_{\alpha\beta\gamma}$) and second ($\gamma_{\alpha\beta\delta}$) dipole hyperpolarizability. In addition to the Cartesian components, properties of interest in this work are the mean ($\bar{\alpha}$) and the anisotropy ($\Delta\alpha$) of the dipole polarizability and the mean of the second hyperpolarizability ($\bar{\gamma}$), defined as,

$$\bar{\alpha} = (\alpha_{xx} + \alpha_{yy} + \alpha_{zz})/3$$

$$\Delta\alpha = (1/2)^{1/2}[(\alpha_{xx}-\alpha_{yy})^2 + (\alpha_{yy}-\alpha_{zz})^2 + (\alpha_{zz}-\alpha_{xx})^2]^{1/2}$$

$$\bar{\gamma} = (1/5)(\gamma_{xxxx} + \gamma_{yyyy} + \gamma_{zzzz} + 2\gamma_{xxyy} + 2\gamma_{yyzz} + 2\gamma_{zzxx}) \tag{3}$$

Our approach to the calculation of electric properties from Eqs 1 and 2 has been presented in rich detail in previous work [17-24]. The theoretical methods used in this paper are self-consistent field (SCF) and second-order Møller-Plesset theory [25].

The interaction electric properties of the $CO_2\cdots Rg$ system are obtained via the well-tested Boys-Bernardi counterpoise-correction (CP) method [26]. The interaction quantity $P_{int}(CO_2\cdots Rg)$ at a given geometric configuration is computed as

$$P_{int}(CO_2\cdots Rg) = P(CO_2\cdots Rg) - P(CO_2\cdots X) - P(X\cdots Rg) \tag{4}$$

The symbol $P(CO_2\cdots Rg)$ denotes the property P for $CO_2\cdots Rg$. $P(CO_2\cdots X)$ is the value of P for subsystem CO_2 in the presence of the ghost orbitals of subsystem Rg. By virtue of Eq 2, the interaction mean polarizability or hyperpolarizability is a linear combination of the Cartesian components of $\alpha_{\alpha\beta}$ or $\gamma_{\alpha\beta\gamma\delta}$. The interaction anisotropy $\Delta\alpha_{int}$ at a given level of theory is computed by inserting in Eq 3 the respective interaction quantities for the Cartesian components of $\alpha_{\alpha\beta}$.

3. Computational Details

We have used flexible, carefully optimized basis sets of Gaussian-type functions (GTF) for all subunits of $CO_2\cdots Rg$. Their construction has been reported in previous work: [5s3p3d1f] for CO_2 [8,27], [6s4p3d] for He [7], [7s5p4d1f] for Ne [7], [8s6p5d4f] for Ar [7], [8s7p6d5f] for Kr [7] and [9s8p7d1f] for Xe [10].

The experimental geometry for CO_2, $R_{C-O} = 2.192$ a_0 [28], has been used in all calculations.

All calculations were performed with GAUSSIAN 94 [29] and GAUSSIAN 98 [30].

Atomic units are used throughout this paper. Conversion factors to SI units are, Energy, 1 E_h = 4.3597482 x 10^{-18} J, Length, 1 a_0 = 0.529177249 x 10^{-10} m, μ, 1 ea_0 = 8.478358 x 10^{-30} Cm, α, 1 $e^2a_0^2E_h^{-1}$ = 1.648778 x 10^{-41} $C^2m^2J^{-1}$, β, 1 $e^3a_0^3E_h^{-2}$ = 3.206361 x 10^{-53} $C^3m^3J^{-2}$, γ, 1 $e^4a_0^4E_h^{-3}$ = 6.235378 x 10^{-65} $C^4m^4J^{-3}$. Property values are in most cases given as pure numbers, i.e. μ/ea_0 or $\alpha_{\alpha\beta}/e^2a_0^2E_h^{-1}$.

4. Results and Discussion

We give in Table 1 the dipole polarizabilities of all subunits of $CO_2 \cdots Rg$, calculated with the basis sets used in this work. We also give reference values of Hartree-Fock quality for He, Ne, Ar, Kr [31], Xe [32] and CO_2 [33]. For all subunits, the SCF value of the polarizability is convincingly close to the reference result.

Table 1. SCF values for the dipole polarizability for all basis sets used in this work (reference values in bold).

System	Basis set	$\bar{\alpha}$	$\Delta\alpha$
He	[6s4p3d]	1.3217	0
	NHF[a]	**1.32223**	**0**
Ne	[7s5p4d1f]	2.3719	0
	NHF[a]	**2.37674**	**0**
Ar	[8s6p5d4f]	10.6556	0
	NHF[a]	**10.758**	**0**
Kr	[8s7p6d5f]	16.4498	0
	NHF[a]	**16.476**	**0**
Xe	[9s8p7d1f]	27.0297	0
	NHF[b]	**27.06**	**0**
CO₂	[5s3p3d1f]	15.7949	11.7839
	(19s14p8d6f1g)[b]	**15.863**	**11.681**

[a] Numerical Hartree-Fock by Stiehler and Hinze [31].
[b] NHF by McEachran et al. [32]. [c] Maroulis [33].

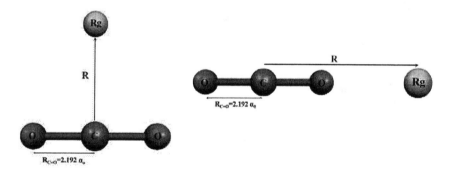

Figure 1. Geometric configuration of the $CO_2 \cdots Rg$ systems.

Table 2. Stable configurations of the $CO_2 \cdots Rg$ systems obtained via second-order Møller-Plesset perturbation theory.

Σύστημα	Διαμόρφωση	R / α_0	E_{int} / E_h
$CO_2 \cdots He$	T	6.019	-0.000155500
	L	8.249	-0.000089372
$CO_2 \cdots Ne$	T	6.206	-0.000290534
	L	8.350	-0.000179960
$CO_2 \cdots Ar$	T	6.605	-0.000814300
	L	8.789	-0.000483379
$CO_2 \cdots Kr$	T	6.846	-0.000982642
	L	9.039	-0.000576846
$CO_2 \cdots Xe$	T	7.332	-0.001005520
	L	9.517	-0.000605612

The definition of the relative position of the subunits is shown in Figure 1. The only parameter defines the distance of the Rg atom from the center of the CO_2 unit. The stable configurations of the

$CO_2\cdots Rg$ systems have been reported in a previous paper [9]. The values used in this work are given in Table 2. They have been determined at the MP2 level of theory. The T-shaped (perpendicular) configuration is the most stable in all cases. The respective distances are systematically shorter that those of the L-shaped (collinear) one.

The interaction dipole moment and dipole (hyper)polarizability is given for all systems and both configurations in Tables 3 and 4. We show in Figures 2-5 the evolution of the interaction properties of $CO_2\cdots Rg$ with the size (atomic number Z) of the Rg atom.

Table 3. SCF and MP2 interaction properties for the T-shaped configuration of $CO_2\cdots Rg$.

System	Method	μ_{int}	$\overline{\alpha}_{int}$	$\Delta\alpha_{int}$	$\overline{\gamma}_{int}$
$CO_2\cdots He$	SCF	0.0070	-0.0417	0.3767	-8.33
	MP2	**0.0063**	**-0.0494**	**0.4375**	**-11.66**
$CO_2\cdots Ne$	SCF	0.0112	-0.0701	0.6248	-16.57
	MP2	**0.0109**	**-0.0897**	**0.8043**	**-25.88**
$CO_2\cdots Ar$	SCF	0.0308	-0.252	2.508	-97.9
	MP2	**0.0282**	**-0.275**	**2.996**	**-108.2**
$CO_2\cdots Kr$	SCF	0.0367	-0.366	3.530	-214
	MP2	**0.0344**	**-0.372**	**4.194**	**-207**
$CO_2\cdots Xe$	SCF	0.0409	-0.52	4.86	-497
	MP2	**0.0397**	**-0.51**	**5.80**	**-460**

Table 4. SCF and MP2 interaction properties for the L-shaped configuration of $CO_2\cdots Rg$.

System	Method	μ_{int}	$\overline{\alpha}_{int}$	$\Delta\alpha_{int}$	$\overline{\gamma}_{int}$
$CO_2\cdots He$	SCF	-0.0032	0.0393	0.3389	-3.39
	MP2	**-0.0025**	**0.0512**	**0.4014**	**-6.58**
$CO_2\cdots Ne$	SCF	-0.0054	0.0676	0.5797	-6.53
	MP2	**-0.0052**	**0.0968**	**0.7533**	**-10.83**
$CO_2\cdots Ar$	SCF	-0.0246	0.263	2.139	9.0
	MP2	**-0.0209**	**0.364**	**2.619**	**33.0**
$CO_2\cdots Kr$	SCF	-0.0359	0.365	2.923	25
	MP2	**-0.0295**	**0.511**	**3.581**	**81.85**
$CO_2\cdots Xe$	SCF	-0.0493	0.51	3.92	61
	MP2	**-0.0401**	**0.72**	**4.83**	**198**

The μ_{int} values in Tables 3 and 4 are practically identical with our previously reported values [9]. We will not repeat the presentation of these results here.

Turning our attention to $\overline{\alpha}_{int}$ we observe that its sign is negative for the T-shaped and positive for the L-shaped configuration. For both configurations, the size of the interaction polarizability increases monotonically with the size of the Rg atom. The effect of electron correlation on $\overline{\alpha}_{int}$ is rather small in absolute terms for all $CO_2\cdots Rg$. Yet, it is important in relative terms for most systems: for $\overline{\alpha}_{int}$ ($CO_2\cdots He$) we obtain -0.0417 (SCF), -0.0494 (MP2) for the T-shaped and 0.0393 (SCF), 0.0512 (MP2) for the L-shaped configuration. These changes correspond to an increase of size by 18.5 and 23.2 %, respectively. Our $\Delta\alpha_{int}$ values show that the interaction is larger by one order of magnitude compared to $\overline{\alpha}_{int}$. $\Delta\alpha_{int}(CO_2\cdots Rg)$ is positive for both configurations. We see that electron correlation effects are sizeable for this property.

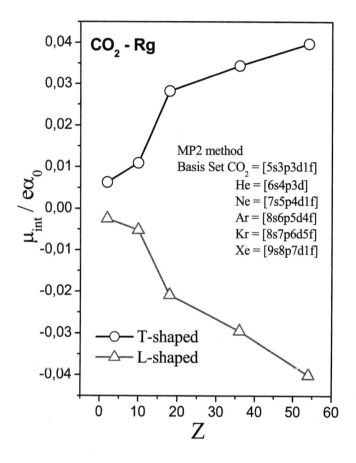

Figure 2. Interaction dipole moment for the T- and L-shaped configuration of CO₂-Rg.

The analysis of the $\bar{\gamma}_{int}$ values requires more attention. For the T-shaped configuration the value of this property is negative at the SCF level. It is also negative at the MP2 level of theory but the second order correction (D2 = MP2 - SCF) is negative for He, Ne, Ar and positive for Kr, Xe. $\bar{\gamma}_{int}$ (CO₂⋯He) = -8.33 at the SCF level of theory. Compared to the respective SCF values $\bar{\gamma}$ (CO₂) = 821 [27] and γ(He) = 36.23 [34] , the $\bar{\gamma}_{int}$ (CO₂⋯He) is certainly small but non negligible. For $\bar{\gamma}_{int}$ (CO₂⋯Xe) we obtain an SCF value of -497, which is significant compared to the respective values of the subunits, $\bar{\gamma}$ (CO₂) = 821 [27] and γ(Xe) = 5364 [34]. For the L-shaped the SCF values of $\bar{\gamma}_{int}$ are relatively smaller in magnitude compared to the T-shaped configuration. Electron correlation has a very strong effect on the SCF values. The second order correction is negative for He and Ne, but strongly positive for Ar, Kr and Xe.

Figures 2-5 show in a very eloquent manner the basic patterns present in the calculated values of the interaction dipole properties for both configurations. For the dipole moment and the mean (hyper)polarizability P_{int}(CO₂⋯Rg) increases in magnitude but so does the absolute difference | P_{int}(CO₂⋯Rg)(T) - P_{int}(CO₂⋯Rg)(L)|. For the anisotropy, $\Delta\alpha_{int}$(CO₂⋯Rg)(T) and $\Delta\alpha_{int}$ (CO₂⋯Rg)(L) are

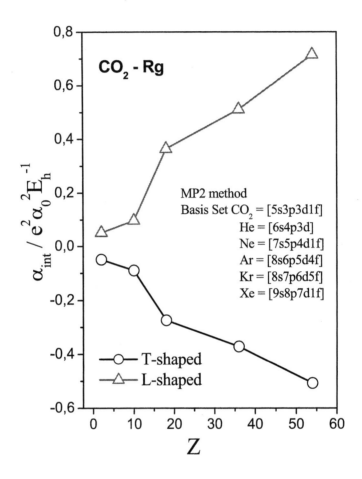

Figure 3. Interaction mean dipole polarizability for the T- and L-shaped configuration of CO_2-Rg.

qualitatively similar.

It is instructive to add at this points a few remarks for the comparison of our results to the available experimental data. The CO_2-He system has been the object of several experimental investigations in late years [35-37] but we are not aware of an experimental value of the interaction dipole moment. Our value $\mu_{int}(CO_2\cdots Ne) = 0.0109$ is very close to the experimental value of 0.0096 referenced by Iida et al. [12]. The agreement is also very good for the remaining systems. We report $\mu_{int}(CO_2\cdots Rg) = 0.0282$ (Ar), 0.0344 (Kr) and 0.0397 (Xe). The respective experimental values are 0.0267 (Ar), 0.0326 (Kr) and 0.0405 (Xe).

5. Conclusions

We have calculated interaction dipole moments and (hyper)polarizabilities for the weakly bound complexes of carbon dioxide with rare gas atoms. The interaction properties have been obtained from finite-field SCF and MP2 calculations with near-Hartree-Fock quality basis sets. We find that the magnitude of the interaction properties increases monotonically with the size of the rare gas atom. The

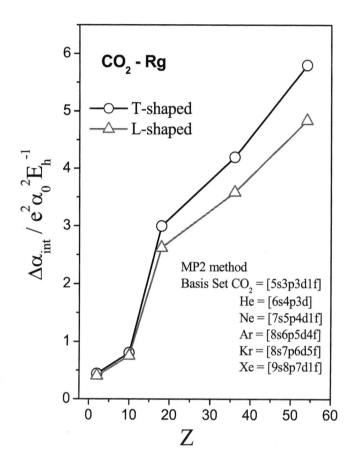

Figure 4. Interaction mean dipole polarizability for the T- and L-shaped configuration of CO_2-Rg.

interaction dipole moments for the T-shaped and L-shaped μ_{int} configurations, $\mu_{int}(CO_2\cdots Rg)(T)$ and $(CO_2\cdots Rg)(L)$, are of the opposite sign. The magnitude of the interaction anisotropy of the dipole polarizability $\Delta\alpha_{int}$ is significantly larger than that of the mean $\bar{\alpha}_{int}$. The magnitude of the interaction hyperpolarizability increases significantly with the size of the rare gas atom. At the MP2 level of theory we find for the T-shaped configuration $\bar{\gamma}_{int}(CO_2\cdots Rg)$ = -11.66 (He), -25.88 (Ne), -108.2 (Ar), -207 (Kr) and -460 (Xe). The magnitude of this quantity for the L-shaped configuration is $\bar{\gamma}_{int}(CO_2\cdots Rg)$ = -6.58 (He), -10.83 (Ne), 33.0 (Ar), 82 (Kr) and 198 (Xe).

Our values for the interaction dipole moment are in very good agreement with the available experimental data.

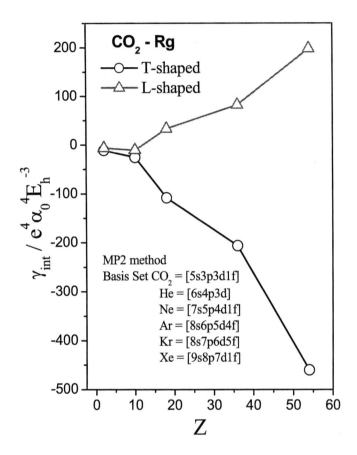

Figure 5. Interaction mean second dipole hyperpolarizability for the T- and L-shaped configuration of CO_2-Rg.

References

[1] A.J.Thakkar, in *Encyclopedia of Chemical Physics and Physical Chemistry*, (eds.) J.Moore and N.Spencer (IOP Publishing, Bristol, 2001), Vol I, pp. 181-186.

[2] G.Birnbaum (Ed.), *Phenomena induced by intermolecular interactions* (Plenum, New York, 1985).

[3] K.L.C.Hunt, in reference 2.

[4] G.C.Tabisz and M.N.Neuman, *Collision- and Interaction-induced Spectroscopy* (Kluwer, Dordrecht, 1995).

[5] T.Bancewicz, Y. Le Duff and J.L.Godet, Adv.Chem.Phys. **119**, 267 (2001).

[6] B.Fernández, C.Hättig, H.Koch and A.Rizzo, J.Chem.Phys. **110**, 2872 (1999).

[7] G.Maroulis, J.Phys.Chem. A **104**, 4772 (2000).

[8] G.Maroulis and A.Haskopoulos, Chem.Phys.Lett. **349**, 335 (2001).

[9] G.Maroulis and A.Haskopoulos, Chem.Phys.Lett. **358**, 64 (2002).

[10] G.Maroulis, A.Haskopoulos and D.Xenides, Chem.Phys.Lett. **396**, 59 (2004).

[11] A.Haskopoulos and G.Maroulis, Chem.Phys. **309**, 271 (2005).

[12] M.Iida, Y.Ohshima and Y.Endo, J.Phys.Chem. **97**, 357 (1993).

[13] A.D.Buckingham, Adv. Chem. Phys. **12**, 107 (1967).

[14] A.D.McLean and M.Yoshimine, J.Chem. Phys. **47**, 1927 (1967).

[15] P.Isnard, D.Robert and L.Galatry, Mol.Phys. **31**, 1789 (1976).

[16] H.D.Cohen and C.C.J.roothaan, J.Chem.Phys **43**, S34 (1965).

[17] G.Maroulis and A.J.Thakkar, J.Chem.Phys. **88**, 7623 (1988).

[18] G.Maroulis and A.J.Thakkar, J.Chem.Phys. **93**, 4164 (1990).

[19] G.Maroulis, J.Chem.Phys. **94**, 1182 (1991).

[20] G.Maroulis, J.Chem.Phys. **101**, 4949 (1994).

[21] G.Maroulis, J.Chem.Phys. **108**, 5432 (1998).

[22] G.Maroulis, J.Chem.Phys. **111**, 6846 (1999).

[23] G.Maroulis, J.Chem.Phys. **118**, 2673 (2003).

[24] G.Maroulis and D.Xenides, J.Phys.Chem. A **107**, 712 (2003).

[25] T.Helgaker, P.Jørgensen and J.Olsen, *Molecular Electronic-Structure Theory* (Wiley, Chichester, 2000).

[26] S.F.Boys and F.Bernardi, Mol.Phys. **19**, 55 (1970).

[27] G.Maroulis, Chem.Phys. **291**, 81 (2003).

[28] G.Graner, C.Rossetti and D.Baily, Mol.Phys. **58**, 627 (1986).

[29] M.J.Frisch, G.W.Trucks, H.B.Schlegel et al., GAUSSIAN 94, Revision E.1 (Gaussian, Inc, Pittsburgh PA, 1995).

[30] M.J.Frisch, G.W.Trucks, H.B.Schlegel et al., GAUSSIAN 98, Revision A.7 (Gaussian, Inc., Pittsburgh PA, 1998).

[31] J.Stiehler and J.Hinze, J.Phys. B **28**, 4055 (1995).

[32] R.P.McEachran, A. D Stauffert and S. Greita, J.Phys.B **12**, 3719 (1979).

[33] G.Maroulis, Chem.Phys.Lett. **396**, 66 (2004).

[34] G.Maroulis, unpublished results.

[35] C.Boulet, J.-P.Bouanich, J.-M.Hartmann, B.Lavorel and A.Deroussiaux, J.Chem.Phys. **111**, 9315 (1999).

[36] J.Tang and A.R.W.McKellar, J.Chem.Phys. **121**, 181 (2004).

[37] J.Buldyreva, S.V.Ivanov and L.Nguyen, J.Raman Spectrosc. **36**, 148 (2005).

Brill Academic Publishers
P.O. Box 9000, 2300 PA Leiden,
The Netherlands

Lecture Series on Computer
and Computational Sciences
Volume 3, 2005, pp. 102-118

The Role of Dynamics in Reactive Processes Relevant to Biophysical Chemistry

Markus Meuwly[1]

Department of Chemistry,
Klingelbergstr. 80
University of Basel,
CH-4056 Basel, Switzerland

Received 24 July, 2005; accepted in revised form 1 August, 2005

Abstract:
The results of molecular dynamics simulations of different biophysically relevant processes are discussed in the light of experimental investigations. Using detailed electrostatic models for ligands it is shown that agreement with a range of experimental data on structural, energetic and spectroscopic questions can be achieved for Mb interacting with CO. The combination of umbrella sampling and stochastic dynamics allows to follow the rebinding dynamics of CO after photodissociation. The time scales depend on one physically relevant parameter only, the asymptotic separation between the bound and the unbound free energy surface. Proper choice of this parameter gives rebinding times in quanitative agreement with experiment. As a last example the electron coupled proton transfer in Ferredoxin is discussed. Computer simulations suggest that a water molecule can be stabilized around the active site and may be involved in the reaction.

Keywords: Molecular Dynamics, Myoglobin, Ferredoxin, Stochastic Modelling, Smoluchowski Equation

1 Introduction

Experimental and simulation techniques have provided a wide range of information to characterize biological systems. Conventional and time-resolved X-ray experiments and NMR spectroscopy have been successfully used in combination with infrared spectroscopy to interrogate systems of biological relevance. Some of the proceses take place on the ns to ps time scale and are thus ideal to be investigated with molecular dynamics (MD) simulations. Examples are ligand diffusion or the infrared spectroscopy of small ligands. Other interesting observations require longer time scales (ms to s) and particular simulation techniques are necessary to better understand them. They include activated processes (larger barriers) or protein folding.

Myoglobin (Mb) is a small globular protein which is typically described as an oxygen reservoir in muscles. More recently, new functional roles have been discovered for this intensively-studied protein, including its role as an intracellular scavenger of nitric oxide (see, for example, Reference [1] and references therein). NO can react with MbO_2, eventually producing ferric oxidised Mb (metMb) and harmless nitrate. O_2 has also been found to react with MbNO to give the same products, although very slowly.[2]

---

[1]Corresponding author. m.meuwly@unibas.ch

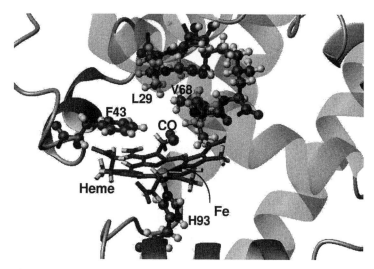

Figure 1: Active site of Mb with the heme-unit (sticks), the dissociated CO molecule, and a few nearby residues shown in ball-and-stick representation.

The binding of small ligands (O_2, CO and NO) to Mb has been widely studied as a model of protein function and ligand discrimination.[3] Recent advances in experimental techniques have led to the direct observation of photodissociation and rebinding of ligands, particularly CO, on the femtosecond to picosecond timescale.[4, 5] Upon photodissociation, CO is found to move to a docking site at the edge of the distal haem pocket, where it lies parallel to the haem plane in two possible (opposite) orientations.[6, 7, 8] The ligand then either rebinds to the haem (so-called geminate recombination) with a time constant of 180 ns[9] or diffuses away to the solvent.

An important aspect in understanding the dynamics of small ligands in the protein interior is to quantify the energetic barriers for rebinding which can be of different origin (steric, electronic, dynamical) [10, 11, 12]. Under favourable conditions, the ligand motion and the associated free energy barriers can be described along suitably chosen (progression) coordinates. Considerable experimental information is available for the migration of CO from the heme binding site (see Fig. 1). The binding site (B) is of major importance since after photodissociation it is rapidly populated. From there, the ligand can either rebind directly or it follows a largely unknown path within the protein to diffuse towards the solvent from where it rebinds at much longer time scales.[10, 13] One possible, secondary binding site in the neighborhood of site B is the Xe4 pocket.[14, 5, 15, 13] For native Mb, experimental evidence for population of the Xe4 pocket by CO after dissociation appears to be sparse. It is only recently that FTIR-TDS experiments have found migration of photolyzed CO from the primary docking site B to the Xe4 cavity.[15] This finding is supported by results from a 90 ns MD simulation of native MbCO which show that CO occupies the Xe4 cavity before migrating further in the protein matrix.[13]

Another process of considerable relevance to biological systems is electron coupled proton transfer (ECPT). Ferredoxin I is the first system for which a microscopically detailed mechanism for ECPT appears to have been observed. Since the reaction takes place on the ms time scale it is difficult to be addressed directly by simulations. Nevertheless, MD simulations have shown to provide insight into more or less probable reaction pathways.

In the present work three different simulation approaches for biologically relevant systems are discussed. In the first example (IR spectrum of photodissociated CO in Mb) the direct sampling of phase space allows to understand the spectroscopy. To describe the ligand migration between different binding sites in Mb a combination of MD and stochastic simulations is used. The last example, ECPT in FdI, is based on MD simulations alone. Thus, the simulations are carried out in such a way as to provide information about the process investigated. The remainder of this contribution is structured as follows. First, some methods are briefly introduced. For more detailed descriptions the reader is referred to the original references. Then, results are presented and discussed in a final section.

2   Methods

2.1   Molecular Dynamics Simulations

MbCO: Molecular dynamics (MD) simulations were carried out with the CHARMM program [16] and the CHARMM22 force field [17]. Additional parameters required to describe the heme–CO interactions were taken from the literature[18, 12].The heme pocket was solvated by a 16 Å sphere of equilibrated water molecules, centered around the heme, and a solvent boundary potential was applied to constrain the water molecules. The "reaction region" of radius 12 Å , inside which the system was propagated with Newtonian dynamics, was centered on the heme while the dynamics of the buffer region between 12 and 16 Å was described using Langevin dynamics. Friction coefficients of 62 ps^{-1} and 250 ps^{-1} were applied to the water oxygens and the remaining non-hydrogen atoms, respectively. The FEPs were calculated using the umbrella sampling method as implemented in CHARMM[19] along escape paths (see Fig. 3) from 1ns molecular dynamics simulations previously calculated.[20]

Ferredoxin: Simulations of the different systems considered (wild type and mutant ferredoxin (Fd)) were carried out using Langevin dynamics (LD) with friction coefficients of 62 kcal/mol for water molecules and mass-scaled friction coefficients (up to 250 kcal/mol) for heavy atoms in the buffer region with a radius of 16 Å and defined around the center of mass of the [3Fe-4S] cluster. In all trajectory calculations that used SHAKE to constrain the hydrogen bond lengths the time step was 1 fs.[21] The total simulation time for each run was 250 ps. For the reduced and the oxidized [3Fe-4S] the charge distributions from density functional theory (DFT) calculations were used. They were calculated at the UB3LYP/6-31G** level of theory. It is of some interest to mention that depending on the method used, the partial charges can differ quite substantially (see Table 1). For all results reported here, NBO charges were used.

Further details on the simulations and individual parameter sets can be found in the original work on MbCO[20, 22, 23] and FdI[24, 25].

2.2   Umbrella Sampling

To calculate the free energy surfaces (FESs) for dissociated CO in native and mutant (L29F) Mb, umbrella sampling was used along the Fe–C coordinate R, where C is the carbon atom of the CO molecule. MD simulations allow to collect statistics n_i for given intervals $\Delta\delta_i$ along a particular reaction coordinate δ by applying a biasing potential of the form $V(\delta) = k_i(\delta - \delta_{0_i})^2$. Here, k_i is the force constant and δ_{0_i} is the value around which δ should be constrained. Associated free energies $G(\delta)$ are extracted from n_i via

$$f(i) = -kT \ln n_i + c,$$

where k is the Boltzmann constant, T is the temperature, and c is an arbitrary constant. Since each interval of the FEP is only defined up to the constant c, the segments are joined to obtain a continuous FEP over the entire range of the coordinate δ. Here, the constant c is calculated as the value which minimizes the functional min $\left[\sum_i (f_2(i) - f_1'(i))^2\right]$ in the overlap region i of two adjoining windows. $f_1(j_1)$ and $f_2(j_2)$ are sets of points in the first and second interval and $f_1'(j_1) = f_1(j_1) - c$, respectively. To find suitable parameters k_i and δ_{0_i} for the biasing potential (see above), a short 5 ps simulation for every interval was run. From this, an approximate form of the FEP was extracted and, if necessary, the parameters k_i and δ_{0_i} were modified for better sampling. Next, statistics were collected during 50 ps for each window $(\delta_l, \delta_h)_i$, from which the final FEP was derived.

For the bound state, $R \in (1.6, 7.0)$ Å , divided into 42 intervals $(\delta_l, \delta_h)_i$ while for the unbound state varied between $R \in (2.4, 10.0)$ Å. Once appropriate parameters k_i and δ_{0_i} were found, the system was equilibrated for 10 ps after which statistics was collected during 60 ps for each of the windows. For the L29F mutant it two umbrella sampling runs were carried out. One included a water molecule in the Xe4 pocket while the other one did not.

2.3 Solving the Smoluchowski Equation

With $G(R)$ calculated at a particular temperature T the relaxation of an initial CO-population $p(R,t)$ along R can be followed by solving the Smoluchowski equation:

$$\frac{\partial p(R,t)}{\partial t} = \frac{\partial}{\partial R} D(R) e^{-\beta G(R)} \frac{\partial}{\partial R} \left[e^{\beta G(R)} p(R,t) \right]. \tag{1}$$

Eq.(1) describes the decay of a nonequilibrium distribution $p(R, t = 0)$ to the equilibrium $p_{eq}(R)$ governed by $G(R)$. To solve eq.(1), the discrete approximation is used [26]. For large R, reflecting boundary conditions were imposed (Fig. 4). $D(R)$ is the (possibly) R-dependent diffusion constant, $G(R)$ is the FEP, and $\beta = 1/kT$ is the Boltzmann factor. In the following, D is position-independent and is set to $D = 2.2$Å ^{2}/ps, which is the value calculated for CO in the distal pocket at 300 K.

3   Results

3.1   Infrared spectrum of photodisociated CO

The infrared spectrum can be related to the time dependence of the dipole moment of the CO molecule, $\mu(t)$ along a trajectory. For this, the real-time dipole–dipole autocorrelation function, $C(t) = \langle M(0)M(t) \rangle$, is accumulated over 2^n time origins, where n is an integer such that 2^n corresponds to between 1/3 and 1/2 of the trajectory, with the time origins separated by 1 fs. The infrared spectrum, $C(\omega)$, is then calculated from the Fourier transform of $C(t)$ [27], using a Fast Fourier Transform with a Blackman filter to minimise noise [28]. The final infrared adsorption spectrum is then calculated by evaluating

$$A(\omega) = \omega \left\{ 1 - \exp\left[-\omega/(kT) \right] \right\} C(\omega), \tag{2}$$

where k is the Boltzmann constant and T is the temperature in Kelvin.

The infrared spectrum (see Figure 2) of the photodissociated CO molecule was calculated using the dipole moment time series from three independent trajectories, each 1 ns in length. Each spectrum was slightly different, reflecting the different trajectories of the CO within the protein. A number of peaks can be seen at different positions between 2170 and 2200 cm^{-1}. The calculated spectra correspond to the signal from a single molecule and are therefore sensitive to the precise position and orientation of the CO molecule, as well as to the surrounding protein environment which is continually changing. In experiments, the spectra recorded correspond to many CO molecules in different environments and therefore give an *average* picture of the CO environment. Compared to the "single molecule spectra" calculated here, some features will move, some will change in magnitude and some will even appear or disappear as the signal is averaged over an ensemble of molecules.

The average spectrum over the 1 ns trajectories (Figure 2) shows two principal bands separated by approximately 8 cm^{-1} with intensities approximately in the ratio 2:1. Both, the splitting and the relative intensities, are in good agreement with the experimental spectra of Lim, Jackson and Anfinrud[6, 4] and gives some confidence in the methods used. In the present case, the time range over which the average is taken is also important, since there are several processes occurring following the photodissociation of CO from Mb: docking, diffusion through the protein and loss of the CO to the solvent. Only the first process is important during the 1 ns trajectories presented here, but if the trajectories are extended, diffusion through the protein environment may become important which changes the nature of the infrared spectra. This is because CO will explore secondary binding sites, such as the Xe4 or the Xe1 pockets.

Detailed analysis of the trajectories for CO motion in the docking site (B) revealed that there are two energetically favoured positions. They include the Fe–CO and the Fe–OC conformation.[4] This is also found from MD simulations with the Fe–OC state favoured over the Fe–CO conformer by 0.3 kcal/mol, in good agreement with experiment.[22] It was argued that each peak in the infrared spectrum can be associated with one of the two states. The original proposal that the peak at higher frequency corresponds to the Fe–OC conformer was recently challenged by additional experiments and density functional theory calculations.[29] From the MD simulations, however, such an identification was not yet possible.

Molecular dynamics simulations on the L29F mutant showed that the fluctuating three point charge model is also capable of capturing the major observables.[5, 22] They include the fact that after photodissociation the CO ligand rapidly escapes to the Xe4 pocket and that the infrared spectrum consists only of one single line. The mutation at position 29 introduces sufficient steric hindrance to force the CO out of the docking site. Thus, the fluctuating three point charge model[22] has been validated with respect to structural data (location of the docking site)[30], the infrared spectra of the photodissociated ligand in native and mutant Mb[4, 5], and the free energy barriers for in-plane rotation in the docking site.[4] In the next section the fluctuating point charge model will be used to investigate free energy barriers between different stable and metastable states.

3.2 Rebinding kinetics of photodissociated CO

Depending on the location from where CO rebinds within the protein matrix this process takes between several hundred ns up to s (rebinding from the solvent).[31, 32] Such time scales preclude the direct use of conventional MD simulations since many events are required to estimate a statistically significant rebinding time. The free energy profiles $G(R)$ calculated from umbrella sampling can be used to follow the time dependence of the rebinding reaction of photodissociated CO. Here, R is the distance between Fe and the C atom of CO.

Free energy curves were calculated along escape paths found in molecular dynamics simulations. On example (for native Mb–CO) is shown in Figure 3. Here, A is the bound state (MbCO), while B and Xe4 are metastable states for dissociated CO. As can be seen, the Fe–C distance R is a meaningful coordinate to calculate $G(R)$. Since the force field for the bound and unbound state of CO are different, one free energy simulation for each state is required (inset of Figure 4). At long range (large R) the bound and unbound states are separated by a constant energy Δ. There are no experimental values for Δ but estimates from DFT calculations yield $\Delta \approx 5$ kcal/mol.[33] The resulting $G(R)$ for the two states are diabatized in the neighborhood of the crossing. This leads to continuous free energy curves as shown in Figure 4. Using such $G(R)$ it is now possible to

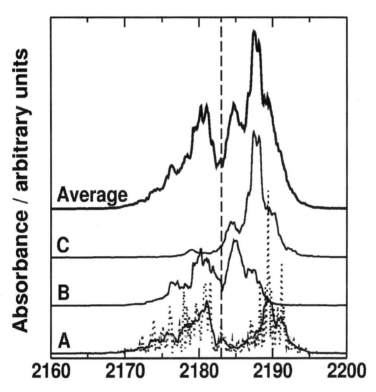

Figure 2: Infrared spectra of CO in Mb calculated over 1 ns from three different trajectories A, B and C at 300 K using the three point fluctuating charge model. The raw and smoothed spectra are shown for trajectory A, and the smoothed spectra from trajectories B and C have been displaced vertically for clarity. The vertical dashed line indicates the calculated position of the absorbance of a free CO from the Fourier transform of $\mu(0)\mu(t)$ at $T = 300$ K. The average spectrum is shown in the top line.

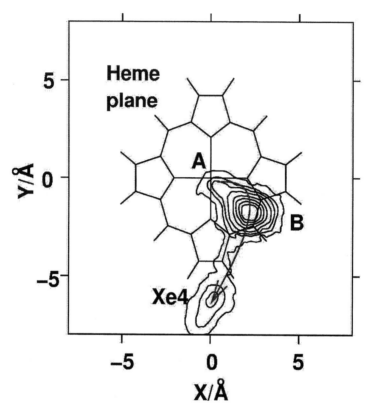

Figure 3: Sample escape path along which FEPs were calculated. The path follows the center of mass of CO from the heme binding site (A) via the docking site (B) to the Xe4 pocket. The contours give the probability (with maximum probability in the B site) from a 1 ns MD simulation[20] for the dissociated state *without* application of an umbrella potential.

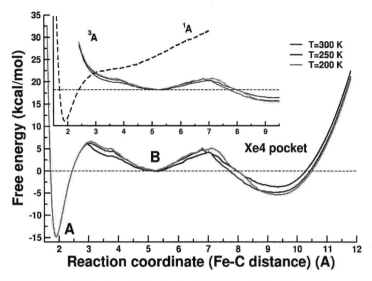

Figure 4: Adiabatic (main figure) and diabatic (inset) FEPs for different temperatures. In the inset, the dashed line corresponds to the bound state FEP while the solid lines are the unbound FEPs for different T. For the ^1A state the FEP becomes repulsive at long range due to interactions with nearby protein residues (see Fig. 1). The energies of the FEPs for the three different temperatures have been shifted vertically to coincide in site B to allow direct comparison of the barriers.

follow a distribution of initial conditions and to calculate the rebinding time from the behaviour of $p(x,t)$. To this end, different $p(x,0)$ are chosen and the rebinding time τ is calculated from $p(x,t)$.[34, 35, 23] For native MbCO a rebinding time of 100 ns is found for $\Delta = 3.7$ kcal/mol. This value of Δ leads to an inner barrier $H_{BA} = 4.3$ kcal/mol, in good agreement with the effective barrier $H_{eff} = 4.5$ kcal/mol previously reported[36]; H_{eff} is the effective barrier in the fully relaxed and fluctionally averaged protein near 300 K.

For the L29F mutant no experimental data on the rebinding times is available. However, femtosecond time resolved infrared spectroscopy and MD simulations have shown that after photodissociation, the CO leaves the docking site on a very short time scale.[5, 22] To more quantitatively describe the rebinding dynamics in the L29F mutant, similar simulations as for native MbCO were carried out. Figure 5 shows two FEPs for the L29F mutant of Mb. In one simulation (red) the Xe4 pocket was occupied by a water molecule while in the other simulation this was not the case. Around the crossing region of the FEP for the bound state and the unbound state the two curves are diabatized. The zero of the two FEPs has been chosen in the minimum of the docking site (B). This allows to directly compare the forward and backward barriers. Using these FEPs and solving the Smoluchowski equation with a diffusion constant $D = 2.2$Å^2/ps, yields rebinding times ≈ 20 ps and ≈ 75 ps for the FEP with and without water in Xe4, respectively. It is instructive to follow the maximum of the probability distribution as a function of time in the three relevant regions (A, B, and Xe4), see figure 6. For the initial condition where only the Xe4 pocket is populated (solid line) a small transient population of the B state is found (between 1 ps and 100 ps). For an equal initial population of B and Xe4 most of the population in B immediately rebinds while a similar portion of the Xe4 population gets trapped in the B state. However, as the solid and dashed black line indicate, the overall rebinding times from the different initial conditions are very similar. The rebinding times reflect the rather low barriers between the Xe4, the B state and the bound A state.

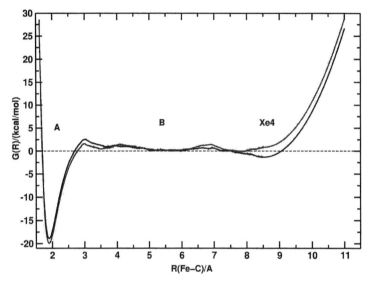

Figure 5: Free energy profiles for CO migration between the bound state (A), the docking site (B) and the Xe4 pocket in the L29F mutant. The black line is for the case where the Xe4 pocket contains no water while the red line is for the case where the Xe4 pocket contains one water molecule.

The combination of molecular dynamics simulations and stochastic modeling provides a detailed understanding of the rebinding dynamics in native and mutant Mb. Since for native Mb the time scale for rebinding is around 100 ns, direct sampling with sufficient statistics is not possible.

3.3   Electron-coupled proton transfer in Ferredoxin

For the electron-coupled proton transfer (ECPT) in Ferredoxin (Fd) less data from experiment suitable for comparison with MD simulations are available[37] than for Mb complexed to CO or NO. The main conclusions from the experimental investigations, together with preliminary molecular dynamics simulations, concerned the role of the Asp15 residue. It was suggested that Asp15 (see Figure 7) serves as the proton relay, i.e. the proton is accepted by the residue Asp15 from the solvent and transported towards the iron-sulphur cluster [3Fe-4S] by rapid movements of the residue.[37] This conclusion was based primarily on site directed mutagenesis and pK_a studies of the native and mutant proteins. It was found that the wild type (Asp15) FdI had a faster rate for protonation (by two orders of magnitude) compared to the D15N and D15E mutants (Asp15 → Asn15 and Asp15 → Glu15, respectively).[37] Cherepanov and Mulkidjanian[38] have reported molecular dynamics simulations on wild type ferredoxin I (PDB code 7FD1) and the Asp15 → Glu mutant (PDB code 1D3W). For each system they analyzed one 1.25 ns trajectory and estimated the free energy along a progression coordinate. To put the calculations in perspective it is useful to recall the putative overall reaction mechanism inferred from the experimental observations.[37] Starting from the oxidized $[3Fe-4S]^+$ cluster electron transfer leads to reduction ($[3Fe-4S]^0$). Next,

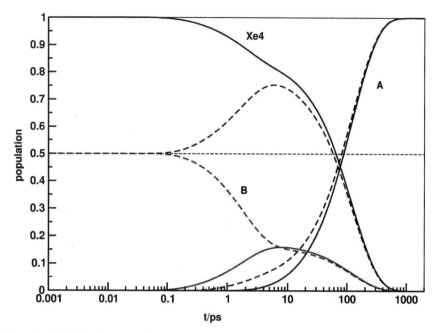

Figure 6: Rebinding dynamics for the L29F mutant using the $G(R)$ withouth water in the Xe4 pocket (black line in Figure 5). The population of the bound (A) state is black, while populations of the metastable B and Xe4 states are in red and blue, respectively. The two initial conditions are 100% population of Xe4 (solid line) and 50% population in B and Xe4 (dashed line). Both initial conditions give very similar rebinding times.

a proton is transferred from the environment and $H^1-[3Fe-4S]^0$ is formed which, after release of the proton, can oxidize back to $[3Fe-4S]^+$.

The results discussed here aim particularly at investigating various aspects of the proton transfer to $[3Fe-4S]^0$, i.e. after electron transfer has occurred. For example, if Asp15 is the source of the proton for $[3Fe-4S]^0$, an important process is the delivery of a proton from the solvent to Asp15, which requires a rotation of the carboxylic group to move the proton close enough to S1 of $[3Fe-4S]^0$. Further, for the observed rate (1300 s^{-1}) the proton on Asp15 (which has a low probability of being there) must spend sufficient time close to S1 of $[3Fe-4S]^0$ to provide a high probability for proton transfer. Thus it is of interest to investigate which protons from the protein framework are sufficiently close to serve as proton donors to the buried [3Fe-4S] cluster.

Possible Proton Donors: Using carefully calculated force field parameters for the buried [3Fe-4S] cluster, molecular dynamics simulations were carried out for reduced and oxidized native ferredoxin and for the D15E mutant. The primary question concerned which protons around the [3Fe-4S] cluster can serve as possible proton donors and what role could a potentially present water molecule in the active site play. MD simulations on native 7FDR reveal a number of hydrogen atoms around the [3Fe-4S] cluster which could, in principle, serve as proton donors. All distribution functions are averages over five independent trajectories, each 250 ps in length. The distribution functions are shown in Figure 8. The most obvious candidate for proton transfer to $[3Fe-4S]^0$ is HD2(Asp15), as

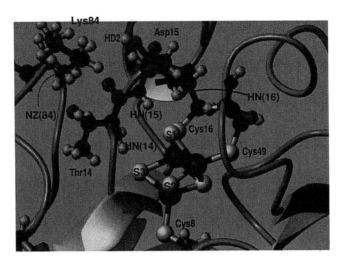

Figure 7: Active site of 7FDR with the central [3Fe-4S] cluster (in ball-and-stick) which is bound to cysteine residues Cys8, Cys16, and Cys49 of the protein. Residues important for the proton transfer (Thr14, Asp15, Cys16) and involved in the salt bridge (Lys84) are also shown and labelled.

was suggested based on the experimental analysis. However there are several other protons in the surroundings. They include HN(Tyr13) (close to S2), HN(Thr14) (close to S3), HN(Asp15) (close to S1), and HN(Cys16) (close to S1).

The profiles indicate that hydrogen atoms other than HD2(15) can approach [3Fe-4S] more closely and thus could serve as possible proton donors. In particular, HN(15)–S1 and HN(14)–S3 have a rather pronounced peak around 2 Å in all simulations. Using ESP or Mulliken charges (see Table 1) instead of the NBO charges has little influence on these results. Also, releasing the constraints on the hydrogen bonds (through SHAKE[21]) does not alter this picture. Since a conventional force field does not allow for bond breaking and bond forming processes to be described, DFT calculations can be used to estimate proton transfer from HN(15) and HD2(15), respectively, to the [3Fe-4S] cluster. In both cases, a double minimum potential with a barrier height of 42 kcal/mol (OD2(15)–HD2(15)–S1) and 44 kcal/mol (N(15)–NH(15)–S1), respectively, was found. The secondary minimum (X–H–S1) has a depth of 10 kcal/mol (OD2(15)–HD2(15)–S1) and 7 kcal/mol (N(15)–NH(15)–S1), respectively. This shows that in the gas phase the proton is almost equally likely to come from the Asp side chain as from the main-chain nitrogen atom while the overall reaction is highly endothermic in both cases. Thus, solvation effects due to the protein and the surrounding water must play a significant role.

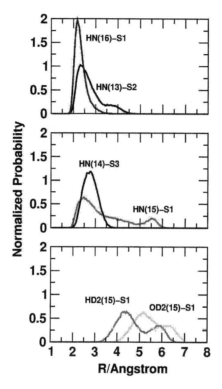

Figure 8: Average of H-bonding distance distributions in 7FDR. All averages are taken over five independent runs, each 250 ps in length. The three panels show possible proton-sulphur distances that would allow proton transfer from the protein to the buried [3Fe-4S] cluster. The lower panel shows the H–S distance (together with the O–S distance) that has been suggested to be involved in proton transfer.

Table 1: Atomic charges using B3LYP/6-31G** for different models for the $[3Fe-4S]^{0/+}$ clusters. S4 signifies the three-fold coordinate inorganic sulphur. All charges are given in units of e.

Model	S4	S	Fe
$[3Fe-4S]^0$			
Mulliken/UHF geometry	−0.630	−0.590	0.800
Mulliken/Exp geometry	−0.510	−0.480	0.650
NBO/UHF geometry	−1.100	−0.980	1.350
NBO/Exp geometry	−1.000	−0.920	1.250
ESP/Exp geometry	−0.700	−0.630	0.863
$[3Fe-4S]^+$			
Mulliken/UHF geometry	−0.560	−0.330	0.850
Mulliken/Exp geometry	−0.360	−0.250	0.700
NBO/UHF geometry	−0.980	−0.590	1.250
NBO/Exp geometry	−0.830	−0.640	1.250
ESP/Exp geometry	−0.538	−0.398	0.910

As indicated above, if Asp15 is the proton transfer agent, transport of the proton from the protein outside to its interior is required. The most obvious path for this is the rotation of COOH around the CB-CG bond (see Figure 7. Umbrella sampling along the the dihedral angle (CA-CB-DG-OD2) yields a barrier for rotation of around 8 kcal/mol which indicates that the rotational motion is an activated process with a time constant of the order of μs.[24]

Comparison of wild type and D15E mutant Fd: The experimental work established that proton transfer in the D15E (Asp → Glu, PDB code 1D3W) mutant is considerably retarded (by two orders of magnitude) compared to native Azotobacter Vinelandii (7FD1).[37] On the basis of this work it was suggested that the carboxylate of Asp15 acts as the proton relay from the solvent to the cluster. For the native 7FD1 it was found in a single 1.25 ns trajectory[38] that the salt bridge between N(Lys84) and O(Asp15) frequently breaks and is restored during MD simulations which should allow the OD2 proton to approach the [3Fe-4S] cluster. In contrast to that, simulations on the D15E mutant apparently also showed that the longer side chain of Glu15 forms a stronger salt bridge than Asp15 with Lys84 through simulations of the D15E mutant (1D3W). It was concluded that "Glu15 is effectively immobilized and cannot serve as a mobile proton carrier" while Asp15 in native 7FDI is able to do so.

Five simulations with varying initial conditions, each 250 ps in length, were carried out and analyzed for both, native and mutant (D15E) ferredoxin where Asp15 is unprotonated. For the native protein (Figure 9 top row) the five runs reveal strong salt bridges which are interrupted only by rotation of the carboxyl group; i.e. when the binding partner to N(Lys84) changes from OD1(Asp15) to OD2(Asp15) in a transition that takes on the order of μs. For 1D3W (see Figure 9 bottom row) two out of five trajectories (black and red) were very similar to those in Ref.[38]. However, the remaining three trajectories showed completely different behaviour. The stable salt bridge between Glu15 and Lys84 (the result found by [38]) the N(Lys84)–O(Glu15) bridge is broken and restored for periods of several ten ps (blue and green) and in one case (yellow) the salt bridge was broken for almost 200 ps. This is shown in Figure 9.

From the time history of the N(Lys84)–O(Asp15) and N(Lys84)–O(Glu15) distances the free energy profile (or potential of mean force) can be estimated according to $-G(q) = kT \log P(q) + G_0$, where G_0 is a constant, $P(q)$ is the distribution function of the time history for the coordinate q in question, k is the Boltzmann constant and T is the temperature. From the OD1(15)–NZ(84) and OD2(15)–NZ(84) distances the coordinate q was determined by choosing the shorter of the two distances along each trajectory (see also [38]). On the right hand side of Figure 9 the $G(q)$

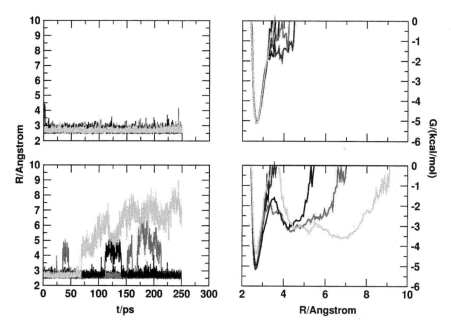

Figure 9: Left hand panels: Time history for the salt bridge in 7FD1 (with unprotonated Asp) and the D15E mutant. The scale for the N(Lys84)–O(X15) is the same, where X is either Asp or Glu, to allow for direct comparison. For Asp only the shorter of the two possible N–O separations is drawn. Right hand panels: $G(q)$ derived from the time histories for the two systems. The color codes correspond to the time history traces. In the upper right figure (native protein) only the red curve suggests breaking of the salt bridge for a short time while in the lower right figure (D15E mutant) the salt bridge is broken in three out of five simulations. The traces of the black and the red simulation are almost identical.

curves are shown for the native and the mutant protein. The $G(q)$ curves reveal that the reported rupture of the salt bridge for the unprotonated wild type protein 7FD1 (RUN) appears to be more the exception than the rule. Because of this apparent disagreement with the results from Ref. [38] five additional runs, each 500 ps in length and with varying initial conditions, were analyzed for RUN. In two cases the disruption of the salt bridge N(Lys84)–O(Asp15) was found as was observed in [38]. For the remaining three runs the salt bridge was again stable over the entire length of the trajectory. For the D15E mutant, however, the salt bridge is present in less than half the cases investigated.

These results suggest that formation and/or destruction of the salt bridge between Asp15 and Lys84 is probably not the primary difference between native and mutant Fd which leads to the observed slowdown of the proton transfer rate. It is, however, possible (and likely) that the salt bridge plays a role in opening and closing the access to the active site. In summary, we find that for the native, unprotonated 7FDI the salt bridge is more stable than previously assumed while for the mutant D15E it is less stable than previously assumed. This will have important consequences for the proposed reaction mechanism which was believed to proceed exclusively via salt-bridge formation between Asp15 and Lys84 in 7FDI.[37, 38]

From the simulations of the wild type protein described with and without Asp protonated and with and without water[24] around the [3Fe-4S] cluster, the following reaction mechanism is proposed: initially, Asp15 is unprotonated which leads to a very stable salt bridge between Lys84 and Asp15 (see Figure 9). Residues Asp15 and Pro50 are moving sufficiently to open and close a cleft which leads towards the [3Fe-4S] cluster.[24, 39] Solvent water can penetrate into the "active region" (the zone around [3Fe-4S]) and protonate Asp15. The salt bridge between N(Lys84) and O(Asp15) now breaks. This is manifest in the simulations. After rotation of Asp15, such that the hydrogen atom is pointing towards the [3Fe-4S] cluster and the enclosed water molecule, proton transfer from COOH to water can occur. This leads to H_3O^+ in the active site. From there, the proton could be transferred directly onto [3Fe-4S] or to the main chain to form NH_2^+ prior to proton transfer onto the cluster. At the same time, the salt bridge N(Lys84)–O(Asp15) is re-established because Asp15 is deprotonated. This in turn opens the cleft. The region between residues Asp15 and Pro50 can open up again and the water molecule is able to leave the active site. This is possible, as shown by simulations of the wild type protein with unprotonated Asp15 with water around the active site.[24]

4   Conclusion

The results of molecular dynamics simulations are summarized which describe a variety of experimental observables. Three cases of different complexity should emphasize what is currently possible with all-atom simulations and where still methodological advances are needed. While structural questions ("where does CO migrate immediately after photodissociation?") do not require particular simulation techniques, illuminating the rebinding dynamics of CO is a rather different issue. Since the time scale for this process is of the order of 100 ns direct sampling with sufficient statistics is no easily possible. The same applies to direct sampling of proton transfer in ferredoxin, the last system described.

Another fact worth mentioning is that mutation studies, such as for the infrared spectrum of photodissociated CO in the native and the L29F mutant, are an invaluable source for rationalizing the results and improving the simulation techniques. With computational methods in general and molecular dynamics simulations in particular it is easier to calculate relative changes in observables. This allows to compare such changes with results from experiments carried out in the same fashion, which is far less sensitive than comparing absolute values. One example are the infrared bands of photodissociated CO, where the difference between the two bands (B_1 and B_2) is 8 cm^{-1} while the absolute values of the absolute values of the observed and calculated frequencies differ by several 10 cm^{-1}.

Finally, it is not always necessary to directly calculate an experimentally measured quantity to provide meaningful insight from simulations. As in the case of ECPT in ferredoxin it is unrealistic to ever directly sample the proton transfer process from the protein to the buried [3Fe-4S] cluster. However, it is possible to investigate whether there are apparent differences between the native and the mutant protein which sufficiently rationalize observed differences between the two. In this context it is worthwhile to mention that the opening motion between residues Asp15 and Pro50 has been found to allow access of water into the protein, as was suggested by the simulations. However, since non-structural water molecules are likely to have a very high motility they are probably not observable by X-ray crystallography. The simulations, however, suggest that water can be readily stabilized around the [3Fe-4S] cluster in the native system.

Acknowledgment

The author acknowledges financial support from the Swiss National Science Foundation through a Förderungsprofessur. He is also grateful to his collaborators, mentioned in the cited references, for their enthusiasm to bring this research forward.

References

[1] Ascenzi, P. & Brunori, M. (2001) *Biochem. Mol. Biol. Edu.* **29**, 183–185.

[2] Cooper, C. E. (1999) *Bioch. Bioph. Acta* **1411**, 290–309.

[3] Olson, J. S. & Phillips Jr., G. N. (1996) *J. Biol. Chem.* **271**, 17593–17596.

[4] Lim, M., Jackson, T. A., & Anfinrud, P. A. (1997) *Nature Structural Biology* **4**, 209–214.

[5] Schotte, F., Lim, M., Jackson, T. A., Smirnov, A. V., Soman, J., Olson, J. S., Phillips, G. N., Jr., Wulff, M., & Anfinrud, P. A. (2003) *Science* **300**, 1944–1947.

[6] Lim, M., Jackson, T. A., & Anfinrud, P. A. (1995) *J. Chem. Phys.* **102**, 4355–4366.

[7] Schlichting, I., Berendzen, J., Phillips, G. N., Jr., & Sweet, R. M. (1994) *Nature* **371**, 808–812.

[8] Teng, T., Šrajer, V., & Moffat, K. (1994) *Nature Structural Biology* **1**, 701–705.

[9] Henry, E. R., Sommer, J. H., Hofrichter, J., & Eaton, W. A. (1983) *J. Mol. Biol.* **166**, 443–451.

[10] Elber, R. & Karplus, M. (1990) *J. Am. Chem. Soc.* **112**, 9161.

[11] Ansari, A., Jones, C. M., Henry, E. R., Hofrichter, J., & Eaton, W. A. (1994) *Biochem.* **33**, 5128.

[12] Meuwly, M., Becker, O. M., Stote, R., & Karplus, M. (2002) *Biophys. Chem.* **98**, 183.

[13] Bossa, C., Massimiliano, A., Roccatano, D., Amadei, A., Vallone, B., Brunori, M., & Nola, A. D. (2004) *Biophys. J.* **86**, 3855.

[14] Tilton, R. F., Jr., Kuntz, I. D., Jr., & Petsko, G. A. (1984) *Biochemistry* **23**, 2849–2857.

[15] Nienhaus, K., Deng, P., Kriegl, J. M., & Nienhaus, G. U. (2003) *Biochem.* **42**, 9647.

[16] Brooks, B. R., Bruccoleri, R. E., Olafson, B. D, States, D. J., Swaminathan, S., & Karplus, M. (1983) *J. Comp. Chem.* **4**, 187–217.

[17] MacKerell, Jr., A. D., Bashford, D., Bellott, M., Dunbrack, Jr., R. L., Evanseck, J. D., Field, M. J., Fischer, S., Gao, J., Guo, H., Ha, S., Joseph-McCarthy, D., Kuchnir, L., Kuczera, K., Lau, F. T. K., Mattos, C., Michnick, S., Ngo, T., Nguyen, D. T., Prodhom, B., Reiher, III, W. E., Roux, B., Schlenkrich, M., Smith, J. C., Stote, R., Straub, J., Watanabe, M., Wiorkiewicz-Kuczera, J., Yin, D., & Karplus, M. (1998) *J. Phys. Chem. B* **102**, 3586.

[18] Kuczera, K., Kuryian, J., & Karplus, M. (1990) *J. Mol. Biol.* **213**, 351.

[19] Kottalam, J. & Case, D. A. (1988) *J. Am. Chem. Soc.* **110**, 7690.

[20] Nutt, D. R. & Meuwly, M. (2003) *Biophys. J.* **85**, 3612–3623.

[21] Van Gunsteren, W. V. & Berendsen, H. J. C. (1977) *Mol. Phys.* **34**, 1311.

[22] Nutt, D. R. & Meuwly, M. (2004) *Proc. Natl. Acad. Sci.* **101**, 5998.

[23] Banushkina, P. & Meuwly, M. (2005) *J. Phys. Chem. B*.

[24] Meuwly, M. & Karplus, M. (2003) *Faraday Discuss. Chem. Soc.* **124**, 297–313.

[25] Meuwly, M. & Karplus, M. (2004) *Biophys. J.* **86**, 1987.

[26] Bicout, D. & Szabo, A. (1998) *J. Chem. Phys.* **109**, 2325.

[27] McQuarrie, D. A. (1976) *Statistical Mechanics* (Harper & Row, New York).

[28] Allen, M. P. & Tildesley, D. J. (1989) *Computer Simulation of Liquids* (Clarendon Press, Oxford).

[29] Nienhaus, K., Olson, J. S., Franzen, S., & Nienhaus, G. U. (2005) *J. Am. Chem. Soc.* **127**, 41.

[30] Vojtechovsky, J., Chu, K., Berendzen, J., Sweet, R. B., & Schlichting, Il (1999) *Biophys. J.* **77**, 2153.

[31] Austin, R.H., Beeson, K.W., Eisenstein, L., Frauenfelder, H., & Gunsalus, I.C. (1975) *Biochem.* **14**, 5355.

[32] Olson, J. & Philipps, G. (1996) *J. Biol. Chem.* **271**, 17593.

[33] Harvey, J. N. (2000) *J. Am. Chem. Soc.* **122**, 12401.

[34] Szabo, A., Schulten, K., & Schulten, Z. (1980) *J. Chem. Phys.* **72**, 4350.

[35] Banushkina, P. & Meuwly, M. (2005) *J. Chem. Theo. Comp.* **1**, 208.

[36] Steinbach, P. J., Ansari, A., Berendzen, J., Braunstein, D., Chu, K., Cowen, B. R., Ehrenstein, D., Frauenfelder, H., Johnson, J. B., Lamb, D. C., Luck, S., Mourant, J. R., Nienhaus, G. U., Ormos, P., Philipp, R., Xie, A., & Young, R. D. (1991) *Biochem.* **30**, 3988.

[37] Chen, K., Hirst, J., Camba, R., Bonagura, C. A., Stout, C. D., Burgess, B. K., & Armstrong, F. A. (2000) *Nature* **405**, 814.

[38] Cherepanov, D. A. & Mulkidjanian, A. Y. (2001) *Bioch. Bioph. Acta* **1505**, 179.

[39] Camba, R., Jung, Y. S., Hunsicker-Wang, L. M., Burgess, B. K., Stout, C. D., Hirst, J., & Armstrong, F. A. (2003) *Biochem.* **42**, 10598–10599.

Brill Academic Publishers
P.O. Box 9000, 2300 PA Leiden
The Netherlands

Lecture Series on Computer
and Computational Sciences
Volume 3, 2005, pp. 119-125

Theoretical Study on Open-Shell Nonlinear Optical Systems

M.Nakano[1]

Department of Materials Engineering Science,
Graduate School of Science,
Osaka University,
Toyonaka, Osaka 560-8531, Japan

Received 1 August, 2005; accepted in revised form 12 August, 2005

Abstract: The static second hyperpolarizabilities (γ) of open-shell organic nonlinear optical (NLO) systems are investigated from a viewpoint of diradical character dependence of γ for diradcical systems using ab initio molecular orbital (MO) and density functional theory (DFT) methods. It is found that neutral singlet diradical systems with intermediate diradical characters tend to enhance γ compared to those with small diradical characters, while the triplet diradical systems tend to significantly reduce the γ due to the Pauli principle. On the basis of these results, we propose a new class of NLO systems, i.e., open-shell NLO systems, which have novel control parameters: the spin multiplicity and diradcial character.

Keywords: Second hyperpolarizability, open-shell molecule, diradical, ab initio MO, density functional
PACS: 31.15.Ar., 31.1.5.Ew, 33.1.5.Kr.

1. Introduction

Although lots of investigations of nonlinear optical (NLO) properties have been performed, most of those studies have been limited to the closed-shell systems. On the other hand, a pioneering study on the open-shell NLO systems has begun with the aim of enhancing the field-induced charge fluctuations in charged radical states [1-3]. These systems are also interesting in designing multi-functional materials, for example, combining magnetic and optical properties [4,5]. On the basis of the perturbational analysis of virtual excitation processes of γ, the charged radical systems having symmetric resonance structure with invertible polarization (SRIP) [6,7] are expected to have unique negative second hyperpolarizability (γ), which is the microscopic origin of the third-order NLO property, in the off-resonant region and to exhibit large electron-correlation dependences [8]. Some crystals composed of these charged radical systems are predicted to present unique third-order NLO properties (negative γ) as well as high electrical conductivity [7]. Further, the open-shell molecules and their clusters involving nitronyl nitroxide radicals, which are important in realizing organic ferromagnetic interaction, also belong to the SRIP systems and are predicted to give negative off-resonant γ values.

The feature of open-shell NLO systems is also described from the viewpoint of the chemical bonding nature [9,10]. We have predicted a remarkable variation in γ of H_2 model with increasing the bond distance and have suggested a significant enhancement of γ in the intermediate correlation (intermediate bond breaking) regime. This feature reflects the fact that the intermediate bonding electrons are sensitive to the applied field, leading to large fluctuation of electrons. Such intermediate bond breaking nature in the intermediate correlation regime is, for example, expected to be realized by the increase in the spin multiplicity in open-shell neutral systems. The investigation of the effect of introducing charge into open-shell systems on γ is also important in view of the spin-control of novel open-shell NLO systems since the control of spin state is often achieved by introducing the charges into systems in molecular magnetism. Recently, we have elucidated the effects of spin multiplicity on γ of open-shell neutral and/or charged π-conjugated systems using simple open-shell model compounds, and have found that the magnitude of γ sensitively depends on the spin and charged states [11].

---

[1] Corresponding author. E-mail: mnaka@cheng.es.osaka-u.ac.jp

Toward pursuiting another way to control γ of open-shell NLO systems, we have considered the diradical systems [12]. The dependence of γ on the diradical character for singlet diradical systems is investigated using diradical model compounds, p-quinodimethane (PQM) models with different both-end carbon-carbon (C-C) bond lengths, by highly correlated *ab initio* molecular orbital (MO) and hybrid density functional theory (DFT) method. The γ value of a real diradical molecule, i.e., 1,4-bis-(imidazole-2-ylidene)-cyclohexa-2,5-diene (BI2Y), which is expected to possess intermediate diradical character, is also examined as well as related molecule (BI2YN) with smaller diradical character. The spin-state dependence of γ for BI2Y has been also investigated. On the basis of these results, we clarify the features of a new class of open-shell NLO systems.

2. Model systems and calculation scheme

As an example, we here show the results of γ for p-quinodimethane models. Figure 1 shows the structure of the p-quinodimethane (PQM) molecule in its singlet ground state. It is described by two resonance forms, i.e., the quinoid and diradical forms. In general, the experimental structures of conjugated diradical systems are well reproduced by the spin-restricted (R)B3LYP and unrestricted (U)B3LYP methods. Indeed, the optimized structure for the present system with D_{2h} symmetry ($R_1 = 1.351$ Å, $R_2 = 1.460$ Å and $R_3 = 1.346$ Å) at the UB3LYP level using 6-311G* basis set is the same as that at the RB3LYP level. The optimized parameters suggest that PQM presents a large degree of quinoid form instead of diradical nature since R_1 and R_3 are like a C-C double bond length and are smaller than R_2 having a C-C single bond nature. In order to increase the degree of diradical nature, we consider several PQM models with bond-length R_1 changing from 1.35 Å to 1.7 Å under the constraint of $R_2 = R_3 = 1.4$ Å. We first determine the diradical character from spin-unrestricted Hartree-Fock (UHF) calculations. The diradical character y_i related to the HOMO-i and LUMO+i is defined by the weight of the doubly-excited configuration in the MC-SCF theory and is formally expressed in the case of the spin-projected UHF (PUHF) theory as [10,13]

$$y_i = 1 - \frac{2T_i}{1+T_i^2},\qquad(1)$$

where T_i is the orbital overlap between the corresponding orbital pairs. T_i can also be represented by using the occupation numbers (n_j) of UHF natural orbitals (UNOs):

$$T_i = \frac{n_{\text{HOMO}-i} - n_{\text{LUMO}+i}}{2}.\qquad(2)$$

Since the PUHF diradical characters amount to 0 % and 100 % for closed-shell and pure diradical states, respectively, y_i represents the diradical character, i.e., the instability of the chemical bond. Table 1 gives the diradical characters y calculated from Eqs. (1) and (2) using HOMO and LUMO of UNOs for PQM systems with various R_1 values. As expected, it is found that the diradical character y increases when increasing the bond length R_1 starting from the optimized (equilibrium) geometry, which possesses a low y value (0.146), i.e., quinoid like structure. Figure 2(a) and 2(b) show the optimized structures BI2Y and BI2YN at the B3LYP/6-31G** level. The two resonance forms, i.e., quinoid and diradical forms, contribute to BI2Y, while the diradical form little contributes to BI2YN because of the lone pairs of N atoms linked with the middle aromatic benzene ring. In fact, BI2Y and BI2YN are turned out to exhibit, respectively, an intermediate ($y = 0.423$) and very small ($y = 0.010$) diradical character.

We use the 6-31G*+p basis set with p exponent of 0.0523 on carbon (C) atoms and 0.0582 on nitrogen (N) atoms since several studies have demonstrated that the use of a split-valence or split-valence plus polarization basis set augmented with a set of p and/or d diffuse functions on the second-row atoms enables us to reproduce the γ of large- and medium-size π-conjugated systems calculated with larger basis sets. We apply the RHF, UHF and UHF-coupled-cluster with single and double excitations as well as a perturbative treatment of the triple excitations (UCCSD(T)). Among the DFT schemes, the R and U BHandHLYP exchange-correlation functionals have been adopted. All calculations have been performed using the Gaussian 98 program package [14]. We confine our attention to the dominant longitudinal components of static γ. The static γ can be obtained by the finite field (FF) approach which consists in the fourth-order differentiation of energy E with respect to different amplitudes of the applied external electric field.

ı *p*-quinodimethane (PQM)

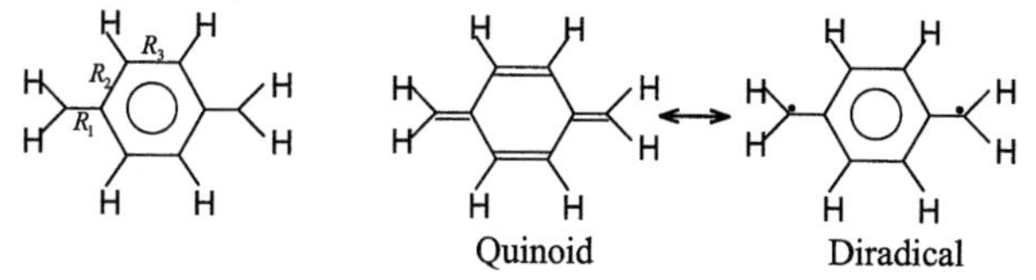

Figure 1: Singlet diradical molecules, *p*-quinodimetahne (PQM) model with several parameters R_1, R_2 and R_3

Table 1: Diradical character y for R_1, R_2 and R_3 in the PQM models[1] shown in Fig. 1.

R_1	R_2	R_3	y
1.351	1.460	1.346	0.146
1.350	1.400	1.400	0.257
1.400	1.400	1.400	0.335
1.450	1.400	1.400	0.414
1.500	1.400	1.400	0.491
1.560	1.400	1.400	0.576
1.600	1.400	1.400	0.626
1.700	1.400	1.400	0.731

[1] The first row corresponds to the equilibrium geometry with D_{2h} symmetry in the singlet ground state optimized at the RB3LYP/6-311G* level

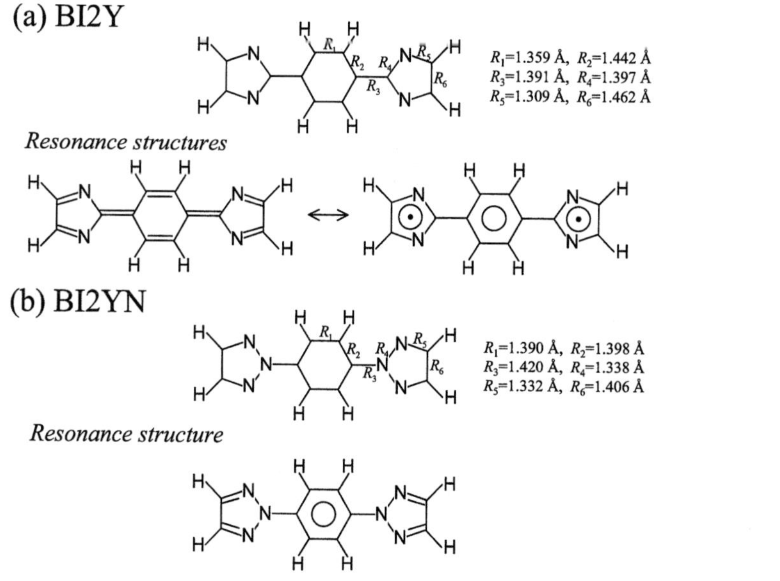

(a) BI2Y

R_1=1.359 Å, R_2=1.442 Å
R_3=1.391 Å, R_4=1.397 Å
R_5=1.309 Å, R_6=1.462 Å

Resonance structures

(b) BI2YN

R_1=1.390 Å, R_2=1.398 Å
R_3=1.420 Å, R_4=1.338 Å
R_5=1.332 Å, R_6=1.406 Å

Resonance structure

Figure 2: Geometrical structures of BI2Y (a) and BI2YN (b) optimized by the B3LYP/6-31G**. Resonance structures are also shown

3. Results and discussion

Figure 3 shows the results calculated by the RHF, UHF and UBHandHLYP methods as well as the UCCSD(T) result, which is considered to be the most reliable for the present systems. It turns out that the γ value at the UCCSD(T) level increases with increasing y in the weak diradical region ($y = 0.1$-0.5), attains a maximum in the intermediate region ($y \approx 0.5$) and then decreases in the region corresponding to large diradical character ($y > 0.5$). The maximum γ value at the UCCSD(T) level (77500 a.u.) for a y value close to 0.5 is 3.3 times as large as the γ value at the equilibrium geometry (23300 a.u.) ($y = 0.146$). The RHF γ value is negative while its magnitude increases strongly with the diradical character. Such incorrect behavior is predicted to originate from the triplet instability of the RHF solution in the intermediate and strong correlation regimes [10]. At the UHF level, γ monotonically decreases with increasing diradical character, so that for y smaller than 0.3 they are larger than the UCCSD(T) reference value whereas they are smaller in the intermediate and large diradical character regions. In contrast, the UBHandHLYP method provides a similar qualitative description of the variation in γ to that at the UCCSD(T) level. Nevertheless, the UBHandHLYP γ value overshoots the UCCSD(T) value in the region with small diradical character ($y < 0.3$) while it is slightly smaller in the region with intermediate and large diradical characters ($y > 0.4$). As a result, it is predicted that hybrid DFT methods with exchange-correlation functionals specifically-tuned for reproducing the NLO properties

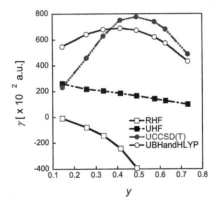

Figure 3: γ value versus diradical character y for PQM models calculated by the RHF, UHF, UCCSD(T) and UBHandHLYP methods.

of small- and intermediate-size diradical species could provide satisfactory results in comparison with the more elaborated UCCSD(T) scheme.

The variation in γ for the PQM models with increasing diradical character can be explained from the analogy to the dissociation of H_2. The increase of γ in the intermediate diradical character region is predicted to be caused by the virtual excitation processes (type III [1]) with zwitterionic contribution between the radicals on both-end carbon atoms, which corresponds to the intermediate dissociation region for the H_2 molecule [9]. In an analogous way, the intermediately spatial polarized wave-functions for α and β electrons, i.e., spin polarization, on both-end carbon sites in the PQM with intermediate diradical character are predicted to contribute to the enhancement of γ through the virtual excitation processes involving the zwitterionic nature as compared to the case with small diradical character (stable bond nature), giving a relatively small polarization. On the other hand, in the region with large radical character, the localized spins on both-end sites exhibit less charge polarization to the applied external electric field due to its strong correlation nature, so that the γ value decreases again.

We next investigate the γ for singlet BI2Y and BI2YN. From the results on the p-quinodimethane models, the UBHandHLYP is shown to give reliable γ values for diradical molecules with intermediate diradical characters, while the RBHandHLYP method gives reliable γ values for diradical molecules with small diradical characters or closed-shell molecules, at least for molecules of the size investigated here. Furthermore, using the 6-31G basis set, the reliability of UBHandHLYP to estimate γ of BI2Y is confirmed. Indeed, its value amounts to 174, -40, 939, 690, 484, and 524 x 103 a.u at the UHF, PUHF

($l = 3$), UMP2, PUMP2 ($l = 3$), UBHandHLYP, and UCCSD(T) levels of approximation, respectively. We therefore apply the R(U)BHandHLYP methods. For comparison, we also apply the UHF and UMP2 methods to the calcuation of γ for BI2Y and the RHF and RMP2 methods to BI2YN. The γ values at the UHF (2002 x 10^2 a.u.) and UMP2 (9962 x 10^2 a.u.) levels are shown to undershoot and overshoot that (6534 x 10^2 a.u.) at the UBHandHLYP level, respectively. These deficiencies are caused by the spin contamination at the UHF and UMP2 solutions. The situation with respect to spin contamination is more cumbersome for BI2YN although the small diradical character ($y = 0.01$). Indeed, the γ value (444 x 10^2 a.u.) at the UHF level is close to that (405 x 10^2 a.u.) at the RHF level, the MP2 values are much different [1001 x 10^2 a.u. (UMP2) versus 617 x 10^2 a.u.(RMP2)] while removal of the spin contamination at the PUHF ($l = 1$) and PUMP2 ($l = 1$) levels of approximation gives γ values of 47 x 10^2 a.u and 642 x 10^2 a.u. . The γ value at the RMP2 [and PUMP2 ($l = 1$)] level is in good agreement with that (654 x 10^2 a.u.) at the RBHandHLYP level, which is known to well reproduce γ value for relatively small size closed-shell π-conjugated molecule at the CCSD(T) level. The UBHandHLYP solution is found to coincide with RBHandHLYP solution for BI2YN as a result of its very small diradical character ($y = 0.01$). Considering the BHandHLYP results, γ increases by about one order of magnitude by going from BI2YN to BI2Y. This trend is in agreement with that observed in *p*-quinodimethane model. The enhancement of γ in the intermediate diradical character regime is predicted to originate from the increase of the field-induced electron fluctuation due to the intermediate bond dissociation nature (p-bond breaking in the present case) in such regime. This fluctuation is expected to be reduced both in the stable bond region with small diradical character and in the complete diradical region where the diradical character is close to 1.

We finally consider the spin-state dependence of γ using singlet and triplet states of BI2Y shown in Fig. 4. The calculated γ values of our model molecules are given in Table 2. For the triplet state, the UBHandHLYP method is also expected to give a reliable γ value (211600 a.u.) since the spin contamination is small ($<S^2> = 2.0475$). In fact, for both singlet and triplet states, the γ values calculated by the UBHandHLYP/6-31G method are similar to those by the UCCSD(T)/6-31G method (Table 2). From these results, the UCCSD(T)/6-31G*+p results are expected to be reproduced well by the UBHandHLYP/6-31G*+p results, which are used in the following discussion. On the other hand, the overshot γ values at the UMP2 (289100 a.u.) and spin projected UMP2 [PUMP2(l=1)] (251800 a.u.) levels of approximation are predicted to be caused by the spin contamination effects. The γ value for the singlet state calculated by the UBHandHLYP method is about three times as large as that for the triplet state as shown in Table 2. This significant difference of γ value between the singlet and triplet states is predicted to be caused by the difference of the field-induced electron fluctuation effect between the singlet and triplet states. This feature can be understood as follows. The large third-order polarization in the singlet state with intermediate diradical character is caused by the intermediate overlap between a pair of radical orbitals. In the triplet state, which corresponds to the highest spin state for diradical system, such polarization is significantly suppressed because of the non-overlap between a pair of radical orbitals due to the Pauli principle. Also such difference between the singlet and triplet states is predicted to be reduced in the region with large y value for the singlet state because the overlap between diradical distributions becomes negligible in such region for both spin states.

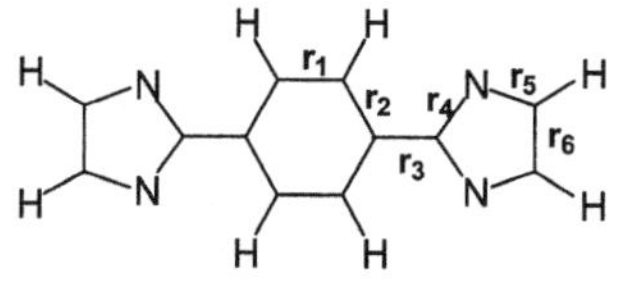

In Å	Singlet	Triplet
r_1	1.359	1.386
r_2	1.442	1.411
r_3	1.391	1.450
r_4	1.397	1.382
r_5	1.309	1.319
r_6	1.462	1.464

Figure 4: Molecular geometries of singlet and triplet BI2Y optimized, respectively, by RB3LYP and UB3LYP methods using 6-31G** basis set.

Table 2. Calculated γ values [a.u.][a] for singlet and triplet BI2Y.

Spin state	Method/Basis-set	$\gamma \square^{\tilde{}} \tilde{\square}\square$
	UBHandHLYP/6-31G*+p	653400
Singlet	UBHandHLYP/6-31G	484360
	UCCSD(T) /6-31G	524400
	UBHandHLYP/6-31G*+p	211600
Triplet	UBHandHLYP/6-31G	196000
	UCCSD(T) /6-31G	192400

a) 1 a.u. = 5.0366 x 10^{-40} esu

4. Summary

We have introduced the concept of open-shell third-order NLO systems and showed some examples. From our previous studies [11,12], the significant dependences of γ of open-shell systems on three parameters, spin multiplicity, charge and diradical character, are found and have focused in this paper on the diradical character dependence of γ. The neutral systems with intermediate diradical character exhibit larger γ value than those with small or large diradical characters. The enhancement of γ in the intermediate diradical character region is predicted to be caused by the virtual excitation processes (type III[1]) involving the zwitterionic nature between both-end radical sites. This model is substantiated by the intermediate spin multiplicity state of neutral π-conjugated systems, e.g., quartet C_5H_7, and neutral singlet diradical molecules, e.g., p-quinodimethane and BI2Y [15]. On the other hand, the introduction of charged defects turns out to cause the relative enhancement of negative contribution (type II[1]), so that charged open-shell symmetric systems satisfying SRIP condition [6], e.g., anion radical pentalene, tend to show negative total γ value. As a result, it is predicted that the sign and amplitudes of γ values of open-shell NLO systems can be controlled by tuning the spin multiplicity, diradical character and charged states. Furthermore, these systems are expected to be candidates for multi-functional materials, e.g., nonlinear optical property combined with magnetic property and/or electric conductivity, in which these properties could be mutually controllable.

Acknowledgments

This work was supported by Grant-in-Aid for Scientific Research (No. 14340184) from Japan Society for the Promotion of Science (JSPS).

References

[1] M. Nakano and K. Yamaguchi, *Chem. Phys. Lett.* **206**, 285-292 (1993).
[2] M.Nakano, I. Shigemoto, S. Yamada and K. Yamaguchi, *J. Chem. Phys.* **103**, 4175-4191 (1995).
[3] S. P. Karna, *J. Chem. Phys.* **104**, 6590, (1996) *erratum* 1996, *105*, 6091.
[4] S. Yamada, M. Nakano, I. Shigemoto, S. Kiribayashi and K. Yamaguchi, *Chem.Phys. Lett.* **267**, 438-444 (1997).
[5] M. Nakano, S. Yamada and K. Yamaguchi, *Bull. Chem. Soc. Jpn.* **71**, 845-850 (1998).
[6] M. Nakano, S. Kiribayashi, S. Yamada, I. Shigemoto, K. Yamaguchi, *Chem. Phys. Lett.* **262**, 66-73 (1996).
[7] M. Nakano, S. Yamada and K. Yamaguchi, *J. Phys. Chem. A* **103**, 3103-3109 (1999).
[8] M. Nakano and K. Yamaguchi, "Analysis of nonlinear optical processes for molecular systems", *Trends in Chemical Physiscs* Vol. 5 (Research trends, 1997) 87-237.
[9] M. Nakano, H. Nagao, K. Yamaguchi, *Phys. Rev. A* **55**, 1503 (1997).
[10] S. Yamanaka, M. Okumura, M. Nakano, K. Yamaguchi, *J. Mol. Structure* **310**, 205 (1994).
[11] M. Nakano, T. Nitta, K. Yamaguchi, B. Champagne and E. Botek, *J. Phys. Chem. A* **108**, 4105-4111 (2004).

[12] M. Nakano, R. Kishi, T. Nitta, T. Kubo, K. Nakasuji, K. Kamada, K. Ohta, B. Champagne, E. Botek, and K. Yamaguchi, *J. Phys. Chem. A*, **109**, 5, 885 - 891 (2005).

[13] K. Yamaguchi, *"Self-Consistent Field Theory and Applications"*, R. Carbo and M. Klobukowski, Eds. (Elsevier: Amsterdam, 1990) 727.

[14] M.J. Frisch, G.W. Trucks, H.B. Schlegel, J.A. Pople, et al. GAUSSIAN 98, Revision A11, Gaussian Inc., Pittsburgh, PA, 1998.

[15] M. Nakano, R. Kishi, T. Nitta, T. Kubo, K. Nakasuji, K. Kamada, K. Ohta, B. Champagne, E. Botek, and K. Yamaguchi, *Mater. Res. Soc. Symp. Proc.* **846**, DD1.4.1- DD1.4.12 (2005).

Brill Academic Publishers
P.O. Box 9000, 2300 PA Leiden
The Netherlands

Lecture Series on Computer
and Computational Sciences
Volume 3, 2005, pp. 126-128

Toxicity of Aliphatic Ethers: A Comparative Study

S. Nikolić,[1] [a] A. Milicević [b] and N. Trinajstić [a]

[a]The Rugjer Bošković Institute, P.O.B. 180, HR-10002 Zagreb, Croatia
[b]The Institute of Medical Research and Occupational Health, P.O.B. 291, HR-10001 Zagreb, Croatia

Received 19 May, 2005; accepted in revised form 30 May, 2005

Abstract: The CROMRsel procedure was used to model the toxicity of aliphatic ethers against mice. The best model obtained is based on three molecular descriptors and is better QSAR model than other models from the literature. Only comparable model is one by Ren, based on four descriptors.

Keywords: aliphatic ethers, CROMRsel procedure, molecular descriptors, QSAR, toxicity

Mathematics SubjectClassification: 05C10, 05C90, 68R10

1. Introduction

There are a number of structure-activity studies on toxicity of aliphatic ethers against mice in the literature [*e.g.*, 1-5]. Authors of these reports used a variety of molecular descriptors: the vertex-connectivity index χ, the first-order and the second-order valence vertex-connectivity indices $^1\chi^v$ and $^2\chi^v$, the weighted identification number WID, the Balaban indices J and J_{het} and the electropy index ε. We collected some of these indices for 21 aliphatic ether and applied to them our CROMRsel modeling procedure [6-8]. This is a multivariate procedure that has been designed to select the best possible model among the set of models obtained for a given number of descriptors, the criterion being the standard error of estimate. The quality of models is expressed by fitted (descriptive) statistical parameters: the correlation coefficient (R_{fit}), the standard error of estimate (S_{fit}) and the Fisher's test (F). The models are also cross(internally)-validated by a leave-one-out procedure. Statistical parameters for the cross-validated models are symbolized by R_{cv} and S_{cv}, where subscript cv denotes the cross-validation.

2. Computational Details

All considered descriptors χ, $^1\chi^v$, $^2\chi^v$, WID, J, J_{het} and ε are well-described in the literature [9-10]. They have been here computed by our own program and are given in Table 1. The experimental toxicities pC are taken from Marsh and Leake [11]. We used these experimental data because they were also used in all previous studies, but by Kier and Hall who used much older data [1].

Table 1: Considered descriptors χ, $^1\chi^v$, $^2\chi^v$, WID, J, J_{het} and ε computed by our own program for 21 ethers

	Ether	pC	χ	χ^v	WID	J	J_{het}	ε	$^2\chi^v$
1	Dimethyl	1.43	1.414	0.816	3.293	1.632	2.016	59.813	0.408

---

[1] Corresponding author. E-mail: sonjai@.hr

2	Methyl ethyl	1.74	1.914	1.404	4.124	1.975	2.352	92.642	0.577
3	Methyl propyl	2.45	2.414	1.904	5.058	2.190	2.480	125.585	0.993
4	Methyl isopropyl	2.26	2.270	1.799	5.074	2.540	2.937	124.263	0.707
5	Methyl butyl	2.70	2.914	2.405	6.032	2.339	2.557	159.393	1.346
6	Methyl isobutyl	2.79	2.770	2.260	6.040	2.628	2.902	158.071	1.272
7	Methyl *sec*-butyl	2.79	2.808	2.337	6.037	2.755	3.053	158.071	1.455
8	Methyl *tert*-butyl	2.79	2.561	2.112	6.050	3.169	3.583	154.164	2.316
9	Methyl pentyl	2.88	3.414	2.905	7.019	2.488	2.615	194.136	1.700
10	Diethyl	2.22	2.414	1.992	5.058	2.190	2.584	128.170	0.781
11	Ethyl propyl	2.60	2.914	2.492	6.032	2.339	2.675	163.300	1.197
12	Ethyl isopropyl	2.60	2.770	2.386	6.037	2.628	3.057	161.978	1.260
13	Ethyl butyl	2.82	3.414	2.992	7.019	2.448	2.722	198.943	1.551
14	Ethyl isobutyl	2.82	3.270	2.847	7.023	2.667	3.007	197.621	2.053
15	Ethyl *sec*-butyl	2.85	3.308	2.924	7.022	2.831	3.205	197.621	1.674
16	Ethyl *tert*-butyl	2.92	3.061	2.700	7.027	3.135	3.619	193.714	2.545
17	Ethyl pentyl	3.00	3.914	3.492	8.013	2.531	2.802	235.269	1.905
18	Ethyl *tert*-pentyl	3.15	3.621	3.261	8.018	3.373	3.779	230.040	2.046
19	Dipropyl	2.79	3.414	2.992	7.019	2.448	2.764	200.265	1.612
20	Propyl isopropyl	2.82	3.270	2.886	7.022	2.677	3.061	198.943	1.920
21	Di-isopropyl	2.82	3.126	2.781	7.025	2.953	3.425	197.621	2.234

3. Results and Discussion

The best single-descriptors model is based on WID (R_{fit}=0.942, R_{cv}=0.920, S_{fit}=0.14, S_{cv}=0.16), the best two-descriptor model is based on χ and J (R_{fit}=0.966, R_{cv}=0.951, S_{fit}=0.10, S_{cv}=0.13), the best three-descriptor model is based on $^1\chi^v$, J and J_{het} (R_{fit}=0.981, R_{cv}=0.968, S_{fit}=0.08, S_{cv}=0.10) and the best four-descriptor model is based on χ, $^1\chi^v$, J and J_{het} (R_{fit}=0.983, R_{cv}=0.969, S_{fit}=0.08, S_{cv}=0.10). These last two models are comparable among themselves – our choice is the model with less descriptors:

$$pC = 0.75\ (\pm0.13) + 0.328\ (\pm0.046)\ {}^1\chi^v + 1.83\ (\pm0.34)\ J - 1.25\ (\pm0.28)\ J_{het}$$

This model is also comparable to the best models from the literature [4,5] that are based on the variable connectivity index $^1\chi^f$ (R_{fit}=0.975, S_{fit}=0.10), and on Xu index [12] and atom type descriptors (R_{fit}=0.987, R_{cv}=0.965, S_{fit}=0.07, S_{cv}=0.12). This latter model is only model, besides our own model, that has been cross-validated.

4. Conclusion

The CROMRsel approach appears to be useful computational procedure for generating reliable QSAR models as was illustrated in this report.

Acknowledgments

This work was supported by Grant No. 0098034 from the Ministry of Science and Technology of Croatia.

References

[1]   L.B. Kier and L.L. Hall, *Molecular Connectivity in Chemistry and Drug Research*, Academic Press, New York, 1976, pp. 202-203.

[2]   O. Mekenyan, D. Boonchev, A. Sabljić and N. Trinajstić, Application of Topological Indices to QSAR. The Use of the Balaban Index and the Electropy Index for Correlations With Toxicity of Ethers on Mice, *Acta Pharm. Yugosl.* **37** (1987) 75-86.

[3]   B. Bogdanov, S. Nikolić, A. Sabljić, N. Trinajstić and S. Carter, On the Use of the Weighted Identification Numbers in the QSAR Study of the Toxicity of Aliphatic Ethers, *Int. J. Quantum Chem.: Quantum Biol. Symp.* **14** (1987) 325-330.

[4]   B. Ren, Novel Atomic Level AI Topological Descriptors for QSPR/QSAR Modeling: Normal Boling Points and Toxicity of Aliphatic Ethers, *unpublished.*

[5]   M. Randić and S.C. Basak, On Use of the Variable Connectivity Index $^1\chi^f$ in QSAR: Toxicity of Aliphatic Ethers, *J. Chem. Inf. Comput. Sci.* **41** (2001) 614-618.

[6]   B. Lučić and N. Trinajstić, Multivariate Regression Outperforms Several Robust Architectures of Neural Networks in QSAR Modeling, *J. Chem. Inf. Comput. Sci.* **39** (1999) 121-132.

[7]   B. Lučić, D. Amić and N. Trinajstić, Nonlinear Multivariate Regression Outperforms Several Concisely Designed Neural Networks on Three QSPR Dana Sets, *J. Chem. Inf. Comput. Sci.* **40** (2000) 403-413.

[8]   A. Milićević and S. Nikolić, On Variable Zagreb Indices, *Croat. Chem. Acta* **77** (2004) 97-101.

[9]   N. Trinajstić, *Chemical Graph Theory*, CRC Press, Boca Raton, 1992, 2nd revised ed., chapter 10.

[10] R. Todeschini and V. Consonni, *Handbook of Molecular Descriptors*, Wiley-VCH, Weinheim, 2000.

[11] D.P. Marsh and C.D. Leake, *Anaesthesiology* **11** (1950) 455-465.

[12] L. Xu, *Chemometrical Method*, Scientific Press of China, Beijing, 1996.

Brill Academic Publishers
P.O. Box 9000, 2300 PA Leiden,
The Netherlands

*Lecture Series on Computer
and Computational Sciences*
Volume 3, 2005, pp. 129-141

Charge instabilities in molecular materials: cooperative behavior from electrostatic interactions

Anna Painelli[*1]**, Zoltán G. Soos**[°]** and Francesca Terenziani**[‡]

*Dip. Chimica GIAF, Parma University & INSTM UdR Parma, 43100 Parma, Italy
°Dep. Chemistry, Princeton University, Princeton, New Jersey 08544
‡ CNRS, UMR 6510, Université de Rennes 1, Institut de Chimie, F-35042 Rennes Cedex, France

Received 15 July, 2005; accepted in revised form 1 August, 2005

Abstract: The interplay between electron transfer and electrostatic interactions leads in molecular materials to interesting charge instabilities. Here we discuss two families of materials where this phenomenon occurs, namely clusters of push-pull chromophores and charge transfer salts with a mixed stack motif. In the first case electrons are localized within each chromophore, whereas in the second case electrons are truly delocalized in one dimension. Cooperative and collective behavior results in the first class of materials from classical electrostatic intermolecular interactions, whereas in CT salts a complex interplay of electrostatic interactions, delocalization and lattice phonons governs the low-energy physics. We underline the similarity in the basic physics of the two systems, including the appearance of discontinuous phase transitions and of multielectron transfer. Collective and cooperative contribution to static polarizabilities are discussed in the two families of materials showing that supramolecular interactions can profoundly alter the material behavior, offering a powerful tool for the optimization of the material properties.

Keywords: electron transfer; electrostatic interactions, push-pull chromophores, charge transfer crystals, neutral-ionic phase transitions; NLO responses; dielectric constant; cooperative and collective behavior; multielectron transfer

PACS: 33.15.Kr; 42.65.-k; 71.10.Fd; 71.35.-y; 77.84.Jd

1   Introduction

Electron-transfer (ET) is a fundamental process in chemistry and biology and is the key process in organic electronics in its wide context. Understanding molecular junctions, photoconversion devices, organic LEDs, just as an example, requires a thorough understanding of ET. ET describes the motion of an electron between different sides of the same molecule (intramolecular ET) or between different molecules (intermolecular ET). In either case an electric charge moves along a sizeable distance: ET processes generate large electric fields and are strongly affected by electric fields. This simple consideration explains why materials with delocalized electrons, i.e materials where ET plays a basic role, are so interesting for non-linear optics (NLO): an applied electric field in fact displaces the delocalized electrons leading to large response fields and hence to non-linear behavior. Electric fields need not to be applied from the outside: large fields can be experienced by electrons inside a material as due to their interaction with surrounding charges. These *environmental* electric fields heavily affect the ET process in the material, leading to important phenomena.

---

[1] Corresponding author. E-mail: anna.painelli@unipr.it

Here we discuss materials where ET processes occurring at different locations interact electrostatically, so that the motion of an electron at a specific location affects ET processes at nearby (and not so nearby) sites. Collective and cooperative phenomena driven by electrostatic interactions in ET-based materials will be the main focus of this work.

Specifically, in the next Section we will discuss the role of intermolecular electrostatic interactions in clusters of push-pull chromophores. Push-pull chromophores are an interesting class of molecules with applications in molecular photonics and electronics [1]. In these molecules an electron donor (D) and an acceptor (A) group are joint together by a π-conjugated bridge and the ET from the D to the A site within each molecule dominates its low-energy physics. In aggregates, films or crystals push-pull chromophores pack with negligible intermolecular overlap, but electrostatic intermolecular interactions are strong. Materials based on push-pull chromophores are then interesting model systems to investigate the role of classical electrostatic interactions on intramolecular ET.

In Section 3 we will discuss charge-transfer (CT) crystals with a mixed stack motif [2]. In these materials electron donors such as tetrathiafulvalene (TTF) and electron acceptors such as chloranil (CA) alternate to form one-dimensional (1D) stacks. The frontier orbitals on adjacent molecules overlap significantly as testified by the fractional average ionicity on the molecular sites. Much as with push-pull chromophores, the low-energy physics of these materials is governed by ET. But in CT salts ET is intermolecular and leads to delocalized electrons in the stack direction. Charge and lattice instability dominate the physics of these materials and require fairly complex models accounting for delocalized electrons in 1D, electrostatic interactions in 3D and electron-phonon coupling.

The detailed phenomenology of the two families of materials described here is different, and the models adopted to describe their physics are very different, however several common features may be recognized in the two systems. Here we will emphasize a few fundamental concepts that underline the complex and interesting behavior of these materials as resulting from the interplay between ET and electrostatic interactions

2 Intramolecular ET and electrostatic interactions: clusters of push-pull chromophores

We describe the electronic structure of push-pull chromophores in terms of a two-state model, originally proposed by Mulliken [3] to describe DA complexes in solution. The model is based on the assumption that the low-energy physics of push-pull chromophores is dominated by the resonance between the neutral and the charge separated (zwitterionic) structures: $DA \leftrightarrow D^+A^-$. Two basis states, $|DA\rangle$ and $|D^+A^-\rangle$, separated by an energy $2z$ and mixed by a matrix element $-\sqrt{2}t$, then define the electronic Hamiltonian. The diagonalization of this Hamiltonian is trivial and was discussed by several authors [4]. The resulting ground and excited states are:

$$
\begin{aligned}
|g\rangle &= \sqrt{1-\rho}|DA\rangle + \sqrt{\rho}|D^+A^-\rangle \\
|e\rangle &= \sqrt{\rho}|DA\rangle - \sqrt{1-\rho}|D^+A^-\rangle
\end{aligned}
\tag{1}
$$

where the ionicity $\rho = (1 - z/\sqrt{z^2 + 2t^2})/2$ measures the weight of $|D^+A^-\rangle$ in the ground state (gs), and is therefore a measure of the molecular polarity. Following Mulliken [3], we recognize that in the adopted basis the dipole moment operator is dominated by μ_0, the dipole moment relevant to $|D^+A^-\rangle$, so that all quantities of interest for spectroscopy can be derived as follows [4]:

$$\mu_G = \langle g|\hat{\mu}|g\rangle = \mu_0 \rho \tag{2}$$

$$\mu_E = \langle e|\hat{\mu}|e\rangle = \mu_0(1-\rho) \tag{3}$$

$$\mu_{CT} = \langle g|\hat{\mu}|e\rangle = \mu_0\sqrt{\rho(1-\rho)} \tag{4}$$

$$\hbar\omega_{CT} = \mathcal{E}_e - \mathcal{E}_g = \sqrt{2}t/\sqrt{\rho(1-\rho)} \tag{5}$$

On this basis, closed expressions for static NLO responses were written, that proved particularly useful since they relate linear and non-linear optical susceptibilities to spectroscopic observables [4].

Optical spectra of push-pull chromophores in solution can be described based on the same model, provided that it is extended to account for the coupling of electrons to slow degrees of freedom, including molecular vibrations and orientational degrees of freedom of polar solvents [5]. Both couplings are important in the definition of optical and static responses and can be described in the framework of Holstein coupling. Here we describe a model for clusters of interacting push-pull chromophores without accounting for slow degrees of freedom. As a matter of fact, in the adiabatic limit, vibrational coupling can be easily accounted for via a renormalization of the model parameters [6]. More generally, the non-adiabatic calculations needed to account for vibrational coupling in molecular clusters are memory and time-consuming, and so far have only been carried out on dimers of push-pull chromophores [7].

The Hamiltonian for a cluster of push-pull chromophores only interacting through electrostatic forces is [8]:

$$H = \sum_i \left(2z\hat{\rho}_i - \sqrt{2}t\hat{\sigma}_{x,i}\right) + \sum_{i,j>i} V_{ij}\hat{\rho}_i\hat{\rho}_j \tag{6}$$

where i and j run on the N molecular sites. The first term above corresponds to the two-state Mulliken model for the isolated chromophore, and the last term accounts for electrostatic inter-chromophore interactions. We have defined $\hat{\rho}_i = (1-\hat{\sigma}_{z,i})/2$ as the operator measuring the amount of CT from D to A in the i-th molecules, and $\hat{\sigma}_{z/x,i}$ as the z/x Pauli matrices defined on the two basis states for the i-th chromophores. Since the expectation value of $\hat{\rho}_i$ fully defines the charge distribution on the i-th chromophore, the operator representing the electrostatic interaction between i and j chromophores is proportional to $\hat{\rho}_i\hat{\rho}_j$ through a proportionality constant, V_{ij}, measuring the interaction energy between the two fully zwitterionic (D^+A^-) molecules. Of course several models are possible for V_{ij}, including dipolar or multipolar approximations, possibly accounting for electrostatic screening.

The Hamiltonian in Eq. 6 is general, here we apply it to describe two one-dimensional (1D) arrays of molecules as sketched in Fig. 1. Moreover, to estimate V_{ij}, we describe the zwitterionic

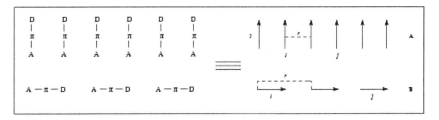

Figure 1: The two one-dimensional array of chromophores discussed in this paper.

molecule as a rigid rod of length l with the positive and negative charges located at the two ends.

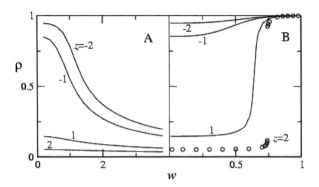

Figure 2: The polarity of chromophores calculated as a function of w for two 1D arrays of 16 chromophores with $v = 2$ and the z marked in the figure. Left panel: repulsive interactions (A geometry); right panel: attractive interactions (B geometry).

This defines the basic unit of electrostatic energy: $v = e^2/l$. In the following we discuss unscreened interactions and introduce the dimensionless inverse intermolecular distance, $w = l/r$, where r is the distance between the chromophores. For B-clusters the condition $w < 1$ applies. For any cluster V_{ij} interactions can be defined in terms of v and w parameters [8].

Intermolecular electrostatic interactions affect the molecular polarity. A molecule inside a cluster feels the electric field generated by the surrounding molecules. In clusters with repulsive interchromphore interactions (A in Fig. 1) each molecule feels an environmental electric field that opposes the charge separation on each molecule, and the average polarity ρ on the molecular sites decreases. Just the opposite occurs for lattices with attractive interactions (B in Fig. 1), where one expects an increase of ρ with intermolecular interactions. This mean-field (mf) description catches most of the gs physics of molecular aggregates, and, specifically, allows one to appreciate the cooperative self-consistent interaction between each molecule and its environment [8].

Fig. 2 shows the variation of the molecular polarity with the inverse intermolecular distance (w) for A and B clusters with different z. For the repulsive A lattice the most interesting results are obtained for negative z for which the isolated molecules at $w = 0$ are in a (zwitter)ionic (I) gs, with $\rho > 0.5$. In that case in fact simply putting the molecules together at a short enough distance turns all the molecule to a neutral (N) gs with $\rho < 0.5$. N and I chromophores have qualitatively different properties and behavior, so intermolecular interactions can profoundly alter the material properties. Of course, for the attractive B-lattice the opposite occurs, and the most interesting case is that of chromophores with positive z. In that case in fact the N isolated molecule can turn to I in the lattice.

Attractive lattices are even more interesting since for large enough v the transition from a N to an I gs becomes discontinuous for suitable z values, as shown in the B-panel of Fig. 2 for $z = 2$. The observation of a discontinuous crossover is the extreme manifestation of cooperative behavior: two large competing interactions, z favoring the N gs, and the intermolecular interactions, favoring the I gs, lead to competing ground states whose energies cross at some special point in the parameter space giving rise to a discontinuous behavior.

The properties of the material in the proximity of the discontinuous N-I crossover are very interesting and unusual [9]. Focusing on the optical excitation with the largest oscillator strength (that always coincides with the lowest excitation in B-lattices) we calculate the number of $|D^+A^-\rangle$ molecules that are created upon photoexcitation, or, equivalently, the number of electrons that

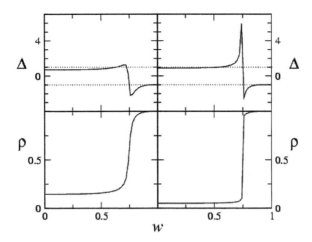

Figure 3: Upper panels: number of zwitterionic species created upon photoexcitation (Δ) vs w. Dotted lines mark the boundary of the $|\Delta| < 1$ region allowed according to the exciton model. Lower panels: $\rho(w)$ curves for the same parameters as for the upper panels. Results are obtained for a 16-site B cluster. Left panels: $v = 1$, $z = 1$; right panels: $v = 2$, $z = 2$

are transferred from D to A: $\Delta = N \left(\langle E_1|\hat{\rho}|E_1 \rangle - \langle G|\hat{\rho}|G \rangle \right)$ (negative Δ means that electrons are transferred from A^- to D^+). In the familiar excitonic approximation [10] the absorption of a photon creates a single excitation, switching a molecule from the local ground to the local excited state: in this approximation $\Delta = 1 - 2\rho$, and the number of transferred electrons upon photoexcitation ranges from 1 for largely N lattices ($\rho \to 0$) to -1 for largely I lattices ($\rho \to 1$). The upper panels in Fig. 3 show the evolution with w of Δ, calculated for a 16-site B lattice near to continuous and a discontinuous neutral to zwitterionic interface (left and right panels, respectively, cf. bottom panels, where the relevant $\rho(w)$ curves are shown). The dotted lines mark the extreme limits of the excitonic approximation for Δ, i.e. $|\Delta| < 1$. The simple excitonic result is spoiled near the charge crossover: deviations are minor near a continuous interface (left panels), but become important near a discontinuous interface: for the parameters in Fig. 3, up to 6 electrons are transferred at a time upon absorption of a single photon.

A detailed analysis of the states involved in absorption leads to a very interesting picture [9]. Upon absorption of a single photon in fact several I chromophores are created on a background of N molecules or vice versa, and these molecules with *reversed* ionicity cluster together forming a droplet of I (N) molecules on a N (I) background. Multi-electron transfer has already been discussed in different contexts, but usually describes a cascading effect related to the relaxation of some slow degree of freedom following a more traditional optical excitation [11, 12]. Here instead multi-electron transfer represents the *primary photoexcitation event*: the absorption of a single photon directly drives the concerted motion of several electrons residing in several nearby molecules.

Important effects from supramolecular interactions are also expected in NLO responses [13]. Fig. 2 shows very clearly that the polarity ρ of a chromophore is strongly dependent on its environment, and, specifically, on the geometry of the cluster, and on intermolecular distances. It is known that linear and non-linear polarizabilities of isolated push-pull chromophores are fixed by ρ [4]. So, when considering clusters of push-pull chromophores one expects that their susceptibilities

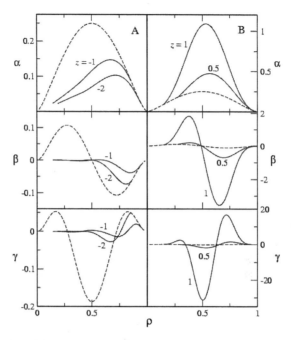

Figure 4: Static polarizability and hyperpolarizabilities per chromophore calculated for clusters of 16 push-pull chromophores with repulsive (A) and attractive (B) interactions and $v = e^2/l = 2$. Results are shown for the selected z marked in the figure and are obtained by varying w as to obtain the ρ values reported in the x-axis. In all panels dashed lines show the (hyper)polarizabilities relevant to the isolated push-pull chromophore: these curves are of course equal in the left and right panels, their different appearance is due to the largely different scales needed to show in the same panel results for the isolated molecule and for the cluster.

vary with the cluster geometry and intermolecular distance, due to the variation of ρ. Indeed the situation is much more complex, with important cooperative and collective contributions to the cluster susceptibility. This is clearly demonstrated by data in Fig. 4 where the linear polarizability α and the first and second hyper-polarizabilities, β and γ, respectively, are shown as a function of ρ. Results are obtained for A (repulsive interactions) and B (attractive interactions) with $v = e^2/l = 2$ and selected z values, while varying w as to span a large interval of ρ (cf. Fig. 2). The dashed lines in the same figure show the linear and non-linear susceptibilities of the isolated chromophore, that only depend on ρ [4]. The deviations of the cluster susceptibilities from the result relevant to the isolated chromophore with the same polarity measure cooperative and collective effects that derive from local-field corrections, as well as from the excitonic and ultraexcitonic mixing of excited states. The role of the different interactions has been recently discussed [1], here we underline that cooperative and collective contributions to the susceptibilities are very large and rapidly increase with the order of non-linearity. In A geometry the cluster response is largely suppressed with respect to the molecular response, whereas in B geometry a very large amplification is observed. A well known result for the isolated chromophore is the symmetry of the responses around $\rho = 0.5$, the so-called cyanine limit. This result is spoiled by electrostatic interactions in the cluster.

Materials based on push-pull chromophores have negligible intermolecular overlap, and here we present a model where electrons are fully localized in the molecular units. In spite of that, electrostatic intermolecular interactions make ET in clusters qualitatively different from ET in the isolated chromophore. In the quest for materials with optimized properties it is then extremely important to properly understand and exploit *supramolecular* structure-properties relationships. This is a challenging task, due to the need to account for complex cooperative and collective phenomena, but it may also be a largely rewarding job as it offers an additional handle to tune, and possibly amplify, the material response.

3 Extended systems and electrostatic interactions: charge-transfer salts with a mixed stack motif

Charge-transfer (CT) crystals have mixed face-to-face stacks of planar π-electron donors and acceptors as sketched in Fig. 5. Intermolecular overlap is negligible between stacks, but not within stacks where $\pi - \pi$ overlap is indicated by less than van der Waals separation between D and A [14]. The gs consequently has fractional charges ρ at D and $-\rho$ at A sites [14]. Just as an example, in the prototypical material, TTF-CA, at ambient conditions about 0.2 electrons are transferred on average from the donor (tetrathiafulvalene, TTF) to the acceptor (chloranil, CA) [2]. We notice that while referring to different physical systems with respect to those described in the previous Section, the parameter ρ corresponds to the same physical quantity. Specifically in both systems ρ measures the fractional charge on D/A sites: in push-pull chromophores the sites corresponds to different chemical groups in the same molecule, in CT salts to different molecular units. In both cases the fractional ionicity is a direct consequence of ET.

As seen in Fig. 5, a regular stack of centrosymmetric molecules is not polar because there is an inversion center at each site, as in fact occurs in the actual structures. The inversion center is lost on dimerization and the gs becomes ferroelectric if dimerization is in the same sense everywhere. Much as it occurs in attractive lattices in Section 2, Madelung interactions favor charge separation,

mixed regular stack

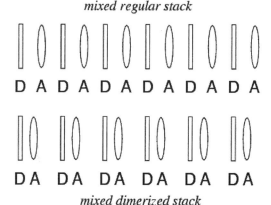

mixed dimerized stack

Figure 5: A schematic view of mixed regular and dimerized stacks.

and a large variation of ρ can be induced by tuning intermolecular distances. At ~ 81 K TTF-CA undergoes a discontinuous phase transition to an I phase with $\rho \sim 0.6$ [2]. Other systems with N-I transitions are known [2]: both continuous and discontinuous transitions have been observed, and in all cases, stack dimerization accompanies the charge crossover. N-I transitions can be induced by temperature, pressure or by absorption of light and represent a complex and interesting phenomenon [15].

Mixed stack CT salts share some physics with B-lattices of push-pull chromophores discussed in the previous Section. Indeed one of the first models for the discontinuous N-I transition in CT salts [16] was based on the description of the stack as a collection of DA pairs only interacting via electrostatic interactions leading exactly to the same Hamiltonian as discussed in the previous Section. More realistic models for the stack must account for delocalized electrons in 1D [17]. Moreover, in order to properly describe structural instabilities, models for CT salts must also include the coupling between electrons and lattice phonons [18, 19]. A minimal model for the gs of CT salts with a mixed stack motif is represented by the Hubbard-Holstein Hamiltonian [18, 19]:

$$H = -\sum_{i,\sigma}[1 + (-1)^i\delta](c_{i,\sigma}^\dagger c_{i+1,\sigma} + H.c.) + \Gamma\sum_{i,\sigma}(-1)^i c_{i,\sigma}^\dagger c_{i,\sigma} + V\sum_i \rho_i\rho_{i+1} + \frac{N}{2\epsilon_d}\delta^2 \quad (7)$$

where $c_{i,\sigma}^\dagger$ creates an electron with spin σ on the i-th site and the number operator, $\hat{n}_{i,\sigma} = c_{i,\sigma}^\dagger c_{i,\sigma}$, is restricted to 0 or 1 on A sites (i even) and to 1 or 2 on D sites (i odd). The first term is the Hückel model for electron transfer between neighbors in the stack; we take $t = -\langle DA|H|D^+A^-\rangle = 1$ as the energy unit, and $t_i = [1 - \delta(-1)^i]$ for dimerized stacks. The second term has site energies $-\Gamma$ at D and $+\Gamma$ at A. The third term is the nearest neighbor Coulomb attraction V that in mf adds $-2V\rho$ to Γ; the charge operator ρ_i is $2 - n_i$ at D sites and $-n_i$ at A sites. Since the full electrostatic (Madelung) energy of the crystal leads to similar modification of Γ in mf theory, at this level V represents any Coulomb or vibronic interaction that modifies sites energies. The last term in the above Hamiltonian measures the bare elastic energy associated with the dimerization mode, with $1/\epsilon_d$ measuring the lattice stiffness. The rigid lattice has $\epsilon_d = 0$.

For $\Gamma >> V$ the Hamiltonian in Eq. 7 describes an almost N lattice ($\rho \to 0$) of donors and acceptors, whereas for $\Gamma << V$ an almost I lattice of spin 1/2 radical ions is obtained ($\rho \to 1$) [17]. The N-I crossover is continuous for small V, but becomes discontinuous at large V [20]. The N and I phases are qualitatively different, with the I lattice being unconditionally unstable to dimerization (spin-Peierls transitions, [21]). The N-I crossover can consequently be identified precisely even for continuous ρ in the rigid regular stack [17]. The possible occurrence of two instabilities leads to a complex phase diagram: soft lattices dimerize in the N phase before reaching the N-I crossover. Harder lattices with sizeable V instead undergo a discontinuous transition from a N regular to an I dimerized phase. Both kinds of transitions have been observed [2].

Fig. 6 summarizes results obtained on a system with large enough V and small enough ϵ_d as to undergo to a discontinuous charge crossover, as evidenced by the $\rho(\Gamma)$ curve in the rightmost panel. The middle panel in the same figure shows the dimerization amplitude: the stack is regular in the N regime and dimerizes in the I phase mimicking TTF-CA. The discontinuous N-I crossover in Fig. 6 closely resembles the discontinuous crossover described in the previous Section for the linear cluster of push-pull chromophores, and, as already noticed, the two systems indeed share some physics. However the physics of the N-I transition in CT salts is much more complex than in push-pull chromophores since electrons are now truly delocalized in 1D, and a lattice instability accompanies the charge instability. In spite of that the intriguing phenomenon of multi-electron transfer survives in CT salts, as shown in the rightmost panel of Fig. 6, where the number of D^+A^- pairs created on photoexcitation is reported.

For the parameters chosen for Fig. 6, that apply to TTF-CA [19], we find that at the N-I interface the absorption of a single photon moves as many as 4 electrons at a time. The analysis of the relevant wavefunction demonstrates that, much as it occurs for clusters of push-pull chromophores, photoexcitation generates droplets of I states in an otherwise N background (or

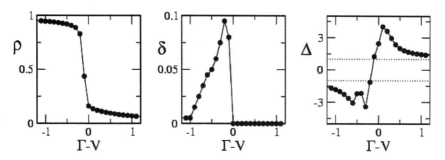

Figure 6: The N-I transition for a CT salt described by the Hamiltonian in Eq. 7 with $V = 3$, and $\epsilon_d = 0.28$. Results are shown for a 12 site chain. Left and middle panels show the evolution with Γ of the ionicity and of the dimerization amplitude, respectively. The right panel shows the number of D^+A^- pairs created upon photoexcitation (result refer to the optical transition with the highest intensity. The two dotted lines in this panel at $\Delta = \pm 1$ mark the limits for Δ in the excitonic approximation.

vice versa). These droplets, created upon vertical phoexcitation, can act as nucleation centers for the photoinduced NIT and can explain the high quantum yield of the process close to the thermally-induced transition [12].

Not only optical spectra, but also gs properties, including linear and non-linear polarizabilities show unusual behavior at the charge crossover. Here we discuss the linear polarizability, that, being related to the static dielectric constant, has an important experimental counterpart. The static susceptibility is the first derivative of the polarization on the applied electric field, $\alpha = dP/dF$, and can be calculated in finite-field approaches from the $P(F)$ dependence, or, in sum-over states approaches as the sum over all excited states of the squared transition dipole moments over the relevant excitation frequencies [22].

The calculation of $P(F)$ and hence of α is straightforward in principle in systems with localized electrons, like the clusters of push-pull chromophores described in Section 2. In that case in fact the linear chain can be naturally partitioned in finite non-overlapping unit cells. The dipole moment operator $\hat{M}$ is then defined as the sum of dipole moments on each cell and enters the F-dependent Hamiltonian as $-\hat{M}F$, allowing the calculation of F-dependent properties. Numerical F-derivatives of P or its perturbative expansion in sum over state approaches are both viable and lead to equivalent results for α. The situation is more complex in extended systems where delocalized electrons make the partitioning of the system in unit cells non-unique. The Berry-phase definition of P for extended systems, or, equivalently for finite systems with periodic boundary conditions (PBC), solves the problem of the calculation of $P(F = 0)$ [23], but leaves open the calculation of F-dependent properties.

In fact, the lack of an explicit definition for the the electric dipole moment operator in PBC systems, hinders both the definition of the F-dependent Hamiltonian $H(F)$ and the calculation of transition dipole moments entering sum over states expressions for the susceptibilities. We have recently solved this problem by introducing an induced dipole operator [19]:

$$\Delta\hat{M} = \frac{N}{2\pi} Im \frac{exp(2\pi i \hat{M}/N)}{Z(F)} \tag{8}$$

where $\hat{M} = \sum_i r_i \rho_i$, is the traditional dipole moment defined for the open chain and $N/2$ is the

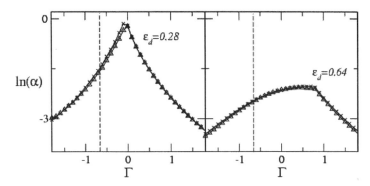

Figure 7: The logarithm of the linear electronic polarizability vs Γ for a mixed stack with $V = 0$ and for the ϵ_d values in the figure. Triangle and crosses refer to $N=12$ and 14, respectively. The dashed line marks in each panel the location of the N-I transition in the rigid ($\epsilon_d = 0$) chain.

number of unit cells. Moreover:

$$Z(F) = \langle G(F)|exp(\frac{2\pi i \hat{M}}{N})|G(F)\rangle \qquad (9)$$

where $|G(F)\rangle$ is the gs of $H(F)$. $\Delta\hat{M}$ enters the definition of an effective Hamiltonian $H(F) = H - F\Delta\hat{M}$, whose diagonalization is equivalent to the minimization of the energy functional $E(F, |G(F)\rangle) = \langle G(F)|H|G(G)\rangle - NFP(F)$, where H is the unperturbed Hamiltonian. The procedure is formally exact when working on a real space basis where $\hat{M}$ is diagonal. Moreover, except for the linear polarizability, which requires $Z(0)$ in Eq. 8, $|G(F)\rangle$ must be found iteratively [19].

Along these lines we have demonstrated that the linear polarizability calculated for the rigid chain ($\epsilon_d = 0$ in Eq. 7) diverges at the continuous NIT of the regular chain with $V = 0$. The dashed line in Fig. 7 locates the electronic transition of the rigid lattice, where the gs is metallic [17]. The divergence is however suppressed by dimerization in soft lattices with $\epsilon_d > 0$ as shown in Fig. 7, where the kink in α marks the Peierls transition.

So far we have discussed the linear electronic polarizability calculated as the first derivative of P on the electric field while keeping the nuclei at their equilibrium location. This electronic polarizability does not exhaust the material polarizability. In fact due to the coupling between electronic and lattice degrees of freedom we expect a vibrational contribution to α, with [18, 19]:

$$\alpha = \frac{\partial P}{\partial F} + \frac{\partial P}{\partial \delta}\frac{\partial \delta}{\partial F} = \alpha_{el} + \alpha_{vib} \qquad (10)$$

The vibrational contribution to the polarizability is governed by the lattice vibration corresponding to the dimerization motion, as described by δ. Specifically α_{vib} goes with the IR intensity of the dimerization mode divided by its frequency. Close to the charge-instability, δ oscillations induce large fluxes of electronic charge from D to A sites and vice versa leading to gigantic IR intensity of the Peierls mode near to the N-I interface [18]. Moreover, the Peierls mode drives the dimerization instability, and its frequency softens in the proximity of the lattice instability. As a consequence, the vibrational contribution to the polarizability is very large and actually dominates over the electronic contribution for systems close to the charge and structural instability [18, 19].

Both α_{el} and α_{vib} contribute to the static dielectric constant, according to:

$$\kappa = \kappa_\infty + \frac{\alpha_{el} + \alpha_{vib}}{\epsilon_0 v_0} \qquad (11)$$

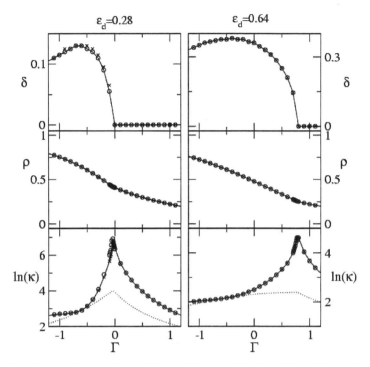

Figure 8: The dimerization amplitude, δ, the ionicity ρ and the logarithm of the static dielectric constant, κ, as a function of Γ calculated for the Hamiltonian in Eq. 7 with $V = 0$ and two different ϵ_d. Circles refer to N = 14, crosses to N = 16. Dotted lines in the bottom panels show the logarithm of the dielectric constant obtained by neglecting the vibrational contribution to α.

where SI units are adopted and ϵ_0 is the vacuum permittivity constant, $\kappa_\infty \sim 3$ is the usual contribution from molecular excited states that are not being modeled, and v_0 is the volume per site. Adopting typical lattice parameters for TTF-CA we obtain the κ values reported in a logarithmic scale in the panels of Fig. 8. A sharp peak in κ is observed with absolute values of the order of 100-1000, in agreement with experimental data [15]. Specifically, the figure shows results obtained for $V = 0$ and two different ϵ_d values. The harder lattice in the left panels dimerizes at $\rho \sim 0.4$, in the near proximity of the charge instability: the coupling between electronic and vibrational degrees of freedom is very large here and the resulting κ is very large. The softer lattice dimerizes far in the neutral regime, at $\rho \sim 0.25$, where charge fluctuations induced by δ are smaller: the calculated peak in the dielectric constant is reduced by almost one order of magnitude with respect to the previous case. In both cases however, κ is dominated by the vibrational contribution: the result obtained by neglecting α_{vib} in Eq. 11 is shown as dotted lines in Fig. 11 and represents a negligible fraction of the dielectric constant.

Results for systems with finite and large V as to describe discontinuous N-I transitions have already been discussed [19]. Here we just notice that $V = 0$ results are enough to describe the behavior of a system with finite V in the mf approximation. Within mf in fact V enters the Hamiltonian via a renormalization of $\Gamma \rightarrow \Gamma - 2V\rho$. For $V > (d\rho/d\Gamma)^{-1}/2 \sim 2.5$ the N-I transition becomes discontinuous and regions of intermediate ρ becomes unaccessible to the system. This leads to a reduction of the κ peak that, for $\epsilon_d = 0.28$ reduces from ~ 1000 at $V = 0$ to ~ 150 at $V = 3$.

4 Conclusions

In this paper we have described the behavior of two interesting classes of materials where ET and electrostatic interactions play an important role. We have focused attention on collective and cooperative behavior appearing as a consequence of the interplay between ET processes occurring at different location in the material as due to electrostatic interactions. Important collective and cooperative phenomena show up clearly in static susceptibilities of clusters of push-pull chromophores. Clamping together local-field corrections, excitonic and ultraexcitonic contributions, classical electrostatic intermolecular interactions lead to a suppression of the responses in the repulsive A-lattice and to an amplification of the responses in the attractive B-lattice, with effects that increase fast with the order of non-linearity. In CT salts it is more difficult to single out the role of electrostatic interactions, due to the complex interplay among delocalization effects, electrostatic 3D interactions and lattice phonons. The electronic contribution to the static linear polarizability in CT salts is enormously amplified by delocalization: in fact, even in the absence of electrostatic V interactions, the polarizability diverges at the continuous NIT of the rigid lattice. However in soft lattices electron-phonon coupling drives the lattice dimerization then reducing the electronic delocalization and strongly suppressing the electronic contribution to the polarizability. At the same time, a vibrational contribution adds to the polarizability, becoming largely dominant over the electronic contribution in the proximity of the dimerization phase transition.

Both in systems with electrons localized within discrete molecular units and in materials with delocalized electrons in 1D, electrostatic interactions can lead to the occurrence of a discontinuous charge crossover, and hence to the appearance of bistable behavior. In the proximity of discontinuous crossovers cooperative and collective phenomena dominate and multielectron transfer represents the most striking demonstration of that. The possibility to induce a concerted motion of several electrons upon absorption of a single photon contrasts sharply with the common excitonic description of optical excitations in molecular materials and opens interesting and new perspectives for the understanding of photoinduced phase transitions and for applications in photoconversion devices.

Acknowledgment

Work in Parma is supported by Italian MIUR through FIRB-RBNE01P4JF and PRIN2004033197-002.

References

[1] A. Painelli and F. Terenziani: in *Nonlinear optical properties of matter: From molecules to condensed phases*, M. G. Papadopoulos, J. Leszczynski and A. J. Sadlej Eds., Kluwer, in press; and references therein.

[2] A. Girlando, A. Painelli, S. A. Bewick, Z. G. Soos, *Synth. Metals* **141** 129 (2004); and references therein.

[3] R. S. Mulliken, J. Am. Chem. Soc. **74** 811 (1952); R. S. Mulliken and W. B. Person, *Molecular Complexes: A Lecture and Reprint Volume*, Wiley, New York, (1969).

[4] see e.g. A. Painelli, *Chem. Phys. Lett.* **285**, 352 (1998); and references therein.

[5] A. Painelli and F. Terenziani *Chem. Phys. Lett.* **312** 211 (1999); A. Painelli and F. Terenziani *J. Phys. Chem. A* **104** 11049 (2000); B. Boldrini, E. Cavalli, A. Painelli, F. Terenziani, *J. Phys. Chem. A* **106** 6286 (2002).

[6] F. Terenziani, A. Painelli, A. Girlando, R. M. Metzger, *J. Phys. Chem. B* **108** 10743 (2004).

[7] F. Terenziani and A. Painelli, *J. Lumin.* **112** 474 (2005); A. Painelli, F. Terenziani L. Angiolini, T. Benelli and L. Giorgini *Chem. Eur. J.* in press.

[8] F. Terenziani, A. Painelli, *Phys. Rev. B* **68**, 165405 (2003).

[9] A. Painelli, F. Terenziani, *J. Am. Chem. Soc.* **125** 5624 (2003).

[10] J. Knoester, in *Organic Nanostructures: Science and Applications*, Proceedings of the International School of Physics *Enrico Fermi*, CXLIX Course, M. Agranovich and G.C. La Rocca Eds, IOS Press, Amsterdam, (2002), and references therein.

[11] H. Tributsch, L. Pohlmann *Science* **279** 1891 (1998).

[12] S. Iwai, S. Tanaka, K. Fujinuma, H. Kishida, H. Okamoto, Y. Tokura *Phys. Rev. Lett.* **88** 057402 (2002).

[13] B. Champagne, D. M. Bishop *Adv. Chem. Phys.* **126** 41 (2003).

[14] Z.G. Soos and D.J. Klein, in: N.B. Hannay (Ed.) *Treatise on Solid-State Chemistry*, Vol. III, Plenum Press, New York (1976), p. 679.

[15] S. Horiuchi, Y. Okimoto, R. Kumai, and Y. Tokura, *J. Am. Chem. Soc.* **123** 665 (2001); S. Horiuchi, Y. Okimoto, R. Kumai and Y. Tokura, *Science* **299** 229 (2003).

[16] Z.G. Soos, H.J. Keller, W. Moroni, and D. Nothe, *Ann. N.Y. Acad. Sci.* **313** 442 (1978).

[17] Y. Anusooya-Pati, Z.G. Soos and A. Painelli, *Phys. Rev. B* **63**, 205118 (2001).

[18] L. Del Freo, A. Painelli and Z.G. Soos, *Phys. Rev. Lett.* **89**, 27402 (2002).

[19] Z.G. Soos, S.A. Bewick, A. Painelli and A. Peri, *J. Chem. Phys.* **120**, 6712 (2004).

[20] A. Girlando and A. Painelli, *Phys. Rev.B* **34**, 2131 (1986); A. Painelli and A. Girlando, *Phys. Rev.B* **37**, 5748 (1988).

[21] J.W. Bray, L.V. Interrante, I.S. Jacobs and J.C. Bonner, in *Extended Linear Chain Compounds*, J.S. Miller, ed. (Plenum, New York, 1983), Vol. 3, p. 353.

[22] D. M. Bishop *Adv. Chem. Phys.* **104** 1 (1998); and references therein.

[23] R. Resta, *J. Phys.: Condens. Matter* **14** R625 (2002).

Brill Academic Publishers
P.O. Box 9000, 2300 PA Leiden
The Netherlands

*Lecture Series on Computer
and Computational Sciences*
Volume 3, 2005, pp. 142-151

Density functional response approach for electric properties of molecules

Sourav Pal[1] and K. B. Sophy

Physical Chemistry Division,
National Chemical Laboratory,
Pune 411 008, India

Received 18 August, 2005; accepted in revised form 18 August, 2005

Abstract: We review in this paper an implementation of the response approach to the Kohn-Sham (KS) density functional theory (DFT) for obtaining the linear and non-linear electric response properties of molecules using Gaussian type orbital basis centered on atoms. We have made a formulation in which the response of the electron density through the solution of the coupled perturbed Kohn-Sham (CPKS) equations has to be obtained only once, instead of iteratively as in the case of completely analytic procedure. Our method is based on a numerical finite-field solution of derivative KS operator, followed by analytic solution of CPKS equation. Further, using the response of the electron density, the dipole moment, polarizability and first-hyperpolarizability of the molecules are evaluated. The method is particularly useful for large systems. We tested our method using HF, BH, H_2O and CO as test molecules, for which, high quality *ab initio* results are available. Further, our study of possible incorporation of non-dynamical electron correlation by studying BH and HF at several internuclear distances is discussed.

Keywords: Density functional theory, Response approach, Molecular properties, polarizability

Mathematics SubjectClassification: Here must be added the AMS-MOS or PACS

PACS: 31.15

1. Introduction

With the advancement in the field of lasers and experimental techniques in the past few years, it has become easy to study the experimental phenomenon namely, the Kerr effect, electro-optical effect, dc-induced second (dc-SHG) harmonic generation, etc. which are of interest to the experimentalists for understanding the non-linear optical (NLO) properties of molecules, and subsequently to identify NLO materials of technological importance. To a theoretical chemist this means computing the response electric properties of a molecule, namely, the dipole moment, polarizabilities and first-hyperpolarizabilities, etc. These properties can be used to understand the response of the atom or molecule to the electric field as external perturbation. Electric response properties [1-8] have been extensively studied using different *ab initio* theories [8-10] as well as density functional theory (DFT) for both atoms and molecules. These properties can be studied using the response theory using different theoretical methods that are available at our disposal. However, these properties are largely affected by, i) the basis set size, and ii) the electron correlation effects. Larger basis sets are required to account for the long-range effects of the response properties, at the outer regions of the molecule. Additionally, the response properties of higher order are sensitive to the inclusion of correlation effects, which leads to improvement in their values.
Density functional theory (DFT) [11,12] includes correlation effects and is applicable to large molecules due to its simple working equations as compared to the *ab initio* theories. Calculations of response electric properties for atoms and molecules have been reported using the time-independent as well as the time-dependent DFT (TDDFT) via different approaches [13-19]. While the time-

---

[1] Corresponding author. E-mail: s.pal@ncl.res.in

independent DFT can give the static electric properties for the ground state of a molecule, the TDDFT can give static as well as dynamic (frequency dependent) polarizabilities for the ground and subsequently, the excited states of the molecule. But, these correlated frequency dependent programs are algebraically complicated and expensive, keeping in mind the large basis sets that are required for obtaining these properties. Therefore, for the static electric properties it is sufficient to stick to the time-independent picture of the DFT, which is simple. These properties can be obtained either as the expectation value of the operator (of the property concerned) or as energy derivatives with respect to the electric field. The energy

Although, the domain of the response properties using KS-DFT has been explored in the past, much of the reported calculations have used the finite field approach [9, 17]. Some others use the least squares fit to a polynomial to obtain the derivatives [9]. Attempts have also been made to obtain the response of the density matrix by solving the CPKS equations [13] for atoms as well as molecules. However, only numerical atomic basis sets have been used in these calculations [17, 18]. These basis sets may suffice for atomic calculations, but for molecules one needs to introduce an atomic orbital basis, and this along with the iterative nature of the CPKS equations makes the entire exercise of obtaining the response, complicated and time consuming. We circumvent this problem through our simplified approach [20, 21].

In Section 2, we discuss the DFT, followed by response approach and CPKS procedure to obtain the response of the electron density. Section 3 gives our method in brief and some technical details of our implementation followed by the results and discussion in the Section 4. The concluding remarks are given in Section 5.

2. Theory

DFT is variational in nature and uses electron density as the basic variable. The ground state electron density of the system is obtained by making the energy stationary with respect to the electron density,

$$\frac{\delta E[\rho]}{\delta\rho(r)} = 0 \tag{1}$$

subject to the normalization of the electron density,

$$\int \rho(r)dr = N \tag{2}$$

where, N is the total number of electrons of the system under study.
The KS-DFT [22] simplifies the theory to a non-interacting picture despite including the effects of electron correlation. The basic working equation in DFT is the Euler-Lagrange equation, which is given by,

$$\mu = \upsilon_{eff}(r) + \frac{\delta T_s[\rho]}{\delta\rho(r)} \tag{3}$$

·where, μ is the Lagrange multiplier associated with the constraint in Eq. (2), $T_s[\rho]$ is the kinetic energy functional which contains the major part of the kinetic energy, the rest being put in the exchange-correlation potential contained in the KS potential, $\upsilon_{eff}(r)$, which can be given as,

$$\upsilon_{eff}(r) = \upsilon(r) + \frac{\delta J[\rho]}{\delta\rho(r)} + \frac{\delta E_{XC}[\rho]}{\delta\rho(r)} \tag{4}$$

where, the first term is the external potential acting on the system, the second term is the coulomb potential and the third term is the exchange-correlation potential.

In the KS method, N one-particle equations are solved to obtain the ground state electron density of the system.

$$[-\frac{1}{2}\nabla^2 + \upsilon_{eff}(r)]\Phi_i(r) = E_i\Phi_i(r) \tag{5}$$

For atomic systems, these N one-particle equations can be solved numerically to a reasonable accuracy, but, for molecules we need to introduce a basis. The electron density can then be written as,

$$\rho(r) = \sum_{i=occ}\sum_s |\Phi_i(r,s)|^2 \tag{6}$$

For molecules, the KS orbitals $(\Phi_i(r)'s)$, are written in terms of contracted Gaussians centered on atoms which can in turn be written in terms of the density matrix, $P_{\mu\nu}$ and the atomic orbitals $(\chi_\mu(r)'s)$ as,

$$\rho(r) = \sum_{\mu}^{N}\sum_{\nu}^{N} P_{\mu\nu}\chi_\mu(r)\chi_\nu^*(r) \tag{7}$$

Eq. (5) is transformed into the matrix equation,

$$FC = SCE \tag{8}$$

where, F is the KS operator in the atom-centered basis, C is the coefficient matrix of KS orbitals in the basis, S is the overlap matrix of the basis and E is the energy matrix. The Eqs. (3), (4), (5), (6) and (8) are solved self-consistently and are the KS equations. The variational nature of the DFT facilitates the use of the (2n+1) rule [23] which arises as a result of the Hellmann-Feynman [24, 25] theorem for calculating the higher order energy derivatives using the nth order response of the electron density.

Using the response theory the energy of the molecule can be expanded as a power series in the electric field, λ, if the field is small.

$$E(\lambda) = E(0) + \sum_i \left(\frac{\delta E(\lambda)}{\delta\lambda_i}\right)_0 \lambda_i + \frac{1}{2}\sum_{i,j}\left(\frac{\delta^2 E(\lambda)}{\delta\lambda_i\delta\lambda_j}\right)_0 \lambda_i\lambda_j$$
$$+ \frac{1}{6}\sum_{i,j,k}\left(\frac{\delta^3 E(\lambda)}{\delta\lambda_i\delta\lambda_j\delta\lambda_k}\right)_0 \lambda_i\lambda_j\lambda_k + \cdots\cdots \tag{9}$$

Eq. (9) defines the static dipole moment, polarizability and first-hyperpolarizability and so on. Alternatively, the above static response properties can also be defined by expanding the field-dependent dipole moment as,

$$\mu_i(\lambda) = \mu_i + \sum_j \alpha_{ij}\lambda_j + \frac{1}{2}\sum_{j,k}\beta_{ijk}\lambda_j\lambda_k + \cdots\cdots \tag{10}$$

To obtain the higher derivative of the energy functional with respect to the electric field, we use the variational principle to calculate the response of the electron density. This is obtained by making the variational DFT equations stationary with respect to the electric field λ.

$$\frac{\delta}{\delta\lambda}\left(\frac{\delta E}{\delta\rho(r)}\right) = 0 \tag{11}$$

The above stationary equation can be transcribed in the KS framework using orbital derivatives with respect to the electric field. The first derivative of density can be written in terms of the KS orbital derivatives. On introduction of the atomic orbital basis, the above equation transforms into the CPKS matrix equation and provides us with the first derivative of the coefficients of KS orbital in the AO basis. This, in turn, gives us the first-order derivative of density or the density response. The CPKS equations are the derivative of the KS matrix equation (8) with respect to external perturbation. The CPKS equation can be written as,

$$F'C^0 + FC' = S'C^0E + SC'E + SC^0E' \tag{12}$$

where, the primes denote the first-order derivatives with respect to the electric field.

Further, the use of the (2n+1) rule simplifies the formula for second and third energy derivatives. The second and third energy derivatives with respect to external fields can be written in terms of only the unperturbed density and first order density response, which are obtained by the first derivatives of the coefficients of the KS orbital in the basis. However, unlike the coupled-perturbed Hartree-Fock (CPHF) [26], the KS operator contains the exchange-correlation part, which can be local or non-local, depending on the form of the exchange-correlation functional. In this procedure, the exchange-correlation part of the potential is fitted as sum of Gaussians over the co-ordinate space and the C' of the above Eq. (12) can be expanded as,

$$C' = C^0 U' \tag{13}$$

Using this above definition for C' and pre-multiplying by transpose conjugate of $C^{0\dagger}$, the CPKS equation (Eq. (12)) takes the form,

$$EU' - U'E = C^{0\dagger}S'C^0E - C^{0\dagger}F'C^0 + E' \tag{14}$$

This equation can now be solved in a self-consistent manner to get the U' matrix and hence the derivative of the coefficient matrix, C'.

3. Methodology and technical details

KS-DFT involves single-particle equations and is variational in nature. We can thus avail the facility of the (2n+1) rule for energy derivatives, which comes as a result of the Hellmann-Feynman theorem. We, therefore, require only the first order response of the electron density to obtain up to the third order response property. To obtain this response of the electron density we solve the analytic CPKS equation, for which we require the derivative KS operator matrices. The exchange-correlation term in the KS operator is obtained by fitting it in terms of Gaussian functions, as already available in the DEMON [27] system of programs. Following this approach, the complete analytic approach would involve fitting the derivative of the KS matrix as well. This step is difficult as well as computationally extensive. As a first step toward this analytic approach, we have devised a method, which is a combination of numerical and analytical procedures [20, 21]. In this, fitting the derivative of exchange-correlation part is by-passed by constructing the derivative of the KS matrix using a finite field approximation. The elements of the derivative KS matrix are computed as a difference between the elements of the KS matrices calculated at suitably chosen electric field values around zero field, namely, +0.002 a.u., +0.001 a.u., -0.001 a.u. and -0.002 a.u. through the DEMON system of programs. Using this derivative KS matrix, the derivative of the molecular orbital coefficients in terms of the atomic basis is obtained analytically by solving the CPKS equation just once. The coefficient derivatives lead to the first-order density. We can then obtain up to the third-order property through the knowledge of the first-order density and the unperturbed density. The dipole moment, which is a first-order property, can be simply calculated through the ground state density. The CPKS code for obtaining the response of the density was written and integrated with the DEMON code. The second-order and third-order properties were obtained using the density response, through complete analytic expressions.

We review the electrical response properties for the hydrogen fluoride molecule at equilibrium as well as stretched geometries [20]. We will also discuss the calculation of properties presented for BH, CO and H_2O in our second paper [21], for which high quality *ab initio* results were available for comparison. The selection of molecules was a test for the method. In addition, we have used the method to examine the properties of BH at different inter-nuclear distances, where non-dynamic electron correlation will be important. The results for hydrogen fluoride molecule for $0.75R_e$, R_e, $1.5R_e$ and $2.0R_e$, are presented in tables 1 through 4 (taken from reference [20]), the results for boron mono-hydride molecule for $.75R_e$, R_e, $1.5R_e$ and $2.0R_e$, are presented in table 5 (taken from reference [21]), the calculations for carbon monoxide and water at R_e are given in table 6 and table 7 respectively (taken from reference [21]). The details of each calculation are given in the respective table. In addition, we have employed auxiliary basis (5,1; 5,1) for hydrogen and (5,2; 5,2) for boron, for fitting the charge density and the exchange-correlation part for the BH molecule. Similarly CO represents an interesting system, where electron correlation plays an important role in determining the sign of the dipole moment. In this case, however, the molecule has been studied only at the equilibrium geometry. An auxiliary basis (4,4; 4,4) has been employed for DFT charge density and exchange correlation fit throughout our calculation for both carbon as well as oxygen in CO. For water, when using the first geometry we have used auxiliary bases (5,1; 5,1) and (5,3; 5,3) for the hydrogen and the oxygen atoms respectively. For the second set of calculations on water molecule we used, (3,1; 3,1) and (4,4; 4,4) auxiliary basis for the hydrogen and oxygen atoms respectively. The exchange-correlation functionals used were all non-local in nature throughout our calculations.

4. Discussion of Results

In this section, we review the results of out method for the systems described above. We see that for HF, the results of dipole moment, polarizability and first-hyperpolarizability at 1.0 R_e (Table 1) are generally in agreement with experiment [1] and other theoretical results [2], wherever available though there is a variation in details. Benchmark FCI [2, 4] and coupled-cluster (CC) [2] results have been tabulated for reference. Obviously, calculations in larger basis sets provide results closer to experiments. In general, though, compared to the dipole moments and polarizabilities, the first-hyperpolarizability values are not reproduced reliably. Dipole moment values come out best either in comparison to the experiments or benchmark calculations. There is less degree of variation in dipole moment results with basis set or the nature of functionals used. However, higher order properties, as expected, are much more sensitive to these variations. We see, however, that BPW91 and BP86 functionals produce better results of the properties, specially the higher-order ones. For HF, we have results at other inter-nuclear distances, namely 0.75 R_e, 1.5 R_e and 2.0 R_e, presented in Tables 2 through 4. We see that quality of results using DFT become progressively worse compared to other benchmark theoretical results [2, 4] in each basis set as the inter-nuclear distance deviates from equilibrium distance. In particular, this is true for first-hyperpolarizability. This is, presumably, due to the increase in multi-reference character of the states at distorted geometry, which the functionals in DFT are not able to incorporate correctly. Comparing across different functionals, however, we still find that BPW91 and BP86 functionals produce relatively higher quality results.

Benchmark calculations were available for comparison of the BH response property values (Table 5) at R_e, but, none were available for comparison with the values at 0.75 R_e, 1.5R_e and 2.0 R_e. We therefore carried out calculations for these geometries using extended coupled-cluster (ECC) response method for comparison [28, 29]. In this case, we used only two functionals BPW91 and PW86PW91. At R_e, we observe that the dipole moment values using both the functionals are as good as the benchmark values from different CC approaches and FCI. For polarizabilities as well as the first-hyperpolarizabilities, the second functional PW86PW91 gives better agreement with the benchmark values. In particular, there is dramatic difference in the results of the two functionals. At different distances, the trend of the hyperpolarizability results is not very clear. However, in general, both functionals provide somewhat better results for dipole moments and polarizability In general, however, quality decreases progressively as geometries differ from equilibrium and the order of properties becomes higher. BH is marked by high degree of quasi-degeneracy and this is reflected in poor quality of DFT results of first-hyperpolarizability.

For CO, in Table 6 we have used three different basis, one extensive [8s6p3d] and other VTZP and VTZP+, which includes s and d field-induced polarization (FIP) functions. BPW91 and PW86PW91 functionals have been used. The DFT results are compared with several benchmark CC results and experiments, wherever available. We find the DFT results in all three basis are in very good agreement with CC results. Results using the first basis are very near the experimental values. It may be noted that the correct sign of dipole moment is produced by the DFT results. Both functional produce good quality results. It is interesting to note that even the first-hyperpolarizability value is in good agreement with CCSD available results. We have also compared with finite-field LDA results [9] using VTZP and VTZP+ basis sets and we observe that our results produce correct trend compared to LDA results.

For H_2O, we have reproduced results in Sadlej, VTZP and VTZP+ basis sets presented in Table 6. We have used BP86 and PW91PW91 functionals for the latter two basis. For these latter two basis, LDA finite-field results were available for comparison [9]. For the first basis, we have used additionally PW86P86 functional. The geometry used for this basis is also slightly different. For this basis, benchmark CCSD results were available for comparison. Comparison with experiments, wherever available, has also been made. Once again, we find that results are, in general, good agreement with the experiments. The trend of results, however, is better than the LDA in VTZP and VTZP+ basis. Sadlej basis results show good agreement with CCSD results, even for first-hyperpolarizability. In general, we observe that the first-hyperpolarizability values are stable with respect to the functional and basis sets, compared to the ones of BH molecule (Table 5).

5. Conclusion

We have reviewed a numerical-analytical DFT response procedure recently developed by us and discussed property results for HF, BH, CO and H_2O using this procedure. While we have reproduced results of CO and H_2O only at equilibrium, HF and BH have been reported at different inter-nuclear distances. In case of water molecule, although we have carried out the study at equilibrium geometry, two slightly varying geometries have been used in the calculations. For CO as well as H_2O molecule, we find that the DFT procedure, as developed by us, produces quite good results. Results of the HF and BH molecules show that the DFT numerical-analytical linear response procedure produces reasonably correct values of dipole moment and polarizability even at stretched geometry. However there is instability in the first-hyperpolarizability values of BH, which should be investigated.

Since DFT is, in principle, a theory for non-degenerate ground state of systems, the encouraging results from our method for the equilibrium geometries can be clearly understood and the fact that the theory includes dynamical correlation can be attributed as a reason for the closeness of our results with those from the correlated *ab initio* theories. We have also explored the applicability of our method for stretched geometries and we feel that the results for these cases could be improved, provided the multi-reference character or the non-dynamical correlation effects are accounted for by the theory.

     Finally, we would like to impress upon the point that our numerical-analytical method, which involves just a single-step solution to the CPKS over the iterative solution, has a great computational advantage, when applied to large molecules. We would like to extend the scope of this method to the calculation of magnetic response properties, particularly the paramagnetic and the diamagnetic susceptibility.

Table 1: Coupled-perturbed Kohn-Sham dipole moment, polarizability and first-hyperpolarizability values of HF Molecule[a] at 0.75R$_e$ (in atomic units); R$_e$ =1.7328 a.u (taken from ref. [20])

DFT	μ_z		α_{XX}		α_{ZZ}		β_{ZZZ}	
	DZ	Sadlej	DZ	Sadlej	DZ	Sadlej	DZ	Sadlej
1	0.798	0.554	0.908	4.952	1.989	5.213	6.79	-3.49
2	0.794	0.555	1.017	5.539	1.989	4.994	5.99	3.62
3	0.787	0.564	0.964	5.002	1.934	4.718	5.54	4.32
4	0.795	0.563	1.010	5.286	1.939	4.655	5.62	9.15
RHF	0.803	0.590	0.883	3.982	1.187	3.981	4.31	1.41
CC[b]	0.801	0.567	-	-	1.899	4.547	-4.929	-2.225

* where, 1. PW91-PW91, 2. PW86-P86, 3. B88-PW91, 4. B88-P86
[a]The molecule in z-direction.
[b]From ref. [2].

Table 2 : Coupled-perturbed Kohn-Sham dipole moment, polarizability and first-hyperpolarizability values of HF Molecule[a] at 1.0R$_e$ (in atomic units); R$_e$ =1.7328 a.u (taken from ref. [20])

DFT	μ_z		α_{XX}		α_{ZZ}		β_{ZZZ}	
	DZ	Sadlej	DZ	Sadlej	DZ	Sadlej	DZ	Sadlej
1	0.868	0.673	0.798	5.792	4.280	6.961	16.61	6.09
2	0.860	0.682	0.846	6.317	4.263	7.017	19.12	12.23
3	0.854	0.677	0.805	5.823	4.177	6.719	18.06	10.88
4	0.868	0.685	0.837	6.170	4.165	6.892	18.67	11.09
RHF	0.936	0.757	0.739	4.457	4.002	5.742	17.59	8.14
FCI	0.898[b]	-	-	-	4.14[b]		-17.32[b]	-
CC[b]	0.896	0.699	-	-	4.179	6.521	-17.513	-9.869
Expt.	0.707[d]		5.08[d]		6.40[d]		-	

* where, 1. PW91-PW91, 2. PW86-P86, 3. B88-PW91, 4. B88-P86
[a]The molecule in z-direction.
[b]From ref. [2]
[d]From ref. [1]

Table 3 : Coupled-perturbed Kohn-Sham dipole moment, polarizability and first-hyperpolarizability values of HF Molecule[a] at 1.5R$_e$ (in atomic units); R$_e$ =1.7328 a.u (taken from ref. [20])

DFT	μ_z		α_{XX}		α_{ZZ}		β_{ZZZ}	
	DZ	Sadlej	DZ	Sadlej	DZ	Sadlej	DZ	Sadlej
1	0.971	0.928	0.572	6.971	11.023	13.379	33.13	19.51
2	0.955	0.940	0.587	7.801	11.061	13.895	38.85	39.94
3	0.935	0.908	0.567	7.201	10.983	13.366	39.49	36.56
4	0.973	0.945	0.583	7.779	10.974	13.573	40.27	40.33
RHF	1.198	1.142	0.564	5.559	12.634	13.488	76.81	68.67
FCI	0.923[b]	-	-	-	12.47[b]	-	-2.39[b]	-
CC[b]	0.926	0.912	-	-	12.383	14.348	-1.012	-24.302

* where, 1. PW91-PW91, 2. PW86-P86, 3. B88-PW91, 4. B88-P86
[a]The molecule in z-direction.

[b]From ref. [2].

Table 4 : Coupled-perturbed Kohn-Sham dipole moment, polarizability and first-hyperpolarizability values of HF Molecule[a] at $2.0R_e$ (in atomic units); R_e =1.7328 a.u (taken from ref. [20])

DFT	μ_z		α_{xx}		α_{zz}		β_{zzz}	
	DZ	Sadlej	DZ	Sadlej	DZ	Sadlej	DZ	Sadlej
1	1.081	1.167	0.488	7.977	19.122	23.334	33.48	28.24
2	1.068	1.166	0.499	9.324	18.688	23.308	38.57	39.09
3	1.013	1.084	0.487	8.233	18.785	22.676	39.31	58.15
4	1.092	1.175	0.497	9.107	18.925	23.179	37.77	65.93
RHF	1.510	1.570	0.500	6.838	26.541	28.746	-	-
FCI	0.605[b]	-	-	-	18.42[b]	-	367	-
CC[b]	0.640	0.820	-	-	18.531	26.189	352.68	279.80

* where, 1. PW91-PW91, 2. PW86-P86, 3. B88-PW91, 4. B88-P86
[a]The molecule in z-direction.
[b]From ref. [2].

Table 5 : Coupled-perturbed Kohn-Sham dipole moment, polarizability and first-hyperpolarizability values of BH molecule[a] (in atomic units); R_e=2.329 a.u. (taken from ref. [21])

Bond length		μ_z	$\alpha_{xx=yy}$	α_{zz}	β_{zzz}	$\beta_{xxz=yyz}$
0.75R_e						
	CPHF	0.966	25.25	15.53	71.5	0.2
	FF (BPW91)	0.910	-	17.95	-	-
	FF (PW86PW91)	0.905	-	16.40	-	-
	BPW91	0.890	17.54	17.98	-12.1	11.1
	PW86PW91	0.896	18.02	16.40	-10.0	36.1
	ECC	0.799	-	16.93	-55.8	-
R_e						
	CPHF	0.686	22.76	22.58	-20.9	52.5
	FF (BPW91)	0.638	-	27.69	-	-
	FF (PW86PW91)	0.621	-	24.11	-	-
	BPW91	0.574	16.57	27.79	406.4	689.2
	PW86PW91	0.580	16.49	24.11	-137.2	51.6
	ECC	0.514	-	23.66	-51.9	-
	CCSD (nonrel)[b]	0.5298	20.433	23.044	-29.04	-
	CCSD (rel)[b]	0.5256	20.308	23.145	-27.70	-
	BCCD[b]	0.5257	20.311	23.143	-27.53	-
	CCSD (T)[b]	0.5190	20.298	23.270	-29.35	-
	BCCD (T)[b]	0.5190	20.295	23.270	-29.20	-
	Full-CI[b]	0.5171	20.289	23.306	-29.25	-
	Expt.[b]	0.4997	-	-	-	-
1.5R_e						
	CPHF	-0.244	22.32	47.51	-118.2	19.6
	FF (BPW91)	-0.184	-	42.42	-	-
	FF (PW86PW91)	-0.179	-	43.40	-	-
	BPW91	-0.278	16.30	42.42	-129.8	9.6
	PW86PW91	-0.279	16.73	43.40	-128.5	9.7
	ECC	-0.344	-	62.97	-57.4	-
2.0R_e						
	CPHF	-1.261	26.94	83.34	56.8	-78.7
	FF (BPW91)	-0.826	-	64.13	-	-
	FF (PW86PW91)	-0.816	-	64.23	-	-
	BPW91	-0.934	18.25	64.13	-150.5	-33.3
	PW86PW91	-0.946	18.75	64.24	33.5	-33.5
	ECC	-0.262	-	30.16	-246.6	-

[a]The molecule in the z direction; B and H atom on the negative and the positive side of the

z axis, respectively.
[b]From ref. [10].

Table 6 : Coupled-perturbed Kohn-Sham dipole moment, polarizability and first-hyperpolarizability values of CO molecule[a] at R_e (in atomic units); R_e=2.1323 a.u. (taken from ref. [21])

Basis Set		μ_z	$\alpha_{xx=yy}$	α_{zz}	β_{zzz}	$\beta_{xxz=yyz}$
[8s6p3d]						
	BPW91	0.075	10.21	15.52	34.7	6.0
	PW86PW91	0.073	10.58	15.66	27.5	5.7
	CCSD (nonrel)[b]	0.070	11.85	15.79	29.4	-
	CCSD (rel)[b]	0.040	11.76	15.50	30.3	-
	BCCD[b]	0.038	11.74	15.52	30.6	-
	CCSD(T)[b]	0.061	11.91	15.61	30.2	-
	BCCD(T)[b]	0.058	11.90	15.62	30.4	-
	Expt.[b]	0.043	11.86	15.51	-	-
VTZP						
	CPHF	-0.120	8.79	13.79	23.0	1.05
	BPW91	0.060	7.97	14.69	19.9	3.1
	PW86PW91	0.052	8.18	14.85	20.3	3.3
	LDA[c]	0.074	9.63	14.87	17.6	2.6
VTZP+						
	CPHF	-0.102	10.93	14.36	30.9	3.9
	BPW91	0.074	9.98	15.39	31.6	4.8
	PW86PW91	0.073	10.35	15.50	39.8	6.7
	LDA[c]	0.095	12.32	15.75	31.3	6.6
	Expt.[c]	0.048	12.1	15.7	-	-

[a]The molecule in the z direction; C and O atom on the negative and the positive side of the z axis, respectively.
[b]From ref. [10]
[c]From ref. [9]

Table 7 : Coupled-perturbed Kohn-Sham dipole moment, polarizability and first-hyperpolarizability values of H_2O molecule (in atomic units) (taken from ref. [21])

Basis Set		μ_z	α_{xx}	α_{yy}	α_{zz}	β_{zzz}	β_{xxz}	β_{yyz}
Sadlej[a]								
	BP86	0.712	10.43	10.64	10.35	-10.9	-2.4	-15.8
	PW86P86	0.707	10.72	10.88	10.66	-12.6	-3.7	-17.3
	PW91PW91	0.689	10.53	10.66	10.40	-6.1	-6.3	-10.3
	SCF/TDHF[b]	0.780	7.84	9.17	8.50	-4.6	-0.6	-9.7
	MBPT (2)[b]	0.730	-	10.05	9.75	-8.7	-2.4	-10.6
	CCSD[b]	0.729	-	9.89	9.49	-7.3	-2.0	-10.8
	Expt.[b]	0.721	-	-	-	-	-	-
VTZP[c]								
	CPHF	0.864	7.55	3.85	6.03	-6.6	-0.8	-15.3
	BP86	0.827	7.95	4.34	6.62	-10.0	-2.5	-17.9
	PW91PW91	0.826	7.98	4.33	6.65	-7.3	-2.1	-14.5
	LDA[d]	0.852	7.93	4.02	6.65	-9.8	-2.3	-16.6
VTZP+[c]								
	CPHF	0.793	9.03	7.64	8.28	-6.8	-0.4	-9.8
	BP86	0.726	10.22	9.76	9.99	-20.7	-8.1	-14.6
	PW91PW91	0.722	10.38	10.29	10.11	-6.6	1.7	-12.0
	LDA[d]	0.743	10.32	9.96	10.12	-17.1	-6.3	-12.9
	Expt.[d]	0.727	10.31±0.09	9.55±0.09	9.91±0.02			

[a]The molecule in the positive z-direction; O atom is placed at the origin and the two H atoms are in the yz plane. The equilibrium geometry used was, R_{OH} = 0.957 Å and $\angle\Theta_{HOH}$ = 104.5°.
[b]From ref. [3].
[c]The molecule in the positive z-direction; O atom is placed at the origin and the two H atoms are in the

xz plane. The equilibrium geometry used was, $R_{OH} = 0.9576$ Å and $\angle\Theta_{HOH} = 104.48°$.
^{d}From ref. [9].

Acknowledgements

The authors acknowledge financial support from the Board of Research in Nuclear Sciences, India, toward this work

References

[1] H. Sekino and R. J. Bartlett, Hyperpolarizabilities of the hydrogen fluoride molecule: A discrepancy between theory and experiment?, *Journal of Chemical Physics* 84 2726-2733(1986).

[2] A. E. Kondo, P. Piecuch, and J. Paldus, Orthogonally spin-adapted single-reference coupled-cluster formalism: Linear response calculation of higher-order static properties, *Journal of Chemical Physics* 104, 8566-8585(1996).

[3] H. Sekino and R. J. Bartlett, Molecular hyperpolarizabilities, *Journal of Chemical Physics* 98, 3022-3037 (1993).

[4] K. B. Ghose, P. Piecuch, S. Pal, and L. Adamowicz, State-selective multireference coupled-cluster theory: In pursuit of property calculation, *Journal of Chemical Physics* 104 6582-6589 (1996).

[5] D.M. Bishop, J. Pipin, B. Lam, Field and field-gradient polarizabilities of BeH, BH and CH$^+$, *Chemical Physics Letters* 127 377-380(1986).

[6] V.E. Ingamells, M.G. Papadopoulos, N.C. Handy, A. Willetts, The electronic, vibrational and rotational contributions to the dipole moment, polarizability and first and second hyperpolarizabilities of the BH molecule, *Journal of Chemical Physics* 109,1845-1859 (1998).

[7] E.A. Salter, H. Sekino, R.J. Bartlett, Property evaluation and orbital relaxation in coupled cluster methods, *Journal of Chemical Physics* 87, 502-509 (1987).

[8] R.J. Bartlett and G. D. Purvis III, Molecular hyperpolarizabilities. I. Theoretical calculations including correlation, *Physical Review* A 20 1313-1321(1979).

[9] J. Guan, P. Duffy, J.T. Carter, D.P. Chong, K.C. Casida, M.E. Casida, M. Wrinn, Comparison of local-density and Hartree-Fock calculations of molecular polarizabilities and hyperpolarizabilities, *Journal of Chemical Physics* 98 4753-4765(1993).

[10] R. Kobayashi, H. Koch, P. Jorgenson, and T. J. Lee, Comparison of coupled-cluster and Brueckner coupled-cluster calculations of molecular properties, *Chemical Physics Letters* 211 94-100 (1993).

[11] R.G. Parr, W. Yang, *Density Functional Theory of Atoms and Molecules*, Oxford University Press, Oxford, 1989.

[12] R. M. Dreizler and E. K. U. Gross, *Density Functional Theory*, Springer, Berlin, 1990.

[13] S. M. Colwell, C. W. Murray, N. C. Handy and R. D. Amos, The determination of hyperpolarizabilities using density functional theory, *Chemical Physics Letters* 210 261-268(1993).

[14] F. Sim, D. R. Salahub and S. Chin, *International Journal of Quantum Chemistry* 43 463-(1992).

[15] M. K. Harbola, Electric dipole polarizabilities for helium-like ions from correlated wavefunctions. A density functional approach, *Chemical Physics Letters* 217 461-465(1994).

[16] G. Senatore and K. R. Subbaswamy, Hyperpolarizabilities of closed-shell atoms and ions in the local-density approximation, *Physical Review* A 34 3619-3629(1986).

[17] P. G. Jasien and G. Fitzgerald, Molecular dipole moments and polarizabilities from local density functional calculations: Applications to DNA base pairs, *Journal of Chemical Physics* 93 2554-2560(1990).

[18] A. Hu, D. M. York, and T. K. Woo, Time-dependent density functional theory calculations of molecular static and dynamic polarizabilities, cauchy coefficients and their anisotropies with atomic numerical basis functions, *Journal of Molecular Structure: THEOCHEM* 591 255-266(2002).

[19] J. Moullet and J. L. Martins, Comparison of self-consistent calculations of the static polarizability of atoms and molecules, *Journal of Chemical Physics* 92 527-535(1990).

[20] K. B. Sophy and S. Pal, Density functional response approach for the linear and nonlinear electric properties of molecules, *Journal of Chemical Physics* 118 10861-10866(2003).

[21] K. B. Sophy and S. Pal, Electric properties of BH, CO and H_2O molecules by density functional response approach, *Journal of Molecular Structue: THEOCHEM* 676 89-95(2004).

[22] W. Kohn and L. J. Sham, Self-consistent equations including exchange and correlation effects, *Physical Review* 140 A1133-(1965)

[23] S.T. Epstein, *The Variation Method in Quantum Chemistry*, Academic Press, New York, 1974.

[24] J. Hellmann, Einfuhrung in die Quantenchemie, Deuticke, Leipzig, 1937.

[25] R.P. Feynman, Forces in molecules, *Physical Review* 56 340-343 (1939).

[26] C. E. Dykstra and P. G. Jasien, Derivative Hartree-Fock theory to all orders, *Chemical Physics Letters* 109, 388-393 (1984).

[27] (a) A. St-Amant, D.R. Salahub, New algorithm for the optimization of geometries in local density functional theory, *Chemical Physics Letters* 169, 387-392 (1990). (b) A. St-Amant, Ph.D. thesis, University of Montreal, 1992. (c) M.E. Casida, C. Daul, A. Goursot, A. Koester, L. Pettersson, E. Proynov, A. St.-Amant, D. Salahub, principal authors, S. Chreties, H. Duarte, N. Godbout, J. Guan, C. Jamorski, M. Leboeuf, V. Malkin, O. Malkina, F. Sim, A. Vela, contributing authors, DEMON-KS version 3.5, DEMON Software, 1998.

[28] N. Vaval, K. B. Ghose and S. Pal, Nonlinear molecular properties using biorthogonal response approach, *Journal of Chemical Physics* 101 4914-4919 (1994).

[29] N. Vaval and S. Pal, Stationary coupled-cluster approaches to molecular properties: A comparative study, *Physical Review* A 54 250-258 (1996).

Brill Academic Publishers
P.O. Box 9000, 2300 PA Leiden
The Netherlands

*Lecture Series on Computer
and Computational Sciences*
Volume 3, 2005, pp. 152-155

On electric polarizabilities and hyperpolarizabilities: The correlation, relativistic and vibrational contributions

M. G. Papadopoulos[a], A. Avramopoulos[a], S. G. Raptis[a] and A. J. Sadlej[b]

[a]Institute of Organic and Pharmaceutical Chemistry, National Hellenic Research Foundation, 48 Vas. Constantinou Ave., Athens 116 35, Greece.
[b]Department of Quantum Chemistry, Institute of Chemistry, Nicolaus Copernicus University, PL-87 100 Toruń, Poland.

Received 1 August, 2005; accepted 12 August, 2005.

Abstract: We present the polarizabilities and hyperpolarizabilities of the hydrides of Cu, Ag and Au and the sulfides of Zn, Cd and Hg. The correlation, relativistic and vibrational contributions to the above properties have been taken into account.

1. Introduction

It is known that the polarizabilities and hyperpolarizabilities are of great scientific and technological interest. Most of the computational works on these properties deal only with the electronic contribution. However, in the recent years an increasing number of articles determine both the electronic and the vibrational contributions to these properties. In most cases the vibrational properties have been computed by employing, in some form, the perturbation method proposed by Bishop and Kirtman. There are relatively few articles, which consider the relativistic contribution to the polarizabilities and/or hyperpolarizabilities. In our presentation we shall review some of our work in this area and in particular our study on the hydrides of Cu, Ag and Au as well as the sulfides of Zn, Cd and Hg. State-of-the-art property values have been reported in which the correlation, relativistic and vibrational contributions have been taken into account.

2.Computational methods

In this section we shall briefly review the methods we have employed for the computation of the vibrational contributions. The clumped nucleus approximation allows the resolution of the electric properties to electronic, P^{el}, and vibrational contributions. The latter have two components, the zero-point vibrational averaging term, P^{ZPVA}, and the pure vibrational contribution, P^{pv}:

$$P = P^{el} + P^{ZPVA} + P^{pv}. \tag{1}$$

Bishop and Kirtman developed a perturbation theory approach (BKPT) for the evaluation of the vibrational corrections (pv and ZPVA) [5]. Within this approach we have:

$$P^{zpva} = [P^e]^{(1,0)} + [P^e]^{(0,1)} \tag{2}$$

$$[P^e]^{(0,1)} = -\frac{\hbar}{4} \sum_\alpha \frac{1}{\omega_\alpha^2} \left(\sum_b \frac{F_{\alpha bb}}{\omega_b} \right) \left(\frac{\partial P^e}{\partial Q_\alpha} \right) \tag{3}$$

and

$$[P^e]^{(1,0)} = \frac{\hbar}{4} \sum_\alpha \frac{1}{\omega_\alpha} \left(\frac{\partial^2 P^e}{\partial Q_\alpha^2} \right), \tag{4}$$

where, ω_α is the harmonic frequency, $F_{\alpha bb}$ is the cubic force constant and Q_α is the normal coordinate. According to BKPT, the pure vibrational contributions to the dipole polarizability and hyperpolarizabilites are given by the following formulas:

$$\alpha^{pv} = [\mu^2]^{(0,0)} + [\mu^2]^{(2,0)} + [\mu^2]^{(1,1)} + [\mu^2]^{(0,2)} \tag{5}$$
$$\beta^{pv} = [\mu\alpha]^{(0,0)} + [\mu\alpha]^{(2,0)} + [\mu\alpha]^{(1,1)} + [\mu\alpha]^{(0,2)} + [\mu^3]^{(1,0)} + [\mu^3]^{(0,1)} \tag{6}$$

e-mail:mpapad@eie.gr;aavram@eie.gr;teoajs@chem.uni.torun.pl

$$\gamma^{pv}=[\alpha^2]^{(0,0)}+[\alpha^2]^{(2,0)}+[\alpha^2]^{(1,1)}+[\alpha^2]^{(0,2)}+[\mu\beta]^{(0,0)}+[\mu\beta]^{(2,0)}+[\mu\beta]^{(1,1)}+[\mu\beta]^{(0,2)}+[\mu^2\alpha]^{(1,0)}$$
$$+[\mu^2\alpha]^{(0,1)}+[\mu^4]^{(2,0)}+[\mu^4]^{(1,1)}+[\mu^4]^{(0,2)} \tag{7}$$

Analytical expressions have been derived by Bishop et al. for the computation of $[A]^{(n,m)}$, where n and m define the orders of the electrical and mechanical anharmonicities, respectively [1].

The vibrational contributions have also been calculated by means of the field-dependent Numerov-Cooley (NC) integration technique. The use of this method has been originally suggested by Dykstra and Malik [2]. For this approach we start with the vibrational nuclear equation

$$[T+E(R,F)]X(R,F)=W(F)X(R,F), \tag{8}$$

where $F=F_z$ is the electric field along the bond axis, $W(F)$ is the energy eigenvalue, T is the vibrational kinetic energy operator, $E(R,F)$ is the field-dependent Born-Oppenheimer electronic energy, R is the distance and $X(R,F)$ is the field-dependent vibrational wave function. Differentiation of eq. (8) with respect to F, leads to the total (i.e. electronic and vibrational) values of the electric properties [3]. For example for the polarizabilities we have:

$$\alpha_{zz}=\alpha=-(\frac{\partial^2 W(F)}{\partial F^2})_{F=0}=\alpha^{pv}+\alpha_{Av}$$

$$\alpha_{Av} = <X(R,0)|\ \alpha(R,0)|\ X(R,0)>$$

$$\alpha^{pv} = 2<(\frac{\partial X(R,F)}{\partial F})_{F=0}|\mu(R,0)|\ X(R,0)>,$$

where $\mu(R,0)$, $\alpha(R,0)$, denote the electronic dipole moment and polarizability functions, respectively, at $F=0$.

3. Results and Discussion

The hydrides of Cu, Ag and Au

We have used a hybrid approach in which the field dependence of the vibrational wave functions has been determined from field dependent CASPT2 calculations and the property functions have been calculated by employing the CCSD(T) method. The valence and sub-valence shells have been correlated. The computation of the *vibrational contributions* is based on the direct Numerov-Cooley (NC) integration of the vibrational equation and the subsequent evaluation of the vibrational corrections from the property curve [4]. The relativistic corrections to the properties of interest have been computed by employing the Douglas-Kroll approximation [5,6].

Table I. The electronic and the vibrational contributions (pv and ZPVA)[a] to the dipole moment (μ), the parallel polarizability (α) and the parallel first hyperpolarizability (β) of the coinage metal hydrides, MeH, Me=Cu,Ag,Au, from relativistic (NR) and relativistic (DK) calculations based on CASPT2 energy curves and CCSD(T) property functions. All property values are in a.u.

	CuH		AgH		AuH	
	NR	DK	NR	DK	NR	DK
μ^{el}	-1.1140	-1.0182	-1.3796	-1.1167	-1.3240	-0.5236
μ^{ZPVA}	-0.0076	-0.0111	-0.0113	-0.0062	-0.0107	-0.0006
α^{el}	37.33	36.42	47.52	44.48	49.84	39.51
α^{ZPVA}	0.97	1.08	1.66	1.13	1.63	0.52
α^{pv}	0.38	0.27	0.60	0.37	0.47	0.00
β^{el}	468	416	734	489	685	166
β^{ZPVA}	20	17	56	23	49	0
β^{pv}	-67	-52	-159	-90	-131	-35

[a] The vibrational contributions (ZPVA and pv) were computed by using the Numerov-Cooley method.

Table I presents the results of the electronic and vibrational contributions (pv and ZPVA) to the dipole moment, dipole polarizability (α_{zz}) and first hyperpolarizability (β_{zzz}) of the coinage metal hydrides

(MH: CuH, AgH and AuH). The ZPVA contribution to the dipole moment is quite small. The relativistic correction reduces (in absolute value) the dipole moment of the AgH and AuH. The NR α^{el} of MH increases as the atomic number of M increases. The results show that α^{el} takes the maximum value for AgH. The relativistic correction increases α^{ZPVA} for CuH. The reverse trend is observed for AgH and AuH. The relativistic correction has a significant effect on α^{ZPVA} of AgH and AuH. The relativistic contribution reduces α^{PV} of MH. The relativistic effect reduces β^{el}, β^{ZPVA} and β^{pv} (the absolute value is considered). We may generalize this observation by noting that all considered property values are reduced by the relativistic corrections except of μ^{ZPVA} and α^{ZPVA} of CuH. The reduction of the vibrational contributions, by the relativistic correction can be attributed to that the mechanical and electrical anharmonicities were diminished by the above correction.

The sulfides of Zn, Cd and Hg

The polarizabilities and first and second hyperpolarizabilities of ZnS, CdS and HgS have been computed by employing a series of high-level methods [7]. The electronic correlation has been taken into account by employing the methods MP2, CCSD and CCSD(T). The relativistic corrections and the interference relativistic-correlation effects have been calculated by using the Douglas-Kroll approximation. The vibrational contributions have been computed by employing the Bishop and Kirtman perturbation theory [1]. Basis sets developed by Sadlej and et al.[8] have been employed for the non-relativistic and the relativistic computations. The results demonstrate (Table II) the importance of (i) the relativistic correction for the computation of accurate polarizability and hyperpolarizability results and (ii) the vibrational contribution (static) for β and γ. This contribution is affected by the relativistic contribution.

Table II. Correlation, relativistic and vibrational contributions to the polarizabilities and first and second hyperpolarizabilities ($\gamma \times 10^3$) of ZnS, CdS and HgS. All property values are in a.u.

	ZnS			CdS			HgS		
	α	β	γ	α	β	γ	α	β	Γ
SCF	64.13	-1231.7	56.3	81.37	-2307.8	101.7	86.26	-2565.6	106.7
El. cor. cont.									
MP2	-0.57	1390.9	-58.3	-3.22	3431.5	-165.8	-5.40	4089.9	-189.8
CCSD	0.32	485.0	-14.4	-0.99	1105.4	-34.0	-2.84	1359.7	-40.2
CCSD(T)	0.53	574.6	-10.4	-1.03	1140.5	-6.4	-3.03	1341.5	-7.2
DK rel. cont.									
SCF	-1.35	148.0	-5.8	-5.30	745.1	-33.4	-19.04	2002.5	-75.4
MP2	-0.79	-108.1	10.7	-3.12	-856.4	78.7	-2.59	-3030.4	178.0
CCSD	-0.64	-23.4	2.1	-1.98	-237.4	14.9	-1.31	-1003.4	38.4
CCSD(T)	-0.64	-50.8	2.1	-1.92	-268.2	1.4	-0.76	-1013.7	10.5
ZPVA cont.									
CCSD[a]	0.03	9.9		0.06	25.6		0.05	21.0	
CCSD(T)[b]	0.02	7.2		0.03	7.9		0.01	-1.1	
PV cont.									
CCSD(T)[a]	0.15	-88.0	2.6	0.22	-143.6	2.4	0.25	-158.9	1.8
CCSD(T)[b]	0.15	-83.0	2.3	0.18	-114.0	2.9	0.03	-31.6	2.3
Total									
CCSD(T)[a]	64.84	-735.2	49.34	80.62	-1348.3	100.3	83.54	-1362.0	103.6
CCSD(T)[b]	62.85	-635.2	45.1	73.33	-859.9	67.0	63.47	-268.0	36.9

[a] Nonrelativistic values at the CCSD(T) level of theory.
[b] Relativistic results which include the relativistic correction and mixed relativistic correlation contributions evaluated at the level of CCSD(T) approximation.

REFERENCES

[1] D. Bishop, J. M. Luis and B. Kirtman, *J. Chem. Phys.*, **108**, 10013 (1998); D. M. Bishop and P. Norman, *"Handbook of Advanced Electronic and Photonic Materials"*, (Editor: H. S. Nalwa) Vol.9, p.1. Academic Press, San Diego, (2001).

[2] C. E. Dykstra and D. J. Malik, *J. Chem. Phys.*, 87, 2806 (1987).

[3] V.E. Ingamells, M. G. Papadopoulos,A. J. Sadlej, *Chem. Phys. Letters,* **316,** 541   (2000).

[4] A. Avramopoulos, V. E. Ingamels, M. G. Papadopoulos and A. J. Sadlej, *J. Chem. Phys.*, **114**, 198 (2001).

[5] M. Douglas and N. M. Kroll, *Ann. Phys.* (N.Y.), **82**, 89 (1974).

[6] B. A. Hess, *Phys. Rev. A*, **33**, 3742 (1986);*ibid.*, **39**, 6016 (1989).

[7]. S. G. Raptis, M. G. Papadopoulos and A. J. Sadlej, *J. Chem. Phys.*, **111**, 7904   (1999).

[8]. V. Kellö and A. J. Sadlej and K. Faegri, *J. Chem. Phys.*, **108**, 2056 (1998); V. Kellö, and A. J. Sadlej, *Theor. Chim Acta*, **91**, 353 (1995); A. J. Sadlej, *Theor. Chim Acta*, **79**, 123 (1991); V. Kellö and A. J. Sadlej, *Theor. Chim Acta*, **94**, 93 (1996).

Brill Academic Publishers
P.O. Box 9000, 2300 PA Leiden,
The Netherlands

Lecture Series on Computer
and Computational Sciences
Volume 3, 2005, pp. 156-167

Molecular Polarizabilities from Density-Functional Theory: From Small Molecules to Light Harvesting Complexes

[a]Mark R. Pederson[1] and [a,b]Tunna Baruah

[a]Center for Computational Materials Science
Naval Research Laboratory Washington DC 20375
and
[a,b]Department of Physics and Astronomy
State University of New York
Stony Brook NY 11794-3800

Received 5 August, 2005; accepted 12 August, 2005

Abstract: We review some recent applications of density-functional theory to molecules and systems of molecules where the role of polarizabilities are particularly relevent. With respect to the implementation of density-functional theory, details related to numerics and basis sets are described. We then describe how self-consistent finite-field calculations may be used to separately extract the electronic polarizability tensor and infrared intensities. We review the relationship between second-harmonic vibrational polarizabilities and molecular infrared intensities. An efficient method for describing the polarization effects in systems of molecules is included and a recent application of this method to a biomimetic light-harvesting complex is discussed.

Keywords: Electronic Structure, Gaussian Basis, Polarizability, Charge Transfer

Mathematics Subject Classification: 81-08, 81V55

PACS: 31.15.EW, 31.15.kr, 34.70.te, 31.15Pf

1 Introduction

The calculation of molecular polarizabilities [1, 2, 3, 4, 5, 6, 7, 8, 9, 10, 11, 12, 13] in molecules remains a problem of great interest due to the important role they play in many physical, chemical, and biological processes. In weakly interacting molecular materials, the polarizabilities are responsible for long-range attractions which may be due to either dipole-/monopole- induced dipole interactions or London interactions. In ionic and molecular crystals, the Clausius-Mossotti relation and extensions relates the materials dielectric constant to the polarizabilities of the molecular systems. In biology or biomimetic materials, light-induced charge transfer interactions are often of interest. Molecular polarization impacts such events both directly and indirectly. First, when a charge-transfer excitation occurs on a given molecule or between two neighboring molecules, all of the spectator molecules surrounding the active molecules respond to the change in electric-fields caused by this excitation. Depending upon how quickly the molecular surroundings respond to such an excitation, several different effects could occur. If the absorption and emission were indeed sudden but long-lived, the emission spectrum would be shifted relative to the absorption specturm.

---

[1]Corresponding author. E-mail: mark.pederson@nrl.navy.mil

At the other limit, where long-range coupling between spectator and active molecules is considered instantaneous, a significant broadening in the absorption and emission spectrum would arise. While the interactions discussed here would often be dominated entirely by electronic polarization effects, dynamical and static coupling between electronic and atomistic degrees of freedom, such as those pioneered by Bishop, complicate the problem further and may be important in some cases. A second, albeit indirect, effect due to molecular polarizability is related to the transition rates that govern the light induced excitations associated with photovoltaic energy conversion. With the possible exception of coherent electron processes, where preparation of an excited electronic configuration is often not considered, the time scale associated with a rattling electron or even an electron-hole pair requires the calculation of transition matrix elements. At the lowest level of approximation, the dipole form of these matrix elements coincides with those requisite for molecular polarizabilities. In this sense, assessing capabilities for calculation of polarizabilities and polarizability derivatives furthers the confidence level in transition rates for excited states.

One method for obtaining information about polarizabilities and polarizability derivatives is the original density-functional theory as well as the various extensions such as time-dependent density functional theory and the hybrid functionals. [14, 15, 16]

In this paper we review some of the recent work in this area that has been primarily accomplished by the NRLMOL suite of density-functional codes. [17, 18, 19, 20] We provide a summary of calculated polarizabilities in many molecules which can will hopefully illustrate general trends.

The paper is organized as follows. Since the methodology utilized here has been developed primarily in the physics and materials-research communities, in Sec. II we briefly discuss the underlying numerical methodologies, the methods for basis-set selection, and provide a few computational benchmarks. In Sec. III we review the calculation of polarizabilities and infrared intensities and discuss how the infrared intensity is related to double-harmonic vibrational polarizabilities. We provide calculated results for several different types of molecules and compare to experimental results as well as other theoretical calculations. In Sec. IV, we review the simple case of an FCC lattice of fullerenes and a simple model for calculating the polarization-induced relaxation associated for a lattice of fullerene molecules. [11] In Sec. V we discuss a problem where all of these capabilities are required. We start by discussing the calculation of the polarizability tensor of a recently proposed light-harvesting molecular triad which is composed of a fullerene molecule, a porphyrin molecule and a carotenoid molecule. [21, 22] The molecule is considered to be interesting because it possesses a charge transfer excited state with a very large dipole moment. We then show that the polarization of surrounding molecules would significantly change the energetics of this excited state compared to excitation in vacuum where charge transfer is not achieved.

2 Computational Methodology

The calculations discussed herein have been performed using the NRLMOL suite of density-functional codes that have been developed by Pederson and collaborators over the last two decades. Except where otherwise noted, we have used the generalized-gradient approximation to exchange-correlation energy [15, 16] for the applications discussed here. For a given molecular geometry, it is necessary to solve a Schroedinger-like equation for each of the electrons. Once the electronic Schroedinger-like equations are self-consistently solved, the forces on each atom are determined from the Hellman-Feynman-Pulay theorem [17][b] and the geometry is relaxed using standard quasi-newtonian approaches such as conjugate gradient.

To solve the Schroedinger-like equations the wavefunctions are expanded in terms of gaussian basis functions. This reduces the problem to the numerical determination of matrix elements, followed by the solution of a standard algebraic secular equation. While the latter task is accomplished using standard linear-algebra codes, there are a few aspects of the former approach that are worth mentioning. In order to obtain the matrix elements, it is necessary to calculate terms such as $< \phi_i | -\frac{1}{2}\nabla^2 + V_{eff}(\mathbf{r}) | \phi_j >$. For gaussian basis functions, the kinetic energy part of this matrix element can be obtained in closed form using standard techniques. However parameterizations of the density functional such as GGA and LDA do not allow the matrix elements for the potential energy to be calculated in closed form. To efficiently and accurately perform the calculation of the potential matrix elements we first determine a variational integration mesh [17]^a that can be used to accurately calculate three-dimensional integrals that are strongly peaked near the positions of nuclei and more slowly decaying in the regions that are far from any atom. This method is based on several different transformations of gaussian quadrature integration strategies and was originally presented in Ref. [17]^a. The general idea is that it is always possible to place a molecule in a parallelpiped that is so large that the basis functions outside the parallelpiped vanish. Then the parallelpiped is iteratively tessalated until one is left with a collection of parallelpipeds that are either empty or have a single atom at the center. The latter case are further reduced to more empty parallelpipeds and an atom at the center of a perfect cube. Three different adaptable and iterative quadrature techniques are then used to develop accurate integration meshes in each of these three regions. Details are given in Ref. [17]^a. Given the integration mesh, it is then only necessary to numerically determine the values of the basis-set pair $[\phi_i(\mathbf{r})\phi_j(\mathbf{r})]$ and the effective potential $(V_{eff}(\mathbf{r}))$. Within the generalized-gradient approximation the exchange-correlation potential depends on the spin densities as well as their first and second derivatives. Numerical determination of the electronic Coulomb potential requires one to solve poisson's equation for each pair of basis functions in the problem. Determination of the GGA potential and the Coulomb potential can in fact be done quasi-simultaneously since the latter case requires algebraic reduction of the two-center density matrices to a superposition of single-center gaussian distributions multiplied by polynomials of degree $2L_{max}$ with L the maximum angular-momentum basis function in the problem. It is a fact that this algebraic reduction can then be used to convert the determination of the GGA potential to order-N complexity. Further this strategy is easily parallelized and can be automatically load balanced. A discussion of load balancing is presented in Ref. [18].

In addition to intrinsic numerical accuracy a second issue is related to the choice of gaussian-basis sets. The basis-sets used here have been specifically optimized for GGA calculations [19]. However, initial tests suggest they perform well for HF calculations also. Issues related to basis-set choice are numerous and some of these must be primarily addressed in a brute-force manner by systematically increasing basis-set size until a calculated quantity stops changing. By performing such calculations on smaller systems, reasonably reliable rules of thumb may be deduced which allow high confidence for calculations on larger systems. However, there are several aspects of basis-set selection that can be rigorously addressed in atomic calculations prior to use in molecules and we first ensure that our basis sets satisfy these requirements. With respect to the cusp condition and selection of shortest-range function in the problem, we rely on a theorem proved in Ref. [19] which we now refer to as the $Z^{10/3}$ theorem. If one requires that the absolute error in energy for a 1s-core electron in a heavy atom be the same as the absolute error in energy for a light atom it can be shown that the ratio of the shortest-range gaussian decay parameters should in fact scale as $(Z_l/Z_h)^{10/3}$ where Z_l and Z_h are the nuclear charges of the lighter and heavier nuclei respectively. We have shown that this condition eliminates the need for counterpoise calculations in systems where basis-set superposition is expected to be an issue. A second issue that can be addressed rigorously at the atomic level is related to the selection of the values of each gaussian decay function. To address this issue we start with a set of N even-tempered decay parameters given by $(\alpha_1, \alpha_2, \alpha_3, ...\alpha_N)$ and perform an SCF calculation on a given atom. When self-consistency is achieved one can determine the derivative of the total energy with respect to each decay parameter $(\frac{dE}{d\alpha_i})$ by noting that the Hellman-Feynman theorem tells us that at self-consistency the total derivative is in fact the partial derivative. In conjunction with the conjugate-gradient algorithm and a series of SCF calculations it is then possible to find the best decay parameters within the constraint of N decay parameters. Finally the total number of decay parameters can be determined by ramping up the value of N until the optimized total energy converges and agrees with a numerical solution to a given accuracy in energy. It was ascertained that the $Z^{10/3}$ scaling law mentioned above is automatically "selected" by the numerical approach discussed above. Also in reference to to basis sets, it is worth mentioning that we find it to be computationally advantageous to use a common set of decay parameters for all angular momentum. However the automated use of more conventional quantum-chemistry basis sets is also an option. The final issue that probably can not be addressed systematically pertains to the total number of contracted orbitals necessary to determine an accurate value. With the exception of the total energy most quantities of interest, polarizabilities being one example, are not variational quantities so care must be exercised in assessing the accuracy of such calculations. In this work the default basis sets employed start with a minimal basis set that exactly reproduces the ground state energies. To this basis set we add 3-4 extra single gaussians of s and p character and 2-3 extra single gaussians of d character. Additional detail can be found in Ref. [19].

3   Polarizabilities in Isolated Molecules and Connections Between Vibrational Spectra and Higher Order Derivatives

When a molecule is placed in an external field each of these molecules polarizes in response to the applied fields. As discussed by Bishop and coworkers the total polarizability of a molecule can be analytically broken up into terms due to different types of interactions. Generally the largest response is the electronic polarizability which is commonly designated as α in the literature. This response is associated field-induced relaxations of the electronic density and the formation of induced dipole moments and is of second-order in the electric field.

The next largest response is often the double-harmonic vibrational polarizability [23, 24, 25, 26, 27, 28] that is designated by $([\mu_2]^{0,0})$ in the literature. This response is due to the following physics. At zero field, the equilibrium geometry of a molecule has no net forces. However when an electric field is applied to a molecule, the forces on each atom become nonzero due to the fact that the nuclear charges feel a direct force from the applied field and also an induced force due to the faster rearrangement of electronic density. The molecule then relaxes its geometry slightly to counteract the appearance of the force and lowers it's energy further. This effect is also second-order in the electric field. It has been shown that double-harmonic vibrational polarizabilities depend on the same quantities that determine the infrared intensities. [29, 30, 31]

While other higher-order effects are also of interest, the discussion in this paper will concentrate entirely on electronic and infrared calculations because these are often the largest contributions to molecular polarization and because they are of immediate interest to the our current studies. Also, these static quantities are indeed ground-state properties and are therefore fully protected by the original formulations of density-functional theory due to Hohenberg, Kohn and Sham. [14] We note that there are many interesting treatises on the more general problem and the interested reader is referred to references at the beginning of the paper.

There are at least two strategies available for calculating the two effects described here. Probably the most general strategy is to determine the ground state of a multiatom system and many of the excited electronic and vibrational states. Given this information, it is possible to develop perturbative expansions of the total energy as a function of applied field and direction and identify the static electronic and double-harmonic vibrational polarizabilities as well as higher-order effects. Such a method also allows for the treatment of dynamical effects. If one is primarily interested in and/or constrained by theoretical formality to look at these two effects, an alternative and conceptually more straightforward approach may be used. The electronic and infrared intensities may be determined directly by mimicing the experimental conditions. In this approach, which is used in entirety in this work, the calculation of the molecular energy and wavefunctions is performed with a small electric field applied for several different orientations. The various contributions to the polarizabilities are then easily determined by observing the calculated change in energy and dipole moment as a function of applied fields and vibrational displacements. A more mathematical discussion of this approach is outlined below.

Within the DFT the total energy (E) of a neutral molecule in an electric field $\mathbf{G}$ is given by

$$E = \quad 2\sum_i < \psi_i | -\frac{1}{2}\nabla^2 + V_{nuc}|\psi_i > +\frac{1}{2}\int d^3r \int d^3r' \frac{\rho(\mathbf{r})\rho(\mathbf{r'})}{|\mathbf{r} - \mathbf{r'}|}$$

$$+ \int d^3r G_{xc}[\rho, \nabla\rho...] + \frac{1}{2}\sum_{\mu\neq\nu} \frac{Z_\mu Z_\nu}{|\mathbf{R}_\mu - \mathbf{R}_\nu|} + \int d^3r \rho_{tot}(\mathbf{r})\mathbf{G}\cdot\mathbf{r}. \qquad (1)$$

For simplicity the above equation assumes a spin unpolarized molecule. The various contributions to the total energy include the electronic kinetic energy, all the coulomb interactions in the problem, the exchange-correlation energy (represented schematically as G_{xc}) and the interaction between the total charge density and the applied electric field. The total charge density includes the charges of the nuclei and is designated as ρ_{tot} whereas the electronic density $\rho = 2\sum_i |\psi_i|^2$ appears without a subscript. At the zero-field equilibrium geometry, the change in energy ($\Delta = E - E_o$) to second-order is given by:

$$\Delta = \sum_i \frac{\partial E}{\partial G_i}G_i + \sum_{\nu ki} \frac{\partial^2 E}{\partial G_i\partial R_{\nu k}}G_i X_{\nu k} + \frac{1}{2}\sum_{\mu i,\nu k} \frac{\partial^2 E}{\partial R_{\mu i}\partial R_{\nu k}}X_{\mu i}X_{\nu k} + \frac{1}{2}\sum_{ik} \frac{\partial^2 E}{\partial G_i\partial G_k}G_i G_k. \qquad (2)$$

Further we note that

$$\frac{\partial p_i}{\partial R_{\nu,k}} = -\frac{\partial^2 E}{\partial G_i\partial R_{\nu k}} = \frac{\partial F_{\nu,k}}{\partial G_i}. \qquad (3)$$

In the above equations, the x-component of the total moment is defined according to $\mathbf{p_x} = \frac{\partial E}{\partial G_x}$. $\mathbf{X}_\nu$ is the displacement of the νth atom. The spring constant matrix is given by $\frac{\partial^2 E}{\partial R_{\mu i}\partial R_{\nu k}}X_{\mu i}X_{\nu k}$. As discussed in Refs. [20], the mixed second derivatives, $\frac{\partial^2 E}{\partial G_i\partial R_{\nu k}}$, when transformed into the space of the normal modes lead to the infrared intensities. Eq. (3) shows that the more conventional representation of this second derivative is in terms of the first derivative of the total dipole moment with respect to an atomic or vibronic displacement. It is both instructive and computationally efficient to notice that the mixed second derivative is also the derivative of the Hellman-Feynman force with respect to applied field. Many authors have noted that this quantity is also responsible for the double-harmonic vibrational polarizability. To see that this is the case, one simply needs to consider the derivative (force) of the above energy with respect to an atomic position:

$$\frac{\partial\Delta}{\partial R_{\nu k}} = \sum_i \frac{\partial^2 E}{\partial G_i\partial R_{\nu k}}G_i + \sum_{\mu i} \frac{\partial^2 E}{\partial R_{\mu i}\partial R_{\nu k}}X_{\mu i} = 0. \qquad (4)$$

From the above equation it is clear that inversion of the spring constant matrix (in the space orthogonal to translational and rotational modes) allows one to find the displacements ($X_{\mu i}$) that minimize the energy. It is also clear that such displacements will be linear in the applied field and that the energy difference for these displacements is then quadratic in the applied field. In Ref. [27] a full derivation is presented and the final result is given in terms of spectroscopically measurable quantities (e.g. frequencies and dynamical dipole moments).

In Table I we present recent NRLMOL-based calculations on static polarizabilities. In Table I we compare results as calculated within the local density approximation and the generalized gradient approximation. Experimental values are also included.

Table 1: Calculated polarizability along principle axes for a selection of molecules. All polarizabilities are reported in $\mathring{A}^3$. Experimental values in parentheses idicate they are average polarizabilities rather than eigenvalues of the polarizability tensor.

Molecule	Method	Ref.	α_1	α_2	α_3
CH_4	LDA	[20]	2.68	2.68	2.68
	GGA	[20]	2.62	2.62	2.62
	Exp	[34]	2.60	2.60	2.60
C_2H_2	LDA	[20]	4.79	2.98	2.98
	GGA	[20]	4.79	2.89	2.89
	Exp	[34]	5.12	2.43	2.43
C_2H_4	LDA	[20]	5.41	3.99	3.45
	GGA	[20]	5.39	3.91	3.44
	Exp	[34]	5.40	3.85	3.40
C_2H_6	LDA	[20]	4.98	4.35	4.35
	GGA	[20]	4.91	4.24	4.24
C_6H_6	LDA	[11]	12.6	12.6	6.75
	Exp	[35]	(10.3)		
NF_3	GGA	[27]	3.40	3.40	2.40
	EXP	[31]	(2.81)		
$TiCl_4$	GGA	[27]	15.03	15.03	15.03
	Exp	[31]	15.00	15.00	15.00
SF_6	GGA	[27]	5.15	5.15	5.15
	Exp	[31]	4.49	4.49	4.49
SiF_4	GGA	[27]	3.72	3.72	3.72
	Exp	[31]	3.32	3.32	3.32
HCN	GGA	[27]	3.42	2.18	2.18
	Exp	[31]	(2.59)		
C_{60}	LDA	[11]	83.5	83.5	83.5
	GGA	[27]	82.9	82.9	82.9
	EXP	[36]	75-83	75-83	75-83
Na_2	GGA	[27]	49.8	28.9	28.9
	Exp	[37]	(38.6)		
Na_8	GGA	[27]	116.2	116.2	116.2
	Exp	[37]	133.5	133.5	133.5
TCNQ $C_6H_4[C(CN]_2)_2$	GGA	[38]	64.9	27.53	12.5
BEDT $[CS_2C_2S_2C_2H_4]_2$	GGA	[38]	75.1	42.4	25.7

In Ref. [27] we have used the second derivatives (Eq. 4) to calculate the second-harmonic vibrational polarizabilities. As is found experimentally [31], the vibrational polarizabilities were found to be relatively large for the halide containing molecules. For example for SiF_4 the (isotropic) vibrational polarizability was found to 2.09 $\mathring{A}^3$ which is on the same order of the electronic polarizability of 3.72 $\mathring{A}^3$. For comparison the experimental values for these quantities are 1.75 and 3.32 $\mathring{A}^3$ respectively. In contrast for a nonpolar tetrahedral molecule (CH_4), the vibrational polarizability was found to be very small (0.04 $\mathring{A}^3$) in good agreement with experiment (0.03 $\mathring{A}^3$). Vibrational polarizabilities for many of the molecules shown in Table I may be found in Ref. [27]. One of the possible trends observed in Ref. [27] was that the vibrational polarizability may be less important (percentage wise) for large molecules. An additional recent calculation [?] that might reinforce this tren is on the TCNQ molecule. This molecule has principal electronic polarizabilities of (61.8, 27.3 15.9) $\mathring{A}^3$. However the principal vibrational polarizabilities are only (1.91, 2.49 and 3.12) $\mathring{A}^3$ respectively.

4 Polarization Effects in Systems of Molecules

In principle, calculation of the polarizabilities of a system of molecules may be accomplished using the formalism of the previous section simply by treating the system of molecules as a single macromolecule with space in between. However, this quickly becomes computationally too expensive. Fortunately, many interesting questions related to polarizability of a collection of nearly nonoverlapping molecules may be described efficiently by first calculating the polarizability tensor for each molecule and then determining how long-range interactions between charged or polarized molecules modify the total response of the system. We review a method for determining the response of a collection of molecules as well as some earlier applications of this method to a lattice of fullerene molecules. This method is conceptually analogous to the Clausius-Mossotti method for extracting dielectric constants from molecular polarizabilities. The primary difference is that it allows one to study such effects for finite collections of polarizable molecules and that the constraint of perfect periodicity is not required. In the next section, we discuss applications of a nonisotropic extension of this method to a more complex biomimetic molecule that has a relatively large polarizability.

Suppose we are given a lattice of N nonoverlapping neutral molecules at sites $(\mathbf{R}_1, \mathbf{R}_2, ...\mathbf{R}_N)$ with polarizabilities of $(\alpha_1, \alpha_2, ...\alpha_N)$. We then consider modifying the charge distribution on these molecules. For example, we could imagine addition or removal of an electron from a given molecule or removing an electron from one molecule and placing it on a second molecule. As a result of the change in charge distribution, electric fields that are spatially slowly varying in space are produced at each lattice site. This causes each molecule to polarize in response to the new electric field. The change in dipole moment at each molecular site further modifies the electric fields in the system which would further change the dipoles at each site. Providing that the original electric fields are small enough, the dipoles on each site would eventually self-consistently adjust themselves so that the induced dipoles are consistent with the external field due to all other charges, dipoles and external fields in the system. In Ref. [11] Pederson and Quong introduced an energy functional for such a lattice of molecules. It is given by:

$$E = \sum_n \left[\frac{|\mathbf{p}_n^{ind}|^2}{2\alpha_n} - \mathbf{p_n} \cdot \mathbf{G} + \Phi(R_n)q_n + T(q_n) \right] + \sum_{n,m} [U^1(p_n, p_m) + U^2(p_n, q_m) + U^3(q_n, q_m)] \quad (5)$$

For a recent article with a similar point of view and a discussion of many other aspects related to Clausius-Mossotti relations, readers are referred to Ref. [32]. In the above equation, $\mathbf{G} = -\nabla\Phi$ is the electric field that is due to an externally applied potential $\Phi(\mathbf{r})$. T(q) is the energy of the particle as a function of charge state. For example, for q=1, T(q) is the ionization energy and for q=-1, T(q) is the electron affinity. The electric field interacts with the total dipole moment on each site and the externally applied field interacts with the charge on each site. In addition, the total dipoles and charges interact with one another in terms of the standard dipole-dipole (U^1), dipole-monopole (U^2) and monopole-monpole (U^3) interactions which can be found in any textbook on Electricity and Magnetism. The first term represents the energy penalty associated with forming an induced dipole on a given site. In the limit of a single site, or well-separated sites, with q_n=0, it is easily verified that the induced dipole moments that minimize the above energy functional satisfy the expected constraint:

$$\mathbf{p}^{ind} = \alpha \mathbf{G} \tag{6}$$

Since the total energy has terms that are both linear and quadratic in the dipole moments, a self-consistent variational principle can be applied if the ratio of polarizability to molecular volume is small enough. Under these conditions one insists that the total energy should be stationary to small varations in the dipole moments. It can be verified that the condition for minimal energy is satisfied when the electric fields, $\mathbf{G}_{tot}(\mathbf{R}_n)$, due to all other dipoles, charges, and the external field lead to fields at each site that satisfy $\mathbf{p_{ind}}(\mathbf{R_n}) = \alpha\mathbf{G_{tot}}(\mathbf{R_n})$.

Since this method, with slight improvements, provides the only computationally feasible method for the molecule discussed in the next session, it is worthwhile discussing some early uses of this method to illustrate its accuracy. In Ref. [11] this method was used to calculate the dielectric constant for an FCC lattice of C_{60} molecules and to calculate the screened Hubbard U parameter for the K_3C_{60} superconductor. Overall the agreement between theory and experiment was found to be reasonably good. For example, within LDA, Pederson and Quong calculated the polarizability of a C_{60} molecule to be approximately 80-82 $\mathring{A}^3$ and found a resulting dielectric constant of 3.66 which is in reasonable agreement with experiment (roughly 4.0). [32]. From the same methodology, it is possible to determine the screened hubbard U which was found to be 1.27 eV in Ref. [11]. This is in reasonably good agreement with experimental measurements in the range of 1.4-1.6 eV. See Ref. [33] for further details about the experiments.

A few remarks on the caveat that the polarizability must be small enough may be useful. If the polarizabilities used in Eq. 5 are too large one finds that a self-consistent solution to this equation is not possible. Instead the energy diverges with the dipoles continuing to grow for each iterative cycle. Clearly if the above methodology leads to a total dipole moment significantly larger than a linear dimension of one of the molecules it is indicative that that the lowest-order penalty function is not enough to describe the system and that higher-order refinement of the energy functional may be required. This behavior is related to the Clausius-Mossotti polarizatin catastrophe. [32]

5 Application to a Light Harvesting Molecule

In this section we illustrate how polarization effects can play an important role in charge transfer processeses. In order to do this it is necessary to modify the form of the energy-penalty function to account for an anisotropic polarizability tensor. We have studied a bio-mimetic molecule that converts solar energy into an electric dipole. The molecule contains a chromophore as well as electron-acceptor and an electron-donor components [21, 22]. The chromophore is a diaryl-porphyrin, the acceptor is a pyrrole-C_{60}, and the donor is a carotenoid connected to the porphyrin through an amide (Fig. 1). Absorption of light at 590 nm by the porphyrin prepares the initial excited state and the charge transfer (CT) occurs through a multistep process resulting in the final charge separated (CS) state with the electron on the C_{60} and the hole on the carotenoid. The CS state which is quite long lived, has a large dipole moment of 153 Debye. For the charge separation process, we have considered a limited number of 100 singly excited states in our density functional calculation [22]. Some of the excited states are localized excitations which have smaller dipole moments but the charge separated states with the electron and hole on different components can have large dipole moments. The experiments on such molecular systems are done in solution and it was determined that the solvent polarity is strongly linked to the occurrence of the CT excitation. It was observed that the CS occurs in the polar benzonitrile and in 2-methyltetrahydrofuran solvents but does not occur in the nearly nonpolar toluene solvent.

The polarizability of the triad is calculated using the finite field method described above and we find that the polarizability of the triad is quite large - the diagonal components of the polarizability

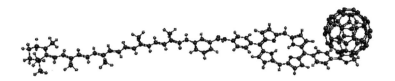

Figure 1: The geometry of the molecular triad shown in the color scheme C=green, H=grey, O=red, N= blue.

tensor are 878, 220, and 199 $\mathring{A}^3$. The largest value is along the molecular axis. The polarization effects become important when we consider a solution or a crystal. For simplicity, if we consider an assembly of the molecule as a 3-dimensional lattice, then when one of the molecule transits to a CS state with a large dipole moment, it will induce dipole moments on other molecules. The field due to the induced dipoles in turn lowers the energy of the CS state. This stabilization energy is larger for excited states with large dipole moments and also depends on the molecular volume. For large molecular volume, the stabilization energy is small while it increases for small volumes. As the volume decreases, it reaches the Clausius-Mossotti dipole catastrophe discussed in the previous section. The stabilization energy as a function of dipole moments and at different volumes is shown in Fig. 2. This stabilization effect has an important consequence in case of the molecular triad. The energies of the CS states in *vacuo* are higher than that of the initial porphyrin excited state. Now, due to the polarization induced lowering of energy of the large dipole states, it also brings the energies of CS states below the porphyrin excited state thus bringing the calculated levels into good agreement with experiment. These results indicate the important role that polarization can play in photo-induced charge transfer reactions.

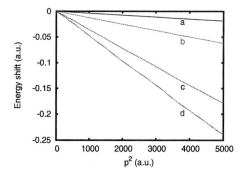

Figure 2: The shift in energy of the excited states as a function of square of dipole moment p and also as a function of molecular volume [$a = 120 \times 42 \times 42, b = 105 \times 36.75 \times 36.75, c = 100 \times 31.5 \times 31.5, d = 100 \times 31 \times 31$ (a.u.).]

6 Summary

In this work we have presented density-functional-based calculations on polarizabilities of several molecules. The results have been generated using reasonably large basis sets using the GGA and LDA versions of DFT. The agreemement with experiment is generally good for the systems presented. In addition we have discussed a methodology for using calculated (or measured) molecular polarizabilities to understand the polarization phenomena in collections of weakly interacting and nearly nonoverlapping molecules. The results in this case, while limited, suggest that polarization phenomena in such systems may be reasonably well accounted for. As a very large scale example we have discussed recent calculations on a light-harvesting organic photovoltaic. [22] Our results show that polarization effects in such charge-transfer systems can be large enough to change the ordering of excited states. While improvements of this method could and will be achieved without significantly increasing the computational burden, the results clearly show that such simplified polarization models allow for reliable computational investigations.

Acknowledgment

T. Baruah was supported in part by NSF grant number NIRT-0304122. M.R. Pederson was supported in part by an HPCMO CHSSI grant. Computational resources for this project were provided the the DoD HPCMO. We thank Prof. P.B. Allen and Dr. R.R. Zope for stimulating and helpful discussions.

References

[1] For a recent survey on many aspects of polarizabilities in molecules and clusters, See *Computational aspects of electric polarizability calculations: Atoms, Molecules and Clusters*, J. Comp. Meth. in Sci. and Eng. (2004) **4**, Issues 3 and 4, Ed. by G. Maroulis.

[2] D.M. Bishop, J. Chem. Phys. (1987) **86** 5613-5616.

[3] D. M. Bishop, Rev. Mod. Phys. (1990) **62**, 343-374.

[4] D.M. Bishop and B. Kirtman, J. Chem. Phys. **95**, 2646-2658 (1991);D.M. Bishop and B. Kirtman, J. Chem. Phys. **97**, (1992) 5255-5256.

[5] G. Maroulis, J. Chem. Phys. (2000) **113**, 1813-1820.

[6] G. Maroulis, J. Chem. Phys. (2004) **121** 10519-10524.

[7] E.A. Perpete, B. Champagne and B. Kirtman, Phys. Rev. B (2000) **61** 13137-13143.

[8] K. Jug, S. Chiodo, P. Calaminici, A. Avramopoulos and M.G. Papadopoulos, J. Phys. Chem. A (2003) **107**, 4172-4183).

[9] P. Caliminici, K. Jug and A.M. Koester, J. Chem. Phys. (1999) **111**, 4613-4620.

[10] A.J. Russell and M.A. Spackman, Mol. Phys. (1996) **88** 1109-1136).

[11] M.R. Pederson and A.A. Quong, Phys. Rev. B **46**, (1992) 13584-13591;A.A. Quong and M.R. Pederson, Phys. Rev. B **46**, R12906 (1992).

[12] S.A. Blundell, C. Guet and R.R. Zope, Phys. Rev. Lett. (2000) **84**, 4826-4829.

[13] J.M. Pacheco and J.L. Martins, J. Chem. Phys. (1997) **106**, 6039-6044.

[14] P. Hohenberg and W. Kohn, Phys. Rev. B **136**, (1964) 864; W. Kohn and L.J. Sham, Phys. Rev. A **140**, (1965) 1133.

[15] J. P. Perdew, J.A. Chevary, S.H. Vosko, K.A. Jackson, M.R. Pederson, D.J. Singh, C. Fiolhais, Phys. Rev. B **46** 6671 (1992).

[16] J. P. Perdew, K. Burke, and M. Ernzerhof, Phys. Rev. Lett. (1996) **77**, 3865-3868.

[17] M.R. Pederson and K.A. Jackson, Phys. Rev. B (1990) **41**, 7453-7461; K.A. Jackson and M.R. Pederson, Phys. Rev. B (1991) **42**, 3276-3281.

[18] M.R. Pederson, Phys. Stat. Solidi (2000) **217** 197-218.

[19] D. Porezag and M. R. Pederson, Phys. Rev. A (1999) **60**, 2840-2847.

[20] D. Porezag and M. R. Pederson, Phys. Rev. B (1996) **54**, 7830-7836.

[21] P.A. Liddell, D. Kuciauskas, J. P. Sumida, B. Nash, D. Nguyen, A.L. Moore, T. A. Moore, D. Gust, J. Am. Chem. Soc. **119**, 1400 (1997).

[22] T. Baruah and M. R. Pederson (submitted to PRL).

[23] D Bishop, L.M. Cheung and A.D. Buckingham, Mol. Phys. (1980) **41**, 1225-1226; D. Bishop Mol. Phys. (1981) **42** 1219-1232.

[24] D.M. Bishop and L.M. Cheung, J. Phys. Chem. Ref. Data (1982) **11**, 119-133.

[25] J. Marti and D.M. Bishop, J. Chem. Phys. (1993) **99**, 3860-3864.

[26] J. Guan, M.E. Casida, A.M. Koester, and D.R. Salahub, Phys. Rev. B (1995) **52**, 2184-2200.

[27] M.R. Pederson, T. Baruah, P.B. Allen and C. Schmidt, J. Chem. Theory. Comput. 1, 590-596 (2005).

[28] O.P. Andrade, A. Aragao, O.A.V. Amaral, T. L. Fonseca, M.A. Castro, Chem. Phys. Lett. (2004) **392**, 270-275.

[29] J. F. Biarge, J. Herranz, and J. Morcillo, An. R. Soc. Esp. Fix. Quiml A (1961) **57**, 81.

[30] W. B. Person and J. H. Newton, J. Chem. Phys. (1974) **61**, 1040-1049.

[31] M. Gussoni, M. Rui, and G. Zerbi, J. Mol. Str. (1998) **447**, 163-215.

[32] P.B. Allen, J. Chem. Phys. **120**, (2004) 2951.

[33] O. Gunnarsson, Rev. Mod. Phys. **69**, (1997) 575.

[34] *Landoldt-Börnstein: Numerical Data and Functional Relationships in Science and Technology*, Springer-Verlag (1982); N. J. Bridge and A. D. Buckingham, Proc. R. Soc. London Ser. A **295**, 334 (1966). G. W. Hills and W. J. Jones, J. Chem. Soc. Faraday Trans. II **71**, 812 (1975).

[35] G. R. Meredith, B. Buchalter and C. Hanzlik, J. Chem. Phys. **78**, 1543 (1983).

[36] K.D. Bonin and V.V. Kresin, *Electric-Dipole Polarizabilities in Atoms, Molecules and Clusters* (World Scientific, Singapore, 1997); P.C. Eklund, A.M. Rao, Y. Wang, P. Zhou, K.A. Wang, J.M. Holden, M.S. Dresselhaus, and G. Dresselhaus, Thin Solid Films, (1995) **257** 211-232; P. Antoine, Ph. Dugourd, D. Rayane, E. Benichou, M. Broyer, F. Chandezon, and C. Guet, J. Chem. Phys. (1999) **110**, 9771-9772; A. Ballard, K. Bonin, and J. Louderback, J. Chem. Phys. (2000) **113**, 5732-5735.

[37] W.D. Knight, K. Clemenger, W.A. de Heer, and W.A. Saunders, Phys. Rev. B (1985) **31**, 2539-2540; R.W. Molof, H.L. Schwartz, T.M. Miller, and B. Bederson, Phys. Rev. A. (1974) **10**, 1131-1140; R. Antione, D. Rayane, A.R. Allouche, M. Aubert-Frecon, E. Benichou, F.W. Dalby, Ph. Dugourd and M. Broyer, J. Chem. Phys. (1999) **110**, 5568-5577.

[38] B. Powell, T.Baruah and M.R. Pederson (unpublished data).

Brill Academic Publishers
P.O. Box 9000, 2300 PA Leiden
The Netherlands

*Lecture Series on Computer
and Computational Sciences*
Volume 3 , 2005, pp. 168-179

Vibrational spectra of medium size molecular systems from DFT and ab-initio quartic force field calculations

C. Pouchan[1], D. Bégué, P. Carbonnière, N. Gohaud

Université de Pau et des Pays de l'Adour
Laboratoire de Chimie Théorique et Physico-Chimie Moléculaire
UMR 5624 – FR CNRS IPREM 2606
IFR Rue Jules Ferry – BP 27540
64075 Pau - France

Received 15 July, 2005; accepted in revised form 12 August, 2005

Abstract: This paper deals with the computation of vibrational transitions by quantum mechanical methods. It shows the capability of a hybrid CCSD(T)/B3LYP approach in which CCSD(T) equilibrium values and harmonic wave-numbers are coupled to B3LYP anharmonic terms to describe the potential energy surface for medium size semi-rigid molecular systems. A general variational scheme to find approximate solutions of the spectral problem for the vibrational Hamiltonian is developed. The efficiency of this new parallel approach is demonstrated for several molecular systems as formaldehyde, acetonitrile, methyl-alkali compounds and its aggregates, vinylphosphine and vinylarsine.

Keywords:
Hybrid CCSD(T)/B3LYP potential; variational method; anharmonic vibrational spectra

PACS:
31.15.-p; 31.25.-v; 31.50.-x; 33.15.-e; 33.20.-t

1. Introduction

Currently there has been intense interest in the study of new computational methods for the calculation of vibrational anharmonic spectra, because IR, Raman and hyper-Raman spectroscopies are among the most powerful and useful techniques for characterizing molecules. While high resolution measurements are often obtained with great success through very sophisticated experimental apparatus, unambiguous assignment of each observed band is never straightforward. In such cases, theoretical study by using quantum mechanical methods becomes an invaluable tool. This theoretical approach requires, in general, two restricted steps: the construction of the potential energy surface (PES) and the solvation of the vibrational equation in order to obtain the anharmonic energy levels and consequently the fundamental, overtone and combination wave-numbers characterizing the vibrational spectra.

For the potential energy surface, in case of semi-rigid molecules, the potential function usually limited to the quartic terms, is based on the electronic energy data (sometimes gradients and Hessian data are concerned) calculated by sophisticated post-Hartree-Fock methods such as CCSD(T); MRCI... in a grid in which the required number of points grows drastically with the increase of the molecular size. Thus, it becomes difficult by using these correlated wave functions to determine a complete anharmonic force field for non-symmetric molecules containing more than four atoms. In order to calculate the potential function for larger molecules, the theoretical chemists have investigated two possible alternatives: the first one relies on the possibility of density functional theory (DFT) methods to approximate high-level post-Hartree-Fock calculations, the second follows the development of generalized least-squares methods in order to fit both energy and its derivatives to reduce the number of required calculations in the grid. Combination of these two possibilities appears as a good way to obtain the anharmonic potential function for medium size and large molecular systems.

---

[1] Corresponding author. E-mail: claude.pouchan@univ-pau.fr

The second step concerns the solving of the vibrational equation. One can approach this problem by using perturbational, variational or mixed perturbation-variation methods. Selection of the method usually depends on the nature of the problem but supplementary restrictions are often imposed by the large size of the molecule under consideration and the difficulty in this case to use a strict application of a variational scheme.

The present contribution focus on both the conditions for obtaining accurate quartic force field for medium size systems and the strategy to calculate anharmonic vibrations by using a parallel approach for a variational algorithm.

2. Method and computational details

2-1. PES and anharmonic force constants

The main difficulty in investigating vibrational spectra of large molecules comes from the determination of reliable quartic force fields, keeping in mind that the CCSD(T) approach is still limited to symmetric penta or hexa-atomic systems, due to its very unfavorable scaling with the number of active electrons.

One of the most significant conclusion of our recent investigation is that spectroscopic properties of semi-rigid molecules can be computed by the latest generation of the density functionals and that the results can be further improved by coupling structures and harmonic force fields computed at higher post HF levels with DFT anharmonic potential [1-3]. Among these hybrid potentials, the CCSD(T)/B3LYP approach is shown to give excellent results for small and medium size molecular systems because it has been observed that B3LYP anharmonicities come in close agreement with their CCSD(T) counterparts. For larger systems this hybrid potential approach can become prohibitive in CPU consumption especially when the system does not posses any symmetry; it is necessary then to use a DFT quartic force field. For organic systems, systematic studies undertaken for small and medium size molecules have already shown that among current DFT methods (LDA, BLYP, BP86, B3LYP, B97-1), the functionals based on the generalized gradient approximation GGA and their hybrid counterparts generally provide accurate results for structural and electronic properties. For vibrational computations our recent investigations point out that GGA functionals generally are not sufficiently accurate and that promising results can be expected from B3LYP and B97-1 hybrid calculations. As, moreover, several studies have shown that diffuse basis functions are much more important for optimizing the DFT performances than larger expansions of valence orbitals and additional polarization functions we conclude that B3LYP (or B97-1) functionals combined with TZ2P+ [as 6-31 + G**] basis sets perform remarkably well the vibrational spectra and present for large systems the best compromise between results quality and computational time [4].

From the knowledge of the potential energy surface, the calculation of the molecular vibration requires an analytical expression for the corresponding potential function. For semi-rigid systems in which anharmonicity is small one of the possible ways to describe modes and their combinations in the mid-infrared region is to write this potential function as a Taylor expansion series in the space of the dimensionless normal coordinate q:

$$V = \frac{1}{2!}\sum_i \omega_i \; q_i^2 + \frac{1}{3!}\sum_{i,j,k} \phi_{ijk} \; q_i \; q_j \; q_k$$

$$+ \frac{1}{4!}\sum_{i,j,k,l} \phi_{ijkl} \; q_i \; q_j \; q_k \; q_l \qquad (1)$$

This dimensionless normal coordinate space is constructed from curvilinear coordinate space or Simons-Parr-Finlan [5] and Morse-like coordinates for strongly anharmonic oscillators.

For small molecular systems, ω_i, ϕ_{ijk}, ϕ_{ijkl} are calculated from a CCSD(T)/cc-pVTZ or cc-pVQZ potential energy surface while a B3LYP/6-31 + G** PES is often used for large molecules.

For medium size systems a hybrid potential in which CCSD(T) equilibrium geometry and harmonic wave-numbers are coupled to B3LYP anharmonic cubic and quartic terms is known to give excellent results [1-3].

Determination of a complete quartic force field is an important step in the calculations. Quadratic, cubic and quartic terms are generally obtained either by fitting the electronic energy data calculated by ab-initio or DFT methods for various nuclear configurations close to the optimized geometry, or by a finite difference procedure for first or second analytical derivatives of the electronic energy with respect to the nuclear coordinates. In both cases the required number of ab-initio data points grows drastically with the increasing size of the molecules which becomes difficult for the determination of a complete quartic force field for medium and large molecular systems without symmetry. In order to reduce the number of calculations we have proposed and implemented in our REGRESS EGH [6] code a generalized least-squares procedure including in the same process of linear regression the value of the energy and the corresponding first analytical derivatives obtained at each point of a well-suited grid [7]. We have shown [7, 8] that the simplex-sum grid of Box and Behnken [9] truncated to the third sum seems the best one by fulfilling the criteria of efficiency and accuracy to determine the coefficients of the Taylor expansion series. The computational gain of this procedure open the door to obtaining accurate force field for larger molecules. For example in the H_2C_nO series by using a B3LYP/6-311 G* method this gain rises to a factor of 3 to 13 when n increases from 1 to 7. This procedure can be easily applied to systems in solution, with the effect of the polarity of the environment being evaluated by using the self-consistent reaction field approach through the Self-Consistent Isodensity Polarized Continuum Model (SCI-PCM) [10], or the so-called polarizable continuum model PCM [11], or the conductor-like polarizable continuum model CPCM [12]. These dielectric continuum theories are now widely used in describing the solvent effects in conjunction with quantum mechanical calculations due to the relative low cost of the calculation in comparison to an explicit solvent model.

2-2. Solution of the vibrational Schrödinger equation

Once the potential energy function is obtained, the second step in the vibrational energy-level calculations consists of solving the vibrational Schrödinger equation. The pure vibrational Hamiltonian can be defined by the following expression:

$$H = \frac{1}{2!}\sum_i \overline{\omega}_i \ p_i^2 + V \qquad (2)$$

where $\overline{\omega}_i$ is the harmonic wave-number for the mode i, p_i the conjugate momentum, and V the potential function expressed in the dimensionless normal coordinates space q and described previously as a Taylor expansion series for semi-rigid molecules.

The vibrational equation can be solved by using the perturbational [13-15], variational [16] or mixed perturbation-variation approaches [17-20].

The perturbational process is the simplest and fastest way to obtain the vibrational energy levels for large molecules, however this method often encounters problems for near-degenerate vibrational states as Fermi and Darling-Dennison resonances. An improvement can be achieved by treating the strongest interactions by diagonalizing the corresponding part of the Hamiltonian matrix [21]. The variational method has often been restricted to small systems due to the size of the matrix to be diagonalized. The techniques used in the variational process are often adapted from expertise acquired in the field of electronic calculations as VSCF [22-24], VMSCF [25], VCASSCF [26] and more recently both the vibrational mean field configuration interaction (VMFCI) [27] and the vibrational coupled cluster (VCC) [28] methods. When a configuration interaction (CI) is guessed by a preliminary perturbative treatment similar to CIPSI [29] as developed in our group [17-20], the term mixed perturbation-variation approach can be used. Indeed, the vibrational Hamiltonian representation is "in fine" diagonalized in a subspace built iteratively by means of a second-order Rayleigh-Schrödinger perturbation theory where only those configurations with weights greater than a given threshold are included in the primary subspace for the following iterations. The multireference vibrational function is then corrected to first order by the remaining states which interact weakly. The eigenvalues and eigenvectors of the corresponding vibrational levels are obtained by diagonalizing the Hamiltonian representation iteratively improved by using a Davidson's procedure [30]. Recently we have proposed a new variational scheme for the treatment of medium-sized systems [3, 31-33]. This approach consists of taking an inventory of the vibrational configurations potentially needed for the description of the problem (for example all fundamentals and some overtones in a spectral region) on the grounds of the potential terms values, taking into account the symmetry of each state and cutting the process into several spectral windows which are dealt in independent processes by utilizing a new parallel software P_Anhar_v1.0 [34].

The major advantage of this algorithm lies in the ability to provide, in several spectral ranges, smaller matrices that contains all the information needed, which makes the execution faster and perfectly adapted to parallel calculations.
In all cases in our applications we have diagonalized several submatrices including 40,000-50,000 configurations, which allows to obtain around 50 converged eigenvalues with an accuracy of 1 cm^{-1} in each spectral window studied.

All results presented in this paper are obtained by using mainly the generalized least-squares procedure, where energy and gradient are both fitted in the same regression process [6] for the determination of the anharmonic force field, and the variational scheme through the P_Anhar_v1.0 software [34] for calculating the vibrational energy levels.

The molecular systems studied, include formaldehyde (H_2CO), acetonitrile (CH_3CN), methyl-alkali compounds (CH_3X; X = Li, Na, K) and their aggregates, ethylene oxide (C_2H_4O), vinyl-phosphine ($CH_2CH\text{-}PH_2$) and vinyl arsine ($CH_2CH\text{-}AsH_2$). All calculated spectra are briefly discussed in light of the experimental data.

3. Results and discussion

3-1. Formaldehyde: gas phase and solution

Formaldehyde is an excellent example to test methods and computational conditions to obtain vibrational spectra. Indeed, experimental data are well known and the small size of this system allows reference calculations by using the most sophisticated methods.

We have reported in Table 1 the fundamentals wave-numbers obtained by using our variational approach from a quartic force field obtained at B3LYP/6-31 + G**; CCSD(T)/cc-pVQZ and CCSD(T)/cc-pVQZ//B3LYP/6-31 + G** (noted CC//B3) levels.

The Coriolis coupling correction is taken into account by a perturbative approach for all calculated wave-numbers for the gas phase.

Table 1: Calculated anharmonic wave-numbers (cm^{-1}) for H_2CO gas at different level of theory

Mode	B3LYP/6-31 + G**	CCSD(T)/cc-pVQZ	CC//B3	Exp.
$\nu_6(b_2)$	1170	1165	1170	1167
$\nu_5(b_1)$	1237	1246	1250	1249
$\nu_3(a_1)$	1499	1499	1506	1500
$\nu_2(a_1)$	1794	1750	1754	1746
$\nu_1(a_1)$	2774	2791	2786	2782
$\nu_4(b_1)$	2832	2848	2847	2843
$\Delta\bar{\nu}$	13.8	4	4.3	
$\Delta\bar{\nu}_{Max}$	48	9	8	

The average and the maximum deviation relatively to experimental data are remarkably good for the CCSD(T)/cc-pVQZ and the hybrid potential CC//B3 approaches. The same is observed for the B3LYP/6-31 + G** method except for the ν_{CO} mode incorrectly reproduced.

For formaldehyde in solvents, the effect of the environment on the vibrational frequencies can be evaluated using the so called polarizable continuum model. The results presented here are obtained with the SCI-PCM model [10]. We have reported in Table 2 the anharmonic spectra calculated for H_2CO in acetonitrile as solvent. As for the gas phase the same conclusions can be made regarding the quality of the results obtained at different levels of theory.

Table 2: Calculated anharmonic wave-numbers (cm^{-1}) for H_2CO in the solvent acetonitrile, at different level of theory

Mode	B3LYP/6-31 + G**	CCSD(T)/cc-pVQZ	CC//B3	Exp.
$\nu_6(b_2)$	1172	1166	1172	
$\nu_5(b_1)$	1235	1243	1249	1247
$\nu_3(a_1)$	1494	1501	1503	1503
$\nu_2(a_1)$	1763	1731	1739	1726
$\nu_1(a_1)$	2804	2812	2817	2808
$\nu_4(b_1)$	2848	2866	2861	2876
$\Delta\bar{\nu}$	15.5	3.75	6	
$\Delta\bar{\nu}_{Max}$	37	5	13	

If we turn our attention to the calculated and observed shifts due to the solvent effect (Table 3) we observe that regardless the level of theory that the sign and the magnitudes are in most cases correctly reproduced with the coupled-cluster approach giving the best results.

Table 3: Calculated and observed shifts (cm^{-1}) for H_2CO in acetonitrile relative to the gas phase

Mode	B3LYP/6-31 + G**	CCSD(T)/cc-pVQZ	CC//B3	Observed
ν_5	-2	-3	-1	-2
ν_3	-5	+2	-3	+3
ν_2	-31	-19	-15	-20
ν_1	+30	+21	+31	+26
ν_4	+16	+18	+14	+33

We have carried out complete anharmonic calculations for numerous solvents in which H_2CO is embedded in a cavity surrounded by a dielectric continuum of permitivity ε, and presented in figure 1 the solvent effect on the carbonyl stretching mode for which experimental data are well known. Solvents chosen are cyclohexane (ε = 2.02), chloroform (ε = 4.81), THF (ε = 7.58), acetonitrile (ε = 37.5), DMSO (ε = 46.7) and water (ε = 78.4). The curves $\Delta\bar{\nu} = f(\varepsilon)$ show the same behavior; B3LYP/6-31 + G** predicts the largest red shift, while CCSD(T)/cc-pVQZ predicts the smallest ones, with experimental values being in between. The shift to lower wave-numbers with increasing solvent polarity can be easily explained by the increase of the C = O bond length due to the induced effect of the solvent.

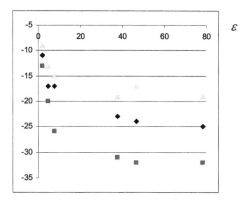

Figure 1: calculated and experimental shifts for $\bar{\nu}_{CO}$ in different solvents. B3LYP/6-31+G** (squares); CCSD (T)/cc-pVQZ (triangles) ;experimental values (rhombuses)

3-2. Ethylene-oxyde

We have calculated for the ethylene oxide C_2H_4O the quartic force field at CC//B3 level of theory. From this quartic force field we have calculated and compared (Table 4) the vibrational spectra obtained by our variational method [31-34] and the vibrational mean-field configuration interaction VMFCI [27] approach proposed by Lievin and Cassam-Chenaï. The results (to be published) are very similar, with the mean deviation found, for fundamentals, to be smaller than 0.2%, and the first harmonic $(2\nu_{10})$, relative to the scissoring mode strongly coupled with the CH_2 stretching vibration to be 1.2%.

Our computational method has produced remarkably accurate results and therefore further suggest reassignment of some of the experimental vibrational modes, in particular the ν_7 mode.

Table 4: Calculated anharmonic wave-numbers (cm^{-1}) for C_2H_4O by using our variational method [31-34] and the VMFCI approach of Lievin and Cassam-Chenaï [27].

Mode	VMFCI	Our results
ν_1	2921	2922
ν_2	1500	1499
ν_3	1273	1271
ν_4	1126	1120
ν_5	879	881
ν_6	3034	3027
ν_7	1151	1152
ν_8	1020	1024
ν_9	2915	2908
ν_{10}	1471	1474
ν_{11}	1128	1130
ν_{12}	823	820
ν_{13}	3044	3041
ν_{14}	1149	1151
ν_{15}	794	800

3-3. Acetonitrile

By using a complete CC//B3 quartic force field and taking into account ten spectral windows and two irreducible representations which lead to twenty submatrices with dimensions included between 10 000 and 12 000 configurations, we have obtained the complete vibrational spectra of acetonitrile [1] in the medium IR region.

For the 31 bands observed in the gas phase our theoretical results give a mean deviation smaller than 0.8%. For fundamentals (Table 5) this mean deviation is only 0.7%, the greater gap being observed for the methyl bending mode v_6 (1.7%).

Table 5: Experimental and calculated (cm-1) anharmonic spectra for fundamentals of acetonitrile

C_3v			Experimental	Theory	Deviation (%)
$v_8^1 (\pm 1)$	CCN bend.	E	365	366	0.2
v_4	CC strech.	A_1	920	916	0.4
$v_7^1 (\pm 1)$	CH$_3$ rocking	E	1041	1038	0.3
v_3	CH$_3$ sym. deformation	A_1	1390	1400	0.7
$v_6^1 (\pm 1)$	CH$_3$ d-deform.	E	1453	1478	1.7
v_2	CN strech	A_1	2266	2271	0.2
v_1	CH$_3$ sym. strech	A_1	2954	2993	1.3
$v_5^1 (\pm 1)$	CH$_3$ d-strech	E	3009	3043	1.1
				Mean deviation	**0.7%**

We note too an excellent agreement for some overtones and harmonics bands expected in the mid and near Infra-Red region (Table 6).

Table 6: Calculated anharmonic spectra for some harmonics and overtones of acetonitrile

	Experimental	Theory	$|\Delta v|$
$v_8^{\pm 1} + v_4 (E)$	1290	1282	8
$2v_8^0 (A_1)$	725	717	8
$v_4 + v_7^{\pm 1} + v_8^{\pm 1} (E)$	2316	2310	6
$v_3 + v_4 (A_1)$	2315	2305	10
$v_2 + v_4 (A_1)$	3182	3178	4
$v_1 + v_7^{\pm 1}$	3989	3980	9
$v_5^{\pm 1} + v_7^{\pm 1} PE$	4089	4049	40
$v_5^{\pm 1} + v_7^{\pm 1} PA$	4094	4083	11
$2v_5^0$	5966	5978	12
$2v_5^{\pm 2}$	6007	6003	4

The two resonances expected near 2300 and 1050 cm^{-1} reported in Table 7 are in very good agreement with experiment. The first one between $v_2(CN)$ and $v_3(\delta_s CH_3) + v_4(CC)$ can be explained by the cubic term W$_{234}$ calculated at 13.3 cm^{-1} and found to be 12.2 cm^{-1} from the experimental data. The second between v_7 (rocking CH$_3$) and $3v_8(\delta_{CCN})$, is due to the quartic term W$_{7888}$ (1.7 cm^{-1}) in agreement with the experimental determination.

Table 7: Fermi resonances calculated for acetonitrile

	Experimental	Theory	Δv
v_2	2266	2271	+5
$v_3 + v_4$	2305	2315	+10
v_7^1	1041	1038	+3
$3v_8^1$	1077	1094	+17

3-4. Methyl-alkali compounds CH$_3$X (X = Li, Na, K)

As the methyl alkali compounds have a strong tendency to aggregate and form oligomers, it presents real difficulty in assigning the IR bands characterizing the monomer. If we consider, as for the previous systems, that the combination of our CC//B3 potential function and our variational process gives the best theoretical results we can expect that the large difference observed between experimental data and theory cannot be explained by only the matrix effects.

We have reported in Table 8 the anharmonic results obtained for CH$_3$Li [31-32] and CH$_3$K [3]. Regarding $v(C - Li)$, and $\delta_s(CH_3)$ for the two systems, a complete disagreement with the experimental assignment is observed. These disagreements are confirmed by the isotopic shifts calculated for the isotopomer forms of the two monomers, and allow us to assign the observed bands at one of the oligomeric forms present in the matrix.

Table 8: Calculated anharmonic wave-numbers and observed data for methylpotassium and methyllithium monomers.

	X = K		X = Li	
	Exp.	Theory	Exp.	Theory
v_{CX}	280	295	530	606
δ_{HCX}	307	304	408	413
$\delta_s CH_3$	1053-1062	931	1158	1069
$\delta_{as} CH_3$	1384	1391	1387	1421
$v_s CH_3$	2732	2824	2780	2823
$v_{as} CH_3$	2775	2874	2820	2861

A complete study of the quartic force field for the dimer (CH$_3$Li)$_2$ [32] confirms that the bands observed at 530 and 1158 cm^{-1} are assignable to the dimer (calculated values 524 and 1174 cm^{-1}).

For the methylpotassium umbrella mode the disagreement is again explained by the presence of the aggregates since the corresponding bands are expected about one hundred cm^{-1} higher in the dimer or the trimer than in the monomer. Differences observed for the stretching vibrations are explained once more by the oligomeric forms for which the stretching modes are found about one hundred cm^{-1} lower than for the monomer.

More complicated is the methylsodium case [35]. Our first results show that, except for δ_{HCN_a} and $\delta_{as} CH_3$, the vibrational spectra must be revisited and the observed bands reassigned.

3-5. Vinylphosphine

Vinylphosphine $CH_2 = CH\ PH_2$ is the phosphorus analog of vinylamine $CH_2 = CHNH_2$. Contrary to the vinylamine where only the gauche form has been observed experimentally, vinylphosphine presents two stable conformers: the syn and the gauche forms. Experimentally the vibrational bands of the two conformers were observed and assigned on the basis of DFT harmonic calculations [36]. From a complete CC//B3 quartic force field we have completely reassigned the IR spectra of the syn and gauche conformers. We have reported in Table 9 our new theoretical results and included the experimental data for comparison.

Table 9: Observed and calculated vibrational wave-numbers for the two conformers of vinylphosphine

	Gauche		Syn	
	Obs.	Calc.	Obs.	Calc.
ν_1	3081	3075	3081	3075
ν_2	3007	3002	3012	3014
ν_3	2966	3018	2971	3013
ν_4	2300	2308	2298	2293
ν_5	2298	2294	2296	2290
ν_6	1603	1606	1605	1606
ν_7	1401	1399	1401	1400
ν_8	1274	1271	1270	1267
ν_9	1074	1074	1093	1088
ν_{10}	1048	1042	1033	1042
ν_{11}	982	981	991	991
ν_{12}	923	918	952	948
ν_{13}	850	847	881	876
ν_{14}	831	830	818	817
ν_{15}	680	648	682	673
ν_{16}	-	504	-	514
ν_{17}	-	327	-	325
ν_{18}	-	143	-	114

It is worth noticing that the agreement between the calculated and the observed data is remarkably good, better than 0.6% (8.5 cm^{-1}) for the gauche conformer and 0.4% ($\approx$ 6.5 cm^{-1}) for the syn. The largest deviations are found for the CH stretching modes ν_3 assigned below 3000 cm^{-1} and for the C-P stretching vibration ν_{15} relative to the gauche conformer. The first disagreement can be explained by the presence of overtone band in this spectral area and the second one to the weakness of the band. Note that the frequency difference observed between the two isomers for ν_{11} (CH bending out of plane), ν_{12} (CH$_2$ wagging), ν_9 (PH$_2$ scissoring), ν_{14} (PH$_2$ wagging) equal respectively to 9, 29, 19, 31 and 13 cm^{-1} are in fair agreement with our calculations which predict these differences to be 10, 30, 14, 29 and 13 cm^{-1}.

On the basis of our results concerning the fundamental bands, the experimental spectrum is actually revisited and calculations of some overtone and harmonic bands are in progress (to be published in collaboration with J.C. Guillemin and A. Benidar – University of Rennes - France).

3-6. Vinylarsine

We present here our first results concerning the vibrational spectra of vinylarsine. Contrary to the previous examples where the quadratic force fields were established at CC//B3 level of theory, the anharmonic potential function is obtained at B3LYP/6-311 + G** level and the resulting vibrational spectra, from a perturbational method.

We have reported in Table 10, only the observed vibrational bands for the two conformers (gauche and syn) and the corresponding calculated wave-numbers. Agreement is fairly good, with the mean deviation for the two conformers being 13 cm^{-1}, as in the vinylphosphine case where the largest deviations are observed for the CH stretching modes.

Table 10: Observed and calculated vibrational wave-numbers for the two conformers of vinylarsine

	Gauche		Syn.	
	Obs.	Calc.	Obs.	Calc.
ν_1	3076	3052	3076	3051
ν_2	3006	3013	3006	3013
ν_3	2959	2997	2959	3008
ν_4	2095	2126	2095	2104
ν_5	2092	2100	2092	2101
ν_6	1601	1615	1601	1617
ν_7	1391	1394	1391	1394
ν_8	1262	1262	1253	1249
ν_9	-	1020	1002	1010
ν_{10}	982	996	989	998
ν_{11}	956	967	976	972
ν_{12}	929	941	945	960
ν_{13}	773	776	786	793
ν_{14}	758	763	737	754

As predicted from our calculations we observe the two AsH_2 stretching (ν_4 and ν_5) modes spaced by 3 cm^{-1} in the case of the syn conformer. This shift predicted to be 26 cm^{-1} for the gauche form is not observed experimentally. For the CH in plane bending mode (ν_8), the CH_2 wagging (ν_{12}) and the AsH_2 twisting vibrations (ν_{13}) the isomeric effects are clearly discerned and confirmed by our calculations. This theoretical confirmation is not assumed for the AsH_2 scissoring (ν_{11}) and AsH_2 wagging (ν_{14}) modes. Our theoretical work is actually in progress to propose an accurate CC//B3 force field open to improve, by using a variational treatment, the description of the vibrational spectra. Supplementary investigations of the IR spectra are also in progress in collaboration with A. Benidar and J.C. Guillemin (to be published).

4. Conclusion

In conclusion we have shown that vibrationnal calculations provide very valuable information for the experimentalists to understand, assign or look for the absorption bands.
For medium size molecular systems it is not as easy to obtain theoretical data with the same accuracy as for a triatomic system but we have shown that the methods presented here for the determination of

the potential function and for the resolution of the vibrational equation provide news insights for the study of larger scale molecular systems.

Acknowledgments

The authors wish to thank D.Y. Zhang for the proof reading manuscript.

References

[1] D. Bégué, P. Carbonnière and C. Pouchan, *J. Phys. Chem. A.* **109** 4611 (2005).

[2] V. Barone, P. Carbonnière and C. Pouchan, *J. Chem. Phys.* **122** 224308/1 (2005).

[3] N. Gohaud, D. Bégué and C. Pouchan, *Int. J. of Quantum Chem.* in press.

[4] P. Carbonnière, T. Lucca, C. Pouchan, N. Rega and V. Barone, *J. of Comput. Chem.* **26** 384 (2005).

[5] G. Simons, R.G. Parr and J.M. Finlan, *J. Chem. Phys.* **59** 3229 (1973).

[6] P. Carbonnière, D. Bégué, A. Dargelos and C. Pouchan, *Regress EGH code.* LCTPCM – UMR 5624 (2000).

[7] P. Carbonnière, D. Bégué, A. Dargelos and C. Pouchan, *Chem. Phys.* **300** 41 (2004).

[8] P. Carbonnière, D. Bégué and C. Pouchan, *Chem. Phys. Lett.* **393** 92 (2004).

[9] G.E.P. Box and D.W. Behnken, *Ann. Mat. Stat.* **31** 838 (1960).

[10] J.B. Foresman, T.A. Keith, K.B. Wiberg, J. Snoonian and M.J. Frisch, *J. Phys. Chem.* **100** 16098 (1996).

[11] V. Barone, R. Improta and N. Rega, *Theor. Chem. Acc.* **111** 237 (2004).

[12] M. Cossi, N. Rega. G. Scalmani and V. Barone, *J. Comput. Chem.* **24** 669 (2003).

[13] D. Papousek and R. Aliev, *Molecular Vibrational Rotational Spectra.* (Elsevier: Amsterdam 1982).

[14] J. Pliva, *J. Mol. Spectrosc.* **139** 278 (1990).

[15] A. Willets and N.C. Handy, *Chem. Phys. Lett.* **235** 286 (1995).

[16] B.T. Sutcliffe, Chem. Modelling: Applications and Theory Vol. 3: Royal Society of Chemistry (2004).

[17] C. Pouchan and K. Zaki, *J. Chem. Phys.* **107** 342 (1997).

[18] C. Pouchan, N. Aouni and D. Bégué, *Chem. Phys. Lett.* **334** 352 (2001).

[19] D. Bégué, P. Carbonnière and C. Pouchan, *J. Phys. Chem. A* **105** 11379 (2001).

[20] D. Bégué, P. Carbonnière and C. Pouchan, *J. Phys. Chem. A* **106** 9290 (2002).

[21] J. M.L. Martin, T.J. Lee, P.R. Taylor and J.P. François, *J. Chem. Phys.* **103** 2589 (1995).

[22] G.D. Carney, L.L. Sprandel and C.W. Kern, *Adv. Chem. Phys.* **37** 37 (1978).

[23] R.B. Gerber and M.A. Ratner, *Chem. Phys. Lett.* **68** 195 (1979).

[24] H. Romanovsky, J.M. Bowman and L.B. Harding, **82** 4155 (1985).

[25] F. Culot and J. Lievin, *Theor. Chim. Acta* **89** 227 (1994).

[26] F. Culot, F. Laruelle and J. Lievin, *Theor. Chim. Acta* **92** 211 (1995).

[27] P. Cassam-Chenaï and J. Lievin, *Int. J. of Quantum Chem.* **93** 245 (2003).

[28] O. Christiansen, *J. Chem. Phys.* **120** 2140 (2004).

[29] B. Huron, J.P. Malrieu and P. Rancurel, *J. Chem. Phys.* **58** 5745 (1973).

[30] E.R. Davidson, *J. Comput. Phys.* **17** 87 (1975).

[31] N. Gohaud, D. Bégué and C. Pouchan, *Chem. Phys.* **210** 85 (2005).

[32] N. Gohaud, D. Bégué, C. Darrigan and C. Pouchan, *J. Comput. Chem.* **26** 743 (2005).

[33] D. Bégué, N. Gohaud and C. Pouchan, *Lecture Series on Computer and Computational Sciences* **817** (2004) and **828** (2004).

[34] D. Bégué, N. Gohaud and C. Pouchan, P_Anhar_v1.0 Code: LCTPCM – UMR 5624 FR CNRS IPREM 2606 (2004).

[35] N. Gohaud, D. Bégué and C. Pouchan, to be published.

[36] A. Benidar, R. Le Douven, J.C. Guillemin, O. Mó and M. Yáñez, *J. Mol. Spectrosc.* **205** 252 (2001).

Brill Academic Publishers
P.O. Box 9000, 2300 PA Leiden
The Netherlands

*Lecture Series on Computer
and Computational Sciences*
Volume 3, 2005, pp. 180-195

Protein Engineering : Role of Computational Methods in Design of Thermally Stable Proteins

V. Renugopalakrishnan,1[1] L. R. Lindvold, 2 S. Khizroev, 3 G. Narasimhan,4 Pingzuo Li 1,5

1. Children's Hospital, Harvard Medical School, 300 Longwood Avenue, Boston, MA 02115, USA
2. Optical Verification Components ApS, CAT Science Park, Frederiksborgvej 399, PO Box 30, DK- 4000 Roskilde, Denmark
3. Center for Nanoscale Magnetic Devices, Florida International University, 10555 W. Flagler St., Miami, FL 33174, USA
4. School of Computer Science, Florida International University, University Park, Miami, FL 33199, USA
5. Shanghai Research Center for Biotechnology, Chinese Academy of Sciences, Shanghai, China 200233

Received 1 August, 2004; accepted 12 August, 2005

Abstract: Our initial focus was on thermal stabilization of bR without loosing its unique photochemistry by stabilizing the photointermediates and enhancing their life times. Due to the unprecedented thermal stability of the genetically engineered bR designed by rational site-directed mutagenesis with physical dimensions in the nanoscale, it is believed that the protein-based devices might become future data storage systems of choice as the aerial densities go beyond 10 Terabit/in^2 mark and has the potential to reach 50 Terabit/in^2. bR mutants developed in our laboratories exhibit excellent physical and optical properties that make them ideal candidates for a storage material. Their naturally evolved properties, along with their genetically engineered variants, make them superior to any magnetic material used in present day memory devices. For example, the bR mutants with < 3 nm in dimension have demonstrated a long-term stability with a shelf life of 10 years at room temperature. In comparison, magnetic grains used in the best hard-drives today become highly unstable if the average grain size is reduced < 3 nm. Among other advantages of bR medium is its unprecedented recyclability and durability including resistance to microbial degradation. Finally, the bR media has demonstrated a faster time response, as compared to the magnetic disks, ps range versus ns range, respectively. The faster time response makes the bR media superiors with respect to the data transfer rate. However, before the protein-based data storage can be finally implemented, suitable methods for immobilizing bR mutants on anchoring platforms including SW CNT and adequate mechanisms for writing and reading information from this type of media should be investigated.

Keywords: Bacteriorhodopsin /Read-Write media/D96N bR/Thermal Stability of Proteins/ Thermodynamic Force Field/Photochemistry

1. Introduction

Proteins are naturally occurring nano systems with nm dimensions endowed with diverse properties. Advances in molecular biology have made it possible to manipulate proteins and selectively modify their properties through rational site-specific mutagenesis. For this, random site-specific mutagenesis has been the method of choice widely used in studies reported in the literature. In our laboratory we have been working on a rational site-specific mutagenesis approach, which rests on the foundation of optimizing specific physical properties by relying upon bioinformatics database. Specifically we have been focusing on the thermal and photochemical properties of light activated proteins.

Due to the unprecedented thermal stability of the genetically engineered bacteriorhodopsin designed by rational site-directed mutagenesis with physical dimensions in the nanoscale, it is believed

---

[1] Address correspondence to: Dr V. Renugopalakrishnan, Enders-12, Children's Hospital, Harvard Medical School, 300 Longwood Avenue, Boston, MA 02115, USA

that the protein-based devices might become future data storage systems of choice as the aerial densities go beyond 10 Terabit/in2mark. Bacteriorhodopsin (bR) mutants developed in our laboratories (Renugopalakrishnan et al., 2003 [1]; Renugopalakrishnan et al., US Patents [2]) exhibit excellent physical and optical properties that make them ideal candidates for a storage material. Their naturally evolved properties, along with their genetically engineered variants, make them superior to any magnetic material used in hard drives. Birge et al.,1999 [3] and Hampp [4] discuss the application of bR in 3D volumetric memory. For example, the bR mutants with less than 3 nanometers in diameter have demonstrated along-term stability with a shelf life of 10 years at room temperature. In comparison, magnetic grains used in the best hard-drives today become highly unstable if the average grain size is reduced below 3 nm. Among other advantages of bR medium is its unprecedented recyclability and durability including resistance to microbial degradation. For example, bR media can be rewritten 10^6 times [3,4,]. Finally, the bR media has demonstrated a faster time response, as compared to the magnetic disks, picosecond range versus nanosecond range, respectively. The faster time response makes the bR media superiors with respect to the data transfer rate. However, before the protein-based data storage can be finally implemented, suitable methods for immobilizing bR mutants on platforms (Tatke et al. [5]) and adequate mechanisms for writing and reading back information from this type of media should be investigated. The study of mechanisms of writing and reading from the protein-based media is the subject of this paper.

Enhancement of thermal stability of proteins has been the Holy Grail in protein engineering studies (Renugopalakrishnan et. al. [6]). Extensive studies of mesophilic, thermophilic, and hyperthermophilic proteins have contributed to a profound understanding of the factors that influence protein thermal stability. With the accumulated data on thermophiles and hyperthermophiles, it has been possible to derive a number of factors that contribute to the structural stability of these species and the proteins contained in them. Some of these factors include (i) amino acid composition and their intrinsic propensity (ii) disulfide bridges (iii) hydrophobic interactions (iv) aromatic interactions (v) hydrogen bonds (vi) ion-pairs. From our studies of thermophilic and hyperthermophilic proteins, we strongly believe that the thermal stability of a protein depends on enhancement of the heat capacity of a protein which in turn is dependent on its primary structure.

2. Random Mutagenesis

The ultimate goal of protein engineering is to design proteins to perform specific functions. However, the exact rules governing protein folding, molecular recognition, and their precise relationship to functions remain to be resolved which renders rational design of proteins quite difficult. Modification of proteins is carried out either via random site-directed mutagenesis or by combinatorial methods that combine random mutagenesis with phage display depending on the desired properties for the candidate proteins. These desired properties include improved bioactivity, photochemical properties, thermostability, etc. Factors that are important for the thermal stability or proteins, e.g.stabilization of α helices, reducing the number of conformations in the unfolded state [7,8]are taken into consideration. There are three widely used strategies of random mutagenesis. The first one is the oligonucleotide-directed mutagenesis. Oligonucleotides can be synthesized to contain mixtures of nucleotides at specific codons which include all the bases in the first two positions and only two bases in the third position, with each combination allowing 20 possible amino acids. The second one is the error-prone polymerase chain reaction (PCR),which uses a DNA polymerase to replicate the target gene. The third method is through DNA shuffling, whereby the gene is first spliced into pieces and then regenerated using a DNA polymerase. Since the pieces of DNA get mixed, mutations from separate copies of the gene can be combined. Because mutagenesis techniques differ in the number and dispersion of mutations introduced into a gene, the appropriate method of mutagenesis depends on the choice of physical properties which are chosen for optimization. The above methods have been used to design and produce a large number of random mutants, but most of them do not possess the desired properties. Therefore, an efficient screening method was introduced, and this was the in vitro phage display which links the protein library to DNA [9], namely, each member of the protein library is connected with its gene, so that the objective mutants can be readily amplified and identified. Specific enzyme inhibitors, increased enzymatic activity, identifying novel peptide ligands and agonists of receptors molecules are examples of the successful use of this method. By sorting these libraries produced by random mutation to select for a predicted function, the small number of active proteins can be separated from the millions of inactive variants.

3. RATIONAL SITE-DIRECTED MUTAGENESIS

Rational site-directed mutagenesis rests on the principle of optimizing a chosen physical property by analyzing the dependence of that property on specific candidate amino acids that are to be substituted by selecting known or putative sites in the protein under investigation. On absorption of light, bR undergoes a cascade of structural changes during its photochemical cycle, see Fig.1.

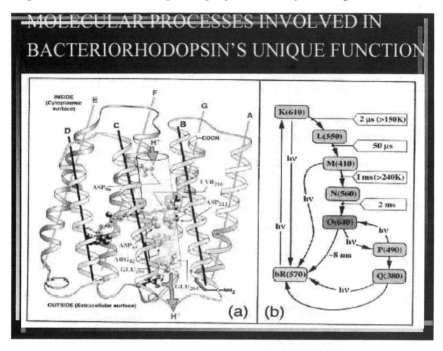

Figure 1: Photochemical Cycle of bR

The cis-trans isomerization, the heart of its photochemistry, triggers an avalanche of structural changes in the bR secondary structure which consists of at least three major steps: (a) the release of proton into the extracellular medium (b) the uptake of a proton from the cytoplasmic medium and (c) the thermal re-isomerization of retinal to the starting all-trans configuration. These structural changes are reflected as changes in the visible region of the spectrum as observed by the sequential formation and decay of the optical intermediates, J, K, L, M, N, and O (Lozier at al., 1975 [10]; EbRey, 1993 [11]).Extensive mutagenesis studies have established that the presence of a carboxylate group at Asp residue 96 is important for rapid reprotonation of the Schiff base (Gerwert et al.,1989 [12]; Marinetti et al., 1989 [13]; Titor et al., 1989 [14]; Otto et al., 1989 [15]; Millerand Osterhelt, 1990[16]). The protein conformational changes in the bacteriorhodopsin photocycle have been discussed by Subramaniam et al. [17]. Reprotonation of the Schiffbase by Asp96 is reflected by the spectroscopically detectable conversion of the Mintermediate to the N intermediate. Asp96 is reprotonated from the cytoplasmic medium, initiating the formation of O-intermediate. The mutation of Asp96 to Asn96 therefore influences the above process and confers long resident time or half-life for the M intermediate. Secondary structure of [D96-N] bR obtained from x-ray crystallography[18] is shown in Fig. 2 below and the intermediate M state (see Fig. 1) is shown in Fig. 3.

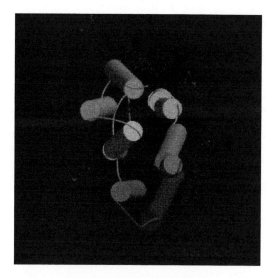

Figure 2: Structure of bR[D96 N]

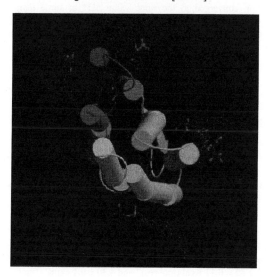

Figure 3: Structure of the excited M state

4. In silico Design of bR Mutants with High Thermal Stability Free Energy of Proteins

The minimum free energy, ΔG^0, can be described as follows:

$$\Delta G^0 = \Delta G^0{}_{\text{intrinsic}} + \Delta G^0{}_{\text{extrinsic}} \tag{1}$$

where ΔG^0 intrinsic is given by the Gibbs–Helmoltz equation:

$$\Delta G^0{}_{\text{intrinsic}} = \Delta_{\text{H0intrinsic}} - T \, \Delta S{}_{\text{intrinsic}} \tag{2}$$

and can be partitioned into individual components arising from constituent amino acid residues as:

$$\Delta G^0{}_{intrinsic} = \sum_{i=1}^{n} \Delta G^0{}_{intrinsic} \qquad (3)$$

where n= is number of residues

One of the fundamental goals in protein engineering is to generate thermally stable protein structures with large free energies of unfolding, ΔF_u, and high mid-point melting temperatures, T_m. The temperature dependence, ΔF_u, can be analyzed by examination of the temperature dependence of ΔH and ΔS. Assuming that the heat capacity, ΔC_p, of a protein is independent of temperature (at least in the range 20-200 ^{0}C) we have

$$\Delta H = \Delta H (T_r) + \Delta C_p (T - T_r) \qquad (4)$$
$$\Delta S = \Delta S (T_r') + \Delta C_p \ln (T / T_r') \qquad (5)$$

where, Tr and Tr' are appropriate reference temperatures. Note that

$$\Delta H (T_r) = \Delta H \text{ of the protein at the temperature } T_r \qquad (6)$$
$$\Delta S (T_r') = \Delta S \text{ of the protein at the temperature } T_r' \qquad (7)$$

The convergence temperatures [19,20] were chosen here because the reference ΔH for different proteins upon derivation converged to the same value ΔH^* at some temperature T_h^*. A similar convergence was also observed for the entropy change per residue, ΔS^*. In physical terms, the convergence temperatures are now believed to be the temperatures at which the polar contribution to ΔH^0 and ΔS^0 reaches zero, respectively.

Therefore,

$$\Delta H = N_{ref} \Delta H^* + \Delta C_p (T - T_h^*) \qquad (8)$$
$$\Delta S = N_{ref} \Delta S^* + \Delta C_p \ln (T / T_s^*) \qquad (9)$$

where N_{ref} = the number of amino acids in the protein. Therefore, the free energy of unfolding, ΔF_u, can be expressed as [19]:

$$\Delta F_u = N_{ref} (\Delta H^* - T \Delta S^*) + \Delta C_p [(T - T_h^*) - T \ln (T / T_s^*)] \qquad (10)$$

The above equation is quite useful when comparing the ΔF between a wild-type protein and its mutants with different values of ΔC_p.

$$\Delta \Delta F_u = \Delta \Delta C_p [(T - T_h^*) - T \ln (T / T_s^*)] \qquad (11)$$

From the above equations, it is evident that for a protein with N_{ref} residues, changes in ΔC_p can result in changes in ΔF_u that can be quantified at all the temperatures. For T_h^* and T_s^* of 100.5 ± 6 ^{0}C and 112 ± 1 ^{0}C, ΔH^* and ΔS^* are 1.35 ± 0.11 Kcal (mol.res)$^{-1}$ and 4.30 ± 0.1 cal K^{-1} (mol.res)$^{-1}$, respectively. Therefore,

$$\Delta F_u = N_{res} (1.35 \pm 0.0043 \text{ T}) + \Delta C_p [(T - 373.6) - T \ln (T / 385.1)] \qquad (12)$$

ΔT_m can be directly estimated from the changes in ΔC_p as follows:

$$\Delta T_m = (3.6 \pm 2.5) \Delta \Delta C_p \qquad (13)$$

5. Calculation of ΔC_p of proteins

For a protein, the quantity ΔCp can be expressed as a function of the change of buried apolar and polar surface areas upon unfolding [20]:

$$\Delta C_p = 0.45 \Delta ASA_{apolar} - 0.26 \Delta ASA_{polar} \qquad (14)$$

where ΔC_p is in calmol^{-1} K^{-1}. ΔASA_{apolar} and ΔASA_{polar} are the changes in solvent accessible surface area (unit Å^2) upon protein denaturation for apolar and polar atoms respectively, The constants are taken from model-compound studies [21] and have been shown to be reasonably accurate in calculating ΔF_u [22]. The above equation shows that increasing buried apolar surface area in the folded state will increase ΔC_p, while increasing buried polar surface area in the folded state will decrease ΔC_p. ΔC_p values of apolar and polar residues are listed in Table1.

Amino Acid	ΔC_p (cal/K/(mol-res))†
Ala	30.15
Ile	63.00
Leu	61.65
Met	41.47
Phe	78.75
Trp	80.73
Tyr	53.62
Tyr‡	70.00
Val	52.65

† The ΔCp values for totally buried hydrophobic amino acids were calculated from reported solvent accessible surface area.
‡ This value is used in special cases when the hydroxyl group of Tyr is solvent exposed.

6. Experimental Determination of ΔC_p

The quantity ΔCp can be experimentally obtained from differential scanning calorimetric studies (DSC). DSC measures the heat capacity of a protein solution as a continuous function of temperature. In a typical DSC measurement, the heat capacity of the sample cell containing the protein solution is measured in relation to that of the reference cell containing only the solvent or buffer. The measured heat capacity Cp (cal /K) can be written as:

$$C_p - m_p\, C_{p,p} + m_b\, C_{p,b} + C_{p,ref} \qquad (15)$$

where, $C_{p,p}$ and $C_{p,b}$ are the heat capacity of the protein and solvent/buffer, while m_b and m_s are the masses of protein and solvent/buffer, respectively. $C_{p,ref}$ includes the heat capacity of the buffer solution in the reference cell. Since the volume of the sample cell is constant, the volume of solvent (or buffer), and consequently its mass, depends on the mass of protein present in the cell, and the above equation can be written as:

$$C_p = m_p\, C_{p,p} + m_p\,.C_{p,b}\, \rho_b\, (V_o - V_p \,.\, m_p) + C_{p,ref} \qquad (16)$$
$$= C_{p,b} \,.\, \rho_b \,.V_o + (C_{p,p} - V_p.\, \rho_b \,.\, C_{p,b}) \,.\, m_p + C_{p,ref} \qquad (17)$$

where, ρ_b is the density of the solution, V_o is the volume of the cell, and V_p is the partial specific volume of the protein. Generally, V_p, ρ_b and $C_{p,b}$ are obtained from standard tables. From the above equation, the slope of a plot of C_p versus the mass of protein in the cell ($\delta C_p / \delta m_p$) provides all the information necessary to calculate the heat capacity, C_p. ΔC_p can be used to calculate T_m.

$$T_m = T_1^* \exp\,[-\Delta S\,(T_1)\,/\,\Delta C_p] \qquad (18)$$

7. Definition of T_{ms}

The temperature of maximal stability of a protein, denoted by T_{ms}, is defined as the midpoint temperature of transition,T_m, during heat denaturation of a protein. Proteins are characterized by two denaturation temperatures (T_d) at which $\Delta G_o = 0$, and Tw as determined by differential scanning calorimetric (DSC) studies. A plot of $\Delta G_o = f(T)$ is usually referred to as the stability curve of a protein. The values of T_{ms} and ΔC_p for a number of proteins are listed in Table 2 (adapted with permission from Ganesh et al., 1999 [22] and shown in Fig. 4).

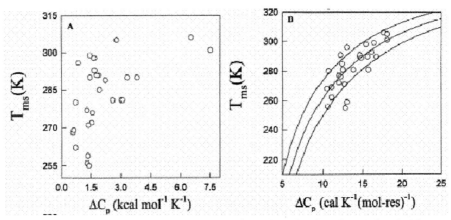

Figure 4: T_m of proteins (whose X-ray structures are known) as a function of ΔC_p.[adapted with permission from Ganesh et al., 1999] [26]

Table 2

Protein	PDB code	N_{s-s}	N_{res}	ΔA^a_{ap}	ΔC^b_p(obs)	ΔC^b_p (calc)	T_{mso}	T_{msc}
Ovomucoid III	1cho	3	56	2712	10.5	10.5	268	257
RNase A	9rsa	4	124	7170	10.6	12.6	256[d]	275
CSP[c]	1csp	0	67	3736	10.7	11.5	280	266
Protein G[c]	1pgb	0	56	2834	11.1	10.5	269	257
CI2[c]	2ci2	0	65	3409	11.1	11.4	262	265
Parvalbumin	5cpv	0	108	6393	12.0	13.5	277	280
Hen Lysozyme	6lyz	4	129	7701	12.0	12.8	272[d]	276
HH myoglobin[c]	1ymb	0	153	9757	12.2	15.3	291	291
Hu lysozyme	1lzl	4	130	8119	12.2	12.8	276	276
Interleukin 1β	5ilb	0	153	10175	12.4	14.4	285	286
Chymotrypsin	4cha	5	241	15383	12.5	14.4	281[d]	286
Iso-1-cytochrome c[e]	1ycc	1	108	6220	12.7	14.4	271	286
Barnase	1rnb	0	110	6166	12.8	13.6	255	281
Sac 7d	1sap	0	66	3446	13.0	11.5	296	265
RNase T1	9rnt	2	104	5847	13.0	12.6	259	274
Trypsin	1tkl	6	223	14161	13.8	14.1	281	284
α-Lactalbumin	1alc	4	123	7404	14.6	12.6	291	274
CAB[c]	2cab	0	260	16850	14.6	15.3	290	291
S. nuclease	2sns	0	149	8360	14.8	14.3	289	286
Thioredoxin	2trx	1	108	6701	15.4	13.1	298[d]	278
Papain	9pap	3	212	13776	15.6	14.5	290	286
T$ lysozyme	1l63	0	164	10024	15.7	14.5	281	287
Cytochrome c[e]	2pcb	1	104	5978	16.1	14.3	293	285
Barstar	1bta	0	89	5770	16.4	12.8	299	276
Hpr[c]	2hpr	0	87	5121	16.7	12.7	290	275
MBP[c]	1omp	0	370	25283	17.6	15.7	306	293
SW myogolbin[c]	5mbn	0	153	9964	18.1	15.3	305	291
PGK[c]	3pgk	0	415	26466	18.1	15.8	301	294

8. Hyperthermophilic Proteins

Hyperthermophilic organisms and the proteins found in them serve as an excellent testbed for the design of thermally stable proteins. Nature has optimized these proteins by the process of evolution, thus providing us with a wealth of principles for the design of thermally stable proteins.

There are currently three proposed models to explain the higher denaturation temperatures of thermophilic proteins. The free energy profile of a mesophilic protein and the proposed models for converting it into a thermophilic protein is shown in Fig. 5. This model was based on the observation that the thermophilic protein shows only a modest increase in ΔG_u at room temperature in comparison to their mesophilic counterparts. However, increase in ΔG_u at room temperature as high as 20 kcal mol^{-1} for thermophilic proteins have been reported [23], which lend support to the "raised" model. There is not much experimental support for the "shifted" model, because decreases in ΔG_u at room temperature have not been observed.

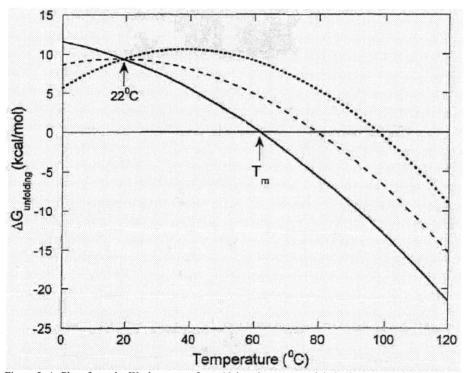

Figure 5: A: Plot of van der Waals energy after a 10-iteration energy minimization vs. van der Waals energy before minimization for random myoglobin structures sampled during a CORE run. B: Van der Waals energy plotted as a function of ΔC_p for these same myoglobin structures, showing no obvious correlation. The shaded region represents the range of ΔC_p values for predicted myoglobin sequences, black circles represent predicted myoglobin sequences, and open circles represent sequences with zero bumps sampled, but not predicted. C: Van der Waals energy after the 10-iteration energy minimization vs. ΔC_p, showing good correlation between these parameters (Jiang et al., Protein Science, 2000, Vol. 9, pages 403-416)

According to the theory presented in this chapter, any one of the three models alone is can not explain the protein thermal stability. First of all, trying to judge the dependence of free energy on temperature by looking at ΔGu values reported (in support of the "raised" model in Fig. 5) by assuming the same ΔCp value for both the thermophilic and mesophilic proteins [24]. The above assumption is incorrect as ΔCp difference is an important factor in determining protein stability.

A quantitative way to look at thermal stability is by taking the derivative of equation (10) with respect to temperature:

$$d(\Delta G)/dT = -N_{res} \times \Delta S^* + \Delta C_p - \Delta C_p \times \ln(T/T_s^*) \qquad (19)$$

Since this is the slope of the ΔG_u vs. T plot at any temperature T, at T_{max} (where ΔG_u reaches maximum), $d(\Delta G)/dT = 0$.

Therefore,

$$T_{max} = T_s^* . \exp(-\Delta S^* . N_{res}/\Delta C_p) = T_s^* \exp(-\Delta S^*/\text{per residue } \Delta C_p) \qquad (20)$$

Because the errors in the values of T_s^* and ΔS^* are relatively small, T_{max} therefore depends mainly on the value of ΔC_p and the nature and number of residues. Because of this conclusion, a thermophilic or hyperthermophilic protein and its mesophilic counterparts differ essentially on ΔC_p and consequently manifest different values of T_m.

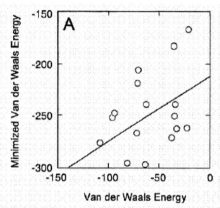

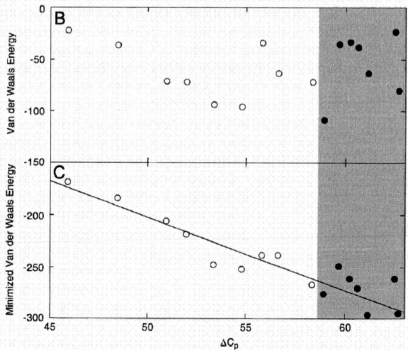

Figure 6: Dependence of ΔF_u on temperature for a hypothetical 100 residue protein at 3 different values of ΔC_p (10 and 20 cal/K/mol. res) are usually lower and upper limit of ΔC_p for most proteins.

9. Conformational Entropy, ΔS_{conf}

The conformational entropy, ΔS_{conf}, arises from the reorganization of hydrophobic amino acid side chains inside the protein and can be calculated by taking the difference of the entropy of the folded and unfolded states from the Boltzman equation:

$$\Delta S_{conf} = S_f - S_u = - R [\ln W^* - \ln W] \qquad (21)$$

where, S_f = entropy of the folded state, S_u = entropy of the unfolded state, W = the number of rotamers in unfolded state, and W^* = the effective number of rotamers in the folded state is given by:

$$W^* = \exp \left(- \sum_{i}^{n} p_i \ln. p_i \right) \qquad (22)$$

where p_i = the fractional population of each rotamer state, i, in the folded state and R = gas constant.
The conformational entropy change for the hydrophobic core of the entire protein, ΔS_{conf}, is defined as the mean of the individual conformational entropies for all core residues. The quantity ΔS_{conf} is an important component of protein stability. Van der waals interactions, E_{vdw}, makes significant contributions to protein stability.
Fig. 6 shows the dependence of ΔC_p on E_{vdw} for several myoglobin mutants. It shows a weak correlation between E_{vdw}, calculated using discrete rotamers library, and ΔC_p. The correlation improves after 10 iterations of minimization. Protein sequences with high ΔC_p values and side-chains compatible with the backbone of the protein, calculated by using a roamer library are stable proteins, are selected.

10. Thermodynamic Force Field for Proteins

The design of a thermally stable protein must include the following criteria:
(a) The side chains of the amino acid substituted to enhance Tm must be sterically compatible within the backbone structure.
(b) The mutated sequence must have a ΔCp value that can be achieved by burying more apolar surface area or by exposing more polar surface area in the folded protein.
(c) The hydrophobic core of the predicted protein must have low conformational entropy of folding to ensure a unique and rigid internal architecture.

11. Compensation and Convergence Temperatures

The convergence of thermodynamic quantities at some temperature will occur when there are two dominant interactions, such as apolar and polar, which independently contribute to the thermodynamic properties. If only apolar contributions are modified by mutation of core residues without significant effect on polar contribution or backbone structure, convergence should not be expected to occur.
It is expected therefore that within a series of proteins in which the backbone structure is maintained and only hydrophobic core residues are altered, the values of ΔH^*, ΔS^*, $T_H *$ and $T_s *$ will remain constant. Under these conditions, increases in ΔC_p will result in increases in T_m and ΔG_u above a fixed temperature.
Let us assume that for a specific protein the apolar contribution to ΔH° and ΔS° are equal to 0 at temperatures T^+ and T_s^+, respectively. Let us also assume ΔH^+ and ΔS^+, respectively. Equation (10) described previously can be used to calculate ΔG_u for any protein where ΔC_p is assumed to be temperature independent. Let us also assume that this specific protein has a compensation temperature of T_c. Based on the enthalpy–entropy compensation relationship, at the compensation temperature, T_c, we have

$$\Delta\Delta H = T. \Delta\Delta S \qquad (23)$$

and

$$(T_c - T_h^+) - T_c \ln (T_c - T_s^+) = 0 \qquad (24)$$

T_c is shared among a native protein and its mutants with approximately the same solution structure, as shown empirically. Based on the above analysis, we deduce T_c is the actual crossing point on the ΔG_u vs T plot for a group of proteins, with T_h^+ and T_s^+ as the temperature at which the apolar contribution to ΔH^o and ΔS^o reaches zero, respectively. The origin of convergence temperatures, T_c, is believed to be hydrophobic effect or the hydration of the apolar groups. Therefore the convergence temperatures (T_h^+ and T_s^+) depend on the composition, structure and the environment of each individual protein.

Enthalpy–entropy compensation is a common property of weak intermolecular interactions. Based on the relationship between compensation and convergence temperatures, it is therefore rational to conclude that T_c is the temperature at which the apolar and polar hydration effects reach certain equilibrium. The value of T_c depends on the primary structure of a protein and hence on its three dimensional structure.

Enthalpy–entropy compensation is a general feature of many chemical reactions and processes in biological systems. The slope of ΔH_o vs. ΔS_o plot is called the compensation temperature, T_c and the values for T_c falls in the range of 260 – 315°K. ΔG_o stays around a constant value. To engineer a protein at a given temperature, T_c, ΔS will also decrease by an equivalent amount. The important points we learn in both components, ΔH_{conf} and ΔS_{conf}, must be optimized simultaneously.

12. Computer Algorithm for in-silico Mutagenesis of Thermally Stable bR

We have developed a program named "Rational site-directed mutagenesis selection algorithm" (RSDMSA) (US Patent Disclosures [2], and Renugopalakrishnan et al. [1]), which was improved from an earlier program (CORE) developed by Jiang et al. [25]. RSDMSA predicts protein hydrophobic core sequences that can fold into a target backbone structure. Basically RSDMSA rejects or eliminates unfavorable mutations starting from an input backbone.

A typical prediction run in RSDMSA starts by randomizing the sequence of hydrophobic core of a selected protein to ensure that no bias is introduced at the start of a run. After this, a simulated annealing run driven by the Metropolis algorithm is initiated. Following the last step, a single mutation of a core residue chosen based on the specific property and nature of the protein is allowed. Residues are allowed to mutate to Ala, Ile, Met, Phe,Tyr, Trp, or Val. RSDMSA then intiates a nested simulated annealing run to determine the best rotamer configuration of all core residues. The simulated annealing run reveals the number of unfavorable van der Waals interactions by calculating the number of hard sphere bumps. If the number of bumps for the simulated annealing run is greater than zero, the sequence is rejected. If the number of bumps is zero, a second simulated annealing on the same sequence is initiated followed by a low temperature Monte Carlo run which yields two parameters :

(i)   mobility of each amino acid in the form of conformational entropy which is averaged to obtain global conformational entropy for the whole protein (ΔS_{conf}) as shown below

$$\Delta S = \sum_1^n \Delta S_{conf}^{residue} \tag{25}$$

(ii)   heat capacity (ΔCP) of the hydrophobic core of the protein calculated based on Table 1.

These two parameters, plus the number of bumps, are used to calculate the "scores" for this particular sequence. This score then drives the main sequence simulated annealing run. After 10 sequences are sampled at each temperature, the Metropolis temperature is gradually decreased during the simulated annealing process. This gradual decrease in the Metropolis temperature ensures that local minima are avoided by slowly lowering the probability that sequences are accepted with scores higher than the previously accepted sequence. The simulated annealing run is automatically terminated when the number of accepted sequences is consistently zero. The temperature at which this occurs, or an arbitrarily low temperature, T, is used for a final sequence Monte Carlo run initiated with the sequence determined from the simulated annealing run. This Monte Carlo run is conducted to sample the sequence space around the simulated annealing sequence. A large values of T allows the program to sample from sequences with wider range of scores around the score for the simulated annealing sequence. Typically these runs generate a family of 100-1000 sequences of proteins with good thermal stability depending on the value of T and the number of core residues mutated.

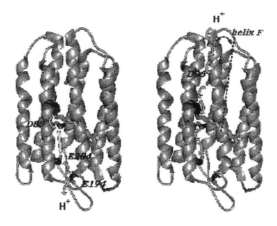

Figure 7: 3D structure of bR showing the trans-membrane helix

13. Design of Thermally Stable bR

Bacteriorhodopsin (bR): Nanoelectronic devices based on bR (see Fig. 7) require thermally robust mutants of bR [1]. Therefore increasing the thermostability of bR has been an area of intense research. One method of choice has been rational site-directed mutagenesis of bR. Advancements in thermostable vectors, antibiotic resistance genes, and the genetic characterization of extreme thermophiles have prompted the development of in vivo thermoselection systems to optimize mesophilic proteins bR for device applications.

Thermus thermophilus has been a useful in vivo screening platform for bR mutants; a versatile, heat-stable expression vector is required. Moreno et al. [26] constructed a bifunctional vector system (pMKE1) capable of expression in both extremely thermophilic (*T. thermophilus*) and mesophilic (*E. coli*) microorganisms. Mutants that retain structural stability at elevated temperatures have been used as starting points for additional rounds of mutagenesis and thermoselection. Several iterations of thermoselection may be required before a bR variant with adequate thermostability can be used for device applications. In a recent study to optimize the heat capacity of bR, eight residues were selected

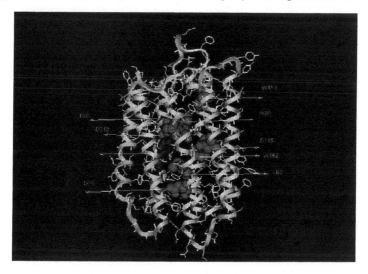

Figure 8: Mutations induced to confer thermal stability

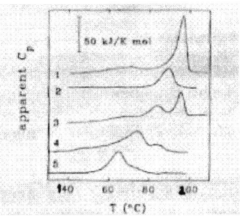

Figure 9: Differential Scanning thermogram of a high Tm mutant of bR.

in bR: D85, W86, L93, D96, D115, W182, W189, D212 (Fig.8). Differential Scanning Calorimetric studies of bR mutants designed and expressed manifest $T_m \sim 192$ °C. shown in Fig. 9.

The proposed mechanisms for recording and retrieving information from the bR medium are both optical. However, unlike previously studied 3-D (see comments) optical systems (holography, Hampp [3] and multi-step two-beam recording and reading systems, Birge et al. [4]), in this work, a two-dimensional (2-D) recording system is chosen by several reasons. There were two difficulties encountered in achieving ultra-density recording using holographic optical mechanisms: (a) Even today, the holographic-based recording methods cannot beat state-of-the-art 2-D recording systems (a recording head and a thinfilm medium) from the perspective of the effectively achievable areal density. (b) It is not trivial to immobilize protein into a 3-D matrix. Therefore, to address and solve these issues, in this project, at least in the preliminary stage it is chosen to use a 2-D implementation. With the 2-D implementation, there is no fundamental difficulty to implement exactly the same recording transducer configuration as used in so called heatassisted magnetic recording (HAMR) (Hergett [27]) . It is believed that HAMR is the most advanced recording system today. In this recording mode, an optical near-field based transducer is used to focus the energy of light (photons) in a spot size substantially than the physical wavelength. In the author's laboratories, a near-field beam spot as small as 30 nm in diameter has been recorded. In the course of this project, a novel recording transducer capable of even smaller beam spots will be demonstrated. For example, a beam spot with a diameter of 8 nm is equivalent to an aerial density of above 10 Terabit/in^2.

As for the manufacturing of a 2-D (thin-film) BR-based recording medium, the 2-D immobilization of bR is much easier to control and more reliable and durable as compared to the 3-D immobilization. As a matter of fact, even the wild-form bR is intrinsically inclined to form a 2-D film because of its molecular structure. As earlier mentioned, bR is found in halobacteria in the form of a 2-D crystalline array integrated into their cell membrane. Up to 80 % of their surface may be covered by one or two of so-called purpose membranes (PM). The purple membranes are made of bR and lipids only. All the bR molecules are uniformly oriented (microscopically yes - but not necessarily in a bulk coated film. Methods for orienting bR molecules are available and described in great detail in literature) with their carboxy termini located in the inner portion of the cell. It is most due to this crystalline nature why bR displays such unprecedented thermal and chemical stability. This naturally occurring 2-D crystalline structure of bR molecules makes it also relatively easy and straightforward to fabricate stable thin-films (2-D recording media).

14. Recording and retrieving information from protein-based media

It is believed that the basic physics behind recording ultra-high density information into protein-based media could be more thoughtfully utilized with a Nanoscale device capable of recording and retrieving information on a single-bit level (instead of a large set (page) of bits used in parallel data access such as in holographic storage). It should be reminded that the purpose of this work is to develop a single-molecule type data recording/retrieving mechanism based on the thorough knowledge of the physics of Nanoscale recording and protein-based media.

Therefore, it is not surprising why in this proposal it has been chosen to use data recording methods very similar to the methods developed for HAMR (a technology previously developed at Seagate Research to focus light at a spot size of a few nanometers). Below, the developed for this purpose experimental setup is described in detail. As for the reading mechanism, it is proposed to use aperturless version (developed in-house) of scattering type of scanning near-field optical microscopy (s-SNOM) very similar to the above-mentioned HAMR system. A novel aperturless transducer was developed for both recording and reading with the purpose to substantially increase the efficiency of the light conduction system.

Finally, the following methods are proposed for writing and retrieving (reading back) information. 1: to record nanoscale bits into the protein media, it is chosen to use a technology similar to the recently developed technology used in heat-assisted magnetic recording (HAMR). 2: to read back nanoscale bits recorded into the protein media, it is chosen to use an improved version of so called aperturless or scattering type of scanning near-field optical microscopy (s-SNOM). (Note: to physically separate the writing process from the reading process, in one implementation of the technology, these processes are performed at two different frequencies.) The improvement is through using apertureless laser at the air bearing surface to maximize the signal-to-noise ratio. This improvement is analogous to the technology used to record information and has been previously developed for different purposes by the investigators during their employment in the industry.

15. Conclusion

In this review, we have discussed the re-optimization of a photoreactive protein, bR, to confer thermal stability for novel bionanotechnological applications. Computational chemistry offers a host of methods that are useful in in-silico design of bacteriorhodopsin for applications in flash memory, generation of hydrogen gas by increasing protonpumping capacity of bR and other proton pumping proteins. The optimization of specific physical properties critical for bionanotechnological applications has several opportunities in other protein systems e.g. cytochrome c superfamily in the design of nitric oxide (NO) and carbon monoxide (CO) sensors. The immobilization of a protein into a stable 2-D recording medium and the development of a transducer capable of recording areal densities of substantially higher values than the areal densities in the current state-of-the-art magnetic recording systems present formidable technological challenges. Nanotechnology and in particular bionanotechnology in turn can spur further developments in the underlying basic science in explaining phenomenon at interfaces and the differences between bulk and interfacial phases.

Acknowledgements

S. Khizroev and V. Renugopalakrishnan express their thanks to National Science Foundation for the award of a MRI grant ECS-0421255. V. Renugopalakrishnan would like to thank Florida International University, Wallace H Coulter Foundation, Office of Naval Research (ONR), and Harvard Medical School for supporting this project. L. Lindvold was originally supported by the European Communities as a part of the POPAM project No. 6863 under the auspices of the ESPRIT Basic Research Program for the work on bR film fabrication. Authors express thanks for copyright permission to the Editor, FEBS Letters, Elsevier, Ganesh, C., et al., FEBS Lett, 1999. 454:31-36 [Ref.22], to reproduce Figs. 4 and Table 1 and Editor, Protein Science for Figs. 5 and 6, Jiang et al., Protein Science, 2000, Vol. 9:403-416 [28]. Research of GN was supported in part by NIH Grant P01 DA15027-01.

References:

[1] V. Renugopalakrishnan, A. Strzelczyk, P. Li, A. A. Mokhnatyuk, S. H. Gursahani, M. Nagaraju, S. L. Lakka, Retroengineering bacteriorhodopsins: Design of smart proteins by bionanotechnology. Int. J. Quantum Chem. 95, (2003), 627-631

[2] V. Renugopalakrishnan, P. Li, Design of a Bacteriorhodopsin, bR 192, with seven point mutations and its expression in P. Pastoris: Stabilization of the M and O photo intermediates with enhanced lifetimes US Provisional Patent (pending).

[3] R. Birge, N. B. Gillespie, E. W. Izaguirre, A. Kusnestow, A. Lawrence, A. F. Singh, D. Singh, Q. W. Song, E. Schmidt, J. A. Stuart, S. Seetharaman, and K. Wise, Biomolecular Electronics: Protein-Based Associative Processors and Volumetric Memories. J.Phys. Chem. B 103, (1999), 10746

[4] N. Hampp, Bacteriorhodopsin as a Photochromic Retinal Protein for Optical Memories. Chem. Rev. 100, (2000), 1755-76

[5] S.S. Tatke, V. Renugopalakrishnan, M. Prabhakaran, Interfacing Biological Macromolecules with Carbon Nanotubes and Silicon Surfaces: A Computer Modeling and Dynamic Simulation study. Nanotechnology 15, (2004), S684-S690

[6] V. Renugopalakrishnan, X. Wei, S.L. Lakka, G. Narasimhan, C. S Verma, P. Li, A. Anumanthan, Enhancement of Protein Thermal Stability : Towards the Design of Robust Proteins for Bionanotechnological Applications. Springer, Dordrecht, The Netherlands, 2004

[7] C. Cohen, D.A. Parry, Alpha-helical coiled coils and bundles: how to designan alpha-helical protein Proteins, 7, (1990) 1-15

[8] R.R. Naik, S.M. Kirkpatrick, M.O. Stone, The thermostability of an alphahelicalcoiled-coil protein and its potential use in sensor applications. Biosens Bioelectron, 16, (2001), 1051-1057

[9] Dennis, M.S., A. Herzka, R.A. Lazarus, Potent and selective Kunitz domain inhibitors of plasma kallikrein designed by phage display.J Biol Chem, 270(43), (1995), 25411-7.

[10] R. H. Lozier, R. A. Bogomolni, W. Stoeckenius, Bacteriorhodopsin: a light-driven proton pump in Halobacterium Halobium. Biophys. J. 15, (1975) 955-963

[11] T. G. Beery, in Thermodynamics of Membrane Receptors and Channels, Ed. M. Jackson, CRC Press, Boca Raton, FL, USA, pp. 353-387, 1993.

[12] K. Gerwert, B. Hess, D. Soppa, D. Osterhelt,. Role of Aspartate-96 in Proton Translocation by Bacteriorhodopsin. Proc. Natl. Acad. Sci. USA 86,(1989), 4943-4947

[13] T. Marinetti, S. Subramaniam, T. Mogi, T. Marti, H. G. Khorana, Replacement of Aspartic Residues 85, 96, 115, or 212 Affects the Quantum Yield and Kinetics of Proton Release and Uptake by Bacteriorhodopsin. Proc. Natl. Acad. Sci. USA 86, (1989), 529-533

[14] J, Tittor, C. Soell, D. Osterhelt, J. J. Butt, E. Bamberg, A defective proton pump, point-mutated bacteriorhodopsin Asp96 Asn is fully reactivated by azide. EMBO J. 8, (1989),3477-3482

[15] H. Otto, T. Mareti, T. Holz, T. Mogi, M. Lindau, H. G. Khorana, M. P. Heyn, Aspartic Acid-96 is the Internal Proton Donor in the Reprotonation of the Schiff Base of Bacteriorhodopsin Proc. Natl. Acad. Sci. USA 86, (1989), 9228-9232,

[16] A. Miller, D. Osterhelt, K, Kinetic optimization of bacteriorhodopsin by aspartic acid 96 as an internal proton donor. Biochim. Biophys. Acta 1020, (1990), 57-64

[17] S. Subramaniam, M. Lindahlr, P. Bullough, A. R. Faruqi, J. Tittor, D. Osterhelt, L. Brown, J. Lanyi, R. Henderson, Protein conformational changes in the bacteriorhodopsin photocycleJ. Mol. Biol. 287, (1989), 145-161

[18] H. Luecke, B. Schobert, H. –T. Richter, J. –P. Cartailler, J. K. Lanyi, Structural Changes in Bacteriorhodopsin During Ion Transport at 2 Angstrom Resolution Science 286, (1999), 255

[19] Murphy, K.P., P.L. Privalov, and S.J. Gill, Common features of protein unfolding and dissolution of hydrophobic compounds Science, 1990. 247: 559-561.

[20] Murphy, K.P. and E. Freire, Thermodynamics of structural stability and cooperative folding behavior in proteins.Adv Protein Chem, 1992. 43: 313-361.

[21] Murphy, K.P. and S.J. Gill, Molecular basis of co-operativity in protein folding, J Mol Biol, 1991. 222: 699-709.

[22] Ganesh, C., Narayanan E, Srivatsava S, Ramakrishnan C., and Varadarajan, R. Prediction of the maximal stability temperature of monomeric globular proteins solely from amino acid sequence FEBS Lett, 1999. 454: 31-36.

[23] Grattinger, M.,Dankesreiter A., Schurig H, and Jaenicke R. Recombinant phosphoglycerate kinase from the hyperthermophilic bacterium Thermotoga maritima: catalytic, spectral and thermodynamic properties J Mol Biol, 1998. 280 :525-533.

[24] Beadle, B. M. Baase, W A, Wilson, D E., Gilkes N R, and Shoichet, B K Comparing the Thermodynamic Stabilities of a Related Thermophilic and Mesophilic Enzyme. Biochemistry 1999, 38: 2570-2576.

[25] Jiang, X., Bishop, E. J., and Farid, R. S. A de Novo Designed Protein with Properties That Characterize Natural Hyperthermophilic Proteins J. Am. Chem. Soc.1997, 119, p 838-839.

[26] Moreno R, Zafra O, Cava F, Berenguer J: *Plasmid* Development of a gene expression vector for Thermus thermophilus based on the promoter of the respiratory nitrate reductase 2003, **49**:2-8.

[27] Herget, P.,T. Rausch, A. C. Shiela, D. D. Stancil, T. E. Schlesinger, J.-G. Zhu, J. A. Bain, Mark shapes in hybrid recording. Appl. Phys. Lett. 80 , (2002), 1835-7

[28] Jiang X, Harid H, Pistor E, Farid RS A new approach to the design of uniquely folded thermally stable proteins. Protein Science, 2000, Vol. 9:403-416

Brill Academic Publishers
P.O. Box 9000, 2300 PA Leiden,
The Netherlands

Lecture Series on Computer
and Computational Sciences
Volume 3, 2005, pp. 196-222

Recent Developments and Challenges in Chemical Simulations

Bernd M. Rode[1], Demetrios Xenides, Thomas S. Hofer, and Bernhard R. Randolf

Theoretical Chemistry Division,
Insitute of General, Inorganic and Theoretical Chemistry
University of Innsbruck,
Innrain 52a, A-6020 Innsbruck, Austria

Received 1 July, 2005; accepted 10 July, 2005

Abstract: The rapid development of statistical chemical simulation methods and its dependence on computational capabilities and performance is outlined, up to the successful combination of classical mechanics and quantum mechanics achieved in the past 15 years, which has given these simulations a level of accuracy comparable to the most advanced experimental techniques. The use of simulation 'data mining' as a universal instrument for obtaining thermodynamic, kinetic and spectroscopic data is described and examplified with a number of applications of the mixed quantum mechanical / molecular mechanical molecular dynamics method. Ions in solution are used as example for the description of methodical achievements but also potential error sources. These studies also demonstrate, where and why experimental techniques could fail in the determination of structures of solvates, and that the inherent limits of experimental measurabilty can be surpassed by accurate simulation techniques, e.g. in the analysis of multiple species forming solutions or in picosecond reaction dynamics. The challenges of further methodical development and the prospective utilisation of further increasing computational power in chemical simulations are addressed.

[1]Corresponding author. E-mail: Bernd.M.Rode@uibk.ac.at

Keywords: Liquids, Electrolyte Solutions, Molecular Dynamics, *ab initio* simulations.

1 Introduction

There is no doubt that any experimental observation and measurement requires a theoretical framework for its interpretation, and the more general and consistent the theoretical framework, the more reliable and relevant are the experimental data. For a long time, chemistry at molecular level has been a domain of rather qualitative models, combined with classical physical theories such as thermodynamics and statistical mechanics, and the quantum mechanical foundations of chemistry, known for almost 80 years from the fundamental equation of Schrödinger [1], have been implemented only very gradually into teaching and research, as up to now only numerical solutions of this equation for many-electron systems can be achieved. Therefore, 'Theoretical Chemistry' is largely synonymous with 'Computational Chemistry', and its progress was and still is bound to progress in computational speed and capacity.

Fortunately, this progress was tremendous in the past decades, both in computational power and in its affordability due to continuously decreasing cost. For this reason, very accurate quantum mechanical calculations of small and medium-sized molecular systems are within the reach of almost every chemist today (provided a solid background knowledge of the theory behind the commercially available programs!), and modelling of large biopolymers and condensed systems in solid and liquid state has become an equally important tool, having found its way not only into academic but also industrial research.

As most chemical reactions and all biochemical processes take place in the presence of a solvent, reliable theoretical approaches to the liquid state are in particular demand. The size of a representative model system for a liquid or a solvated species, however, has prohibited until recently its investigation by means of quantum mechanical calculation methods, and at present still demands a compromise combining classical physics with quantum chemistry in various forms of 'simulations'.

The term 'simulation' implies the attempt to construct a model as similar as possible to reality, i.e. including as much as possible of contemporary theoretical knowledge and all known contributions to the physical and chemical behaviour of the species contained in the system to be studied.

On the other hand, there are so many systems, where even the most modern and expensive experimental methods reach their limits in determining structural details and/or dynamical properties, and thus the reasons for their specific reactions at molecular level. Many reactions in solution and a large number of biologically important reactions take place within the pico- or even femtosecond range. Only recently laser pulse spectroscopy [2] has given access to measure such processes, but hitherto only for very simple systems such as pure liquids. For these ultrafast processes and all related structural changes, species formation and interchanges and underlying reaction mechanisms, chemical simulations have become one of the most important tools, sometimes the only way of access. This explains, why so much emphasis and interest is focussed on the development of increasingly accurate simulation methods. The accuracy of the simulations is of particular importance, when it can be verified only partially by the reproduction of experimental data, transgreding the limits of all experimental methods and thus demanding a method-inherent control mechanism.

Statistical mechanics simulations of the Monte Carlo (MC) and Molecular Dynamics (MD) type are well-established methods for almost 50 years. However, their sophistication and applicability has been continuously increasing: while in the early time interactions within the simulation box were mostly described by empirical potentials, the advances in computational chemistry successively enabled the calculation of complete energy surfaces, providing thus the material for the construction of *ab initio* generated analytical interaction potentials. Such potentials, often modified by empirical parameters or variational parts to include effects such as mutual polarisation, are still an important tool in MC and MD simulations of liquids today. On the other hand, for organic and biomolecules, numerous empirical force fields have been developed, e.g. AMBER [3], GROMACS [4], GROMOS [5], COSMOS [6], FOCUS [7], CHARMM [8] consisting of force constants, electrostatic and Van-der-Waals components. Most of these methods allow the inclusion of water as solvent and thus high-quality modelling of biopolymers within reasonable computing times. By this approach it is also possible to simulate molecular interactions such as drug-receptor docking, and the scanning of potential interaction sites and modes by Monte Carlo methods represents another important application of chemical simulations. It should not be concealed, however, that these force field methods are mostly suited only for the simulation of large organic molecules, and that the investigation of simple electrolyte solutions already calls for the employment of much more sophisticated potentials or, as will be shown later, for the combination of quantum chemical methods with the classical simulation techniques.

The continuous improvement of computer speed and capacities has given the opportunity to significantly advance the methodical development and the applicability of much more accurate computations, in particular in the past decade. Classical MC and MD simulations, including the construction of *ab initio* generated potential functions - ten years ago still a major effort for mainframe computers - have become a desktop task and often just serve for the generation of good starting configurations for more sophisticated simulations.

These more sophisticated simulations all have one idea in common, namely to utilise the superiority of quantum mechanics in the description of molecular systems in statistical simulations. It had been recognised quite soon that classical simulations based on pair potentials lead to roughly wrong results, whenever interactions between species are strong, e.g. in the case of a solvated ion with higher charge [9, 10, 11, 12, 13, 14, 15, 16]. The addition of 3-body correction potentials could partly correct this failure, but it had to be expected that higher correction terms might be of importance as well. To construct energy surfaces for such corrections, however, is a tedious and often unresolvable problem [15, 17]. Even if such correction functions were available, they still would not take into account properly all typical quantum effects, in particular charge transfer and mutual polarisation, nor could they reflect well-known phenomena such as the Jahn-Teller distortion [18, 19].

To perform a chemical simulation of an elementary box representative for a liquid system, treating all the hundreds of species in it by *ab initio* quantum mechanics, is still a task beyond the capacity of even the best supercomputers available. Therefore, two basic simplifications in applying quantum mechanics had to be thought of. The first one was to reduce the quantum mechanical calculations to semiempirical levels, the second one to reduce the quantum mechanical treatment to a subregion of the system and to retain the classical simulation by potential functions for the rest of the elementary box. Attempting the first approach one very soon recognised that typical semiempirical MO methods failed in the prediction of correct structures [41], whereas density functional methods (DFT) looked more promising. In the case of solvated ions it was shown, however, that simple density functionals, e.g. BLYP [43, 45] employed in Car-Parrinello simulations (CP, [37, 38, 39, 42]), the RIDFT method [47] or the BP86 [43, 46], do not correctly reproduce even basic properties such as coordination numbers [39, 42], and that the more suited hybrid functional B3LYP [44, 45] does not lead to any advantage in computational effort compared to *ab initio* calculations with the same basis sets.

The second approach, where the quantum chemical subsystem (QM) is reduced to a manageable dimension while treating the remaining simulation box by classical molecular mechanics (MM), has been successfully applied to large biomolecules, for which it had originally been developed [20, 21, 22], and is used for such systems with varying levels of quantum mechanics until today [24, 25, 26, 27, 28, 29]. A similar approach was also made for solvated ions [41, 48, 49] and further developed to treat more complex solvates [50, 51, 52, 53, 54, 55, 56] and pure liquids as well [32, 59]. All of these methods are generally termed by the acronym 'QM/MM', reflecting the combination of QM and MM treatment. Again it is the available computation power, which determines the manageable size of the QM region and the level of quantum mechanical methods this region can be treated. A further criterion to choose these parameters and thus the computational effort to be spent is the level of accuracy required for specific data. To evaluate only structural data may allow a somewhat simpler approach than the analysis of dynamics and reaction pathways. A short overwiew, which data can be obtained from simulations, will not only show the universality of this computational chemistry instrument, but also give an idea of the sensibility of results to the methodical framework employed.

2   'Data Mining' in Simulation Trajectories

The 'history' files of MC and MD simulations contain a wealth of information about the simulated systems. In both cases structural details and microspecies distributions can be extracted as well as thermodynamic quantities. The history file of MD simulations, representing the trajectories of all particles, additionally allows the evaluation of time-dependent quantities, from rovibrational spectra to reaction rate constants. With some exceptions, e.g. the scanning of binding sites of large molecules, where an easily adjustable size of molecular displacements is needed, preference is given to MD simulations, therefore, as their trajectories give access to many more data than MC history files.

The analysis of trajectories with appropriate instruments, i.e. evaluation programs, largely has the character of 'data mining': from Gigabytes of coordinates and velocities, specified atom-atom or molecule-molecule radial distribution functions (RDF), complemented by various angular distribution functions (ADF) can be extracted to give a complete picture of the average structure of a liquid system, and the changes of structures in time leads to detailed knowledge of microspecies present in the system, ranging from coordination number distributions to the various solvate complexes realised in solution and thus forming the basis for reactivity and reaction pathways of the solutes.

Whereas all of this data can be extracted from an MC history file as well, only an MD trajectory provides time dependent data and thus gives access to velocity autocorrelation functions (VACF) and through these by Fourier transformation to the rovibrational spectra observable by IR and Raman techniques. Further dynamics-related quantities available from the trajectories are diffusion constants, reaction rates for exchange processes and average life-times of bonds and species. By the successive application of the appropriate 'mining tools' one can, therefore, fully characterise a liquid system in all of its aspects, which would have required numerous different experimental techniques with enormous technical and financial implications. For some of the data there even are no suitable experimental methods available yet, and thus make high-quality trajectories of MD simulations the only resource to access them.

It should be mentioned at this point that the evaluation of data from trajectories is not a trivial task and that there are numerous error sources in chemical simulations, which have gradually been recognised (and probably new ones will be found!) in the course of methodical improvements of the simulations. Ion solvates are a very good example for this process and will, therefore, be used to demonstrate capabilities, limits and possible future developments in the process of 'data mining' in simulation trajectories.

3 QM/MM methods in simulations

Three crucial parameters determine the success of a QM/MM simulation, the first being the size of the QM region, the second the quality of the quantum mechanical method applied in this region including the choice of basis sets, and the handling of the transition region between QM and MM region. The latter is more a problem in biopolymers, where the separation line runs through chemical bonds, demanding an artificial breaking and compensation for the broken bonds, but it calls for care also in solvated systems, where solvent molecules migrate through this line, turning from a part of the quantum chemical system into 'classical', force-field determined molecules. Any rigid solvent model would create problems in this case, as the molecule would have to jump into its 'standard' configuration and immediately freeze in this conformation. On the other hand, the larger the solvent molecules are, the more difficult it is to handle transitions, where half of the molecule is within and the other half outside the defined QM region. For small solvent molecules like the most important water, these problems are easier to manage, but even in this case a flexible description of the solvent molecules in the MM region and a careful smoothing algorithm for the transitions are mandatory.

Classical simulations with pair and 3-body potentials are the common starting point of QM/MM simulations, in order to obtain a suitable starting configuration for the much more time-consuming simulation including quantum mechanically calculated interactions, but also in order to determine the optimal size of the QM region from the classical RDF data. In the most common QM/MM formalism the classical description is continuously needed for the whole system, as the influence of the quantum effects is introduced by a subtraction of QM and MM forces in the QM region, i.e. as a correction of the classical simulation by quantum mechanical effects in the selected subregion. This need for a permanent classical evaluation of all classical forces in the whole simulatin box implies the availability of appropriate potential functions for all species present, in the case of solvated species the functions describing solute-sovent and solvent-solvent interactions. For the more complicated solute-solvent interactions, frequently adequate 3-body correction functions have to be constructed, which makes the performance of reliable conventional QM/MM simulations strongly dependent on the quality of these functions. A possible way to solve this problem will be shown shortly below.

As mentioned above, the choice of a proper quantum mechanical level of calculation is a crucial factor determining the accuracy, sometimes even the qualitiative correctness of QM/MM results. The influence of the QM level has been investigated for solvated ions from semiempirical [40, 41] and DFT level [37, 38, 39, 42, 47] to *ab initio* methods with varying basis set quality [60] at Hartree-Fock and MP2 level [62]. Summarising the results of these investigations it can be stated that the *ab initio* HF method with DZP basis sets is - at present - the best compromise between accuracy and affordable computational effort, and that an enlarged QM region at this level improves the results more than the introduction of correlation effects [32].

The common simulation protocol for QM/MM simulations (cubic box with ≈ 500 molecules, periodic boundary condition, NVT ensemble temperature controlled by the Berendsen algorithm [57], time-step of 0.2 fs, flexible BJH model for water [58]) for all simulations given as examples and illustrations below, has been detailed described in numerous publications and review articles [31, 64, 65], where also the selection and the quantum chemical treatment of the QM subregion and the smoothing procedure for the transition to the MM region are outlined. In this presentation it seems sufficient, therefore, to say a few words about the general 'philosophy' behind the practical design of a QM/MM simulation.

As previously outlined, one has to select an appropriate subregion of the system either on physical arguments or by chemical 'intuition', where one expects higher n-body and quantum effects to play a significant role, large enough to produce erroneous results upon their neglect. In the case of solvated ions one had expected ion plus first hydration shell to be a sufficiently large region (easily determined from a classical ion-solvent RDF) for this purpose. In the course of simulations it was found, however, that if one needs more than mere structural data for the solvate, one would have to include parts, if not the whole second hydration shell into the QM region [63, 65]. Even for a good description of the pure solvent water, a QM/MM approach containing a central water plus 2 coordination shells proved essential [32, 59]. With same basis sets such an amplification from a 1-shell to a 2-shell *ab initio* QM/MM simulation boosts the computational demand by a factor of ≈ 5, which explains the burning desire of computer chemists for ever faster processors. At present, an average *ab initio* QM/MM MD simulation of 10 ps (50.000 time steps) of a hydrated ion on 6-8 Opteron 64 bit processors consumes about 1-2 months for a 1-shell, and 3-5 months for a 2-shell system, depending on the number of solvent molecules residing in average in the shells.

Anticipating the expected rapid progress in computer hardware, leading to a significantly increased speed of processors every year, the design of further improved simulation techniques relying on this technical progress is one of the main challenges in computational chemistry. It has to be said that it is mainly the scalar processing capacity that determines the possible progress in QM/MM methods, as parallelisabity of quantum mechanical force calculations - which consume 95 % of the simulation time - is rather limited, with an optimum reached already with 6 to 10 processors. However, the favourable prospects for scalar processing speed to come are very encouraging for setting new horizons in the treatment of solvated systems. We have developed, therefore, a new approach to deal with composite so-

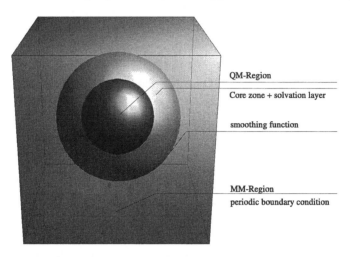

Figure 1: Schematic presentation of the QMCF approach.

lutes, which introduces a refined interplay between QM and MM regions and, most important, allows to perform *ab initio* QM/MM simulations without the need of solute-solvent interaction potentials. This advantage gives a wide access to complex species in solution, for which the construction of such potentials would be a very difficult endeavour, sometimes even leading to hardly resolvable problems. This new approach will be briefly outlined in the following, and some tests performed for it will be presented at the end of the 'Examples' section.

4   *ab initio* Quantum Mechanical Charge Field Molecular Dynamics (QMCF-MD)

As a comparison of 1- and 2-shell *ab initio* QM/MM simulations has already shown that quantum effects can play a significant role for various solvate properties, the new approach started with the principle condition to include at least 2 solvation shells. This means a strongly increased computational effort, but in the treatment of interactions between solute and solvent it is also the key factor for avoiding the construction of analytical potential functions.

The basic principle of the QM/MM formalism, i.e. the partition into QM and MM regions is maintained. The main difference of the QMCF framework lies in the definition and size of the QM region, which now consists of 2 subregions ('core' and 'layer') and the treatment of interactions/forces between the 'core' QM subregion and the MM region.

The 'core' QM region contains the solute, which can be any type of molecule or composite species, e.g. a metal ion with identical or different ligands, and the 'layer' QM region, a complete layer of solvent molecules surrounding the solute (cf. Fig. 1). In the simple case of a hydrated ion, this would correspond to an ion plus 2 complete hydration spheres in the QM region, an example for a complex species would be $[Fe(CN)_6]^{3-}$ (core region) with one full hydration layer (layer region). Outside these QM regions, the remaining solvent is treated by molecular mechanics, i.e. by suitable potential functions, which are already available for many common solvents. A smoothing function as being used in conventional QM/MM simulations [8, 31, 66] ensures a continuous and steady transition between the quantum mechanically treated outer solvation layer and the MM region. Transitions between core and layer QM regions are fully included in the quantum mechanical description, thereby avoiding artifacts possibly arising from the QM/MM transition and associated smoothing procedure as in 1-shell QM/MM simulations.

Inside the QM region, all interactions are evaluated by means of quantum mechanics, in the examples presented here at *ab initio* Hartree-Fock level with DZP basis sets. Interactions between the 'layer' QM region and solvent molecules in the MM region, and interactions between solvent molecules within the MM region are treated by the solvent-solvent potential functions, making use of the smoothing function at the transition between MM and 'layer' QM region. For the Coulombic interactions between solvent molecules in the QM region and those in the MM region, however, the actual quantum mechanically evaluated partial charges inside the QM region are used. Transitions between QM and MM regions are allowed and, due to the smoothing function and the relatively large distance of the transition region from the solute, occur without any discontinuity of energies and forces.

The main difference to conventional QM/MM techniques concerns the evaluation of interactions between solvent molecules in the MM region and the solute, i.e. the 'core' QM region. Due to the distance of this core region to the MM region, all short-range terms of any type of interaction potentials, which usually consist of a Coulombic term plus a series of r^{-n} (n>3) terms, sometimes also an exponential term, become negligibly small, and in most cases are already eliminated by the usual cut-offs in typical simulation protocols. This means that only Coulombic interactions will contribute to such interactions, which can be evaluated simply from the point charges assigned to the molecules/atoms in the MM region and the partial charges on the atoms of the solute. This corresponds to an electrostatic description by a dynamically changing field of point charges, which varies according to the movements of atoms inside the QM region(s) and molecules in the MM region in the course of the simulation. As a further improvement of this charge field, point charges assigned to atoms in the QM region will be redefined in every step of the simulation according to the new configuration, by population analysis within the quantum mechanical calculation step. This ensures the continuous adaptation of the dynamical charge field to all polarisation and charge transfer effects within solute and surrounding solvent layers according to the dynamical changes of the solute's structure (which are much more significant than the small changes of charges eventually occurring at the solvent molecules in the MM region, which are not considered in this formalism). For the QM regions, the influence of the charges of molecules in the MM region is taken into account by incorporating these charges as additional, perturbational term into the core Hamiltonian, representing the outside solvent influence at molecular level according to the actual location and orientation of the solvent molecules. The most important solvent exchange processes take place within the QM region, and thus the relevant dynamics are evaluated with the accuracy of quantum mechanically calculated forces. By this approach, the treatment of solutes of very different nature becomes a much simpler and straightforward process, as only potential functions for solvent-solvent interactions are needed, whereas all other interactions are dealt with at quantum mechanical level, which also means at much higher accuracy than molecular mechanics potentials could achieve. The application of all other conditions of a MD simulation (large number of solvent molecules, periodic boundary condition, minimal image convention and long-range forces corrections by reaction field) ensure that the system corresponds to the condensed, liquid state. The flexible treatment of charges inside the QM region and the incorporation of the MM charges in the

Table 1: Multiple species forming ions: ion–O distances r_{ion-O} in Å, most prominent coordination numbers CN (> 10 %), and mean residence times τ_{MRT}^a of water ligands in ps.

Ion	r_{ion-O}	CN(%)	τ_{MRT}^a	Ref.
K(I)	2.81	7(24), 8(43), 9(29)	2.0	[67, 68]
Rb(I)	2.95	6(21), 7(35), 8(29)	2.0	[69]
Cs(I)	3.25	6(16), 7(26), 8(32), 9(15)	1.8	[70]
Ba(II)	2.86	9(68), 10(24)	18.6	[71]
Ag(I)	2.6	4(23), 5(62), 6(12)	5.5	[72]
Au(I)	2.45	4(40), 5(42), 6(16)	3.1	[73]
Hg(II)	2.42	6(84), 7(16)	23.6	[74, 75]

^{a} t^*=0.5 ps; based on the number of water molecules leaving/entering the coordination shell for at least t>t*, and thus performing a sustainable exhange process, obtained with the direct method [75].

core Hamiltonian can be seen as an additional advantage for systems, where charge transfer and polarisation effects play an important part, e.g. in solvates of highly charged solutes. The QMCF method is very new and thus not much experience has been collected with it so far - an average simulation of a small solute like an ion takes again some months of computing time! - but the few examples reported below seem to confirm that the results, achieved without the use of any intermolecular potential functions except for the solvent itself, are of at least the same quality as those of conventional 2-shell *ab initio* QM/MM MD simulations. For some aspects, e.g. the second solvation shell, QMCF seems to perform even better. Last, but not least, by treating most of the solute-solvent interactions by *ab initio* quantum mechanics this method further enhances the nonempirical character of chemical simulations. By its independence of specific potential functions it further opens a wide field of new applications in solution chemistry and biochemistry.

Table 2: Distribution of second shell coordination numbers CN (> 10 %) of selected ions, mean residence times of water ligands in the second shell τ_{MRT}^a in ps, and the number of accounted exchange events N_{ex} in 10 ps (Values in parentheses denote percent).

Ion	CN	τ_{MRT}^a	N_{ex}/10ps	Ref.
Ba(II)	22(16), 23(22), 24(20), 25(18)	5.5	42.7	[71]
Ag(I)	21(14), 22(18), 23(20), 24(16)	2.6	65.6	[72]
Au(I)	30(12), 31(15), 32(22), 33(20), 34(14)	4.6	72.1	[73, 75]
Hg(II)	20(14), 21(18), 22(26), 23(20), 24(12)	4.8	45.7	[74, 75]
Ca(II)	17(13), 18(22), 19(20), 20(23), 21(13)	4.4	43	[61]
Cu(II)	10(12), 11(26), 12(35), 13(18)	7.7	15.3	[78]

[a] t^*=0.5 ps; based on the number of water molecules leaving/entering the coordination shell for at least $t > t^*$, and thus performing a sustainable exhange process, obtained with the direct method [75].

5   Examples

In order to demonstrate the high instrumentality of *ab initio* QM/MM MD simulations, a few typical examples of solvated ions have been selected, where experimental studies are difficult or even not feasible yet. The first example refers to ions forming multiple species with different coordination numbers and geometries in aqueous solution. When structural data from diffraction methods have to be fitted to models, usually a single species model is taken as a basis for this fitting process. It is clear that in the case of such ions the obtained data can only be a very rough approximation to the actual structures of the various species present, and that bond lengths and angles thus obtained will not coincide with the variety of bonds and

Table 3: One-shell and two-shell Hartree-Fock simulations of highly charged metal ions: average ion–O distances $r_{\text{ion}-O}$ in Å for Al(III), Cu(II), and Ti(III) (first and second shell values are denoted with 1 and 2, respectively).

Ion	$r^1_{\text{ion}-O}$	$r^2_{\text{ion}-O}$	CN_1	CN_2	Ref.
Al-1HF	1.8	4.2	6.0	13.8	[65, 86]
Al-2HF	1.9	4.1	6.0	12.2	[65, 66]
Al-Exp[b]	1.87-1.90	3.98-4.04	6.0	12-14	[40]
Cu-1HF	2.07/2.2	4.6	6.0	11.7	[47]
Cu-2HF	2.03/2.15/2.3	4.2	6.0	12.7	[47]
Cu-Exp[b]	1.95/2.29	-	6.0	-	[80]
Cu-Exp[b]	2.04/2.29	-	6.0	-	[81]
Cu-Exp[b]	1.95/2.6	-	6.0	-	[82]
Cu-Exp[b]	1.98/2.39(2.34)	-	6.0	-	[83]
Cu-Exp[b]	1.96	-	4.1	-	[84]
Cu-Exp[b]	1.97	-	6	-	[85]
Ti-1HF	2.08/2.11/2.15	4.7	6.0	8.0	[79]
Ti-2HF	2.03/1.99/2.06/ 2.12/2.14	4.2	6.0	11.0	[79]

[a] t^*=0.5 ps; based on the number of water molecules leaving/entering the coordination shell for at least $t > t^*$, and thus performing a sustainable exhange process, obtained with the direct method [75]. [b] Experimental value.

angles actually existing. Ag (I), Au(I), Hg(II), and some of the alkali and alkaline earth metal ions are representative for this situation, where the first solvation shell is flexible enough to produce species with varying coordination numbers and thus different conformations within the picosecond scale (cf. Table 1). If properties of the second coordination shell are to be determined, the ambiguity of experimental investigations becomes even much larger, as most ions exchange ligands in this shell extremely fast (cf. Table 2), which leads to a rather wide coordination number distribution. As among all spectroscopic techniques only femtosecond laser pulse spectroscopy could access the picosecond dynamics of these ligand exchange processes, and as this technique has not been extended to study ion-ligand processes so far, simulations are at present the only method to obtain detailed information about such systems. The comparison of *ab initio* QM/MM data for the pure solvent water, where femtosecond laser pulse spectroscopic data are available [77], has shown that this type of simulations has reached the level of accuracy to provide reliable results [32, 59]. Therefore, the results of such simulations can be used to construct better models for the interpretation of diffraction measurements, and simulations should thus become a most useful tool supporting and stimulating experimental solution chemistry. This applies in particular to the case of mixed solvents, where ligands are competing for the solute's coordination sites, often leading to a variety of species with very similar bond distances but entirely different composition, thus making a unique intrepretation of experimental measurements almost impossible [53]. On the

Table 4: Hartree-Fock and electron correlated values of Cu–O and O–O distances r in Å, average coordination numbers CN and mean residence times τ_{MRT} in ps of Cu(II) and pure solvent H_2O (first and second shell values are denoted with 1 and 2, respectively).

	r^1	r^2	CN_1	CN_2	$\tau_{MRT,1}^a$	$\tau_{MRT,2}^a$	Ref.
Cu-1HF	2.07/2.2	4.6	6.0	11.7	-	7.7	[47]
Cu-1MP2	2.07/2.35/2.5	4.6	6.0	10.4	-	-	[62]
Cu-2HF	2.03/2.15/2.3	4.2	6.0	12.7	-	21	[47]
H_2O-1HF	2.97	4.4	4.9	21.2	1.6	2.4	[32]
H_2O-1MP2	2.87	4.7	4.7	22.8	2.5	2.4	[32]
H_2O-2HF	2.92	$-^b$	4.2	19.3	1.5	1.7	[32]

[a] t^*=0.5 ps; based on the number of water molecules leaving/entering the coordination shell for at least $t>t^*$, and thus performing a sustainable exhange process, obtained with the direct method [75]. [b] There is no clear indication of a unique second shell peak maximum in the gO–O graph.

other hand, the need to rely on simulation data for the determination of structural and dynamic data makes the methodical error sources of simulations a most crucial question. The comparison of 1-shell and 2-shell *ab initio* QM/MM MD simulation results collected in Table 3 illustrates that the extension of the QM region can be quite essential. In the case of many group metal ions, it just leads to a 'fine-tuning' of the data, but in the case of the Jahn-Teller distorted ions Ti(III) and Cu(II), the delicate interplay of 1st and 2nd shell processes apparently requires a 2-shell QM region to reproduce the properties of the hydrated ion. A comparison of a 2-shell HF level simulation with a 1-shell MP2 level simulation of Cu(II) even shows that the extension of the QM region is more important than the improvement of the accuracy for the first-shell description (cf. Table 4). A similar conclusion was derived in studies of pure water as well [32]. These findings, therefore, suggest that methodical controls will have to accompany simulation work whenever new types of solutes are to be investigated, and that not only computational improvements, but also methodical development will play an important role in the future of chemical simulations. In this context the first results achieved with the new QMCF MD approach can be seen as a meaningful indication to how chemical simulations will be able to tackle with new challenges, in particular with questions related to structure and reactivity of more complex solutes and active metal centres in biological processes. So far only hydrated ions have been treated by this approach, but these simulations allow an evaluation of the method in comparison to conventional *ab initio* QM/MM simulations. 3 ions have been selected for this comparison, namely Mn(II), Cu(II) and Al(III). Table 5 lists some characteristic data obtained by both methods, and Figure 2 shows the differences in the RDFs. Experimental values have been added, where available. Generally it can be stated that the QMCF method

Table 5: QMCF simulation of higly charged metal ions: Maximum ion–O distances r_{ion-O} in Å, average coordination numbers CN of Al(III), Mn(II), and Cu(II) (first, second, and third shell values are denoted with 1, 2 and 3, respectively).

	$r^1_{ion(O)-O}$	$r^2_{ion(O)-O}$	$r^3_{ion(O)-O}$	CN_1	CN_2	CN_3	Ref.
Al(III)	1.9	4.2	6.4	6.0	11.8	31.7	[87]
Mn(II)	2.2	4.6	-	6.0	26.1	-	[87]
Cu(II)	2.03/2.25/2.35	4.5	-	6.0	13.5	-	[87]

leads to results of at least the same quality as the conventional approach, with the great advantage that no metal-solvent potentials (pair + 3-body) had to be used. At

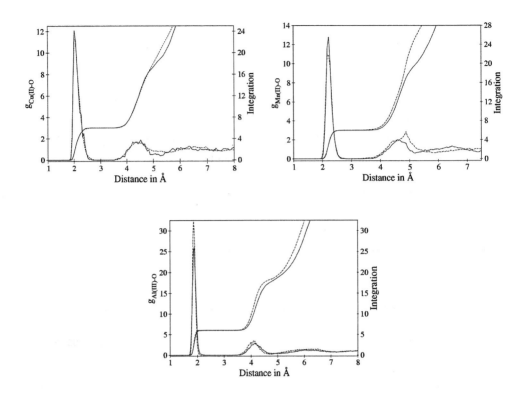

Figure 2: Ion–O RDFs of highly charged metal ions obtained by both QM MM/MD (dashed line) and QM CF/MD (solid line) approaches.

the same time, the computational effort remained almost identical. When looking at the data in more detail it can be recognised that the more sophisticated treatment of electrostatic interactions implemented in the QMCF method apparently influences structure and dynamics of the 2nd shell, improving shape and 'smoothness' of the RDFs and 'fine-tuning' dynamical data.

6   Conclusion

Chemical simulations have established themselves as a most useful and flexible tool for the study of liquid systems and biomolecules. The incorporation of quantum mechanics into this tool has given them the necessary accuracy to become a tool equivalent, sometimes superior to experiment in chemical research on these systems. Therefore, they have become one of the most important branches of computational chemistry and at the same time a practical method in increasing demand by chemical industry. The expected further increase of computer performance will enhance this trend and continuously enlarge the fields of application. New methodical improvements will assist this process, which in a not too distant future will have a strong impact on the working style and abilities of chemists both in basic and applied research.

Acknowledgment

Financial support from Austrian Science Foundation (FWF, Project 16221) and an Ernst Mach grant from the Austrian Ministry for Education, Science and Culture for D.X. are gratefully acknowledged.

References

[1] E. Schrödinger, An Undulatory Theory of the Mechanics of Atoms and Molecules, *Phys. Rev.* **28**, 1049-1070(1926).

[2] M. Dantus, M. Rosker, and A. H. Zewail, Real-Time Femtosecond Probing of 'Transition States' in Chemical Reactions, *J. Chem. Phys.* **87**, 2395-2397(1987).

[3] D. A. Pearlman, D. A. Case, J. W. Caldwell, W. S. Ross, T. E. Cheatham, III , S. DeBolt, D. Ferguson, G. Sebel, and P. Kollman, AMBER, a package of computer programs for applying molecular mechanics, normal mode analysis, molecular dynamics and free energy calculations to simulate the structural and energetic properties of molecules, *Comp. Phys. Commun.* **91**, 1-41(1995).

[4] H. J. C. Berendsen, D. van der Spoel and R. van Drunen, GROMACS: A message-passing parallel molecular dynamics implementation, *Comp. Phys. Commun.* **91**, 43-56(1995).

[5] W. F. van Gunsteren and H. J. C. Berendsen, Computer Simulation of Molecular Dynamics: Methodology, Applications and Perspectives in Chemistry, *Angew. Chem. Int. Ed. Engl.* **29**, 992-1023(1990).

[6] A. D. MacKerell Jr., Atomistic Models and Force Fields *in Computational Biochemistry and Biophysics*, O.M. Becker, A.D. MacKerell, Jr., B. Roux and M.Watanabe (Eds.), Marcel Dekker, Inc. New York, 7-38(2001).

[7] A. P. Lemon, P. Dauber-Osguthorpe, and D. J. Osguthorpe, FOCUS: a molecular dynamics analysis program New features for the characterisation of lipid bilayers and solvated systems, *Comp. Phys. Commun.* **91**, 97-109(1995).

[8] B. R. Brooks, R. E. Bruccoleri, B. D. Olafson, D. J. States, S. Swaminathan, and M. Karplus, CHARMM: A program for macromolecular energy, minimisation and dynamics calculations, *J. Comput. Chem.* **4**, 187-217(1983).

[9] H. Kistenmacher, H. Popkie, and E. Clementi, Study of the structure of molecular complexes. VIII. Small clusters of water molecules surrounding Li^+, Na^+, K^+, F^-, and Cl^- ions, *J. Chem. Phys.* **61**, 799-815(1974).

[10] I. Ortega-Blake, O. Novaro, A. Les, and S. Rybak, A molecular orbital study of the hydration of ions. The role of nonadditive effects in the hydration shells around Mg^{2+} and Ca^{2+}, *J. Chem. Phys.* **76**, 5405-5413(1982).

[11] I. Ortega-Blake, J. Hernández, and O. Novaro, The role of nonadditive effects in the second hydration shell of Mg^{2+} and Ca^{2+}. A molecular orbital study of the three-body potential energy surface, *J. Chem. Phys.* **81**, 1894-1900(1984).

[12] T. P. Lybrand and P. A. Kollman, Water-water and water-ion potential functions including terms for many body effects, *J. Chem. Phys.* **83**, 2923-2933(1985).

[13] N. R. Texler and B. M. Rode, Monte-Carlo simulations of Cu(II) in water with 3-body potential, *J. Phys. Chem.* **99**, 15714-15717(1995).

[14] G. W. Marini, N. R. Texler, and B. M. Rode, Monte Carlo simulations of Zn(II) in water including three-body effects, *J. Phys. Chem.* **100**, 6808-6813(1996).

[15] S. Hannongbua, The role of nonadditive effects in the first solvation shell of Na^+ and Mg^{+2} in liquid ammonia: Monte Carlo studies including three-body corrections, *J. Chem. Phys.* **106**, 6076-6081(1997).

[16] H. D. Pranowo and B. M. Rode, Solvation of Cu^{2+} in liquid ammonia: Monte Carlo simulation including three-body corrections, *J. Phys. Chem. A* **103**, 4298-4302(1999).

[17] S. Hannongbua, On the solvation of lithium ions in liquid ammonia: Monte Carlo simulations with a three-body potential, *Chem. Phys. Lett.* **288**, 663-668(1998).

[18] L. Curtiss, J. W. Halley, and X. R. Wang, Jahn-Teller effect in liquids: General principles and a molecular-dynamics simulation of the cupric ion in water, *Phys. Rev. Lett.* **69**, 2435-2438(1992).

[19] I. B. Bersuker, Modern aspects of the Jahn-Teller effect theory and applications to molecular problems, *Chem. Rev.* **101**, 1067-1114(2001).

[20] A. Warshel and M. Levitt, Theoretical studies of enzymic reactions: Dielectric, electrostatic and steric stabilization of the carbonium ion in the reaction of lysozyme, *J. Mol. Biol.* **103**, 227-249(1976).

[21] J. Gao, Hybrid quantum and molecular mechanical simulations: An alternative avenue to solvent effects in organic chemistry, *Acc. Chem. Res.* **29**, 298-305(1996).

[22] K. P. Eurenius, D. C. Chatfield, B. R. Brooks, M. Hodoscek, Studying enzyme mechanism with hybrid quantum mechanical and molecular mechanical potentials. I. Theoretical considerations, *Int. J. Quantum Chem.* **60**, 1189-1200(1996).

[23] D. Bakowies and W. Thiel, Hybrid models for combined quantum mechanical and molecular mechanical approaches, *J. Phys. Chem.* **100**, 10580-10594(1996).

[24] M. Sironi, A. Fornilia, and S. L. Fornili, Water interaction with glycine betaine : A hybrid QM/MM molecular dynamics simulation, *Phys. Chem. Chem. Phys.* **3**, 1081-1085(2001).

[25] L. Ridder, A. J. Mulholland, Modeling biotransformation reactions by combined quantum mechanical/molecular mechanical approaches: From structure to activity, *in Current Topics in Medicinal Chemistry* **3**, 1241-1256(2003).

[26] K. J. Jalkanen, Energetics, structures, vibrational frequencies, vibrational absorption, vibrational circular dichroism and Raman intensities of Leu-enkephalin, *J. Phys.: Condens. Matter* **15**, S1823-S1851(2003).

[27] C. Peter, W.F. van Gunsteren, and P.H. Hünenberger, A fast-Fourier-transform method to solve continuum-electrostatics problems with truncated electrostatic interactions: algorithm and application to ionic solvation and ion-ion interaction, *J. Chem. Phys.* **119**, 12205-12223(2003).

[28] C. Peter, C. Oostenbrink, A. van Dorp, and W.F. van Gunsteren, Estimating entropies from molecular dynamics simulations, *j. Chem. Phys.* **120**, 2652-2661(2004).

[29] J. Dolenc, C. Oostenbrink, J. Koller and W.F. van Gunsteren, Molecular dynamics simulations and free energy calculations of netopsin and distamycin binding to an AAAAA DNA binding site, *Nucl. Acids Res.* **33**, 725-733(2005).

[30] R. Baron, D. Bakowies, and W.F. van Gunsteren, Principles of Carbopeptoid Folding: A Molecular Dynamics Simulation Study, *J. Peptide Science* **11**, 74-84(2005).

[31] B. M. Rode, C. F. Schwenk, and A. Tongraar, Structure and dynamics of hydrated ions-new insights through quantum mechanical simulations, *J. Mol. Liq.* **110**, 105-122(2004).

[32] D. Xenides, B. R. Randolf, and B. M. Rode, Structure and ultrafast dynamics of liquid water: A quantum mechanics/molecular mechanics molecular dynamics simulations study, *J. Chem. Phys.* **122**, 174506(2005).

[33] M. Sprik, J. Hutter, and M. Parrinello, *ab initio* molecular dynamics simulation of liquid water: Comparison of three gradient-corrected density functionals, *J. Chem. Phys.* **105**, 1142-1152(1996).

[34] P. L. Silvestrelli and M. Parrinello, Structural, electronic, and bonding properties of liquid water from first principles, *J. Chem. Phys.* **111**, 3572-3580(1999).

[35] S. Izvekov and G. A. Voth, Car-Parrinello molecular dynamics simulation of liquid water New results, *J. Chem. Phys.* **116**, 10372-10376(2002).

[36] V. Dubois, P. Archirel, and A. Boutin, Monte carlo simulations of Ag^+ and Ag in aqueous solution. Redox potential of the Ag+/Ag couple, *J. Phys. Chem. B* **105**, 9363-9369(2001)

[37] L. M. Ramaniah, M. Bernasconi, and M. Parrinello, *ab initio* molecular-dynamics simulation of K^+ solvation in water, *J. Chem. Phys.* **111**, 1587-1591(1999).

[38] A. P. Lyubartsev, K. Laasonen, and A. Laaksonen, Hydration of Li^+ ion. An *ab initio* molecular dynamics simulation, *J. Chem. Phys.* **114**, 3120-3126(2001).

[39] I. Bakó, J. Hutter, and G. Pálinkás, Car-Parrinello molecular dynamics simulation of the hydrated calcium ion, *J. Chem. Phys.* **117**, 9838-9843(2002).

[40] H. Ohtaki and T. Radnai, Structure and dynamics of hydrated ions, *Chem. Rev.* **93**, 1157-1204(1993).

[41] T. Kerdcharoen, K. R. Liedl, and B. M. Rode, A QM/MM simulation method applied to the solution of Li^+ in liquid ammonia, *Chem. Phys.* **211**, 313-323(1996).

[42] A. Pasquarello, I. Petri, P. S. Salmon, O. Parisel, R. Car, É. Tóth, D. H. Powell, H. E. Fischer, L. Helm, and A. Merbach, First Solvation Shell of the Cu(II) Aqua Ion: Evidence for Fivefold Coordination, *Science* **291**, 856-859(2001).

[43] A. D. Becke, Density-functional exchange-energy approximation with correct asymptotic behavior, *Phys. Rev. A* **38**, 3098-3100(1988).

[44] A. D. Becke, Density-functional thermochemistry. III. The role of exact exchange, *Phys. Rev. A* **98**, 5648-5652(1993).

[45] C. Lee, W. Yang, and R. G. Parr, Development of the Colle-Salvetti correlation-energy formula into a functional of the electron density, *Phys. Rev. B* **37**, 785-789(1988).

[46] J. P. Perdew, Density-functional approximation for the correlation energy of the inhomogeneous electron gas, *Phys. Rev. B* **33**, 8822-8824(1986).

[47] C. F. Schwenk and B. M. Rode, Extended *ab initio* quantum mechanical/molecular mechanical molecular dynamics simulations of hydrated Cu^{2+}, *J. Chem. Phys.* **119**, 9523-9531(2003).

[48] T. Kerdcharoen, K. R. Liedl, and B. M. Rode, A QM/MM simulation method applied to the solution of Li^+ in liquid ammonia, *Chem. Phys.* **211**, 313-323(1996).

[49] A. Tongraar, K. R. Liedl, and B. M. Rode, Solvation of Ca^{2+} in water studied by Borm-Oppenheimer *ab initio* QM/MM dynamics, *J. Phys. Chem. A* **101**, 6299-6309(1997).

[50] A. Tongraar and B. M. Rode, Preferential solvation of Li^+ in 18.45 aqueous ammonia: A Born-Oppenheimer *ab initio* quantum mechanics/molecular mechanics MD simulation, *J. Phys. Chem. A* **103**, 8524-8527(1999).

[51] A. Tongraar and B. M. Rode, A Born-Oppenheimer *ab initio* QM/MM Molecular Dynamics Simulation on Preferential Solvation of Na+ in aqueous ammonia solution, *J. Phys. Chem. A* **105**, 506-510(2001).

[52] A. Tongraar, K. Sagarik, and B. M. Rode, Effects of many-body interactions on the preferential solvation of Mg^{2+} in aqueous ammonia solution: A Born-Oppenheimer *ab initio* QM/MM dynamics study, *J. Phys. Chem. B* **105**, 10559-10564(2001).

[53] A. Tongraar, K. Sagarik, and B. M. Rode, Preferential solvation of Ca^{2+} in aqueous ammonia solution: Classical and combined *ab initio* quantum mechanical/molecular mechanical molecular dynamics simulations, *Phys. Chem. Chem. Phys.* **4**, 628-634(2002).

[54] A. Tongraar and B. M. Rode, *ab initio* QM/MM molecular mynamics simulation of preferential solvation of K^+ in aqueous ammonia solution, *Phys. Chem. Chem. Phys.* **6**, 411-416(2004).

[55] C. F. Schwenk, T. S. Hofer, B. R. Randolf, and B. M. Rode, The influence of heteroligands on the reactivity of Ni^{2+} in solution, *Phys. Chem. Chem. Phys.* **7**, 1669-1673(2005).

[56] R. Armunanto, C. F. Schwenk, and B. M. Rode, *ab initio* QM/MM simulation of Ag^+ in 18.6% aqueous ammonia solution: Structure and dynamics investigations, *J. Phys. Chem. A*, accepted.

[57] H. J. C. Berendsen, J. P. M. Postma, W. F. van Gunsteren, A. DiNola, and J. R. Haak, Molecular dynamics with coupling to an external bath, *J. Chem. Phys.* **81**, 3684-3690(1984).

[58] P. Bopp, G. Jancsó, and K. Heinzinger, An improved potential for non-rigid water molecules in the liquid phase, *Chem. Phys. Lett.* **98**, 129-133(1983).

[59] D. Xenides, B. R. Randolf, and B. M. Rode, Hydrogen bonding in liquid water: An *ab initio* QM/MM MD simulation study, *J. Mol. Liq.*, accepted.

[60] C. F. Schwenk, H. H. Loeffler, and B. M. Rode, Structure and dynamics of metal ions in solution: QM/MM molecular dynamics simulations of Mn^{2+} and V^{2+}, *J. Am. Chem. Soc.* **125**, 1618-1624(2003).

[61] C. F. Schwenk, H. H. Loeffler, and B. M. Rode, Molecular dynamics simulations of Ca^{2+} in water: Comparison of a classical simulation including three-body corrections and Born-Oppenheimer ab initio and density functional theory quantum mechanical/molecular mechanics simulations. *J. Chem. Phys.* **115**, 10808-10813(2001).

[62] C. F. Schwenk and B. M. Rode, Influence of electron correlation effects on the solvation of Cu^{2+}, *J. Am. Chem. Soc.* **126**, 12786-12787(2004).

[63] H. H. Loeffler and B. M. Rode, The hydration structure of the Lithium ion, *J. Chem. Phys.* **117**, 110-117(2002).

[64] C. F. Schwenk and B. M. Rode, Cu(II) in liquid ammonia - An approach by hybrid quantum mechanical/molecular mechanical molecular dynamics simulation, *Chem.Phys.Chem* **5**, 342-348(2004).

[65] T. S. Hofer, B. R. Randolf, and B. M. Rode, Influence of polarization and many body quantum effects on the solvation shell of Al(III) in dilute aqueous solution?extended *ab initio* QM/MM MD simulations, *Phys. Chem. Chem. Phys.* **7**, 1382-1387(2005).

[66] B. M. Rode, C. F. Schwenk, T. S. Hofer, and B. R. Randolf, Coordination and ligand exchange dynamics of solvated metal ions, *Coord. Chem. Rev.* **xx**, xxx-xxx(2005).

[67] A. Tongraar, K. R. Liedl, and B. M. Rode, Born-Oppenheimer *ab initio* QM/MM molecular dynamics simulations of Na^+ and K^+ in water: From structure making to structure breaking effects, *J. Phys. Chem. A* **102**, 10340-10347(1998).

[68] A. Tongraar, K. R. Liedl, and B. M. Rode, Dynamical properties of water molecules in the hydration shells of Na^+ and K^+: ab initio QM/MM molecular dynamics simulations, *Chem. Phys. Lett.* **102**, 378-373(2004).

[69] T. S. Hofer, B. R. Randolf, and B. M. Rode, Structure-breaking effects of solvated Rb(I) in dilute aqueous solution - An *ab initio* QM/MM MD approach, *J. Comp. Chem.* **26**, 949-956(2005).

[70] C. F. Schwenk, T. S. Hofer, and B. M. Rode, The structure breaking effect of hydrated Cs^+, *J. Phys. Chem. A* **108**, 1509-1514(2004).

[71] T. S. Hofer, B. M. Rode, and B. R. Randolf, Structure and dynamics of solvated Ba(II) in dilute aqueous solution - an *ab initio* QM/MM MD approach, *Chem. Phys.* **312**, 81-88(2005).

[72] R. Armunanto, C. F. Schwenk, and B. M. Rode, Structure and dynamics of hydrated Ag^+: *ab initio* quantum mechanical - molecular mechanical molecular dynamics simulation, *J. Phys. Chem. A* **107**, 3132-3138(2003).

[73] R. Armunanto, C. F. Schwenk, H. T. Tran, and B. M. Rode, Structure and dynamics of hydrated Au^+: *ab initio* QMMM/MD simulations, *J. Am. Chem. Soc.* **126**, 2582-2587(2004).

[74] C. Kritayakornupong and B. M. Rode,Molecular dynamics simulations of Hg^{2+} in aqueous solution including N-body effects , *J. Chem. Phys.*, **118**, 5065-5070(2003).

[75] T. S. Hofer, H. T. Tran, C. F. Schwenk, and B. M. Rode, Characterization of dynamics and reactivities of solvated ions by *ab initio* simulations, *J. Comp. Chem.* **25**, 211-217(2004).

[76] C. Kritayakornupong, K. Plankensteiner, and B. M. Rode, The Jahn-Teller effect of the Ti(III) ion in aqueous solution: Extended *ab initio* QM/MM molecular dynamics simulations, *Chem.Phys.Chem.*, **5**, 1499-1506(2004).

[77] A. J. Lock, S. Woutersen, and H. J. Bakker, Vibrational relaxation and energy equilibration in liquid water *in Femtochemistry and Femtobiology*, A. Douhal and J. Santamaria (Eds.), World Scientific, New Jersey, 234-239(2002).

[78] C. F. Schwenk, T. S. Hofer, and B. M. Rode, New insights into the Jahn-Teller effect through *ab initio* quantum mechanical - molecular mechanical molecular dynamics simulations in Cu(II) in water, *Chem.Phys.Chem.* **4**, 931-943(2003).

[79] C. Kritayakornupong, K. Plankensteiner, and B. M. Rode, The Jahn-Teller effect of the Ti(III) ion in aqueous solution: Extended *ab initio* QM/MM molecular dynamics simulations, *Chem.Phys.Chem.* **5**, 1499-1506(2004).

[80] M. Nomura and T. Yamaguchi, Concentration dependence of EXAFS and XANES of copper(II) perchlorate aqueous solution: comparison of solute structure in liquid and glassy states, *J. Phys. Chem.* **92**, 6157-6160(1988).

[81] B. Beagley, A. Eriksson, J. Lindgren, I. Persson, L. G. M. Pettersson, M. Sandstrom, U. Wahlgren, and E. W. White, A computational and experimental study on the Jahn-Teller effect in the hydrated copper (II) ion. Comparisons with hydrated nickel (II) ions in aqueous solution and solid Tutton's salts, *J. Phys.: Condens. Matter* **1**, 2395-2408(1989).

[82] T. K. Sham, J. B. Hastings, and M. L. Perlman, Application of the EXAFS method to Jahn-Teller ions: static and dynamic behavior of $Cu(H_2O)_6^{2+}$ and $Cr(H_2O)_6^{2+}$ in aqueous solution, *Chem. Phys. Lett.* **83**, 391-396(1981).

[83] M. Magini, Coordination of copper(II). Evidence of the Jahn-Teller effect in aqueous perchlorate solutions, *Inorg. Chem.* **21**, 1535-1538(1982).

[84] P. S. Salmon and G. W. Neilson, The coordination of Cu(II) in a concentrated copper nitrate solution, *J. Phys.: Condens. Matter* **1**, 5291-5295(1989).

[85] G. W. Neilson, J. R. Newsome, and M. Sandström, Neutron diffraction study of aqueous transition metal salt solutions by isomorphic substitution, *J. Chem. Soc., Faraday Trans. 2* **77**, 1245-1256 (1981).

[86] T. S. Hofer, B. R. Randolf, and B. M. Rode, *unpublished.*

[87] B. M. Rode, T. S. Hofer, B. R. Randolf, C. F. Schwenk, D. Xenides, and V. Vchirawongkwin, *submitted.*

Brill Academic Publishers
P.O. Box 9000, 2300 PA Leiden
The Netherlands

*Lecture Series on Computer
and Computational Sciences*
Volume 3, 2005, pp. 223-228

Density Functional Calculations for Static Dipole Interaction Polarizabilities of Ne_N Cluster up to $N < 30$.

Ralf Tonner[1] and Peter Schwerdtfeger[2]

Theoretical and Computational Chemistry Research Centre,
Bldg.44, Institute of Fundamental Sciences, Massey University (Albany Campus),
Private Bag 102904, North Shore MSC, Auckland, New Zealand

Received 1 June, 2005; accepted in revised form 15 June, 2005

Abstract: Accurate two-body interaction potentials of extended Lennard-Jones type for neon are used to determine the global minimum structures for clusters up to Ne_{29}. These structures are then used to calculate ionization potentials, triplet excitation energies and static dipole polarizabilities by density functional theory. The basis set dependence on the polarizability is discussed. The cluster polarizabilities increase monotonically with increasing cluster size N and extrapolation to the infinite system is made. The anisotropy component of the polarizability goes through a maximum in-between the magic cluster numbers N_{magic}.

Keywords: Neon Cluster, Dipole Polarizability, Ionization Potential, Singlet-Triplet Gap

PACS: 36.40.Cg, 32.10.Dk

1. Introduction

The prediction of accurate polarizabilities for rare gas cluster is a non-trivial task. First, it is not easy to accurately describe Van der Waals interactions by quantum theoretical methods. Density functional theory (DFT) is currently not reliable for such systems even if long-range corrections are introduced [1], and sophisticated *ab-initio* correlation procedures such as multi-reference configuration interaction or coupled cluster are simply too computer time consuming and not feasible for large cluster sizes. Second, the lighter rare gas element helium, and neon to some extent, are considered to be quantum-like in their atomic movements leading to a break-down of the Born-Oppenheimer approximation [2,3]. This is realized in rather large vibrational energy contributions for the liquid or solid state in these systems [4]. Third, for the heavier rare gas elements relativistic effects become important [5], and one needs to consider the spatial expansion of the occupied $p_{3/2}$ spinor. Fourth, special care has to be taken for basis sets in polarizability calculations, as more diffuse and high-angular momentum functions are required [6]. And last, vibrational contributions to polarizabilities have to be considered as well.

The static dipole polarizability of neon has been investigated quite recently [5-7], and accurate relativistic coupled cluster values are in good agreement with the experimental result of 2.670 ± 0.005 a.u. [8]. Nicklass *et al.* pointed out that scalar relativistic effects increase the polarizability by only 0.007 a.u. [9] in agreement with recent Douglas-Kroll calculations of Nakajima and Hirao [5]. Direct relativistic perturbation theory used by Klopper *et al.* gives a correction of 0.00443 a.u. [6]. They report a most accurate static dipole polarizability of Ne of 2.66312 a.u. at the CCSD(T) level of theory [6]. Accurate calculations for the static second hyperpolarizability of neon are also available [10].

The static or dynamic polarizabilities of rare gas clusters have not been studied extensively in the past [11]. In this preliminary study we consider dipole polarizabilities, ionization potentials and singlet-triplet gaps of neon clusters up to Ne_{29} calculated at the density functional level at optimized cluster geometries obtained from extended Lennard-Jones two-body potentials. Basis set effects and the

[1] Fachbereich Chemie, Philipps-Universität Marburg, Hans-Meerwein-S trasse, 35032 Marburg, Germany.

[2] Corresponding author. E-mail: p.a.schwerdtfeger@assey.ac.nz .

accuracy of the density functional approximation are discussed in detail for the atom and the neon dimer, and compared to accurate coupled cluster results. An attempt is made to extrapolate to the infinite (bulk) system.

2. Method

The neon cluster geometries are obtained from an accurate two-body potential of extended Lennard-Jones type [12],

$$V^2(r) = \sum_{i=1}^{6} c_{2i+4} r^{-2i-4} \tag{1}$$

with the coefficients c_{2i+4} obtained from coupled cluster calculations for Ne_2 by Cybulski and Toczylowski [13]. The coefficients are given in ref.12. This potential gives a minimum distance of 3.0985 Å and a potential depth of 28.62 cm⁻¹. This potential has the advantage that it has the correct behaviour for $r \to \infty$ and $r \to 0$, and solid-state structures and cohesive energies can be determined analytically [12]. Three-and higher body effects are neglected, as are vibrational and relativistic effects in this study.

The global minimum geometries for all the clusters up to $N=29$ were obtained from using a simulated annealing procedure [14] in conjunction with a conjugate gradient local optimization routine as incorporated in our cluster structure program MAMBO [15]. The optimized geometries for the elemental rare gas clusters were all true minima, established by a harmonic vibrational analysis. Structures of the optimal configurations resembled those of typical Lennard-Jones clusters.

Density functional calculations [16] for the polarizability, the singlet-triplet gap and ionization potential of Ne_N clusters with $N < 30$ were performed at the optimized global geometries as obtained from eq.(1). A modified aug-cc-pVDZ basis set [17] was used as discussed below. For cluster polarizabilities it is convenient to define the interaction polarizability (per atom),

$$\Delta\alpha(N) = N^{-1}\alpha(N) - \alpha(1) \tag{2}$$

with $\alpha(1)$ being the atomic static dipole polarizability of atomic Ne, and $\alpha(N)$ is the total isotropic static polarizability of the neon cluster of size N,

$$\alpha(N) = tr[\alpha_{ij}(N)]/3 \tag{3}$$

with α_{ij} being the Cartesian polarizability tensor. It is clear that for the atom we have $\Delta\alpha(1)=0$. The polarizability anisotropy is defined as,

$$\beta(N) = +\left(\frac{1}{2}\left[\left(\alpha_{xx}(N) - \alpha_{yy}(N)\right)^2 + \left(\alpha_{xx}(N) - \alpha_{zz}(N)\right)^2 + \left(\alpha_{yy}(N) - \alpha_{zz}(N)\right)^2\right]\right)^{1/2} \tag{4}$$

where the tensor α_{ij} is diagonal. For a linear cluster this leads to the well known expression of $\beta(N) = \alpha_p(N) - \alpha_\perp(N)$. Ionization potentials and triplet excitation energies were obtained for clusters up to $N = 20$ only, since the open-shell calculations started to become prohibitive in computer time and the SCF convergence was rather slow.

3. Discussion

We briefly analyze the accuracy of the density functional approximation for the smallest systems, Ne and Ne_2. Table 1 shows the dependence of the ionization potential and dipole polarizability for atomic neon on various well-known density functionals [16] and basis sets applied, using Dunning's series of correlation consistent basis sets [17]. The ionization potential E_{IP} converges nicely with increasing size of the basis set, while the dipole polarizability is very sensitive to the basis set applied as one expects. The different DFT methods all give similar values for the polarizability. The aug-cc-pvdz set gives polarizabilities far below the basis set limit. This is not only due to the neglect of additional diffuse s or p-functions. For example, there is little change at the LDA level of theory using a (doubly augmented, one extra diffuse sp-set) d-aug-cc-pvdz basis set which gives 2.1155 a.u. for the Ne polarizability, and

2.1161 a.u. with a (triply augmented, two extra diffuse sp-sets) t-aug-cc-pvdz basis set. However, adding more d-polarization functions extracted from pvtz and pvqz sets we now get much improved values of 2.6086 a.u. for the aug-cc-pvdz+1d set, and 2.8490 a.u. for the aug-cc-pvdz+2d set. These basis sets are however by far too large for larger cluster calculations. We therefore studied the basis set dependence on the interaction polarizability and anisotropy of the polarizability tensor for Ne$_2$.

Table 1: Ionization potentials E_{IP} (in eV) and static dipole polarizabilities α (in a.u.) for the Ne atom using different density functionals and Dunning's correlation consistent aug-cc-pvMz basis set (M=D, T, Q and 5).

b_1 M=	D	T	Q	5	D	T	Q	5
Method	E_{IP}	E_{IP}	E_{IP}	E_{IP}	α	α	α	α
LDA	22.86	22.76	22.74	22.74	2.1012	2.6206	2.8659	2.9620
Xα	21.11	21.05	21.03	21.03	2.1213	2.6453	2.9004	3.0079
PW91	21.84	21.77	21.76	21.76	2.1497	2.6361	2.8589	2.9651
PBE	21.75	21.68	21.66	21.65	2.1391	2.6351	2.8797	2.9927
BLYP	21.75	21.69	21.67	21.67	2.1448	2.6498	2.9016	3.0180
B3LYP	21.81	21.73	21.71	21.71	2.0520	2.5187	2.7349	2.8238
HF	19.85	19.69	19.66	19.66	1.8319	2.1939	2.3290	2.3631
MP2	21.62	21.68	21.75	21.79	1.9884	2.4345	2.6159	2.6696
CCSD(T)	21.42	21.47	21.55	21.58	1.9800	2.4184	2.5901	2.6389
Exp.		21.56 [18]				2.670±0.005 [19]		

Table 2: Isotropic interaction polarizability $\Delta\alpha$ and anisotropic component β of the static dipole polarizability (in a.u.) of Ne$_2$ using different density functionals and correlation consistent basis sets. A bond distance of 3.0985 Å is used.

b_1	aug-cc-pvdz	aug-cc-pvtz	aug-cc-pvqz	aug-cc-pvdz	aug-cc-pvtz	aug-cc-pvqz
Method	$\Delta\alpha(2)$	$\Delta\alpha(2)$	$\Delta\alpha(2)$	$\beta(2)$	$\beta(2)$	$\beta(2)$
LDA	0.0702	0.0559	0.0365	0.4150	0.4011	0.3603
Xα	0.0737	0.0593	0.0391	0.4397	0.4247	0.3792
PW91	0.0731	0.0580	0.0374	0.4499	0.4207	0.3773
PBE	0.0703	0.0555	0.0355	0.4295	0.4115	0.3698
BLYP	0.0694	0.0537	0.0323	0.4349	0.4111	0.3653
B3LYP	0.0539	0.0403	0.0226	0.3422	0.3276	0.2948
HF	0.0249	0.0164	0.0061	0.1711	0.1782	0.1698
MP2	0.0354	0.0250	0.0113	0.2353	0.2406	0.2252
CCSD(T)	0.0351	0.0249	0.0115	0.2324	0.2392	0.2247

Table 2 shows that while the basis set dependence is less severe for $\beta(2)$, it is still important for $\Delta\alpha(2)$. Moreover, all DFT approximations significantly overestimate the $\Delta\alpha(2)$ and $\beta(2)$ values as compared to the more accurate CCSD(T) results. This is clearly not a satisfying situation and one may have to introduce larger basis sets and long-range corrections into the DFT procedure to obtain more satisfying results. Nevertheless, the B3LYP functional shows the best results of all functionals and we will use this one for the following. However, for the calculation of the cluster polarizabilities we slightly improved the aug-cc-pvdz basis set. A diffuse p function of exponent 0.05 was added and the hard   d-function of exponent 2.202 was replaced by a most diffuse d-function of exponent 0.213. For the Ne atom this leads to considerable improvements, i.e. polarizabilities of 2.7538 a.u. (B3LYP) and 2.8892 a.u. (LDA) are now obtained in much better agreement with the experimental value (Table 1). The ionization potential with 21.89 eV (B3LYP) is also quite reasonable. We note that the experimental (spin-orbit averaged) value for the $^1S \rightarrow ^3P$ transition is 16.675 eV [19] and B3LYP therefore overestimates this transition by about 2 eV. The results for the Ne_N cluster up to N <30 are shown in Table 3.

Table 3: Isotropic interaction polarizabilities $\Delta\alpha$, anisotropic component β of the static dipole polarizabilities (in a.u.), vertical ionization potentials E_{IP} and triplet excitation energies E_{ST} (in eV) for the Ne_N cluster using a B3LYP functional and a modified aug-cc-pvdz basis set. The cluster geometries are taken from optimizations using an extended Lennard-Jones two-body potential as shown in eq.(1).

N	$\Delta\alpha(N)$	$\beta(N)$	$E_{IP}(N)$	$E_{ST}(N)$	N	$\Delta\alpha(N)$	$\beta(N)$	$E_{IP}(N)$	$E_{ST}(N)$
1	0	0	21.830	18.696	16	0.0500	1.043	17.084	14.535
2	0.0167	0.281	19.545	16.757	17	0.0507	1.236	17.042	14.510
3	0.0282	0.404	18.801	16.260	18	0.0517	0.872	17.011	14.495
4	0.0409	0.0	18.421	15.951	19	0.0508	1.395	16.987	14.478
5	0.0407	0.436	18.111	15.627	20	0.0511	1.305	16.953	14.457
6	0.0432	0.0	17.955	15.591	21	0.0504	1.314	-	-
7	0.0454	0.774	17.759	15.229	22	0.0515	1.048	-	-
8	0.0468	0.766	17.602	15.098	23	0.0522	1.287	-	-
9	0.0471	0.873	17.501	14.907	24	0.0525	0.943	-	-
10	0.0509	0.881	17.427	14.787	25	0.0521	0.564	-	-
11	0.0488	0.618	17.347	14.708	26	0.0537	0.0	-	-
12	0.0496	0.382	17.309	14.678	27	0.0539	1.987	-	-
13	0.0495	0.0	17.282	14.666	28	0.0531	1.695	-	-
14	0.0498	0.583	17.184	14.626	29	0.0532	1.291	-	-
15	0.0495	0.917	17.133	14.578					

Figure 1 shows the B3LYP interaction polarizability increasing sharply up to N=4 then leveling off with increasing N and converging quite smoothly towards the infinite system. This increase is easily explained; the highest occupied σ^* molecular orbital consisting mainly of Ne 2p increases in energy with increasing atomic interaction, thus increasing the polarizability and decreasing both the ionization potential and the singlet-triplet gap (Table 3). LDA most likely overestimates $\Delta\alpha$ as binding energies for Ne cluster are overestimated as well. Extrapolation to $N \rightarrow \infty$ using a $f(N^{-1/3})$ fit leads approximately to $\Delta\alpha (\infty)$= +0.07 a.u., $\Delta E_{IP}(\infty)$ = -6.5 eV and $\Delta E_{ST}(\infty)$ = -5.2 eV at the B3LYP level of theory. This gives an ionization potential for solid neon of approximately 15 eV, and a single-particle excitation energy of about 11.5 eV using experimental values for the atomic properties [19]. Note that the latter value is far below the estimated band gap for solid neon, which at the Γ point is 21.42 eV [20]. We find this discrepancy difficult to explain. Since the interaction between the neon atoms will change significantly by removal of an electron, and large reorganization effects are expected for the positively charged system, the adiabatic ionization potentials (and singlet-triplet excitation energies) are expected to be significantly lower compared to the vertical values, especially for the smaller cluster.

Figure 2 shows that the anisotropy β is zero at the magic cluster numbers N_{magic}= 4, 6, 13 and 26. The corresponding high-symmetry structures are shown in Figure 3. In-between the magic clusters, β goes almost smoothly through a maximum. This very much reflects shell-structure effects well-known in cluster physics and usually explained through the jellium model. The LDA and B3LYP results for the anisotropy closely agree with each other.

Figure 1: The dipole interaction polarizability, $\Delta\alpha(N)$, as a function of cluster size N for neon at the LDA and B3LYP level of theory.

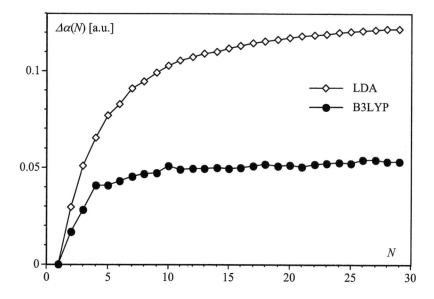

Figure 2: The anisotropy component of the dipole polarizability, $\beta(N)$, as a function of cluster size N for neon at the LDA and B3LYP level of theory.

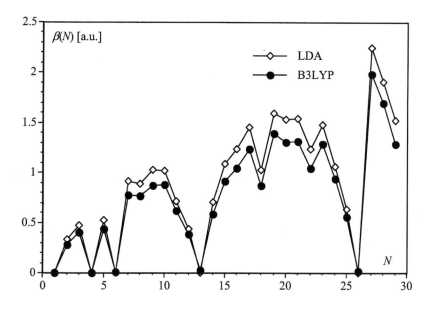

Figure 3: Optimized structures for the highly symmetric (Magic) neon cluster with the anisotropic polarizability component $\beta = 0$. In sequence: Ne_4 (T_d), Ne_6 (O_h), Ne_{13} (I_h) and Ne_{26} (T_d).

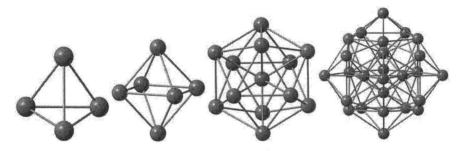

To conclude, more work has to be done for the accurate determination of rare gas cluster polarizabilities. Accurate density functionals with long-range corrections [21] including the Van der Waals tail have to be developed, and tailored basis sets have to be constructed to further improve on our results. Other possible errors here are the neglect of spin-orbit coupling and vibrational effects, both will increase the dipole polarizability for the cluster. Further, the cluster geometries were obtained from a two-body potential neglecting higher-body effects.

Acknowledgments

This work was supported by the Marsden Fund administered by the Royal Society of New Zealand. We are grateful to the Allan Wilson Centre for large amounts of computer time on their high-performance parallel computer HELIX.

References

[1] J. Tao and J. Perdew, *J. Chem. Phys.* <u>122</u>, 114102 (2005).

[2] M. L. Klein and J. A. Venables, *Rare Gas Solids*, Vol. 1, Academic Press, London, 1976.

[3] C. Predescu, P. A. Frantsuzov, and V. A. Mandelshtam, *J. Chem. Phys.* <u>122</u>, 154305 (2005).

[4] K. Rosciszewski, B. Paulus, P. Fulde, and H. Stoll, *Phys. Rev. B* <u>62</u>, 5482 (2000).

[5] T. Nakajima and K. Hirao, *Chem. Lett.* <u>30</u>, 766 (2001).

[6] W. Klopper, S. Coriani, T. Helgaker, and P. Jørgensen, *J. Phys. B: At. Mol. Opt. Phys.* <u>37</u>, 3753 (2004).

[7] R. Franke, H. Müller, and J. Noga, *J. Chem. Phys.* <u>114</u>, 7746 (2001).

[8] R. H. Orcutt, and R. H. Cole, *J. Chem. Phys.* <u>46</u>, 697 (1967).

[9] A. Nicklass, M. Dolg, H. Stoll, and H. Preuss, *J. Chem. Phys.* <u>102</u>, 8942 (1995).

[10] F. Pawlowsky, P. Jørgensen, and C. Hättig, *Chem. Phys. Lett.* <u>391</u>, 27 (2004).

[11] K. D. Bonin and V. V. Kresin, *Electric-Dipole Polarizabilities of Atoms, Molecules and Clusters*, World Scientific, Singapore, 1997.

[12] P. Schwerdtfeger, N. Gaston, R. Krawczyk, M. Lein, R. E. Tonner, G. E. Moyano, to be published.

[13] S. M. Cybulski and R. R. Toczylowski, *J. Chem. Phys.* <u>111</u>, 10520 (1999).

[14] W. Press, S. Teuklosky, W. Vetterling and B.P. Flannery, *Numerical Recipes in Fortran 77*, Cambridge University Press, New York (1992).

[15] G. Moyano, M. Pernpointner, and P. Schwerdtfeger, Program MAMBO: A multitask simulated annealing many-body potential program for the optimization of clusters, Massey University, Auckland, 2004.

[16] M.J. Frisch, G. W. Trucks, H. B. Schlegel *et. al.* GAUSSIAN 03, Gaussian Inc. Pittsburgh, PA, 2004; and references therein.

[17] D. E. Woon and T. H. Dunning, Jr., *J. Chem. Phys.* <u>98</u>, 1358 (1993).

[18] R. H. Orcutt, and R. H. Cole, *J. Chem. Phys.* <u>46</u>, 697 (1967).

[19] C. E. Moore, *Atomic Energy Levels*, US GPO, Washington, 1958.

[20] E. Boursey, J.-Y. Roncin, and H. Damany, *Phys. Rev. Lett.* <u>25</u>, 1279 (1970).

[21] T. Sato, T. Tsuneda, and K. Hirao, *Mol. Phys.* <u>103</u>, 1151 (2005).

Brill Academic Publishers
P.O. Box 9000, 2300 PA Leiden,
The Netherlands

*Lecture Series on Computer
and Computational Sciences*
Volume 3, 2005, pp. 229-233

Polarizability of Carbon Nanotubes

Gustavo E. Scuseria[1] and Edward N. Brothers

Department of Chemistry,
Mail Stop 60,
Rice University,
Houston, Texas 77005-1892

Received 15 July, 2005; accepted 12 August, 2005

Abstract: We have calculated both the transverse and longitudinal polarizability of single wall carbon nanotubes. The results were then fit to curves based on simple physical interpretations.

Keywords: Density functional theory, polarizability, carbon nanotubes.

1 Introduction

One of the earliest studies of carbon nanotube polarizabilities was carried out by Benedict, Louie, and Cohen in 1995 using the tight binding approximation.[1] In that study, they predicted that transverse polarizability per unit length would be proportional to the square of the radius, and longitudinal polarizability per unit length[2] would be proportional to radius over the square from the gap. However, subsequent papers on polarizability have disputed these trends; for example, there are other tight binding predictions of transverse polarizability that argue for dependence of band gap.[3, 4]

We are currently able to calculate the polarizability of nanotubes and other infinite systems using the *Gaussian*[5] suite of programs and periodic boundary conditions (PBC)[6], with code modified in our research group to calculate response to electric fields.[7] Note that this code permits calculation of both HF[9] and DFT[8] polarizabilities.

There are two methods of calculating polarizability that will be discussed in this paper. The first method is approximate polarizability which uses a sum over states formula:

$$\alpha_{zz}^A = \frac{2a}{\pi} \sum_a^{occ} \sum_r^{unocc} \int_{BZ} \frac{|\Omega_{ar}(k_z)|^2}{\epsilon_r(k_z) - \epsilon_a(k_z)} dk_z. \tag{1}$$

where a and r are occupied and unoccupied crystal orbitals (CO), $\epsilon(k)$ is CO energy at a given **k** point, and Ω describes the CO's mixing in response to an electric field. This method is referred to as approximate as it contains zero field orbitals, *i.e.* no density relaxation, and thus is expected to be limited in its ability to model physical systems.

The second method we currently use is the calculation of polarizabilities through numerical second derivatives of the energy with respect to an electric field.[10] Formally:

$$\alpha_{zz} = \frac{\partial E^2}{\partial^2 F_z} \tag{2}$$

---

[1] Corresponding author. E-mail: guscus@rice.ed

Table 1: **Transverse polarizabilities per unit length.** Labels m and s refer metallic and semiconducting respectively. All values were calculated with LSDA/6-31G(d).[14, 15]

tube(n_1,n_2)	Conductivity	Radius(Å)	α/L (Å^2)
(5,0)	s	1.98	3.76
(3,3)	m	2.06	3.80
(6,0)	m	2.37	4.80
(4,4)	m	2.74	5.59
(7,0)	s	2.76	5.64
(8,0)	s	3.15	6.92
(5,5)	m	3.41	7.77
(9,0)	m	3.54	8.17
(10,0)	s	3.93	9.73
(6,6)	m	4.09	10.32
(11,0)	s	4.32	11.38
(12,0)	m	4.71	12.94
(7,7)	m	4.76	13.24

Where F_z is an electric field in the z direction and E is the SCF energy. This method is more computationally expensive than approximate calculations as it requires two SCF calculations instead of one,[11] but as this method includes density relaxation, it is expected to produce more accurate results, especially in systems where the polarization of one part of the system aids or inhibits polarization of another part of the system.

The next sections will provide a brief overview of nanotube polarizability trends calculated in our research group. The LSDA[14] functional is used throughout, and was chosen here solely to illustrate trends; a larger study with more advanced functionals is currently in preparation.

2   Transverse Polarizability

We have recently published a study of transverse polarizability utilizing several functionals that concurs with the conclusions of Benedict, Louie, and Cohen,[13] *i.e.* we found that transverse polarizability depends only on tube radius and not conductivity, *i.e.* metallic and semiconducting polarizabilities follow the same trend. A brief illustration of this can be found it Table 1.

The physical explanation for transverse polarizabilities independence of the gap is quite simple. The conductivity of the nanotube is in the longitudinal direction, *i.e.* the smaller the band gap, the easier it is for an electron to travel down the nanotube. Band gap is not an issue in electron mobility around the outside of the tube. Thus transverse polarizability depends solely on nanotube radius.

3   Longitudinal Polarizability

The picture for longitudinal polarizability becomes significantly more complicated. Since the conductivity of a nanotube is physically expected to be in the direction of the tube, band gap is expected to play an important role. In addition, there are numerical issues with the calculation of polarizability which must be considered first. These issues can be seen by examining the predictions in Table 2. Note that the polarizability determined by the curve fit is expected to be the most reliable as it utilizes the most data.

Table 2: **Effect of field strength on predicted longitudinal polarizability of a (10,0) nanotube.** The entry labeled "Curve" is polarizability determined from a second order polynomial fit. Field strength has units of 1.0×10^{-4} a.u, while polarizability has units of Å^3. All calculations were performed with LSDA/3-21G. [14, 16]

Field	LSDA
1	622.4
2	621.4
3	619.7
4	617.2
5	614.1
Curve	614.2

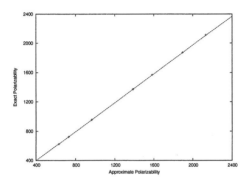

Figure 1: **Comparison of longitudinal polarizabilities with the approximate longitudinal polarizabilities.** All values are calculated with LSDA/3-21G.[14, 16]

The computational expense of repeating the work in Table 2 for a large range of nanotubes with a variety of functionals would be prohibitive. One argument that could be made is that the polarizabilities calculated using only one field strength are accurate enough, with the difference between the curve-predicted value and the prediction furthest from it in the example being $\sim 1\%$. We decided instead to compare the polarizabilities to the approximate polarizabilities for a series of nanotubes to see if the results were similar enough that the approximate method could be used reliably. As can be seen by Figure 1, the approximate values are comparable to the longitudinal polarizabilities for single-wall nanotubes.

Approximate longitudinal polarizabilities were then calculated for a set of semiconducting nanotubes. Note that the data set included chiral as well as zigzag semiconducting nanotubes in order to ensure a diverse test set.[12] Figure 2 contains the LSDA results plotted using the same trends as Benedict, Louie, and Cohen, *i.e.* polarizability per unit length is plotted versus radius over gap squared. This fit is numerically acceptable, but further attempts at analyzing the data resulted in a better fit.

The physical expectation for longitudinal polarizability in carbon nanotubes is that smaller band gaps lead to large polarizability. It is also necessary that the nanotubes are somehow normalized for size, such as considering polarizabilities per length or per atom rather than considering

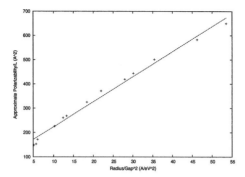

Figure 2: **Longitudinal polarizability per unit length versus radius over band gap squared.** All values are calculated with LSDA/3-21G.[14, 16]

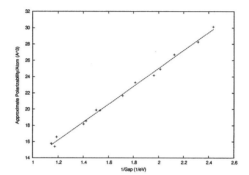

Figure 3: **Longitudinal polarizability per atom versus inverse band gap.** All values were calculated with LSDA/3-21G. [14, 16]

raw polarizabilities in order to get a clear picture of what is occurring. Based on these considerations, we plotted polarizability per atom versus inverse band gap in Figure 3 This fit is simpler than the previous curve and is numerically at least as good. Thus at present we prefer this curve for prediction of longitudinal polarizability.

4   Conclusion

The polarizability of carbon nanotubes is still an active area of research, especially in the case of longitudinal polarizabilities. The response of nanotubes to electric fields will be of great importance to the emerging field of nanoelectronics. While further work with additional functionals is necessary to resolve certain issues, we are able to fit nanotube polarizabilities to very simple curves based on physical expectations, and believe we have a reasonable understanding of nanotube polarizability.

Acknowledgments

The authors wish to thank K.N. Kudin for his insightful comments during the course of this work.

This work was supported by the Nanoscale Science and Engineering Initiative of the National Science Foundation under NSF Award Number EEC-0118007, NSF-CHE-0457030, and the Welch Foundation.

References

[1] L. X. Benedict, S. G. Louie, and M. L. Cohen, Phys. Rev. B **52**, 8541 (1995)

[2] Longitudinal polarizability was considered for semiconducting nanotubes only; for metallic nanotubes, the longitudinal polarizability would be infinite.

[3] J. O'Keeffe, C. Wei, and K. Cho, App. Phys. Lett. **80**, 676 (2002)

[4] Y. Li, S. V. Ritalin, and U. Ravaioli, Nanolett., **3**, 183, (2003)

[5] Gaussian Development Version (Revision B.07) M. J. Frisch, G. W. Trucks, H. B. Schlegel *et al.*, Gaussian, Inc., Pittsburgh, PA, 2004.

[6] K. N. Kudin and G. E. Scuseria, Phys. Rev. B, **61**, 5141, 2000.

[7] K. N. Kudin, and G. E. Scuseria, J. Chem. Phys. **113**, 7779 (2000)

[8] P. Hohenberg and W. Kohn. Phys. Rev. B, **136**, 864 (1964); W. Jon and L. J. Sham, Phys. Rev. A, **140**, 1133 (1965).

[9] C. C. J. Roothan, Rev. Mod. Phys., **23**, 69, (1951).

[10] The first derivative of the dipole with respect to electic field perturbations could also have been used.

[11] The two SCF cycles (one for no field, and one with field) are not independent as the converged density is necessary to form the dipole matrix, and the zero field density usually provides a reasonable guess for the field perturbed density. See the Kudin and Scuseria paper above.[6]

[12] Specifically, the zigzag (n,0) nanotubes (10,0), (11,0), (13,0) (14,0), (16,0), (17,0), (19,0), (20,0), (22,0), and (23,0) were used, along with chiral nanotubes (8,4), (10,5), (12,4) and (14,7).

[13] E. N. Brothers, K. N. Kudin, G. E. Scuseria, and C. W. Bauschlicher, Jr., Phys. Rev. B In Press

[14] S. H. Vosko, L. Wilk, and M. Nusair, Can. J. Phys. **58**, 1200, (1980).

[15] M. J. Frisch, J. A. Pople, J. S. Binkley, J. Chem. Phys. **80**, 3265 (1984) and references therein.

[16] J.S. Binkley, J.A. Pople and W.J. Hehre, J. Am. Chem. Soc. **102**, 939 (1980).

Brill Academic Publishers
P.O. Box 9000, 2300 PA Leiden,
The Netherlands

*Lecture Series on Computer
and Computational Sciences*
Volume 3, 2005, pp. 234-247

Understanding local dielectric properties of water clusters

P. Senet[1]‖, **M. Yang**†, **P. Delarue**‖, **S. Lagrange**‖, and **C. Van Alsenoy**‡

‖Théorie de la matière condensée,
CNRS-UMR 5027, Laboratoire de Physique,
Université de Bourgogne, 9 Avenue Alain Savary - BP 47870,
F-21078 Dijon Cedex, France

†Department of Physics,
Central Michigan University,
Mt. Pleasant,MI 48859, USA

‡Structural Chemistry Group,
Department of Chemistry,
University of Antwerp, Universiteitsplein 1,
B-2610 Antwerp, Belgium

Received 1 August, 2005; accepted in revised form 12 August, 2005

Abstract: The dipolar polarisabilities of the most stable geometries of water clusters W_n ($n = 1 - 20$) are computed by using Density Functional Theory with the B3LYP exchange-correlation functionals and the second order Møller-Plesset perturbation theory. The effects of the size and of the structure of a cluster on its global polarisability are discussed. On the other hand, using the linear response theory, the Hirshfeld partitioning of the electronic density, and an electrostatic model for the local electric field, we define a screened molecular polarisability for a molecule in a cluster. One shows that this local polarisability varies with the size of the cluster, the location of the molecule, and the hydrogen bond network. In particular, one establishes the correlation between the charge redistribution occurring in hydrogen bonding and the molecular polarisabilities.

Keywords: Hydrogen bonds, Hirshfeld charges, Water clusters, Polarisability

1 Introduction

Hydrogen bonding in water is a kind of dipole-dipole electrostatic interactions as suggested a long time ago by Pauling.[1] It is certainly true that the interactions between two water molecules are dominated at large distances by this sole physical Coulomb interaction. But the molecules are not inert. Even at several angstroems apart from the equilibrium distance, one observes of course a shift of the molecular electronic densities due to the mutual polarisation of the water molecules by their dipolar electric fields. This trivial interaction is there well represented by local molecular polarisabilities equal to the polarisability of an isolated molecule. But at closer distances, where the hydrogen bond forms, a more drastic change of the electronic density occurs (see below). As already mentioned by many authors,[2],[3],[4],[5],[6],[7],[8] and in particular by Weinhold and collaborators,[7],[8] the electronic charge redistribution upon bonding is responsible for a net charge

---

[1]Corresponding author. E-mail: psenet@u-bourgogne.fr

transfer in the water dimer. Because of this effect, the qualitative nature of hydrogen bonding can be viewed also as a kind of resonance.

The intermolecular charge transfer varies with the application of an electric field along the bond in the water dimer.[9] What are the effects of these charge redistributions on the polarisability of a molecule in a larger cluster ? How to define a local molecular polarisability ? How these quantities vary with the cluster size ? How to include these effects in polarizable force fields ? The present lecture attempt to answer these questions by investigating in detail the polarisabilities of stable water clusters (up to 20 molecules) and some of their isomers using Density Functional Theory (DFT) and second order Møller-Plesset perturbation theory (MP2). Results obtained for these structures, which can be produced experimentally,[10] are also relevant for liquid water. Indeed, the hydrogen bond network formed in liquid water is dynamic and can be thought in terms of metastable water clusters.[11]·[12]

The *molecular polarisability* or the polarisability of "*a water molecule in a cluster*" has no unique definition. Several approaches in quantum chemistry are available to partition the cluster polarisability in molecular ones.[9],[13],[14] In the present work, one uses the density of the electrons as the basis of the partitioning of the cluster electronic responses.[3], Molecular polarisabilities are defined by using electrostatics, the linear response theory and the Hirshfeld (also named stockholder) electronic density partition.[15],[16] Evaluation of such a molecular polarisability permits a description of the local cluster dielectric properties relevant for its interactions with other water molecules and with solutes. Anticipating the results presented below, one observes that this molecular polarisability decreases with the cluster size n and varies with the position within the structure. The molecular polarisability is different from the *mean polarisability of a water molecule* defined as the polarisability of a water cluster W_n divided by the number of molecules n. Recent first principles calculations of water cluster polarisabilities, including results presented below, show indeed that the mean polarisability varies weakly with the cluster size n.[3],[17],[18],[19] One aim of the present paper is to show that the different variation of *molecular* polarisabilities can be understood in terms of charge redistributions upon hydrogen bonding.

The paper is organized as follows. The local polarisability and its computational scheme is explained in next Section. In the third Section, the variations of the mean and molecular polarisability with the size of the cluster and the hydrogen bond network are reported and discussed. In the fourth Section, one establishes relations between the local dielectric properties of the cluster and its charge distribution. Conclusions are presented in the last Section.

2 Mean and molecular polarisabilities and stockholder scheme

2.1 Stockholder partition of polarisabilities

Mean and molecular polarisabilities may be defined and related to each other by using the stockholder or Hirshfeld partitioning of the electronic density[15] as follows. In the stockholder scheme, the electronic density of an atom in a molecule ρ_L may be defined as the expectation value of an atomic charge operator $\hat{\omega}_L$ by the following equation

$$\rho_L(\mathbf{r}) = \left\langle \hat{\omega}_L \right\rangle = \omega_L(\mathbf{r})\rho(\mathbf{r}), \tag{1}$$

$$\hat{\omega}_L \equiv \sum_{i=1}^{N} \omega_L(\mathbf{r})\delta(\mathbf{r} - \mathbf{r}_i), \tag{2}$$

where $\rho(\mathbf{r})$ is the molecular electronic density and $N = \int d\mathbf{r}\, \rho(\mathbf{r})$ its integral whereas $\mathbf{r}_i$ is the *ith* electron coordinate. The scalar function $\omega_L(\mathbf{r})$ is defined by

$$\omega_L(\mathbf{r}) \equiv \frac{\rho_L^0(\mathbf{r})}{\sum_{K=1}^{M} \rho_K^0(\mathbf{r})}, \tag{3}$$

in which the numerator ρ_L^0 is the electronic density of the atom L when it is free and neutral and the denominator is the sum of the electronic densities of all the atoms of the molecule. The superposition of the atomic densities of the free neutral atoms located at the positions they have in the molecular geometry is called the density of the promolecule and serves to define the weighting function ω_L. In the present work, one will be interested by the atomic charge q_L and dipole moment $\mathbf{p}_L$ (in units of elementary charge $e = 1.602\ 10^{-19}C$) obtained from Eq. (1)

$$q_L = Z_L - \int d\mathbf{r}\ \rho_L(\mathbf{r}), \tag{4}$$

$$\mathbf{p}_L = -\int d\mathbf{r}\ \mathbf{r}\ \rho_L(\mathbf{r}) + Z_L \mathbf{R}_L, \tag{5}$$

where Z_L and $\mathbf{R}_L$ are the atomic number and position, respectively. From these atomic quantities, one computes the effective dipole moment of the water *molecule I* in the cluster

$$\mathbf{P}_I = \sum_{K(I)} \mathbf{p}_K, \tag{6}$$

where the sum is restricted to the three atoms K of the molecule I.

Local molecular polarisabilities can be computed from the variations of the molecular dipole moments Eq. (6) induced by an external electric field $\mathbf{E}$. One distinguishes qualitatively two types of polarisabilities for a molecule in a cluster. First, the molecular polarisability corresponds to the dipolar response of the molecule to the field *applied* : it is a *global* property. Its value is expected to be non transferable from one molecule to another or from a given environment to another. Second, the molecular polarisability corresponds to the dipolar response of the molecule to the *total (screened)* field: it is mainly a *local* property. Its value is expected to be more or less constant from molecules *in the same local environment*. This screened molecular polarisabilty is the one used in polarizable force fields.[20] In these models, all the molecules are equivalent and the effects of the environment is taken into account only through the electrostatic field induced by the other molecules. The polarisability corresponding to the response of the molecule to the screened field is the quantity in which we are interested here. *Both* types of polarisabilities can be defined from Eq. (6) as follows.

The Cartesian components of the molecular dipole moment induced by the field applied is given to the first-order by

$$\delta P_{I\alpha} = \sum_{\beta} \frac{\partial P_{I\alpha}}{\partial E_{\beta}} E_{\beta}. \tag{7}$$

From Eq.(7), one defines of course the molecular polarisability (response to the applied field) as

$$\widetilde{\Omega}_{I\alpha\beta} \equiv \frac{\partial P_{I\alpha}}{\partial E_{\beta}}. \tag{8}$$

For a cluster of size n, these molecular polarisabilities are related to the *mean* polarisability $\overline{\Omega}$ by the relation

$$\frac{\sum_I \widetilde{\Omega}_{I\alpha\beta}}{n} \equiv \overline{\Omega}_{\alpha\beta}. \tag{9}$$

On the other hand, the *screened* electric field e_I can be defined by using a point dipole model by the following relation[21]

$$e_I = E + \sum_{J \neq I} \frac{3(\delta P_J . R_{JI}) R_{JI} - R_{JI}^2 \delta P_J}{R_{JI}^5},$$ (10)

where $R_{JI} = R_J - R_I$ and R_I is the position of molecular dipole defined for instance as the center of mass or as the geometrical center of the Ith water molecule.

Expanding the molecular dipole in terms of the local field Eq. (10), one finds to the first-order

$$\delta P_{I\alpha} = \sum_\beta \frac{\partial P_{I\alpha}}{\partial e_{I\beta}} e_{I\beta},$$ (11)

where we assume that the induced dipole moment of a molecule depends solely of the local electric field at the position of that molecule (the so-called Lorentz approximation).[22] The screened molecular polarisability is of course

$$\Omega_{I\alpha\beta} \equiv \frac{\partial P_{I\alpha}}{\partial e_{I\beta}}.$$ (12)

The relation between the mean and the screened molecular polarisability is deduced from Eqs. **(7-12)**,

$$\sum_I \sum_\beta \Omega_{I\alpha\beta} e_{I\beta} \equiv n \sum_\beta \overline{\Omega}_{\alpha\beta} E_\beta.$$ (13)

The importance of the local fields can be appreciated by using the following approximation often used in force fields:

$$\Omega_{I\alpha\beta} \simeq < \Omega > \delta_{\alpha\beta},$$ (14)

where $< \Omega >$ is an effective isotropic screened polarisability *identical* for all molecules. Validity of the approximation (14) will be discussed below. Eq. (13) becomes

$$< \Omega > \sum_I e_{I\alpha} = n \sum_\beta \overline{\Omega}_{\alpha\beta} E_\beta.$$ (15)

By choosing to apply an external electric field along one direction, say x, one finds from Eq. (15)

$$\frac{< \Omega >}{\overline{\Omega}_{xx}} = \frac{E_x}{\sum_I e_{Ix}/n}.$$ (16)

Because the applied field is screened by the response field (Eq. (10)), one expects the average local field (denominator of Eq. (16)) to be lower than the applied field (numerator of Eq. (16)). One deduces that mean polarisability $\overline{\Omega}$ should be larger than the screened polarisability $< \Omega >$: this is exactly what we find numerically. Finally, it is worth noting that the well-known distributed atomic and molecular polarisabilities[9],[15] correspond to a partitioning of the cluster electronic response to the applied field and should be compared to $\overline{\Omega}$ instead of $< \Omega >$ present definitions.

2.2   Intermolecular charge transfers and corrected polarisabilities

The induced dipole moment evaluated from Eq. (11) is coordinate-dependent because water molecules are in general charged in the cluster due to small intermolecular charge transfers [2],[3],[4], [5],[6],[7],[8] as shown below. To remove the coordinate-dependent terms, $\delta \mathbf{P}_I$ in Eqs. (10) and (11) must be replaced by a "corrected" induced dipole defined by

$$\delta \mathbf{P}_I^c = \delta \mathbf{P}_I - \frac{Q_I}{3} \sum_{K(I)} \mathbf{R}_K = \Omega_I^c \, \mathbf{e}_I^c, \tag{17}$$

$$Q_I \equiv \sum_{K(I)} q_K, \tag{18}$$

where the net charge of molecule is Q_I. From Eq. (17), one evaluates an average dipolar polarisability of the molecule I (trace of Ω_I^c) in the cluster that we called the *"corrected screened molecular polarisability"* or more simply *"screened molecular polarisability"*. Eq. (17) is independent of the choice of coordinates because it defines the induced dipole moment relative to the geometrical center of the molecule. Only Ω_I^c is expected to be a transferable property of a water molecule in a condensed phase. On the contrary, Ω_I deduced from Eq. (11) is coordinate-dependent because it contains a charge dependent contribution which is a global cluster property, we name it as *"uncorrected screened molecular polarisability"*.

2.3   Computational details

The initial Hartree-Fock optimized geometries for W_1 to W_{20}[23] are taken from the Cambridge Cluster Database (CCD)[24] and optimized using the B3LYP functional and the 6-31G(d,p) basis set using the Gaussian98 program.[25] The structures are similar to the initial ones .[4],[23] On the other hand, we dot find major differences between the present DFT geometries and those computed at the MP2/aug-cc-pvDZ level for the smallest clusters.[26]

   The electronic polarisabilities are calculated using a finite field method[27] with DFT (up to 20 molecules) and MP2 (up to 15 molecules) using the B3LYP-optimised structures. The applied field strength is set to 0.001au and the SCF convergence criteria are tighten to 10^{-9}. Only the traces of the polarisability tensors are presented and discussed. For the calculation of the polarisabilities, one chooses a basis set 6-31++G(d,p). Larger basis sets become computationally demanding for water clusters as large as W_{20}. In order to achieve a full picture of the variations of water polarizabilities with cluster size, the basis set 6-31++G(d,p) is selected as a compromise between reliability of the results and the computational efficiency. This choice is based on a careful numerical analysis of the basis set effects for cluster size up to $n = 6$ which demonstrates that the values obtained with the 6-31++G(d,p) calculations reproduce about 70-80% of the values obtained with large basis sets such as aug-cc-pVDZ and aug-cc-pVTZ .[26] Our best estimation in the present work using a moderate 6-31++G(d,p) basis set of the polarisability of an isolated molecule is only 73 % of the experimental value[28] and is obtained with DFT-B3LYP. This inaccuracy compared to experiment does not modify the main conclusions described below as they concern not the absolute values of the cluster polarisabilities but their variations at a given approximation level, with the size and the geometry.

   The calculations of polarisabilities and dipole moments are also performed with Gaussian98 program[25] and the stockholder multipolar expansion of the electronic density is computed with the program STOCK[16] which is part of the BRABO package as in Ref. [29].

3   Size and hydrogen bonding effects

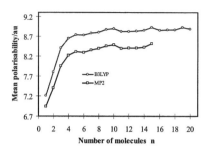

Figure 1: Mean molecular polarisability in water clusters W_n. Calculations are performed at DFT-B3LYP (circles) and MP2 (squares) levels with basis set 6-31++G(d,p). All values are in a.u..

The mean polarisabilities of water clusters computed with DFT-B3LYP and MP2 are shown in Fig. (1). Both methods, as well as HF and DFT-B3PW91,[3],[4] give similar trends. One remarkable property deduced from the Fig. (1) is that the mean polarisability is rather constant for $n > 6$, equal to about 8.9 a u in DFT-B3LYP calculations corresponding to an *increase* of 23 % relative to the value of the isolated molecule calculated with the same approximations.[3],[4] Fig. (1) means that the polarisability of water clusters scales linearly with their size. This has been found also by Ghanty and Ghosh[19] in a recent study. For the smallest size ($n < 6$), the mean polarisability varies with the size but this variation depends also on the basis set.[4] Different results are obtained for the screened molecular polarisabilities. In Fig. (2), we report the variation of the screened molecular polarisability, as averaged on all the molecules of a cluster, with the cluster size. Uncorrected molecular polarisabilities follow the same trend[3],[4]: a decrease of the molecular polarisability as the cluster size increases. We may extrapolate from the Fig. (2) that the screened polarisability per molecule in large water clusters is approximately 35% *smaller* than the polarisability of an isolated molecule. For the largest clusters studied here and at the same level of theory, this corresponds to a value twice smaller than the mean polarisability. This reduction of the molecular polarisability is the effect of screening described by Eq. (16).

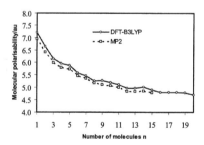

Figure 2: Screened molecular polarisabilities in water clusters W_n. Calculations are performed at DFT-B3LYP (circles) and MP2 (dashed line and squares) levels with basis set 6-31++G(d,p). All values are in a.u.

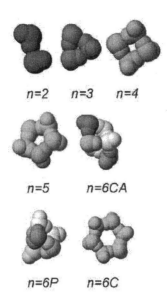

Figure 3: Color maps of molecular polarisabilities in small water clusters W_n ($n = 1 - 6$). 6CA, 6P, 6C are respectively the cage, prism and cyclic isomers of W_6. The values increase from white (low values) to black (high values). Calculations are performed using B3LYP/6-31++G(d,p).

Although the average molecular polarisability represented at Fig. (2) varies monotonously with n, the molecular polarisability in a cluster fluctuates with the molecule position and depends on the hydrogen bond network. In Fig (3) we have represented the screened polarisability according to a red color scheme with a hue proportional to the polarisability value. In general, as stated above, the value is smaller in large clusters compared to the dimer. We can observe also that all the molecules are not equivalent. For a given cluster size, the polarisability has a tendency to be smaller for molecules forming more hydrogen bonds. In cyclic structures, one observes no variation of the molecular polarisability with the position as the molecules are nearly identical forming two hydrogen bonds, one as proton donor and the other as proton acceptor.

The relation between the hydrogen bond network and the polarisability is best studied in water cluster isomers. In Fig. (3), the W_6 most stables isomers: the cage (CA), the cyclic structure (C) and the prism (P) structure are represented. All have an average polarisability which is very similar with a standard deviation of 0.43%.[4] Although the cluster polarisability does not vary with the isomeric structure, the polarisability of a molecule within the cluster depends on its location and more precisely of its hydrogen bonds with the other molecules. There are three kinds of molecules in the cage structure. The two molecules donating one proton and receiving one proton have greatest polarizabilities, followed by the two donating two protons and receiving one proton. The left two molecules donate one proton and receive two protons and have the smallest polarizabilities. All the six molecules in the cycle isomer are almost equivalent. Each of them donates one proton and receives one proton, and therefore have nearly equivalent polarizabilities. The molecules in the prism structure form two groups. Three of them with two protons donated and one proton received have greater polarizabilities than the other three molecules with one proton donated and two protons received. On concludes that the polarizability of a water molecule is directly related to its proton acceptor/donor character in its bonding with the surrounding molecules and decreases with the number bonds it does. This tendency is also confirmed in the largest clusters.[3],[4] The variation of the screened molecular polarisability in the fused cube structure of W_{20} is shown in Fig. (4A). Twelve molecules in W_{20}, having four hydrogen bonds and located in the body of the cluster, have smaller polarisabilities than the others. On the contrary, molecules at both ends in W_{20} having dangling bonds, possess larger polarisabilities than those in the body. As in ice, molecules in the bulk of W_{20}, accept and give two hydrogen atoms although the molecules in W_{20} are not in a tetrahedral network. On may expect the screened polarisability of water in condensed phase to be also smaller than in gas phase. Variations of the screened molecular polarisability with the position can be also related to intermolecular charge transfers as explained next.

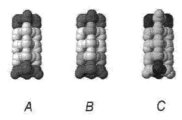

$$A \qquad B \qquad C$$

Figure 4: Molecular polarisability (A), net charge (in absolute value) (B) and HOMO charge (C) in W_{20}. The values increase from white (low values) to black (high values). Calculations are performed using B3LYP/6-31++G(d,p).

4   Relations between molecular polarisability and charges

Electron density has been recognized as a fundamental variable in chemical bonding in the early days of quantum chemistry: Hellman and Feynman demonstrated that bonding may be explained by electrostatic forces.[31] In particular, the variations of the density with the bond lengths contain all we need to evaluate the variations of the intermolecular forces. DFT tells us that these variations of the density are those which conduct to a minimum total energy for the distorted geometry.[32] In the present work, the density variations in the formation of a hydrogen bond can be simply analyzed in terms of stockholder atomic charges and their variations when a bond is elongated.

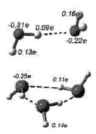

Figure 5: Stockholder charges computed with MP2/aug-cc-pVDZ for water dimer and trimer optimized geometries.

One consider here the more accurate MP2/aug-cc-pVDZ optimized geometries of the water dimer and trimer.[26] The charges are evaluated at the same level of the theory which give the distributions represented at the Fig. (5) for the most stable dimer and trimer geometries.[26] The atomic charges in the water monomer are : $q_O = -0.28e$, $q_H = 0.14e$. The major changes occur for the hydrogen (Hb) of the proton donor molecule and the oxygen (Oa) of the proton acceptor. Hb is less positive when bonded by 36 % and Oa is less negative when bonded by about 21 %. The next important change is observed on the other oxygen atom which becomes more negative by about 10%. These local changes are observed in all water clusters.[33] For instance, for the trimer, where all the molecules are both proton donor and acceptor, the charge of Hb and O are reduced by 21 % and 11 %, respectively. In the trimer, all water molecules are equivalent and the net charge of the molecules are nearly zero. It is not the case for the dimer, where we observe a net charge transfer of 0.09 electrons from the proton acceptor molecule to the proton donor molecule at the equilibrium distance. The net charge of the molecules decays exponentially with the hydrogen bond length as it is shown for the proton donor molecule in Fig. (6). The geometries of the molecules are kept rigid when the Hb...Oa distance is increased. Additional test done using DFT-B3LYP, DFT-PBE, CCSD with medium and large Gaussian basis sets as well as plane-wave basis sets do not modify the conclusions reported here.[33]

To the best of our knowledge is the first application of the stockholder scheme to analyze the charges in water clusters. It is interesting to compare these findings with other analysis. Using Natural Population Analysis (NPA), Weinhold and collaborators[8] found a net charge transfer from one molecule to the other in the water dimer using various basis sets and methods. The electron deficit is on the proton acceptor molecule as in Fig. (6). The most important changes for Hb and Oa are in fair agreement with those found here with the Hirshfeld partitioning but the small modifications of the charges of the other atoms are different. Recently, Chelli et al

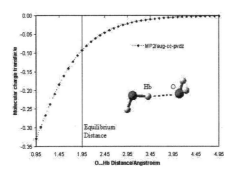

Figure 6: Variation of the stockholder net charge of the proton donor molecule in the water dimer with the hydrogen bond length Hb...Oa. Values are computed with MP2/aug-cc-pVDZ.

using NPA reported a behavior similar to the Fig. (6) for the variation of the charge transferred with the water-water molecules distance.[2] On the other hand, Galvez and collaborators have used recently the Atom In Molecule (AIM)[14] partitioning of the electronic density to analyse the charge redistributions in several hydrogen bonded systems, including the water dimer.[6] Their results for the charge transferred and the charges of Oa and Od are also in agreement with the trends observed in Figs. (5) and (6). Mulliken charge analysis, although less accurate than AIM and NPA, gave also a net negative charge for the molecule donating a proton in the water dimer.[34] On the contrary, the so-called Atomic Polar Tensors (APT),[35] which are *different* quantities because they are related to the vibrational properties of the molecule, shown an opposite behavior: the APT charge is negative for the molecule accepting the proton.[34] Finally, one should note that empirical models based on fluctuating charges gave the wrong charge transfer.[33],[36] Being based on electronegativities differences, they predict a transfer of electrons from the hydrogen Hb to the oxygen Oa on the contrary to what is found in density partitioning *ab initio* methods.

It is interesting to note that the reduction of *intramolecular* charges in the trimer correspond also to a reduction of the molecular polarisability. These charge redistributions as well as the *intermolecular* charge transfer is cooperative[8] and depends on the donor/acceptor character of the molecules. On contrast to ice where each molecule bonds to other four molecules and acts as both electron donor and acceptor, molecules in water clusters are unsaturated in their bonding, especially for the molecules at corners or at surface of the clusters which might serve only as either electron donor or acceptor. Therefore the molecules at corners are charged highest. This can be seen in Fig. (4B), where the molecules in W_{20} at end possess greater amount of net charge (absolute value) than the molecules in body which indicates a *close relation between the molecular polarisability and net charge* (compare Figs. (4A) and (4B)). It is interesting to note that these end regions are also the most reactive towards an electrophilic agent as it can be seen from the molecular HOMO charge map (Fig. 4C). A correlation between molecular charge distribution and molecular polarisability is found for other water clusters. Fig. (7) summarizes the variations of the molecular polarisabilities and of molecular charges for all the molecules in the stable geometries.[4] It is obvious that molecular polarisabilities fluctuate in accordance with molecular net charge for all molecules in a cluster independent of its size.

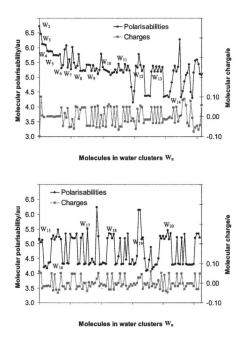

Figure 7: Correlation between molecular polarisability and molecular net charge in water clusters. Calculations are performed at B3LYP level with basis set 6-31++G(d,p).

5   Conclusion

In spite of an amazing amount of works devoted to the intermolecular interactions between water molecules in condensed phases, we have not yet reached a consensus on an accurate model and understanding of water-water bonds. For instance, the transferability of empirical polarizable models is now questioned in the literature using water chains as test cases.[37] There it is demonstrated that the (mean) polarisability of a monomer leads to overpolarization when used in a cluster. Study of the less symmetrical water clusters may open new roads to understand the electronic properties of water and of the hydrogen bonds cooperativity.[8] Using first-principle computations of molecular charge distributions and polarisabilities of stable water clusters, we arrive in the present paper to several conclusions. The screened molecular polarisability, the relevant quantity to use in a polarizable model, decreases with the size of the water cluster. Its value depends also on the local bonding of a molecule and of the cluster charge distribution. A molecule having more hydrogen bonds has also a smaller polarisability. On the other hand, we find a close relation between polarisability and net molecular charge. We can be confident in the qualitative (if not quantitative) intermolecular charge redistributions found here using the Hirshfeld partitioning of the electronic density. For the dimer, it is shown to be in qualitative agreement with other methods (NPA, AIM). A systematic study of other electronic responses and of the charges using both the Gaussian accurate basis sets and plane-waves basis sets is now undertaken for water clusters, ice and liquid water in order to build a transferable water-water potential model.[33]

Acknowledgment

M. Y. thanks the "Conseil Régional de Bourgogne" for financial support during his visit in Dijon. The authors acknowledge the use of the computer resources of the "Centre de Ressources Informatiques" of the "Université de Bourgogne".

References

[1] L. Pauling: *The Nature of the Chemical Bond.* Cornell University Press, 2nd ed., Ithaca, 1940.

[2] R. Chelli, M. Pagliai, P. Procacci, G. Cardini, and V. Schettino, *Journal of Chemical Physics* **122**, 74504 (2005).

[3] M. Yang, P. Senet ,and C. Van Alsenoy, *International Journal of Quantum Chemistry* **101**, 535 (2005).

[4] P. Senet, M. Yang, and C. Van Alsenoy *in : Electric polarisability.* Imperial College Press, London (to be published).

[5] A. van der Vaart, and J. K. M. Merz, *Journal of Chemical Physics* **116**, 7380 (2001).

[6] O. Gálvez, P. C. Gómez, and L. F. Pacios, *Journal of Chemical Physics* **115**, 11166 (2001); ibid *Journal of Chemical Physics* **118**, 4878 (2003).

[7] F. Weinhold, *Journal of Molecular Structure (Theochem)* **398-399**, 181 (1997).

[8] A. E. Reed, L. A. Curtiss, and F. Weinhold, *Chemical Review* **88**, 899 (1988).

[9] M. in het Panhuis, P. L. A. Popelier, and R. W. Munn, J. G. Ángyán, *Journal of Chemical Physics* **114**, 7951 (2001).

[10] U. Buck, F. Huisken, J. Schleusener, J. Schaefer, *Journal of Chemical Physics* **72**, 1512 (1980).

[11] J. D. Bernal and H. R. Fowler, *Journal of Chemical Physics* **1**, 515 (1933).

[12] F. Weinhold, *Journal of Chemical Physics* **109**, 373 (1998).

[13] A. J. Stone, *Chemical Physics Letters* **83**, 233 (1981).

[14] R. F. W. Bader, *Chemical Review* **91**, 893 (1991).

[15] F. L. Hirshfield, *Theoretica Chimica Acta (Berl)* **44**, 129 (1977).

[16] B. Rousseau, A. Peeters, C. Van Alsenoy, *Chemical Physics Letters* **324**, 189 (2000).

[17] G. Maroulis, *Journal of Chemical Physics* **113**, 1813 (2000).

[18] L. Jensen, M. Swart, P.T. van Duijnen, J. G. Snijders, *Journal of Chemical Physics* **117**, 3316 (2002).

[19] T. K. Ghanty, S. K. Ghosh, *Journal of Chemical Physics* **118**, 8547 (2003).

[20] See for instance the following recent review on water potentials, B. Guillot, *Journal of Molecular Liquids* **101**, 219 (2002).

[21] R. Kubo, T. Nagamya: *Solid State Physics*. Mc Graw-Hill, New York, 1969.

[22] P. Senet, L. Henrard, Ph. Lambin, and A. A. Lucas in *"Electronic Properties of Novel Materials"*. Eds. H. Kuzmani, J. Fink, M. Mehring, and S. Roth, World Scientific Publishing, Singapore, 393, 1994.

[23] S. Maheshwary, N. Patel, N. Sathyamurthy, A. D. Kulkarni, S. R. Gadre, *Journal of Physical Chemistry A* **105**, 10525 (2001).

[24] D. J. Wales, J. P. K. Doye, A. Dullweber, M.P. Hodges, F. Y. Naumkin, F. Calvo, J. Hernández-Rojas, T. F. Middleton. The Cambridge Cluster Database. URL: //http://www-wales.ch.cam.ac.uk/CCD.html.

[25] M. J. Frisch, G. W. Trucks, H. B. Schlegel, G. E. Scuseria, M. A. Robb, J. R. Cheeseman, V. G. Zakrzewski, J. A. Montgomery Jr., R. E. Stratmann, J. C. Burant, S. Dapprich, J. M. Millam, A. D. Daniels, K. N. Kudin, M. C. Strain, O. Farkas, J. Tomasi, V. Barone, M. Cossi, R. Cammi, B. Mennucci, C. Pomelli, C. Adamo, S. Clifford, J. Ochterski, G. A. Petersson, P. Y. Ayala, Q. Cui, K. Morokuma, D. K. Malick, A. D. Rabuck, K. Raghavachari, J. B. Foresman, J. Cioslowski, J. V. Ortiz, A. G. Baboul, B. B. Stefanov, G. Liu, A. Liashenko, P. Piskorz, I. Komaromi, R. Gomperts, R. L. Martin, D. J. Fox, T. Keith, M. A. Al-Laham, C. Y. Peng, A. Nanayakkara, M. Challacombe, P. M. W. Gill, , B. Johnson, W. Chen, M. W. Wong, J. L. Andres, C. Gonzalez, M. Head-Gordon, E. S. Replogle, J. A. Pople, 1998, Gaussian 98, Revision A.11, Gaussian, Inc., Pittsburgh PA.

[26] M. Yang, P. Senet and C. Van Alsenoy, to be published.

[27] H. A. Kurtz, J. J. P. Stewart , K. M. Dieter, *Journal of Computational Chemistry* **11**, 82 (1990).

[28] W. F. Murphy, *Journal of Chemical Physics* **67**, 5877 (1977).

[29] C. Van Alsenoy, A. Peeters, *J. Mol. Struct. (THEOCHEM)* **286**, 19 (1993).

[30] H. Hellman, *Einfuhrung in die Quantumchemie*, Deuticke, Leipzig, 1937, P. 285.

[31] R. Feynman, *Physical Review* **56**, 340 (1939).

[32] P. Hohenberg, and W. Kohn, *Physical Review* **136**, B864 (1964).

[33] P. Senet et al., unpublished.

[34] P.-O. Åstrand, K. Ruud, K. V. Mikkelsen, and T. Helgaker, *Journal of Physical Chemistry A* **102**, 7686 (1998).

[35] J. Cioslowski, *Physical Review Letters* **62**, 1469 (1989).

[36] Z.-Z. Yang, Y. Wu and D.-X. Zhao, *Journal of Chemical Physics* **120**, 2541 (2004).

[37] T. J. Giese and D. M. York, *Journal of Chemical Physics* **120**, 9903 (2004).

Brill Academic Publishers
P.O. Box 9000, 2300 PA Leiden,
The Netherlands

*Lecture Series on Computer
and Computational Sciences*
Volume 3, 2005, pp. 248-253

Structural and Electronic Properties of Metal Clusters

Michael Springborg,[1] Valeri G. Grigoryan[2] Yi Dong,[3] Denitsa Alamanova,[4]
Habib ur Rehman,[5] and Violina Tevekeliyska[6]

Physical and Theoretical Chemistry,
University of Saarland,
66123 Saarbrücken,
Germany

Received 10 July, 2005; accepted in revised form 15 July, 2005

Abstract: In order to study the properties of a whole series of metal clusters, for which the structure has been optimized in an optimized way, we have used various parameterized methods for the calculation of the total energy for a given structure in combination with different methods for determining the structure of the global total-energy minimum. Specifically, we have used the embedded-atom method as well as a parameterized density-functional method. The calculations give first of all the total energy and the structure at the global total-energy minimum for clusters with up to 150 atoms. In order to extract information from these numbers we have constructed a number of different descriptors from which stability, overall shape, radial distribution of the atoms, growth patterns, similarity with finite pieces of the crystal, etc. can be analysed. We present these for clusters of Na, Cu, Ni, and Ag atoms and in some cases we also compare the results from the different types of calculations.

Keywords: Metal clusters, structure, stability, density-functional calculations, embedded-atom calculations

PACS: 36.40.-c, 36.90.+f, 61.46.+w, 73.22.-f

1   Introduction

Clusters are intermediates between small, finite molecules and infinite, periodic crystals. The finite size or, equivalently, the relative large number of surface atoms compared with the total number of atoms makes it possible to tune their properties simply through the variation of the size. Experimental studies on these systems are non-trivial due to difficulties in uniquely identifying the size of a given clusters, in obtaining sufficiently large amounts of a specific (monodisperse) cluster size, and in identifying the structure of a given cluster size (compared to a distribution of clusters with the same size but different structures). On the other hand, theoretical studies are non-trivial due to the fact that the clusters are large but finite together with the essentially exponential growth of the number of meta-stable structures as a function of cluster size. Ultimately, this means that theoretical studies have to consider either single clusters with an essentially known structure when

---

[1] Corresponding author. e-mail: m.springborg@mx.uni-saarland.de
[2] e-mail: vg.grigoryan@mx.uni-saarland.de
[3] e-mail: y.dong@mx.uni-saarland.de
[4] e-mail: deni@springborg.pc.uni-sb.de
[5] e-mail: haur001@rz.uni-saarland.de
[6] e-mail: vili@springborg.pc.uni-sb.de

using accurate methods or, alternatively, to use less accurate methods on a whole class of clusters and cluster sizes.

The purpose of this contribution is to present some of our results on metal clusters where we have used the second strategy, i.e., we have studied a whole class of clusters for which we have, for each single cluster size, optimized the structure using some unbiased approach. In some cases we have used two different parameterized methods, making it possible to address the accuracy of the two. The calculations yield first of all the total energy as a function of cluster size as well as the coordinates of the nuclei at the structure of the lowest total energy. In order to extract useful information from these data we have constructed various descriptors that we also shall present and discuss.

2   Total-Energy Methods

As one method we used the embedded-atom method (EAM) in the parameterization of Voter and Chen [1, 2, 3]. According to this method, the total energy for a system of N atoms is written as a sum of atomic components, each being the sum of two terms. The first term is the energy that it costs to bring the atom of interest into the electron density provided by all other atoms, and the second term is a pair-potential term. Both terms are assumed depending only on the distances between the neighbouring atoms, and do therefore not include any directional dependence. Accordingly, the EAM emphasizes geometrical effects, whereas electronic effects are included only very indirectly.

Furthermore, we used the density-functional tight-binding method (DFTB) as developed by Seifert and coworkers [4, 5]. With this method, the binding energy is written as the difference in the orbital energies of the compound minus those of the isolated atoms augmented with pair potentials. In calculating the orbital energies we need the Hamilton matrix elements $\langle \chi_{m_1 n_1} | \hat{H} | \chi_{m_2 n_2} \rangle$ and the overlap matrix elements $\langle \chi_{m_1 n_1} | \chi_{m_2 n_2} \rangle$. Here, χ_{mn} is the nth atomic orbital of the mth atom. The Hamilton operator contains the kinetic-energy operator as well as the potential. The latter is approximated as a superposition of the potentials of the isolated atoms, $V(\vec{r}) = \sum_m V_m(|\vec{r} - \vec{R}_m|)$, and subsequently we assume that the matrix element $\langle \chi_{m_1 n_1} | V_m | \chi_{m_2 n_2} \rangle$ vanishes unless at least one of the atoms m_1 and m_2 equals m. Finally, the pair potentials U_{m_1, m_2} are obtained by requiring that the total-energy curves from parameter-free density-functional calculations on the diatomics are accurately reproduced.

3   Structure Optimization

In the EAM calculations we optimized the structure using our own *Aufbau/Abbau* method [6, 7, 8]. The method is based on simulating experimental conditions, where clusters grow by adding atom by atom to a core. By repeating this process **very** many times and in parallel also removing atoms from larger clusters, we can identify the structures of the lowest total energy.

Finally, in the DFTB calculations we used two different approaches. In one approach the structures of the EAM calculations were used as input for a local relaxation, i.e., only the nearest local-total-energy minimum was identified. In another set of calculations, we optimized the structures using the so-called genetic algorithms [9, 10, 11]. Here, from a set of structures we generate new ones through cutting and pasting the original ones. Out of the total set of old and new clusters those with the lowest total energies are kept, and this process is repeated until the lowest total energy is unchanged for a large number of generations.

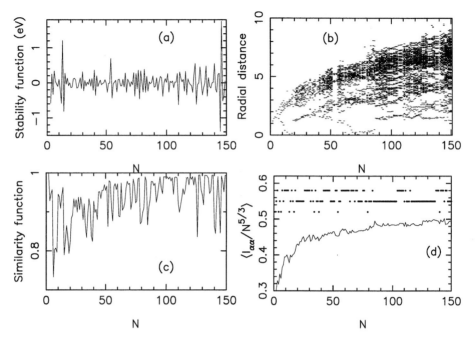

Figure 1: Properties of Au$_N$ clusters from the EAM calculations. The four panels show (a) the stability function, (b) the radial distribution of atoms, (c) the similarity function, and (d) the shape-analysis parameters, respectively. Lengths and energies are given in Å and eV, respectively. In (d) the upper rows show whether the clusters have an overall spherical shape (lowest row), an overall cigar-like shape (middle row), or an overall lens-like shape (upper row).

4   An Example: Au Clusters

Gold clusters constitute on of the most intensively studied classes of clusters [12]. Therefore, we shall in this short contribution present results for only this class of clusters.

Figs. 1, 2, and 3 show various properties from the EAM and DFTB calculations on Au$_N$ clusters. With the EAM method in combination with our *Aufbau/Abbau* method in optimizing the structure we studied clusters with N up to 150, giving the results of Fig. 1. Moreover, we used these optimized structures as input for the DFTB calculations for N up to 40, giving the results of Fig. 2. Finally, we also optimized the structures for N up to 40 with the DFTB method in combination with the genetic algorithms, giving the results of Fig. 3.

In order to identify the particularly stable clusters we have introduces the stability function,

$$\Delta_2 E(N) = E_{\text{tot}}(N+1) + E_{\text{tot}}(N-1) - 2E_{\text{tot}}(N) \tag{1}$$

where $E_{\text{tot}}(K)$ is the total energy of the Au$_K$ system. $\Delta_2 E(N)$ has local maxima when Au$_N$ is particularly stable and it is shown in Figs. 1(a), 2(a), and 3(a) and shows clearly different behaviours from the three approaches. In particular, inclusion of electronic effects leads to much larger oscillations in $\Delta_2 E(N)$.

Information on the structure can be obtained from the radial distances of the atoms, defined

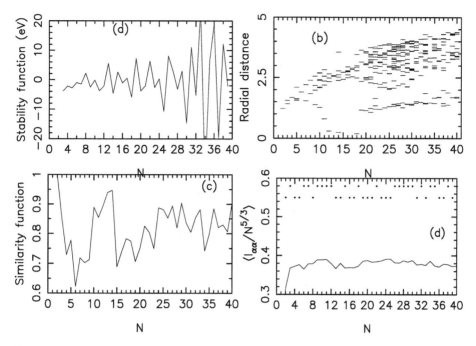

Figure 2: As Fig. 1, but from the DFTB calculations with the structures of the EAM calculations.

as follows. First, we define the center of the Au_N cluster, $\vec{R}_0 = \frac{1}{N} \sum_{i=1}^{N} \vec{R}_i$ and, subsequently, we define for each atom its radial distance $r_i = |\vec{R}_i - \vec{R}_0|$. In Figs. 1(b), 2(b), and 3(b) we show the radial distances for all atoms and all cluster sizes. Each small line shows that at least one atom for the given value of N has exactly that radial distance. The figure shows that somewhere around $N = 10$ a second shell of atoms is being built up, with a central atom for $N = 13$. Around $N = 54$, there are only few values of the radial distance, i.e., the clusters have a high symmetry. Around $N = 75$ we see that a third atomic shell is being formed.

We have earlier found [8] that it was useful to monitor the structural development of the isomer with the lowest total energy through the so-called similarity functions. We shall study how clusters grow and, in particular, if the cluster with N atoms can be derived from the one with $N - 1$ atoms simply by adding one extra atom. In order to quantify this relation we consider first the structure with the lowest total energy for the $(N - 1)$-atom cluster. For this we calculate and sort all interatomic distances, d_i, $i = 1, 2, \cdots, \frac{N(N-1)}{2}$. Subsequently we consider each of the N fragments of the N-cluster that can be obtained by removing one of the atoms and keeping the rest at their positions. For each of those we also calculate and sort all interatomic distances d_i', and calculate, subsequently, $q = \left[\frac{2}{N(N-1)} \sum_{i=1}^{N(N-1)/2} (d_i - d_i')^2 \right]^{1/2}$. Among the N different values of q we choose the smallest one, $q_{\min}$, and calculate the similarity function $S = \frac{1}{1+q_{\min}/u_l}$ ($u_l = 1$ Å) which approaches 1 if the Au_N cluster is very similar to the Au_{N-1} cluster plus an extra atom. This function is shown in Figs. 1(c), 2(c), and 3(c). We see that the structural development is very irregular over the whole range of N that we have considered here, with, however, some smaller

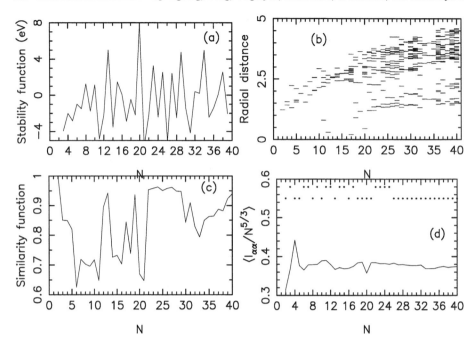

Figure 3: As Fig. 1, but from the DFTB calculations with the genetic-algorithm optimization of the structures.

intervals where S is relatively large, for instance for N slightly above 20.

Finally, we shall consider the overall shape of the clusters. As we showed in our earlier report on Ni clusters [8], it is convenient to study the 3×3 matrix containing the elements $I_{st} = \frac{1}{u_i^2} \sum_{n=1}^{N} (R_{n,s} - R_{0,s})(R_{n,t} - R_{0,t})$ with s and t being x, y, and z. The three eigenvalues of this matrix, $I_{\alpha\alpha}$, can be used in separating the clusters into being overall spherical (all eigenvalues are identical), more cigar-like shaped (one eigenvalue is large, the other two are small), or more lens-shaped (two large and one small eigenvalue). The average of the three eigenvalues, $\langle I_{\alpha\alpha} \rangle$, is a measure of the overall extension of the cluster. For a homogeneous sphere with N atoms, the eigenvalues scale like $N^{5/3}$. Hence, we show in Figs. 1(d), 2(d), and 3(d) quantities related to $I_{\alpha\alpha}$ but scaled by $N^{-5/3}$. In the figures we also mark the overall shape of the clusters through the upper points with the lowest row meaning spherical, the middle row meaning cigar-shaped, and the upper row meaning lens-shaped clusters. Some clusters with an overall spherical shape can be recognized, that, simultaneously, are clusters of particularly high stability according to Figs. 1(a), 2(a)3(a).

5 Conclusions

This short presentation shows clearly that it is possible to address the properties of a whole class of clusters using theoretical methods without making severe approximations on the structure of the clusters but, on the other hand, only when using approximate theoretical methods. Therefore, comparing the results of different theoretical approaches, as done here, is important when estimating the accuracy of the results.

All methods yield similar results with, however, a clear difference depending on whether electronic effects are included directly or indirectly. The former case leads to much more pronounced peaks in the stability function than the latter case.

As we shall in more detail directly at the conference, interesting general trends for different metals but also material-specific differences can be identified from our studies on several different types of metal clusters.

Acknowledgment

We gratefully acknowledge *Fonds der Chemischen Industrie* for very generous support. This work was supported by the SFB 277 of the University of Saarland and by the German Research Council (DFG) through project Sp439/14-1.

References

[1] A. F. Voter and S. P. Chen, in *Characterization of Defects in Materials*, edited by R. W. Siegal, J. R. Weertman, and R. Sinclair, MRS Symposia Proceedings No. 82 (Materials Research Society, Pittsburgh, 1987), p. 175.

[2] A. Voter, Los Alamos Unclassified Technical Report No LA-UR 93-3901 (1993).

[3] A. F. Voter, in *Intermetallic Compounds*, edited by J. H. Westbrook and R. L. Fleischer (John Wiley and Sons, Ltd, 1995), Vol. 1, p. 77.

[4] G. Seifert and R. Schmidt, New J. Chem. **16**, 1145 (1992).

[5] G. Seifert, D. Porezag, and Th. Frauenheim, Int. J. Quant. Chem **58**, 185 (1996).

[6] V. G. Grigoryan and M. Springborg, Phys. Chem. Chem. Phys. **3**, 5125 (2001).

[7] V. G. Grigoryan and M. Springborg, Chem. Phys. Lett. **375**, 219 (2003).

[8] V. G. Grigoryan and M. Springborg, Phys. Rev. B **70**, 205415 (2004).

[9] J.-O. Joswig, M. Springborg, and G. Seifert, Phys. Chem. Chem. Phys. **3**, 5130 (2001).

[10] J.-O. Joswig and M. Springborg, Phys. Rev. B **68**, 085408 (2003).

[11] Y. Dong and M. Springborg, in *Proceedings of 3rd International Conference "Computational Modeling and Simulation of Materials"*, Ed. P. Vincenzini *et al.*, Techna Group Publishers, p. 167 (2004).

[12] F. Baletto and R. Ferrando, Rev. Mod. Phys. **77**, 371 (2005).

Brill Academic Publishers
P.O. Box 9000, 2300 PA Leiden,
The Netherlands

Lecture Series on Computer
and Computational Sciences
Volume 3, 2005, pp. 254-263

Structures and energetics of hydrogen-bonded clusters

Ajit J. Thakkar[1]

Department of Chemistry,
University of New Brunswick,
Fredericton, New Brunswick E3B 6E2,
Canada

Received 1 August, 2005; accepted 12 August, 2005

Abstract: Our laboratory has been involved in the quantum chemical determination of the structures and energetics of hydrogen-bonded clusters of acid and water molecules using semiempirical, density functional, and ab initio Møller-Plesset perturbation theory and coupled cluster methods. Our work on clusters involving glycolic acid and water molecules is reviewed.

Keywords: Hydrogen-bonded clusters, energy minimization, Gaussian-type functions, Hartree-Fock method, density functional theory, Møller-Plesset perturbation theory, coupled cluster approach.

PACS: 36.40.-c, 31.25.Qm, 31.15.Ew

1 Introduction

The properties of a piece of bulk crystal do not change dramatically as we repeatedly subdivide it until the piece reaches molecular dimensions or, in other words, the nanometer scale. Particles of a material consisting of a few to a few thousand atoms are called clusters. The properties of clusters often show dramatic size and shape dependence. Clusters of metals, semiconductors, ionic solids, rare gases, and small molecules have been studied using both theoretical and experimental methods. Intense interest in clusters arises because they can be used to investigate surface properties including mechanisms of heterogeneous catalysis [1], and because clusters can serve as building blocks for new materials and electronic devices. An outstanding example of the fruits of cluster chemistry is fullerene chemistry [2,3] which grew out of the study of carbon clusters, and has become important in nanotechnology [4]. Recent books on clusters include a monograph on metal clusters [5], and edited collections on molecular clusters [6] and metal nanoparticles [7] .

Molecular clusters are held together by relatively weak intermolecular forces [8,9] or by hydrogen bonds [10]. Hydrogen-bonded clusters are an important class of molecular clusters. Small clusters of water molecules have received a lot of attention; see, for example, the experimental work of Saykally and coworkers [11], and the computational investigations of Xantheas and coworkers [12].

In my laboratory, we have studied several types of hydrogen-bonded clusters involving simple acids and water molecules. We examined the dimers of glycolic acid [13] and nitric acid [14]. We have performed density functional theory (DFT) computations to study the structures of trimers [15], tetramers [16,17], pentamers [18], and hexamers [19] of formic acid. Semiempirical PM3 computations have been used for exploratory work on dodecamers of formic acid [20]. More

---

[1]E-mail: ajit@unb.ca, Web page: http://www.unb.ca/chem/ajit/

recently, 2nd-order Møller-Plesset (MP2) and CCSD(T) calculations have been carried out [17] to probe more deeply the nature of formic acid tetramers.

The computational challenges involved in these studies include the choice of computationally efficient but sufficiently accurate quantum chemical methods and basis sets. Since the number of structural isomers grows exponentially with cluster size, a major challenge is to find all low-lying structures including the global minimum. In this talk I will review the work done in my laboratory on gw_n clusters consisting of one glycolic acid (g) molecule and n water (w) molecules with $n = 1, 2$ [21], $n = 3, 4, 5, 6$ [22], and $n = 16, 28$ [23].

2 Glycolic acid

Glycolic acid ($CH_2OHCOOH$), see Fig. 1, is the simplest α-hydroxy carboxylic acid, and plays an important role in dermatology and the cosmetics industry [24, 25]. It has some biological significance because it is involved in several life processes [26] and because it has a structure very similar to that of glycine, the simplest α-amino acid. Glycolic acid has rich functionality that allows it to simultaneously form intra- and intermolecular hydrogen bonds, and also allows for weaker C-H$\cdots$O interactions. The labels for the atoms in glycolic acid shown in Fig. 1 are used throughout.

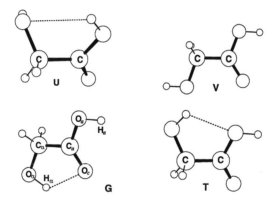

Figure 1: The four most stable rotamers of glycolic acid. Adapted from Thakkar et al. [21].

Eight rotamers of glycolic acid have been identified using experimental [27–31] and theoretical [28, 30–35] techniques. Four of the five lowest-energy rotamers are shown in Fig. 1 as G, T, U, and V. The fifth low-energy rotamer, U' (not shown), is very close in energy to U and differs from U mainly by a smaller $C_aC_\alpha O_\alpha H_\alpha$ torsional angle. The G rotamer is the most stable, the T rotamer lies about 2.7 kcal/mol above G, the U rotamer and its near-degenerate partner U' lie about 3.2 kcal/mol above G, and the V rotamer is about 1.2 kcal/mol higher still [35]. The most stable rotamer, G in Fig. 1, involves an $O_\alpha H_\alpha \cdots O_c$ intramolecular H-bond, as do T, U, U', and one other rotamer of higher energy.

The structure of crystalline glycolic acid, established by neutron diffraction and X-ray analysis [36, 37] consists of loose three-dimensional H-bonded networks in which the intramolecular H-bond is broken in favor of stronger intermolecular H-bonds. On the other hand, the most stable dimer of glycolic acid has the intramolecular H-bond intact as shown by ab initio calculations [13]. This cyclic C_{2h} dimer consists of two G rotamers held together by two short, linear hydrogen bonds between the carboxylic groups. The next two dimer structures of higher-energy are of the same type except that one or both of the monomers are in the T rotameric form.

Raman and infrared spectroscopy studies [38] suggest that glycolic acid exists in a monomeric form in dilute aqueous solution. A first step towards understanding such a solution is to consider small gas-phase clusters consisting of a glycolic acid (g) molecule and a few water (w) molecules. Our goal has been to locate as many local minima as possible to give ourselves a fighting chance of finding the global minimum and all the local minima within 3 kcal/mol of the global minimum.

3 Tiny glycolic-acid-water gw_n clusters with $n = 1, 2$

Thakkar et al. [21] used MP2/6-311+G(2d,2p) calculations to locate 12 and 17 local minima on the potential energy surfaces of the gw ($CH_2OHCOOH–H_2O$) and gw_2 ($CH_2OHCOOH–H_2O–H_2O$) complexes respectively. Single-point MP2/cc-pVQZ energies were computed for these structures. Higher-order correlation corrections were estimated using CCSD(T)/cc-pVDZ single-point energies. Zero-point corrections were also computed. All but two of the reported structures had the glycolic acid in G, T, or V conformations. The three lowest-energy isomers of gw and gw_2 found in that work [21] are shown in Fig. 2.

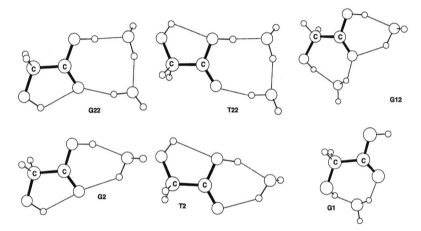

Figure 2: The three most stable structures of the complexes of glycolic acid with one and two water molecules. Adapted from Thakkar et al. [21].

In the lowest-energy gw structure G2, the water forms two H-bonds with the carboxylic group of the glycolic acid. The lowest-energy structure G22 of gw_2 consists of a trans water dimer attached to the carboxylic group by two H-bonds. The structures of second-lowest energy, T2 (2.3 kcal/mol) and T22 (2.1 kcal/mol), are analogous with the glycolic acid being in the T form. The intramolecular H-bond remains intact in the glycolic acid moiety of the two lowest-energy structures of both complexes. The intramolecular H-bond is broken in the third-lowest-energy structures G1 (2.9 kcal/mol) and G12 (4.1 kcal/mol). The gw structure G1 has the intramolecular H-bond replaced by two H-bonds to the water molecule. The gw_2 structure G12 can be thought of as a water molecule bound by two H-bonds to the carboxylic group of the G1 complex.

Binding energy calculations [21] revealed that g$\cdots$w H-bonds are stronger than w$\cdots$w H-bonds. The counterpoise-corrected binding energy of G2 relative to separated, undistorted monomers is

10.17 kcal/mol. At the same level of theory, the binding energy of the water dimer is 4.70 kcal/mol. Two H-bonds are formed in G2 and the binding energy per H-bond is 5.08 kcal/mol which is 0.39 kcal/mol more than the binding energy of the water dimer. The synergistic effect of forming a cyclic hydrogen-bonded network in G2 probably accounts for the added stability. The binding energy of G22 relative to three separated, undistorted monomers is 20.99 kcal/mol [21]. It is also interesting to consider G22 as a complex formed between glycolic acid and a water dimer. The binding energy of G22 relative to glycolic acid and $(H_2O)_2$ is 16.38 kcal/mol [21]. This is 6 kcal/mol more than the binding energy of G2 perhaps because the larger hydrogen-bonded network is sterically more relaxed in the G22 cluster allowing nearly linear hydrogen bonds to be formed.

Another useful result [21] was that the B3LYP/6-311G** method was as accurate as MP2/6-31G** for the clusters with two water molecules, establishing it as the method of choice for calculations on larger clusters.

4 Small glycolic-acid-water gw_n clusters with $3 \le n \le 6$

A four-stage search was used to explore the potential energy surfaces of the gw_n complexes with $3 \le n \le 6$ [22]. In each stage, trial structures generated by the previous stage were subject to symmetry-unconstrained, geometry optimization at a level of theory higher than that used in the previous stage. Then an energy threshold was used to select a subset of the optimized structures to serve as trial structures for the next stage. The methods used at the first to fourth stages were PM3, HF/4-31G, B3LYP/6-31G** and B3LYP/6-311G** respectively. In this way 5, 14, 29 and 30 local minima were identified at the fourth and final stage for the gw_n complexes with $n = 3, 4, 5, 6$ respectively. Single-point B3LYP energies using large polarization-consistent basis sets [39–42], up to the quadruple-zeta pc3 set, were computed at all the local minima found. Zero-point vibrational corrections and solvation corrections were also computed.

Consider the complexes with three water molecules first. The three lowest B3LYP/pc3//6-311G** structures of the gw_3 complexes are shown in Fig. 3. The lowest energy clusters G301 and T302 (2.0 kcal/mol) have a water trimer H-bonded to the carboxylic group of the acid to form a cyclic network of H-bonds. The w_3 moiety in G301 and T302 is an open form whereas the lowest-energy water trimer is cyclic [12, 43, 44]. The intramolecular H-bond in glycolic acid is intact in all three complexes. Each water molecule in these three structures serves both as a proton donor (d) and acceptor (a), or in a da role for short.

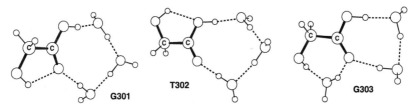

Figure 3: The three lowest-energy structures of gw_3 clusters. Adapted from Roy et al. [22].

There are eight gw_4 complexes with B3LYP/pc3 energies less than 3 kcal/mol. The three lowest gw_4 structures are shown in Fig. 4. The lowest-energy structure of the isolated water tetramer is cyclic with all four waters in a da role [12, 45]. The four lowest-energy gw_4 structures G401, G402 (0.5 kcal/mol), G403 (0.9 kcal/mol), and G404 (1.5 kcal/mol) all consist of a cyclic water tetramer bound by two H-bonds to the carboxylic group of the glycolic acid moiety in the G conformation. The four water tetramers differ only in the relative arrangement of the H atoms not taking part in the H-bonding. In each of these structures, the two water molecules not linked to the acid serve in a *da* role, one of the waters linked to the acid serves a *ada* role, and the remaining water H-bonded to the acid serves a *dad* role. The intramolecular H-bond in glycolic acid is intact in all eight structures.

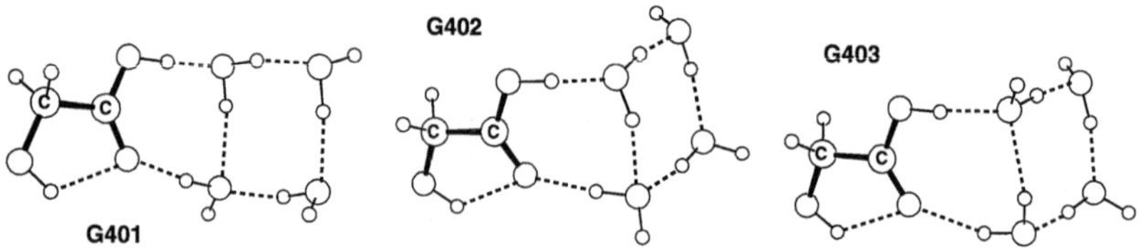

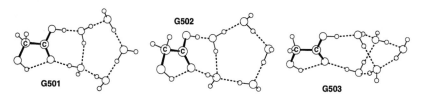

Figure 4: The three lowest-energy structures of gw_4 clusters. Adapted from Roy et al. [22].

There are 17 gw_5 structures with B3LYP/pc3 energies less than 3 kcal/mol. The three lowest gw_5 structures are shown in Fig. 5. The lowest-energy structure of the isolated water pentamer is cyclic with all five waters in a *da* role [12, 45]. Seven of the 17 gw_5 clusters, including five of the eight most stable, consist of a cyclic water pentamer attached to the carboxylic group of the glycolic acid moiety by two H-bonds. Five complexes including G503 have a water pentamer with a bridged cyclic structure bound to the carboxylic group. The intramolecular H-bond in glycolic acid is intact in all but two clusters.

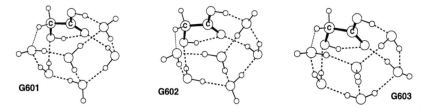

Figure 5: The three lowest-energy structures of gw_5 clusters. Adapted from Roy et al. [22].

18 gw_6 structures with B3LYP/pc3 energies less than 3 kcal/mol were found by Roy et al. [22]. Their three lowest gw_6 structures are shown in Fig. 6. The lowest-energy structures of the isolated water hexamer are the prism, cage, book and S_6 ring which all lie within 1 kcal/mol of each other [12]. At the B3LYP/pc3//6-311G** level, the lowest-energy water hexamer is a ring and not a prism.

Figure 6: The three lowest-energy structures of gw_6 clusters. Adapted from Roy et al. [22].

A striking difference between gw_6 and smaller gw_n clusters is that 14 of the 18 lowest-energy

gw_6 structures do not involve a water hexamer bound to the carboxylic group of the acid. In these 14 structures including all those seen in Fig. 6, both the carboxylic and α-hydroxy groups of the acid are attached to water molecules. The intramolecular H-bond in glycolic acid is intact in 10 of the 18 gw_6 clusters including the five most stable ones.

The interaction energy $\delta E_b(gw_n)$ of a gw_n cluster with respect to a glycolic acid molecule and the lowest-energy w_n cluster at their equilibrium geometries was found to be -9.60, -15.60, -14.43, -11.57, -11.75 and -11.66 kcal/mol for $n = 1$, 2, 3, 4, 5, and 6 respectively [22]. These values differ by less than 0.9 kcal/mol from the more accurate values, 10.17 and 16.38 kcal/mol for $n = 1$ and $n = 2$ respectively, obtained previously [21]. The noticeably higher values of $\delta E_b(gw_n)$ for $n = 2$ and $n = 3$ are probably due to the synergistic effect of forming a cyclic H-bond network at the acid. For larger values of n, $\delta E_b(gw_n)$ seems to remain between -11.8 and -11.5 kcal/mol.

The effects of aqueous solvation, as approximated by the Kirkwood-Onsager reaction field [46–49] and COSMO [50] models, do change the lowest-energy structure of gw_6 but not the smaller complexes. Unfortunately, the two solvation models disagree in their detailed predictions for gw_6 [22].

5 Larger glycolic-acid-water gw_n clusters with $n = 16, 28$

The four-stage strategy used for the smaller gw_n clusters with $n \leq 6$ could not be used for the larger clusters with $n = 16$ and $n = 28$ due to limitations of the computer resources available to us [23]. The first three stages were used for $n = 16$ and only the first two stages for $n = 28$.

The lowest energy B3LYP/pc2//B3LYP/6-31G** structure of gw_{16} is shown in Fig. 7. Note that the water molecules appear in the form of fused pentagonal biprisms just as they do in w_{20} clusters [51]. Addition of solvation corrections using the Onsager reaction field does not change the lowest-energy structure.

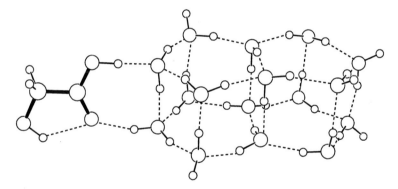

Figure 7: The lowest-energy gw_{16} cluster. Adapted from Roy et al. [23].

The results for gw_{28} are only qualitative because the geometry optimizations were not taken past the HF/4-31G level, and because the sampling of structures was much less thorough. As an example of our results, a quasi-spherical isomer of a cluster of glycolic acid and 28 water molecules is shown in Figure 8.

The cluster in Fig. 8 fits standard conceptions of a solvated glycolic acid molecule. However, it is higher in energy than the most stable cluster that we have found so far. The lowest-energy B3LYP/pc1a//HF/4-31G structure of gw_{28} is shown in Figure 9. Note that some of the waters

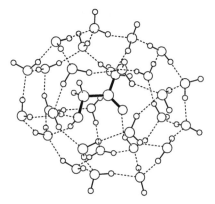

Figure 8: A quasi-spherical, structural isomer of a cluster of glycolic acid and 28 water molecules. Adapted from Roy et al. [23].

are again arranged in pentagonal biprismatic form.

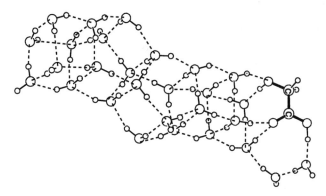

Figure 9: The lowest-energy gw_{28} structure. Adapted from Roy et al. [23].

We do not think that gw_{28} structures with the acid surrounded by a cage of water molecules can be the most stable ones even when entropic effects are taken into account. Instead, we expect a structure in which the acid is bound at one spot to a very stable w_{28} cluster. We guess that about 70 water molecules are required before a cage-like structure becomes the most stable one.

Acknowledgments

This lecture would not have been possible without the contributions of my coworkers: Amlan K. Roy, Shaowen Hu, Noureddin El-Bakali Kassimi, and James R. Hart. I thank the Natural Sciences and Engineering Research Council of Canada for their continuing support.

References

[1] G. Ertl and H. J. Freund. Catalysis and surface science. *Physics Today* **52**, 32–38 (1999).

[2] R. Taylor. *Lecture Notes on Fullerene Chemistry: A Handbook for Chemists* (Imperial College, London, 1999).

[3] A. Hirsch. *The Chemistry of the Fullerenes* (Wiley-VCH, New York, 2002).

[4] E. Osawa. *Perspectives of Fullerene Nanotechnology* (Kluwer Academic, New York, 2002).

[5] W. Ekardt. *Metal Clusters* (Wiley, New York, 1999).

[6] M. Driess and H. Nöth (Eds.). *Molecular Clusters of the Main Group Elements* (Wiley, New York, 2004).

[7] D. L. Feldheim and C. A. Foss, Jr. (Eds.). *Metal Nanoparticles* (Marcel Dekker, New York, 2001).

[8] A. J. Stone. *The Theory of Intermolecular Forces* (Oxford, New York, 1996).

[9] A. J. Thakkar. Intermolecular interactions. In *Encyclopedia of Chemical Physics and Physical Chemistry*, (eds.) J. Moore and N. Spencer (Institute of Physics Publishing, Bristol, 2001), vol. I. Fundamentals, chap. A1.5, pp. 161–186.

[10] S. Scheiner. *Hydrogen bonding: A theoretical perspective* (Oxford, New York, 1997).

[11] F. N. Keutsch and R. J. Saykally. Water clusters: Untangling the mysteries of the liquid, one molecule at a time. *Proc. Nat. Acad. Sci. U. S.A.* **98**, 10533–10540 (2001).

[12] S. S. Xantheas, C. J. Burnham, and R. J. Harrison. Development of transferable interaction models for water. II. Accurate energetics of the first few water clusters from first principles. *J. Chem. Phys.* **116**, 1493–1499 (2002).

[13] N. E.-B. Kassimi, E. F. Archibong, and A. J. Thakkar. Hydrogen bonding in the glycolic acid dimer. *J. Mol. Struct. (Theochem)* **591**, 189–197 (2002).

[14] J. R. Hart and A. J. Thakkar. Nitric acid dimers. *J. Mol. Struct. (Theochem)* **714**, 217–220 (2005).

[15] A. K. Roy and A. J. Thakkar. Structures of the formic acid trimer. *Chem. Phys. Lett.* **386**, 162–168 (2004).

[16] A. K. Roy and A. J. Thakkar. Formic acid tetramers: A structural study. *Chem. Phys. Lett.* **393**, 347–354 (2004).

[17] A. Karpfen and A. J. Thakkar. Tetramers of formic acid: A closer look. *To be published* (2005).

[18] A. K. Roy and A. J. Thakkar. Pentamers of formic acid. *Chem. Phys.* **312**, 119–126 (2005).

[19] A. K. Roy, S. P. McCarthy, and A. J. Thakkar. Hexamers of formic acid. *To be published* (2005).

[20] S. A. Blair and A. J. Thakkar. Dodecamers of formic acid. *To be published* (2005).

[21] A. J. Thakkar, N. E.-B. Kassimi, and S. Hu. Hydrogen-bonded complexes of glycolic acid with one and two water molecules. *Chem. Phys. Lett.* **387**, 142–148 (2004).

[22] A. K. Roy, S. Hu, and A. J. Thakkar. Clusters of glycolic acid with three to six water molecules. *J. Chem. Phys.* **122**, 074313 (2005).

[23] A. K. Roy, J. R. Hart, and A. J. Thakkar. Clusters of glycolic acid with 16 and 28 water molecules. *To be published* (2005).

[24] L. S. Moy, K. Howe, and R. L. Moy. Glycolic acid modulation of collagen production in human skin fibroblast cultures in vitro. *Dermatal. Surg.* **22**, 439–441 (1996).

[25] R. G. Males and F. G. Herring. A ^{1}H-NMR study of the permeation of glycolic acid through phospholipid membranes. *Biochim. Biophys. Acta-Biomembr.* **1416**, 333–338 (1999).

[26] I. Zelitch. Plant respiration. In *McGraw-Hill Encyclopedia of Science and Technology* (McGraw-Hill, New York, 1992), vol. 13, pp. 705–710. 7th ed.

[27] C. E. Blom and A. Bauder. Structure of glycolic acid determined by microwave spectroscopy. *J. Am. Chem. Soc.* **104**, 2993–2996 (1982).

[28] H. Hollenstein, T.-K. Ha, and H. H. Günthard. IR induced conversion of rotamers, matrix spectra, ab initio calculation of conformers, assignment and valence force field of *trans* glycolic acid. *J. Mol. Struct.* **146**, 289–307 (1986).

[29] K. Iijimaa, M. Katoa, and B. Beagley. Molecular structure and barrier to internal rotation of gaseous glycolic acid. *J. Mol. Struct.* **295**, 289–291 (1993).

[30] P. D. Godfrey, F. M. Rodgers, and R. D. Brown. Theory versus experiment in jet spectroscopy: glycolic acid. *J. Am. Chem. Soc.* **119**, 2232–2239 (1997).

[31] I. D. Reva, S. Jarmelo, L. Lapinski, and R. Fausto. First experimental evidence of the third conformer of glycolic acid: combined matrix isolation, FTIR and theoretical study. *Chem. Phys. Lett.* **389**, 68–74 (2004).

[32] M. D. Newton and G. A. Jeffrey. Stereochemistry of the α-hydroxycarboxylic acids and related systems. *J. Am. Chem. Soc.* **99**, 2413–2421 (1977).

[33] T.-K. Ha, C. E. Blom, and H. H. Günthard. A theoretical study of various rotamers of glycolic acid. *J. Mol. Struct. (Theochem)* **85**, 285–292 (1981).

[34] M. Flock and M. Ramek. Ab-initio SCF investigation of glycolic acid. *Int. J. Quantum Chem.* **S26**, 505–515 (1992).

[35] F. Jensen. Conformations of glycolic acid. *Acta Chem. Scand.* **51**, 439–441 (1997).

[36] R. D. Ellison, C. K. Johnson, and H. A. Levy. Glycolic acid: direct neutron diffraction determination of crystal structure and thermal motion analysis. *Acta Cryst.* **B27**, 333–344 (1971).

[37] W. P. Pijper. The molecular and crystal structure of glycolic acid. *Acta Cryst.* **B27**, 344–348 (1971).

[38] G. Cassanas, M. Morssli, E. Fabreque, and L. Bardet. Étude spectrale de l'acide glycolique, des glycolates et du processus de polymérisation. (Spectral study of glycolic acid, glycolates and of the polymerization process). *J. Raman Spectrosc.* **22**, 11–17 (1991).

[39] F. Jensen. Polarization consistent basis sets: Principles. *J. Chem. Phys.* **115**, 9113–9125 (2001). Erratum: J. Chem. Phys. 116, 3502 (2002).

[40] F. Jensen. Polarization consistent basis sets. II. Estimating the Kohn-Sham basis set limit. *J. Chem. Phys.* **116**, 7372–7379 (2002).

[41] F. Jensen. Polarization consistent basis sets. III. The importance of diffuse functions. *J. Chem. Phys.* **117**, 9234–9240 (2002).

[42] F. Jensen. Polarization consistent basis sets. IV. The basis set convergence of equilibrium geometries, harmonic vibrational frequencies, and intensities. *J. Chem. Phys.* **118**, 2459–2463 (2003).

[43] O. Mó, M. Yáñez, and J. Elguero. Cooperative (nonpairwise) effects in water trimers: An ab initio molecular orbital study. *J. Chem. Phys.* **97**, 6628–6638 (1992).

[44] S. S. Xantheas and T. H. Dunning, Jr. The structure of the water trimer from ab initio calculations. *J. Chem. Phys.* **98**, 8037–8040 (1993).

[45] S. S. Xantheas and T. H. Dunning, Jr. Ab initio studies of cyclic water clusters $(H_2O)_n$, $n = 1$–6. I. Optimal structures and vibrational spectra. *J. Chem. Phys.* **99**, 8774–8792 (1993).

[46] J. G. Kirkwood. Theory of solutions of molecules containing widely separated charges with special application to zwitterions. *J. Chem. Phys.* **2**, 351–361 (1934).

[47] L. Onsager. Electric moments of molecules in liquids. *J. Am. Chem. Soc.* **58**, 1486–1493 (1936).

[48] O. Tapia and O. Goscinski. Self-consistent reaction field-theory of solvent effects. *Mol. Phys.* **29**, 1653–1661 (1975).

[49] M. W. Wong, M. J. Frisch, and K. B. Wiberg. Solvent effects. 1. The mediation of electrostatic effects by solvents. *J. Am. Chem. Soc.* **113**, 4776–4782 (1991).

[50] V. Barone and M. Cossi. Quantum calculation of molecular energies and energy gradients in solution by a conductor solvent model. *J. Phys. Chem. A* **102**, 1995–2001 (1998).

[51] G. S. Fanourgakis, E. Aprà, and S. S. Xantheas. High-level ab initio calculations for the four low-lying families of minima of $(H_2O)_{20}$. I. Estimates of MP2/CBS binding energies and comparison with empirical potentials. *J. Chem. Phys.* **121**, 2655–2663 (2004).

Brill Academic Publishers
P.O. Box 9000, 2300 PA Leiden,
The Netherlands

*Lecture Series on Computer
and Computational Sciences*
Volume 3, 2005, pp. 264-281

Multi-scale approaches in Computational Materials Science

P. Weinberger[1]

Center for Computational Materials Science,
Technical University of Vienna,
Gumpendorferstr. 1a, A-1060 Vienna, Austria

Received 1 June, 2004; accepted 10 June, 2004

Abstract: Multi-scale approaches in computational materials science are classified according to (1) their use, namely either as iterative or "one-shot" procedures, and (2) with respect to the need of using different levels in physics such as e.g. by combing quantum mechanical with phenomenological or thermodynamically defined descriptions. Furthermore, it is pointed out that multi-scale procedures not only apply to length scales but also to times scales if integrations over characteristic times are required.

1 Introduction

The term "multi-scale" is presently very much *en vogue,* one almost gets the impression that this term very often is simply used to emphasize the "importance" of a particular scheme or to impress an audience with a "buzz word". In the applied mathematics literature- [1] it seems that essentially two types of multi-scale schemes are in discussion, namely "one shot" schemes in which one approach is combined in a consecutive manner with another one of different mathematical origin, and procedures by intertwining two such approaches "iteratively" (or to use a term more common in physics and chemistry, namely "selfconsistently").

Clearly enough the easiest way to define multi-scale procedures in particular in the realm of physics and chemistry would be to state that a combination of say two different kinds of differential equations is required. Although this in principle would be a valid definition it is too narrow, since, e.g., any use of density functional theory (DFT) requires already the application of two differential equations of different kind, namely the Kohn-Sham equations (effective Schrödinger or Dirac equation) and the Poisson equation, in an "iterative" manner. Surely enough nobody would call ab-initio type calculations in terms of the DFT a "multi-scale" procedure. This simple counter-example indicates that it is perhaps quite appropriate to discuss the concept of "multi-scale" only in the context of a particular field of research or discipline. In the present paper such a discussion is devoted to computational physics, in particular to computational materials science, since this is a well-established field of research in which many different types of computer simulations are performed.

2 Classification of multi-scale schemes

Suppose that instead of narrowing down multi-scale schemes to a combination of two different kinds of differential equations, a combination of different levels in physics as indicated in Fig. 1 is meant such as e.g. combining quantum mechanical approaches with phenomenological approaches.

[1]Corresponding author. E-mail: pw@cms.tuwien.ac.at

Clearly enough one could also say by combining microscopical with macroscopical schemes. A "one shot" multi-scale procedure would then consist of a quantum mechanical calculation (e.g. within the framework of the DFT) followed by a phenomenological one, in which the results of the former are used; an "iterative" procedure combines both in a kind of selfconsistent manner. In the latter case of course great care has to be taken that fundamental concepts are not violated (microcosmos versus macrocosmos), i.e., that only quantities can be varied that are well-defined on both "conceptual" levels.

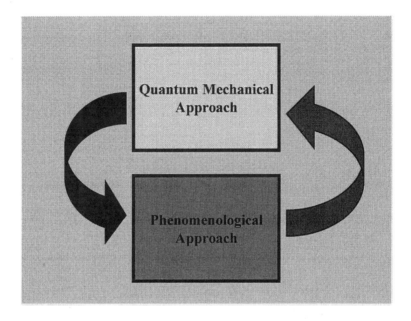

Figure 1: If the computational path follows only the left arrow, a "one shot" multi-scale approach is used. An "iterative" multi-scale procedure implies a "selfconsistent" procedure via the computational return path indicated by the right arrow.

In Fig.2 two typical situations are depicted, namely augmenting a time-independent quantum mechanical scheme with (1) the concept of time (e.g., as indicated there in terms of the phenomenological Landau-Lifshitz-Gilbert equation), and (2) with a method typical for statistical mechanics (e.g., Monte Carlo simulations based on ab-initio determined parameters), both schemes in fact can be operated in an iterative manner.

More frequently, however, are "one-shot" multi-scale procedures, in which mostly physical properties of materials are calculated in terms of the results of ab-initio approaches. In order to be classified as a multi-scale approach the evaluation of these properties has to be based on a scheme, which by definition is different from a typical DFT method such as, e.g., the Kubo equation for evaluating electric and (magneto-) optical transport, i.e., by requiring physical separate program

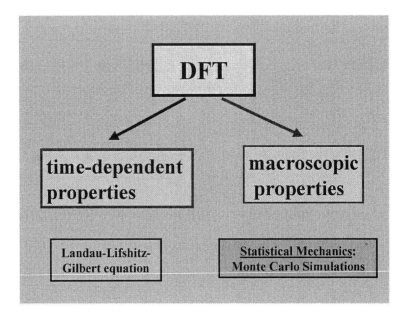

Figure 2: Typical combinations of ab-initio type quantum mechanical approaches with phenomeno-logical schemes either by describing time evolutions (left path) or in terms of ensemble averaging and making use of methods typical for statistical mechanics (right path).

packages that in the end provide macroscopic quantities. Properties derived directly from DFT results such as derivatives of the total energy with respect to atomic coordinates do not fall in this category. Very often also combinations of "one shot" multi-scale procedures are used as for example indicated in Figs. 3 and 4. The example in Fig. 3 refers to an evaluation of critical currents in current induced switching, that in Fig. 4 to an ab-initio determination of Kerr rotation and ellipticity angles.

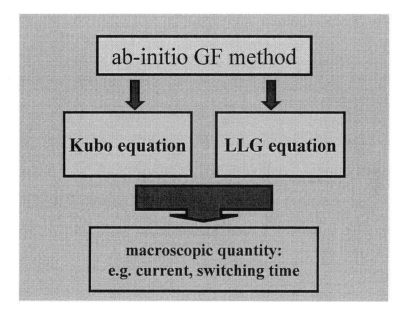

Figure 3: A "one shot" multi-scale procedure consisting of a quantum mechanically defined physical property and a time evolution scheme.

3 Examples

3.1 Iterative multi-scale procedures

3.1.1 Spin dynamics

The quantum mechanical step

In order to deal with exchange splitting and relativistic scattering on equal theoretical footing a first principles spin dynamics scheme can be based on a constrained Kohn–Sham–Dirac equation [2, 3],

$$\mathcal{H}^{con}(\mathbf{r}) = c\,\alpha\cdot\mathbf{p} + \beta mc^2 + V(\mathbf{r}) + \mu_B\beta\,\Sigma\cdot(\mathbf{B}^{xc}(\mathbf{r}) + \mathbf{B}^{con}(\mathbf{r})) \;, \tag{1}$$

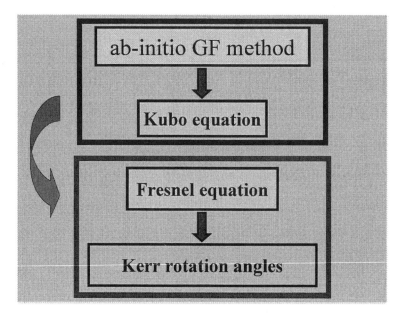

Figure 4: A combination of a quantum mechanical package with a classical physics (classical optics) package.

Table 1: Different types of Bloch walls considered, L'=L-2b.

N	L_1, L_2	type
0	$L_1 = L_2 = L'/2$	180^0 Bloch wall
0	$L_1 = L'$, $L_2 = 0$	90^0 Bloch wall
$\neq 0$	$L_1 = L_2 < L'/2$	two 90^0 Bloch walls

where α and β are the (usual) Dirac matrices, Σ is the spin operator, $V(\mathbf{r})$ stands for the Hartree and the exchange–correlation potential, while within the local spin density approximation (LSDA) $\mathbf{B}^{xc}(\mathbf{r})$ is an effective exchange field.

The phenomenological step

A numerically efficient tool to search for an equilibrium "spin" arrangement is then to trace the time evolution of the orientation of spin-only magnetic moments until a stationary state is achieved [2]. Suppose $\{\mathbf{e}_i(t)\}$ denotes the orientations of a system of well–defined local (magnetic) moments, $\{\mathbf{m}_i(t) = m_i \mathbf{e}_i(t)\}$, then the time evolution of these orientations can be described by the following "quasi-classical" equation of motion (Landau-Lifshitz-Gilbert equation),

$$\frac{d\mathbf{e}_i}{dt} = \gamma \mathbf{e}_i \times \mathbf{B}_i^{eff} + \lambda \left[\mathbf{e}_i \times (\mathbf{e}_i \times \mathbf{B}_i^{eff}) \right] , \qquad (2)$$

where $\mathbf{B}_i^{eff}$ is an effective magnetic field averaged over cell i, γ is the gyromagnetic ratio and λ is a damping (Gilbert) parameter. In this equation at any moment of time the orientational state has to be evaluated within a constrained density functional theory (DFT) in which a local constraining field, $\mathbf{B}_i^{con}$ ensures the stability of a non-equilibrium orientational state. This implies that the internal effective field that rotates the spins in the absence of a constraint is just the opposite of the constraining field.

The iterative use of Eqs. (1) and (2) forms the very basis of a relativistic *spin–only* dynamics. For finite magnetic (nano-) structures on top of semi-infinite substrates usually the so-called Embedded Cluster Method is used, which easily can be implemented within the Screened Korringa-Kohn-Rostoker method (for both schemes, see Ref. [3]). The example in Fig. 5 shows the artificial time evolution of the angles θ and the ϕ angles defining the orientation of the spin moments for seven Co atoms in a finite chain located at a step edge of Pt(111).

3.1.2 Segregation phenomena

The quantum mechanical step

The total energy of one configuration of a binary alloy can be expressed [4] in the form of an effective Hamiltonian of Ising type as

$$H = E_0 + \sum_{RQ} D_{\mathbf{R}}^Q \eta_{\mathbf{R}}^Q + \frac{1}{2} \sum_{RQ,R'Q'} V_{\mathbf{RR'}}^{QQ'} \eta_{\mathbf{R}}^Q \eta_{\mathbf{R'}}^{Q'} + \cdots \quad . \qquad (3)$$

The parameters of the Hamiltonian, see, e.g., Ref. [5], are the configurationally independent part of the alloy internal energy E_0, the on-site energies $D_{\mathbf{R}}^Q$, the interatomic pair interactions $V_{\mathbf{RR'}}^{QQ'}$, and generally, interatomic interactions of higher order. A particular configuration of the alloy is

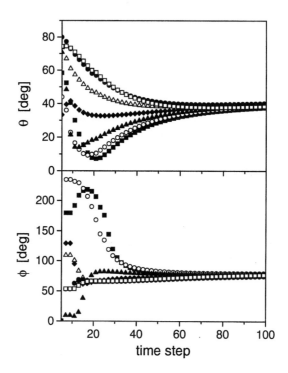

Figure 5: Artificial time evolution of the angles θ (top) and the ϕ (bottom) angles defining the orientation of the spin moments for seven Co atoms in a finite chain located at a step edge of Pt(111). The symbols refer to the following Co atoms numbered from the left to the right: full squares 1, open circles 2, full triangles 3, full diamonds 4, open triangles 5, full circles 6, open squares 7. Shown are the results for only the first 100 time steps. During the next 900 time steps the angles converge very smoothly. From Ref. [2].

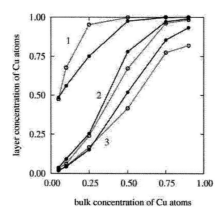

layer concentration of Cu atoms

bulk concentration of Cu atoms

Figure 6: Concentration x of Cu in the first three surface layers as a function of the bulk concentration for an $Cu_x Ni_{1-x}(001)$ surface as calculated from an iterative (selfconsistent) multi-scale procedure (full line and full symbols) and a "one-shot" (non-selfconsistent) multi-scale scheme (dotted line and open circles). From Ref. [4].

characterized by a set of occupation indices $\eta_{\mathbf{R}}^Q$, where $\eta_{\mathbf{R}}^Q = 1$ if site $\mathbf{R}$ is occupied by an atom of the type Q ($Q = A, B$), and $\eta_{\mathbf{R}}^Q = 0$ otherwise. It is assumed that $V_{\mathbf{RR}'}^{QQ'} = 0$ if $\mathbf{R} = \mathbf{R}'$.

The Hamiltonian in Eq. (3) can be modified in many ways, e.g., in terms of a lattice gas model using the relations $\eta_{\mathbf{R}}^A = \eta_{\mathbf{R}}$, $\eta_{\mathbf{R}}^B = 1 - \eta_{\mathbf{R}}$ such that

$$H = E_0' + \sum_{\mathbf{R}} D_{\mathbf{R}}\, \eta_{\mathbf{R}} + \frac{1}{2} \sum_{\mathbf{RR}'} V_{\mathbf{RR}'}\, \eta_{\mathbf{R}}\, \eta_{\mathbf{R}'} + \cdots \quad , \qquad (4)$$

where

$$D_{\mathbf{R}} = D_{\mathbf{R}}^A - D_{\mathbf{R}}^B + \sum_{\mathbf{R}'} (V_{\mathbf{RR}'}^{AB} - V_{\mathbf{RR}'}^{BB}) \quad , \qquad (5)$$

and

$$V_{\mathbf{RR}'} = V_{\mathbf{RR}'}^{AA} + V_{\mathbf{RR}'}^{BB} - V_{\mathbf{RR}'}^{AB} - V_{\mathbf{RR}'}^{BA} \quad . \qquad (6)$$

The quantities $D_{\mathbf{R}}$ and $V_{\mathbf{RR}'}$ can be evaluated on an ab-initio level by using, e.g., the Tight-Binding Linearized Muffin-Tin-Orbital method (TB-LMTO, see Refs. [4, 6]).

The statistical mechanics step

A starting configuration $\Gamma^{(1)}$ is constructed in a random way with respect to the bulk composition. In a repetitive manner, one first picks randomly an atom in a given configuration $\Gamma^{(n)}$ and select, again randomly, one of its nearest neighbors. Second, one considers the variation of the total energy of the system, ΔE, due to the interchange of these two atoms. Two cases are possible according to the sign of ΔE, namely, (i) $\Delta E < 0$: the interchange of two atoms is energetically favorable and represents a new configuration $\Gamma^{(n+1)}$, and (ii) $\Delta E \geq 0$: the interchange is not necessarily energetically favorable. However, the two atoms are only interchanged to form a new configuration $\Gamma^{(n+1)}$ if $\exp\left(-\Delta E/kT\right) > \tau$, where τ is a random number ($\tau \in [0, 1]$), for a review of Monte Carlo simulations, see, e.g., Ref. [7].

This procedure is repeated by choosing a new central atom and the system approaches the thermodynamical equilibrium if the number of interchanges is sufficiently large. In order to avoid correlations in the various steps of the Markov chain the output parameters can be defined as

$$\langle c_p \rangle = \frac{1}{m} \sum_{j=\mu_0}^{\mu_0+m-1} c_p(\nu_j) , \tag{7}$$

where c_p is the concentration of A-atoms in the p-th plane of a tow-dimensionally invariant system, and $\nu_j = jN$, with N being the number of atoms in the computational cell.

The value of μ_0 must be large enough in order to eliminate the influence of the initial configuration and m, the number of iterations, is chosen as large as possible so as to reduce the standard deviation defined by

$$(\Delta c_p)^2 = \frac{1}{m(m-1)} \sum_{j=\mu_0}^{\mu_0+m-1} (c_p(\nu_j) - \langle c_p \rangle)^2 . \tag{8}$$

As an example for this kind of procedure in Fig. 6 the case of surface segregation for $Cu_x Ni_{1-x}(100)$ is displayed [4]. In this particular example eight surface layers have been used each containing $N_p = 15 \times 17$ atoms (the total number of atoms in the computational cell is $N = 8N_p$). The values of m and μ_0 used in the cited calculations were $m = 200N$ and $\mu_0 = 80N$. It should be noted that this choice of parameters depends on the temperature at which the simulation is performed, whereby near the critical temperature, i.e., at temperatures close to a phase transition, these values have to be substantially increased.

3.2 One-shot multi-scale procedures

3.2.1 Magnetic domain walls

Phenomenological description

The phenomenological Landau-Ginzburg theory. see Ref. [8], predicts the following dependence of the Bloch wall energy $E(L)$ on the width of the Bloch Wall L,

$$E(L) = \alpha I_1 \frac{1}{L} + \beta I_2 L , \tag{9}$$

where α and β are related to the exchange coupling (spin-stiffness) and the (in-plane) magnetic anisotropy, respectively, while the constants I_1 and I_2 are integrals depending on the profile of the magnetization direction, $\phi_0(\xi)$ ($0 \leq \xi \leq 1$) defined as

$$\phi(z) = \phi_0(\frac{z}{L}) , \quad 0 \leq z \leq L , \tag{10}$$

$\phi(z)$ denoting the polar angle of the magnetization direction as a function of the position z across the Bloch wall. Obviously, the function in Eq. (9) has a minimum at the thickness

$$L_{BW} = \sqrt{\frac{\alpha\, I_1}{\beta\, I_2}} , \tag{11}$$

associated with the (equilibrium) thickness of the Bloch wall. Although the energy has to be minimized with respect to the profile $\phi_0(\xi)$, Eq. (11) suggests that any profile taken in actual calculations is suitable to predict L_{BW} once the quantity $\sqrt{I_1/I_2}$ is known. It is easy to show that in the case of 90° Bloch walls of cubic ferromagnets (e.g., bcc Fe) or 180° Bloch walls of ferromagnets with uniaxial magnetic anisotropy (e.g., hcp Co) the assumption of a simple linear profile

$$\phi_0(\xi) = \begin{cases} 0 & \xi < 0 \\ \xi \Delta\phi & 0 \le \xi \le 1 \\ \Delta\phi & \xi \ge 1 \end{cases} \quad , \quad \Delta\phi = \pi \text{ or } \pi/2 \quad , \tag{12}$$

results into an enhancement of L_{BW} by a factor of $\sqrt{2}$ with respect to the exact (soliton) solution.

Ab-initio domain wall formation energies

Let $\Delta E(\mathcal{C}_i(L))$ denote the energy difference of a particular magnetic configuration $\mathcal{C}_i(L)$ of L atomic layers (properly embedded inbetween two semi-infinite systems) with respect to a given reference configuration $\mathcal{C}_0(L)$,

$$\Delta E(\mathcal{C}_i(L)) = E(\mathcal{C}_i(L)) - E(\mathcal{C}_0(L)) \quad . \tag{13}$$

By definition [3] $\Delta E(\mathcal{C}_i(L)) > 0$ implies that $\mathcal{C}_0(L)$ is the preferred configuration, whereas for $\Delta E(\mathcal{C}_i(L)) < 0$, $\mathcal{C}_i(L)$ is preferred.

For a study of 90^0 and 180^0 Bloch walls [9] it is useful to define [10] the following magnetic configurations ($\hat{\mathbf{x}}$ and $\hat{\mathbf{y}}$ refer to the in-plane coordinate vectors, $\hat{\mathbf{z}}$ is normal to this plane),

$$\mathcal{C}_0(L) = \{\underbrace{\hat{\mathbf{x}}, \hat{\mathbf{x}}, \ldots, \hat{\mathbf{x}}}_{L}\} \quad , \tag{14}$$

and,

$$\begin{aligned} \mathcal{C}_i(L) &\equiv \mathcal{C}_i(L_1; N; L_2; b) \\ &= \{\underbrace{\hat{\mathbf{n}}_l, \ldots, \hat{\mathbf{n}}_l}_{b}, \hat{\mathbf{n}}_1, \hat{\mathbf{n}}_2, \ldots, \hat{\mathbf{n}}_{L_1}, \underbrace{\hat{\mathbf{y}}, \ldots, \hat{\mathbf{y}}}_{N}, \hat{\mathbf{n}}'_1, \hat{\mathbf{n}}'_2, \ldots, \hat{\mathbf{n}}'_{L_2}, \underbrace{\hat{\mathbf{n}}_r, \ldots, \hat{\mathbf{n}}_r}_{b}\} \quad , \end{aligned} \tag{15}$$

where $L = N + L_1 + L_2 + 2b$. For matters of simplicity in Eq.(14) it was assumed that the orientation of the magnetization in the two domains that form the semi-infinite systems is pointing along $\hat{\mathbf{x}}$. The orientation of the magnetization in the individual layers $\hat{\mathbf{n}}_k$ is given by

$$\hat{\mathbf{n}}_{k_1} = R(\Theta_{k_1})\hat{\mathbf{x}} \quad , \quad \hat{\mathbf{n}}'_{k_2} = R(\Theta'_{k_2})\hat{\mathbf{y}} \quad , \tag{16}$$

where $R(\Theta_{k_1})$ and $R(\Theta'_{k_2})$ are (clock-wise) rotations around the z-axis, and $k_1(k_2)$ is an integer between 1 and L_1 (L_2). In using a simple model of a "linear" Bloch wall, the angles Θ and Θ' are simply given by

$$\Theta_{k_1} = k_1 \frac{90^0}{L_1} \text{ and } \Theta'_{k_2} = k_2 \frac{90^0}{L_2} \quad . \tag{17}$$

In Eq. (15) the index i denotes different configurations, i.e., different choices of L_1, L_2 and N. This set of atomic layers contains b "buffer-layers" at each end of the wall with orientations $\hat{\mathbf{n}}_l = \hat{\mathbf{x}}$ on the left and $\hat{\mathbf{n}}_r = -\hat{\mathbf{x}}$ or $\hat{\mathbf{n}}_r = \hat{\mathbf{y}}$ on the right. Eq. (14) refers to a collinear magnetic configuration with the magnetization oriented uniformly in-plane along $\hat{\mathbf{x}}$, while Eq. (15) specifies a non-collinear magnetic structures in which, however, in each (atomic) plane the orientation of the magnetization is uniform (two-dimensional translational symmetry), but the orientation between different planes can vary in an arbitrary manner.

If in Eq. (13) $C_0(L)$ refers to the magnetic ground state configuration and $C_i(L)$ to a particular choice in Eq. (15) $\Delta E(C_i(L))$ can be regarded as the energy of formation for a domain wall; it is a kind of twisting energy that is needed to form non-collinear structures.

In Fig. 7 the (band) energy difference of the magnetic configuration defined in Eq. (17) with respect to the ferromagnetic configuration – to be associated with the energy of the Bloch wall – is shown [9]. As can be seen from this figure the calculated energy data can nicely be fitted to the function in Eq. (9) (solid lines) giving thus evidence from first principles that the phenomenological Landau-Ginzburg theory applies fairly well to these systems, i.e., suggest a useful (multi-scale) combination of these two schemes. The Bloch walls investigations shown in Fig. 7 comprise about 150 (hcp Co) as well as 800 (bcc Fe) atomic planes. In the case of technologically important ferromagnetic alloys such as FePt or CoPt, however, due to the enhanced magnetic anisotropy the width of the Bloch wall is expected to be of the order of a few tens of atomic planes. This in turn raises the question of the validity of the Landau-Ginzburg theory in such a regime of widths and therefore the validity of a multi-scale calculational scheme based on this equation. It should be noted that the quantum mechanical part can for example be calculated in terms of the fully relativistic spin-polarized Screened KKR method [3].

3.2.2 Current-induced switching

The quantum mechanics related part

In defining the twisting energy $\Delta E(\Theta; m)$ in a typical spin valve system, see Fig. 8, as follows [12]

$$\Delta E(\Theta; m) = E(\Theta; m) - \min[E(\Theta; m)] , \tag{18}$$

where for matters of simplicity m denotes the number of spacer layers, this quantity is positive definite for all Θ. It should be noted that the difference $E(\pi; m) - E(0; m)$ is nothing else but the well-known interlayer exchange coupling energy, which specifies whether the coupling of two magnetic slabs separated by a non-magnetic spacer couples parallel (ferromagnetic) or antiparallel (antiferromagnetic). The energy $\Delta E(\Theta; m)$ has to be regarded as a kind of excitation energy, which, e.g., by means of an external magnetic field, has to be supplied in order to move the system from the ground state, $\min[E(\Theta; m)]$, to a state referring to a different – in general – non-collinear magnetic configuration.

Furthermore, since in principle the spacer can be inhomogeneously disordered, in the following in general the notation $\Delta E(\Theta; \mathbf{x}, m)$ is used, where $\mathbf{x}$ is an m-dimensional vector whose elements specify the concentrations of two chosen constituents in each spacer layer (inhomogeneous binary alloying). In the case of a homogeneous alloy forming the spacer $\mathbf{x} = x\mathbf{I}$, $\mathbf{I} = (1, 1, \ldots, 1)$, for an ordered system $\mathbf{x} = \mathbf{I}$. If interdiffusion at the interfaces between the magnetic slabs and the spacer occurs then $\mathbf{x}$ has to include also a few layers of the magnetic slabs.

Making use of complex Fermi energies, $\mathcal{E}_F = \epsilon_F \pm i\delta$, the sheet resistance [11] for a given magnetic configuration characterized by a particular value of Θ is defined by

$$r(\Theta; \mathbf{x}, m) = \lim_{\delta \to 0} r(\Theta, \delta; \mathbf{x}, m) , \tag{19}$$

where

$$r(\Theta, \delta; \mathbf{x}, m) = \sum_{i,j=1}^{N} \rho_{ij}(\Theta, \delta; \mathbf{x}, m) , \tag{20}$$

$$\sum_{k=1}^{N} \rho_{ik}(\Theta, \delta; \mathbf{x}, m)\sigma_{kj}(\Theta, \delta; \mathbf{x}, m) = \delta_{ij} , \tag{21}$$

and N consists of m spacer layers and a sufficient number of layers of the lead material. The sheet resistance $r(\Theta; \mathbf{x}, m)$ is related to the resistance $R(\Theta; \mathbf{x}, m)$ via the relation $r(\Theta; \mathbf{x}, m) = A_0 R(\Theta; \mathbf{x}, m)$, in which A_0 denotes the unit area, and to $\rho_{CPP}(\Theta; \mathbf{x}, m)$, the resistivity in the current-perpendicular to the planes of atoms geometry (CPP), via $r(\Theta; \mathbf{x}, m) = L\rho_{CPP}(\Theta; \mathbf{x}, m)$, where L is the overall length of the structure for which the conductivity is calculated.

The functional form of the actually calculated sheet resistance with respect to the imaginary part of the Fermi energy can by the way be used to qualitatively interpret the underlying type of conductance, since

$$\frac{d\,[r(\Theta; \mathbf{x}, m; \delta)]}{d\delta} = \begin{cases} > 0; & \text{"metallic"} \\ < 0; & \text{"non-metallic"} \end{cases} . \tag{22}$$

The last equation means *inter alia* that in the case of a negative slope of the sheet resistance with respect to δ "tunneling" might occur. The parameter δ obviously acts like a (small) constant selfenergy: in the regime of metallic conductance an increase of the selfenergy implies an increased resistivity (sheet resistance); in the non-metallic regime an increase of δ reduces the resistance, the system becomes more metallic. In this part of the "one-shot" multi-scale approach the relevant quantities can be obtained using for example the Screened KKR method [3] and the Kubo equation [11].

The macroscopic part

In terms of the twisting energy $\Delta E(\Theta; \mathbf{x}, m)$, see Eq. (18), and the sheet resistance $r(\Theta; \mathbf{x}, m)$ a corresponding current $I(\Theta; \mathbf{x}, m)$ can be defined [12] as

$$I(\Theta; \mathbf{x}, m) = \sqrt{\frac{A_0 \Delta E(\Theta; \mathbf{x}, m)}{\tau(\Theta; \mathbf{x}, m) r(\Theta; \mathbf{x}, m)}} \tag{23}$$

$$= \sqrt{\frac{\langle A_0 \rangle}{\langle \tau(\Theta; \mathbf{x}, m) \rangle}} I_0(\Theta; \mathbf{x}, m) ,$$

where $\tau(\Theta; \mathbf{x}, m)$ is the time needed to accomplish such a rotation by Θ. In Eq. (23) $\langle A_0 \rangle$ and $\langle \tau(\Theta; \mathbf{x}, m) \rangle$ denote the magnitude of the corresponding quantities within the international system of units; $I_0(\Theta; \mathbf{x}, m)$ is the so-called reduced current that just depends on the twisting energy and the sheet resistance. It should be noted that Eq. (23) is based on a mapping of the energy flux corresponding to a rotation by Θ, namely the twisting energy per time needed for the twisting, onto a (macroscopic) magnetic Joule's heat

$$Q(\Theta; \mathbf{x}, m) = \frac{\Delta E(\Theta; \mathbf{x}, m)}{\tau(\Theta; \mathbf{x}, m)} , \tag{24}$$

such that

$$Q(\Theta; \mathbf{x}, m) = R(\Theta; \mathbf{x}, m) I^2(\Theta; \mathbf{x}, m) , \tag{25}$$

with

$$R(\Theta; \mathbf{x}, m) = \frac{r(\Theta; \mathbf{x}, m)}{A_0} , \tag{26}$$

The quantity $I(\Theta; \mathbf{x}, m)$ therefore refers to a macroscopic current.

At zero temperature (the temperature at which ab-initio calculations are performed) the twisting energy corresponds to the free energy such that for determining the times $\tau(\Theta; \mathbf{x}, m)$ the

Landau-Lifshitz-Gilbert equation can be applied by using the following k-th order power series in $\cos(\Theta)$,

$$\Delta E^{(k)}(\Theta; \mathbf{x}, m) = \sum_{s=0}^{k} a_s(\mathbf{x}, m) \left(\cos(\Theta)\right)^s \quad , \tag{27}$$

since the internal field that appears in the Landau-Lifshitz-Gilbert equation is nothing but the derivative of the twisting energy with respect to the orientation of the magnetization in the rotated magnetic slab. The only quantity in Eq. (23) that cannot be determined theoretically is the unit area A_0, since it is an experimental parameter, which of course depends very much on the design of the prepared samples.

In Fig. 9 the twisting energy, sheet resistance, reduced current and magneto-resistance defined as follows

$$MR(\Theta, m) = \frac{r(\Theta, m) - r(0, m)}{r(\Theta, m)} \quad , \tag{28}$$

are shown [13] for a permalloy ($Ni_{85}Fe_{15}$) related spin valve with 20 monolayer Cu serving as spacer. The switching times from the ground state (non-collinear!) to the collinear final states (parallel or antiparallel alignment of the orientations of the magnetization) are displayed together with the critical currents in Fig. 10. The critical reduced current is that reduced current $I_0(\Theta; \mathbf{x}, m)$ that at least has to be applied to perform switching into a well-defined final state.

It should be noted that in this particular example not only a mapping onto a macroscopical quantity is involved, but also that once again a phenomenological approach, namely the Landua-Lifshitz-Gilbert equation is employed in order to describe the time dependence and time averaging.

4 Discussion

In this contribution an attempt was made to classify typical multi-scale procedures used in computational materials science, in particular since frequently used phrases such as e.g. "bridging the time and length scale" do not mean a lot. It was pointed out that multi-scale procedures essentially serve two types of purposes, namely (1) describing mesoscopic or even macroscopic physical properties in terms of quantum mechanically well-defined parameters and (2) to perform averaging over characteristic volumes or times. Mesoscopic in this context means extending computational schemes to physical properties characteristic for length scales of several hundred to a few thousand nanometers, macroscopic refers to thermodynamically defined quantities, while time averaging refers to time scales of nano- pico- and femtoseconds. Very often, but wrongly, also schemes to evaluate ground state geometries in terms of molecular dynamics approaches are labelled as multi-scale schemes, since the two kinds of differential equations - the Schrödinger equation for the motion of the nuclei and that for the electrons, refer both to a quantum mechanical level. It was hopefully pointed out bluntly enough that in combining two computational schemes one of them has to be "borrowed" either from classical physics such as e.g. optics or statistical mechanics (thermodynamics) in order to generate a multi-scale computational scheme.

5 Acknowledgement

Financial support from the Austrian Ministry for Economics and Labour (Zl 98.366) is gratefully acknowledged.

References

[1] *Multiscale Modelling and Simulation*, Lecture Notes in Computational Science and Engineering, (Eds. S. Attinger and P. Koumoutsakos) Vol. 39, Springer-Verlag, 2004.

[2] B. Újfalussy, B. Lazarovits, L. Szunyogh, G. M. Stocks, and P. Weinberger, Phys. Rev. B **70**, 1(R) (2004).

[3] J. Zabloudil, R. Hammerling, L. Szunyogh, P. Weinberger, *Electron Scattering in Solid Matter*, Springer Verlag, Heidelberg, 2004.

[4] V. Drchal, J. Kudrnovský, A. Pasturel, I. Turek, and P. Weinberger, Phys. Rev. B **54**, 8202 (1996).

[5] F. Ducastelle, *Order and Phase Stability* (North-Holland, Amsterdam, 1991).

[6] I. Turek, V. Drchal, J. Kudrnovský, M. Šob and P. Weinberger, *Electronic Structure of Disordered Alloys, Surfaces and Interfaces,* Kluwer Academic Publishers, 1997

[7] K. Binder and D.W. Heermann, *Monte Carlo Simulation in Statistical Physics* (Springer, Berlin, 1994).

[8] A. M. Kosevich, in *Modern Problems in Condensed Matter Sciences*, Vol. 17, edited by V. M. Agranovich and A. A. Maradudin (North Holland, Amsterdam, 1986), p. 495.

[9] J. Switalla, B. L. Gyorffy, and L. Szunyogh, Phys. Rev. B **63**, 104423 (2001).

[10] S. Gallego, P. Weinberger, L. Szunyogh, P. M. Levy, and C. Sommers, Phys. Rev. B **68**, 054406/1-8 (2003)

[11] P. Weinberger, Physics Reports **377**, 281 - 387 (2003)

[12] P. Weinberger, A. Vernes, B. L. Gyorffy, and L. Szunyogh, Phys. Rev. B **70**, 094401/1-13 (2004)

[13] A. Vernes, P. Weinberger, and L. Szunyogh, Phys. Rev. B, MS# BAR1062, in press (2005)

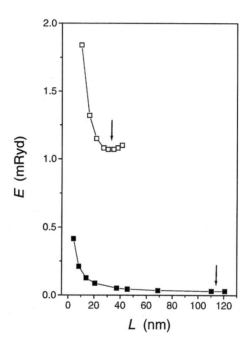

Figure 7: Energy of a 90° Bloch Wall in bcc Fe and of a 180° Bloch Wall in hcp Co as a function of the width L of the Bloch Walls. A linear magnetization profile for the Bloch Wall as described in the text is used. The equilibrium width L_{BW} of the Bloch Wall determined by the minimum of $E(L)$ (labelled by vertical arrows), see also Eq. (9), is about 113 nm for bcc Fe (experiment: ~ 40 nm) and about 35 nm for hcp Co (experiment: ~ 15 nm). From Ref. [9].

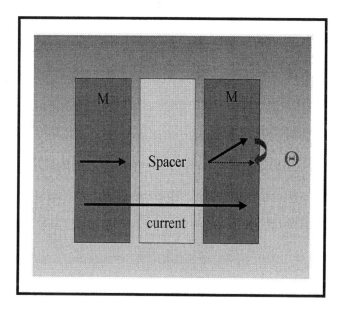

Figure 8: Typical spin valve system: M denotes the magnetic slabs with orientations of the magnetization as indicated by (short) arrows. The current is applied perpendicular to the planes of atoms (CPP).

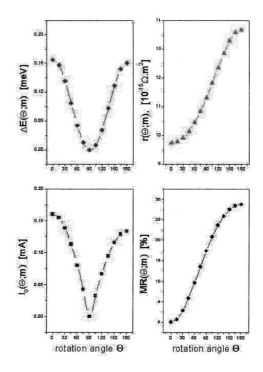

Figure 9: Twisting energy $\Delta E(\Theta; m)$, sheet resistance $r(\Theta; m)$ (top row), reduced current $I_0(\Theta; m)$ and magnetoresistance $MR(\Theta; m)$ (bottom row) for $m = 20$ ML of Cu in Py/Cu/Py. From Ref. [13].

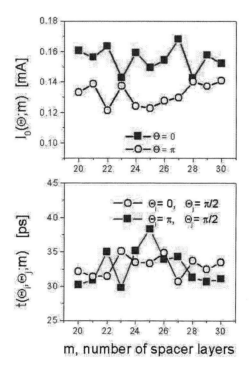

Figure 10: Reduced critical current $I_0(\Theta; m)$ (top) and minimal (switching) time $t(\Theta_i, \Theta_j; m)$ (bottom) versus the number of Cu spacer layers in Py/Cu/Py. From Ref. [13].

Brill Academic Publishers
P.O. Box 9000, 2300 PA Leiden,
The Netherlands

*Lecture Series on Computer
and Computational Sciences*
Volume 3, 2005, pp. 282-288

Adaptive grid technique for computer simulations of condensed matter using orbital-free embedding formalism.

M. Dułak and T.A. Wesołowski[1]

Department of Physical Chemistry,
30, quai Ernest-Ansermet, CH-1211 Genève 11, Switzerland

Received 5 July, 2005; accepted in revised form 10 Aug 2005.

Abstract: Adaptive grid technique, introduced originally for Kohn-Sham equations, is applied here to one-electron equations for embedded orbitals [Wesolowski and Warshel, J. Phys. Chem., **97** (1993) 8050]. For two embedded systems (solvated acetone and a water molecule in beryl crystal), it is shown that the adaptive grid technique provides an efficient strategy to reduce the size of the grid.

Keywords: orbital-free embedding, non-additive kinetic energy functional, one-electron equations, multi-scale simulations, electron density partitioning, numerical integration

PACS: 31.15.Ew, 31.15.Bs, 71.10.Ca, 71.70.-d

1 Introduction

In orbital-free embedding computer simulations, embedded orbitals ($\{\phi_i^A\}$) are derived from the following equations [1]:

$$\left[-\frac{1}{2}\nabla^2 + V_{eff}^{KSCED}\left[\rho_A, \rho_B; \vec{r}\right] \right] \phi_i^A = \epsilon_i^A \phi_i^A \quad i = 1, N^A,$$
(1)

where $\rho_A = 2\sum \left|\phi_i^A\right|^2$ is the electron density associated with the investigated system (embedded molecule) and ρ_B is a given function $\rho_B = \rho_B(\vec{r})$ representing the electron density of the environment.

The effective potential expressed by means of universal density functionals: $E_{xc}[\rho]$ (exchange-correlation energy) and $T_s^{nad}[\rho_A, \rho_B]$ (non-additive kinetic energy - $T_s^{nad}[\rho_A, \rho_B] \equiv T_s[\rho_A + \rho_B] - T_s[\rho_A] - T_s[\rho_B]$, where $T_s[\rho]$ is the kinetic energy in a reference system of non-interacting electrons) reads:

$$V_{eff}^{KSCED}\left[\rho_A, \rho_B; \vec{r}\right] = V_{eff}^{KS}\left[\rho_A; \vec{r}\right] + V_{eff}^{emb}\left[\rho_A, \rho_B; \vec{r}\right],$$
(2)

where

$$V_{eff}^{KS}\left[\rho_A; \vec{r}\right] = \sum_{i_A}^{N_{nuc}^A} -\frac{Z_{i_A}}{|\vec{r} - \vec{R}_{i_A}|} + \int \frac{\rho_A(\vec{r}')}{|\vec{r}' - \vec{r}|} d\vec{r}' + \left.\frac{\delta E_{xc}[\rho]}{\delta\rho}\right|_{\rho=\rho_A},$$
(3)

---

[1]Corresponding author. E-mail: tomasz.wesolowski@chiphy.unige.ch

and

$$V_{eff}^{emb}[\rho_A, \rho_B; \vec{r}] = \sum_{i_B}^{N_{nuc}^B} -\frac{Z_{i_B}}{|\vec{r} - \vec{R}_{i_B}|} + \int \frac{\rho_B(\vec{r}')}{|\vec{r}' - \vec{r}|} d\vec{r}' \qquad (4)$$

$$+ \left.\frac{\delta E_{xc}[\rho]}{\delta \rho}\right|_{\rho = \rho_A + \rho_B} - \left.\frac{\delta E_{xc}[\rho]}{\delta \rho}\right|_{\rho = \rho_A} + \frac{\delta T_s^{nad}[\rho_A, \rho_B]}{\delta \rho_A},$$

and where $V_{eff}^{KS}[\rho_A; \vec{r}]$ is the Kohn-Sham effective potential [2] for the *isolated* subsystem A. The above one-electron equations will be referred to as Kohn-Sham Equations with Constrained Electron Density (KSCED).

The key differences between the Kohn-Sham effective potential for the whole system and the effective potential for the embedded orbitals given in Eq. 1 are underlined below:

- $V_{eff}^{KSCED}[\rho_A, \rho_B; \vec{r}]$ is used in Eq. 1 to perform Euler-Lagrange minimization of the total-energy functional with respect to variation of the electron density in one subsystem (ρ_A) while keeping ρ_B frozen. Therefore, Eq. 1 can lead to the sum of ρ_A and ρ_B which equals to the *ground-state* electron density (ρ_0) only for such ρ_B that $\rho_0 - \rho_B$ is *pure-state non-interacting v-representable* [3]. In Kohn-Sham calculations, the minimization is free from such a constraint and can lead to ρ_0 if it is pure-state non-interacting v-representable [4].

- Compared to $V_{eff}^{KS}[\rho_A + \rho_B; \vec{r}]$, $V_{eff}^{KSCED}[\rho_A, \rho_B; \vec{r}]$ comprises an additional term arising form non-additivity of the kinetic energy in the reference system of non-interacting electrons ($T_s[\rho]$).

- $V_{eff}^{KSCED}[\rho_A, \rho_B; \vec{r}]$ is a functional of *two* electron densities ρ_A and ρ_B whereas $V_{eff}^{KS}[\rho_A + \rho_B; \vec{r}]$ depends on *one* electron density.

In the orbital-free embedding formalism, the total energy E of the system is an explicit functional of the orbitals $\{\phi_i^A\}$ in one subsystem and the electron density of the other one (ρ_B):

$$E[\{\phi_i^A\}, \rho_B] = 2\sum_{i=1}^{N^A} < \phi_i^A| -\frac{1}{2}\nabla^2|\phi_i^A > +T_s[\rho_B] + T_s^{nad}[\rho_A, \rho_B]$$

$$+ V[\rho] + J[\rho] + E_{xc}[\rho],$$

where $\rho = \rho_B + 2\sum_{i=1}^{N^A} |\phi_i^A|^2$, and $V[\rho] = -\int \sum_{i=1}^{N_{nuc}} \frac{Z_i \rho(\vec{r})}{|\vec{r} - \vec{R}_i|} d\vec{r}$.

Multi-level simulations based on Eqs. 1-5 involve three issues: *i*) adequacy of the approximations to the functionals $E_{xc}[\rho]$ and $T_s^{nad}[\rho_A, \rho_B]$, *ii*) CPU effort in generating the electron density of the environment (ρ_B), and *iii*) accuracy of the numerical integration - the main issue of this work.

The system-independent form of the effective potential in one-electron equations for embedded orbitals (Eq. 1) is the same as the one in the subsystem formulation of density functional theory introduced by Cortona [5]. In Cortona's formalism, all components of the total electron density (ρ_A and ρ_B in our case) are subject to variational calculations i.e. none of them is subject to additional simplifications/approximations applied multi-level type of simulations considered in this work.

2   Numerical Integration with Adaptive Grid

Adaptive grid techniques, originally proposed for Kohn-Sham calculations [7, 8] aim at reducing the number of grid points needed for numerical integration with a given accuracy, as the time spent on numerical integration is proportional to the number of grid points. This technique applicable for the case of the grid constructed as a superposition of atomic spherical grids will be outlined below.

If each orbital ϕ_i in the standard Kohn-Sham calculations is constructed as a linear combination of atomic orbitals ($\{\chi_j\}$), the following matrix elements are evaluated numerically [4]:

$$\left(V_{eff}^{KS}\right)_{ij} = \left(\chi_i(\vec{r}) \left| V_{eff}^{KS} [\rho_A; \vec{r}] \right| \chi_j(\vec{r})\right) \tag{5}$$

For atom-centered grids, each matrix element can be partitioned into atomic contributions (I_A):

$$(V_{eff}^{KS})_{ij} = \sum_A I_A. \tag{6}$$

Each atomic integral I_A can be further partitioned into contributions from radial shells.

In the *adaptive grid* technique by Krack and Köster [8], the value of the 'grid generating function' (f_t^{KS}) defined for each diagonal matrix element as:

$$f_t^{KS} = \left(\chi_i(\vec{r}) \left| v_{xc}[\rho_A] \right| \chi_i(\vec{r})\right). \tag{7}$$

is used as a criterion in selecting the angular grid for each radial shell in the atomic grid. Starting from the smallest angular grid, it grows until the difference between the values of f_t^{KS} for two consecutive grids falls below the user-determined threshold t.

3   Numerical Integration with Adaptive Grid for the Orbital-Free Embedding Calculations

In this work, we analyze the applicability of the *adaptive grid* technique, used for the purpose of optimizing the grid for the KSCED calculations.

If each embedded orbital ϕ_i^A is constructed as a linear combination of atomic orbitals ($\{\chi_j\}$), the following matrix elements are evaluated numerically in the self-consistent procedure to solve Eq. 1:

$$\left(V_{eff}^{KSCED}\right)_{ij} = \left(\chi_i(\vec{r}) \left| V_{eff}^{KSCED} [\rho_A, \rho_B; \vec{r}] \right| \chi_j(\vec{r})\right) \tag{8}$$

$$= \left(\chi_i(\vec{r}) \left| V_{eff}^{KS} [\rho_A + \rho_B; \vec{r}] \right| \chi_j(\vec{r})\right) + \left(\chi_i(\vec{r}) \left| \frac{\delta T_s^{nad}[\rho_A, \rho_B]}{\delta \rho_A} \right| \chi_j(\vec{r})\right).$$

For local (Local Density Approximation, LDA) and semi-local (Generalized Gradient Approximation, GGA) approximations to $E_{xc}[\rho]$ and $T_s^{nad}[\rho_A, \rho_B]$ (for details of these approximations see Ref. [6]), $V_{eff}^{KSCED}[\rho_A, \rho_B; \vec{r}](\vec{r})$ cannot be neglected anywhere where either $\rho_A(\vec{r})$ or $\rho_B(\vec{r})$ is non-zero. The *supermolecular grid* i.e. the one applied in conventional Kohn-Sham calculations for the whole system seems, therefore, indispensable. It is worthwhile to notice, however, that V_{eff}^{KSCED} is multiplied by the products of localized basis functions in Eq. 8. Therefore, if the two electron densities ρ_A and ρ_B overlap only in some regions in the real space, the grid can be reduced further.

Application of the adaptive grid technique in KSCED equations follows the similar lines as the one outlined for the Kohn-Sham calculations. Since the effective potentials in KSCED and Kohn-Sham equations differ, the 'grid generating function' applicable in KSCED (f_t^{KSCED}) reflects this fact. It reads:

$$f_t^{KSCED} = \left(\chi_i(\vec{r}) \left| v_{xc}[\rho_A + \rho_B] + \frac{\delta T_s^{nad}[\rho_A, \rho_B]}{\delta \rho_A} \right| \chi_i(\vec{r})\right). \tag{9}$$

4   Results

Three types of supermolecular grids of different quality are considered: COARSE (pruned (50,194) grid), MEDIUM (pruned (75,302) grid) , or FINE (pruned (99,590) grid). More details concerning such grids can be found in Ref. [9, 10]. For each considered system, a series of adaptive grids differing in the value of the threshold parameter t are tested.

The first considered system, the acetone molecule surrounded by 57 water molecules, was used in our study on the solvatochromic effect [11]. The other one, comprising a water molecule inside the $Al_6Be_6Si_{12}O_{24}H_{12}$ cage, provides a model of the water molecule (in two conformations denoted as I and II) in the beryl crystal [12]. The $Al_6Be_6Si_{12}O_{24}H_{12}$ model of the cage was cut from the bulk crystal and the cut bonds were saturated by hydrogens. The free space in the center of this molecule is large enough to contain one water molecule at various possible orientations. A different set of approximations was used in each case for $E_{xc}[\rho]$ and $T_s^{nad}[\rho_A, \rho_B]$: LDA for solvated acetone, and LDA and GGA for water trapped in beryl. In the embedded acetone case, the electron density of the environment was obtained from Kohn-Sham LDA calculations applying the STO-3G basis set and the A2 auxiliary basis set for fitting [10]. For the second system, the electron density of the $Al_6Be_6Si_{12}O_{24}H_{12}$ cage was obtained from Kohn-Sham PW91 calculations applying the DZVP basis set and the A2 auxiliary basis set for fitting [10]. The quantities shown in Tables 1- 3 are chosen as indicators of any inaccuracy in the numerical integration: i) $T_s[\rho]$ evaluated analytically using embedded orbitals derived from Eq. 1, ii) the energy of the highest occupied embedded orbitals derived from Eq. 1 (ϵ_{HOMO}), iii) $T_s^{nad}[\rho_A, \rho_B]$ evaluated numerically using a given ρ_B and ρ_A derived from Eq. 1, and iv) $\Delta E = E[\rho_A, \rho_B] - E[\rho_B]$ which involves both analytical and numerical components. Numerical values of $T_s[\rho]$ and ϵ_{HOMO} depend on the quality of the integration of the potential in Eq. 1 whereas that of $T_s^{nad}[\rho_A, \rho_B]$ and ΔE depend additionally on the quality of the integration of the energy components (Eq. 5).

Table 1: Dependence of the results on the grid type: a) supermolecular (COARSE, MEDIUM, FINE) and b) adaptive (see text for description of the parameter t). Calculations are made for the acetone molecule embedded in a cluster of 57 H_2O molecules. LDA approximations were applied for $E_{xc}[\rho]$ and for $T_s^{nad}[\rho_A, \rho_B]$. $T_s[\rho_A]$ is the analytically calculated kinetic energy of the embedded acetone molecule. $\Delta E = E[\rho_A, \rho_B] - E[\rho_B]$.

Grid type	Grid Size (total)	Grid Size (isol.)	$T_s[\rho_A]$ [Hartree]	ΔE [Hartree]	$T_s^{nad}[\rho_A, \rho_B]$ [Hartree]	ϵ_{HOMO} [eV]
COARSE	430657	24431	189.906565	-191.562056	0.063474	-6.14964
MEDIUM	1147000	65683	189.906473	-191.562111	0.063468	-6.14941
FINE	3254093	186183	189.906367	-191.562128	0.063470	-6.14961
t=1E-03	23759	5230	189.905711	-191.565089	0.063124	-6.15438
t=1E-04	56220	13318	189.906435	-191.562024	0.063447	-6.14913
t=1E-05	130248	31373	189.906205	-191.562139	0.063501	-6.15002
t=1E-06	275201	63340	189.906367	-191.562129	0.063467	-6.14960
t=1E-07	527083	113797	189.906356	-191.562129	0.063470	-6.14963
t=1E-08	912952	198063	189.906359	-191.562129	0.063470	-6.14962
t=1E-09	1464124	298721	189.906360	-191.562129	0.063470	-6.14962

Table 1 shows clearly that the adaptive grid technique allows one to select more efficiently the grid points. For instance, the stability at the eighth significant digit in $T_s[\rho]$ of the solvated acetone is reached with 275201 integration points generated using the adaptive grid technique whereas the same numerical stability requires about ten times more integration points if the supermolecular grid is used.

Table 1 shows also that using the adaptive grid corresponding to $t = 1E - 06$ leads to the numerical stability of ϵ_{HOMO} at fifth significant digit, ΔE at ninth significant digit, and $T_s^{nad}[\rho_A, \rho_B]$ at fourth significant digit. It is worthwhile to notice that the numerical values of ΔE stabilize at significantly smaller grids than those of $T_s[\rho_A]$.

Table 2: Dependence of the results on the grid type: a) supermolecular (COARSE, MEDIUM, FINE) and b) adaptive (see text for description of the parameter t). Calculations are made for the geometry I of the H_2O molecule embedded in the $Al_6Be_6Si_{12}O_{24}H_{12}$ cage. LDA approximations were applied for $E_{xc}[\rho]$ and for $T_s^{nad}[\rho_A, \rho_B]$. $T_s[\rho_A]$ is the analytically calculated kinetic energy of embedded H_2O molecule. $\Delta E = E[\rho_A, \rho_B] - E[\rho_B]$.

Grid type	Grid Size (total)	Grid Size (isol.)	$T_s[\rho_A]$ [Hartree]	ΔE [Hartree]
COARSE	148923	7699	75.477426	-75.900758
MEDIUM	399998	20672	75.477168	-75.900803
FINE	1131630	58618	75.477753	-75.900767
t=1E-03	9444	1603	75.473965	-75.898348
t=5E-04	10262	1929	75.474778	-75.898264
t=1E-04	20007	4213	75.476369	-75.900882
t=5E-05	23521	4652	75.477063	-75.900821
t=1E-05	47666	8608	75.477474	-75.900758
t=5E-06	61977	10138	75.477410	-75.900768
t=1E-06	120768	16377	75.477471	-75.900758
t=5E-07	143920	18149	75.477463	-75.900758
t=1E-07	245498	27840	75.477454	-75.900761

For the other embedded system considered in this work (water molecule trapped in a cage formed in the beryl crystal), a more detailed study of the effect of the grid on the calculated properties was made. Two geometries of the water molecule were considered: one referred to as *geometry I* in which the C_2 axis of the water molecule is oriented perpendicularly to the axis of the cage in the beryl crystal whereas and the other one for which the orientation is parallel (referred to as *geometry II*). As in the case of solvated acetone, the adaptive grid technique leads to the more efficient choice of the integration points for both considered geometries. (see Table 2). For geometry I, the adaptive grid corresponding to $t = 1E - 06$ is sufficient to stabilize the numerical values of ΔE at the seventh significant digit. Such an adaptive grid uses only 120768 integration points. For comparison, supermolecular grid FINE, which leads to the similar accuracy of ΔE, comprises as much as 1131630 integration points.

The above results indicate clearly that the adaptive grid is much better suited for KSCED calculations then the supermolecular grid. In both cases, the adequate grid (FINE and $t = 1E - 06$, in the supermolecular and the adaptive case, respectively) is significantly larger than the corresponding grid for the isolated molecule under investigation. In the supermolecular grid case, the number of additional points is proportional to the number of atoms in the environment. In the adaptive grid case, the number of additional points is significantly smaller because only the ones, where the two electron densities (ρ_A and ρ_B) do overlap, are included. This result could be expected due to the fact that calculations were performed using local density approximations for both $E_{xc}[\rho]$ and $T_s^{nad}[\rho_A, \rho_B]$.

Table 3: Dependence of the results on the grid type: a) supermolecular (COARSE, MEDIUM, FINE) and b) adaptive (see text for description of the parameter t). Calculations are made for the geometry I of the H_2O molecule embedded in the $Al_6Be_6Si_{12}O_{24}H_{12}$ cage. GGA approximations were applied for $E_{xc}[\rho]$ and for $T_s^{nad}[\rho_A, \rho_B]$. $T_s[\rho_A]$ is the analytically calculated kinetic energy of embedded H_2O molecule. $\Delta E = E[\rho_A, \rho_B] - E[\rho_B]$.

Grid type	Grid Size (total)	Grid Size (isol.)	$T_s[\rho_A]$ [Hartree]	ΔE [Hartree]
COARSE	148923	7699	75.622660	-76.435163
MEDIUM	399998	20672	75.621765	-76.435251
FINE	1131630	58618	75.622210	-76.435202
t=1E-03	9564	1688	75.621494	-76.432515
t=5E-04	10646	1929	75.620176	-76.432725
t=1E-04	21278	4238	75.621339	-76.435293
t=5E-05	25169	4626	75.622484	-76.435124
t=1E-05	53140	9205	75.622092	-76.435203
t=5E-06	67860	10329	75.622105	-76.435192
t=1E-06	135661	17470	75.622087	-76.435195
t=5E-07	156225	20109	75.622090	-76.435192
t=1E-07	264393	31165	75.622089	-76.435195

Table 3 collects the results obtained for the same geometry (I) but using generalized gradient approximation for the relevant functionals. The same trends as the ones discussed previously for the LDA case hold for gradient-dependent functionals.

Table 4: Dependence of the results on the grid type: a) supermolecular (COARSE, MEDIUM, FINE) and b) adaptive (see text for description of the parameter t). Calculations are made for two geometries (I and II) of the H_2O molecule embedded in the $Al_6Be_6Si_{12}O_{24}H_{12}$ cage. GGA approximations were applied for $E_{xc}[\rho]$ and for $T_s^{nad}[\rho_A, \rho_B]$.

Grid type	$\Delta E(II)$ [Hartree]	$\Delta E(I)$ [Hartree]	$\Delta E(II) - \Delta E(I)$ [kcal/mol]
COARSE	-76.420072	-76.435163	9.4696
MEDIUM	-76.420164	-76.435251	9.4673
FINE	-76.420152	-76.435202	9.4443
t=1E-03	-76.417537	-76.432515	9.3990
t=5E-04	-76.417425	-76.432725	9.6010
t=1E-04	-76.419943	-76.435293	9.6326
t=5E-05	-76.420009	-76.435124	9.4850
t=1E-05	-76.420168	-76.435203	9.4343
t=5E-06	-76.420167	-76.435192	9.4284
t=1E-06	-76.420149	-76.435195	9.4415
t=5E-07	-76.420149	-76.435192	9.4397
t=1E-07	-76.420152	-76.435195	9.4393

Finally, we analyze the performance of the adaptive grid for determination of *relative energies* of embedded molecules. Table 4 collects the interaction energies at two considered geometries of the water molecule in the beryl cage (I and II) obtained using different considered grids. The adaptive grid corresponding to $t = 1E - 06$, is sufficient to stabilize numerically the relative energy ($\Delta E(II)$ - $\Delta E(I)$) within the 0.01 kcal/mol range.

5 Conclusions

In the orbital-free embedding calculations, the embedded orbitals do not usually extend over the whole system under investigation but are localized. An optimized grid can be constructed using the *adaptive grid technique*. This technique, proposed originally for Kohn-Sham calculations, is based on a single-number simple criterion (threshold t on the grid generating function f_t^{KSCED}). We have shown, using two cases of embedded systems, that the size of the adequate *adaptive grid* is significantly smaller than *supermolecular grid* used in all our previous studies of embedded system by means of Eq. 1. The reduction of the grid is of key importance for multi-level simulations because it can lead to significant savings of the CPU time. Currently, we are working on a method for fast generation of the adaptive grid.

Acknowledgment

This work is supported by the Swiss National Science Foundation (Project 200020-107917/1) and Federal Office for Science and Education (COST). The calculations were made using our implementation [13] of the KSCED formalism into the code deMon2K [10].

References

[1] T. A. Wesolowski and A. Warshel, *J. Phys. Chem.*, **97** (1993) 8050.

[2] W. Kohn and L. J. Sham, *Phys. Rev.* **140** (1965) A1133.

[3] T. A. Wesolowski, *One-electron equations for embedded electron density: challenge for theory and practical payoffs in multi-level modelling of soft condensed matter.* In: *Computational Chemistry: Reviews of Current Trends*, Vol. XI, World Scientific, 2005 *in press*.

[4] R. G. Parr, W. Yang, *Density Functional Theory of Atoms and Molecules*, Oxford University Press, New York, 1989.

[5] P. Cortona, *Phys. Rev. B.*, **44** (1991) 8454.

[6] T. A. Wesolowski and F. Tran, *J. Chem. Phys.*, **118** (2003) 2072.

[7] J. Baker, J. Andzelm, A. Scheiner, and B. Delley, *J. Chem. Phys.*, **101** (1994) 8894.

[8] M. Krack and A. M. Köster, *J. Chem. Phys.*, **108** (1998) 3226.

[9] P. M. W. Gill, B. G. Johnson, and J. A. Pople, *Chem. Phys. Lett.*, **209** (1993) 506.

[10] A. M. Köster, R. Flores-Moreno, G. Geudtner, A. Goursot, T. Heine, J. U. Reveles, A. Vela, D. R. Salahub, deMon 2003, NRC, Canada.

[11] J. Neugebauer, M. J. Louwerse, E. J. Baerends, and T. A. Wesolowski, *J. Chem. Phys.*, **122** (2005) 094115.

[12] M. Prencipe, *Phys. Chem. Minerals*, **29** (2002) 552.

[13] M. Dulak and T. A. Wesolowski, *Int. J. Quant. Chem.*, **101** (2005) 543.

Brill Academic Publishers
P.O. Box 9000, 2300 PA Leiden
The Netherlands

Lecture Series on Computer
and Computational Sciences
Volume 3, 2005, pp. 289-297

High-level calculations of electronic optical and structural properties of small silicon nanocrystals and nanoclusters

Aristides D. Zdetsis

Department of Physics,
University of Patras,
GR-26500 Patras, Greece

Received 1 August, 2005; accepted 12 August, 2005

Abstract: High-level accurate *ab initio* calculations for silicon nanocrystals and nanoclusters of sizes up to a few hundred silicon atoms are now possible with current computational resources. Such calculations provide accurate structural, electronic and optical properties. The simpler and most economical and consistent method to study both structural and electronic properties of silicon nanoclusters is the density functional theory (DFT), especially by the use of the hybrid nonlocal exchange and correlation functional of Becke and Lee, Yang and Parr (B3LYP). It is demonstrated, using the example of the Si_6 cluster that the DFT/B3LYP method works very well even in cases of very flat energy hyper-surfaces, where simple perturbation theory fails. The accurate description of the optical properties and in particular of the absorption spectrum demands high-level correlation of both ground and excited states. The most economical way of studying the absorption spectrum is through time-dependent density functional theory (TDDFT). It is shown that the use of hybrid nonlocal B3LYP functional substantially improves the efficiency and accuracy of the TDDFT calculations. The high accuracy of the TDDFT/B3LYP calculations is verified by high-level multi-reference second order perturbation theory (MRMP2). In particular, the accuracy of the optical gap is better than 0.3 eV. Such high accuracy allows safe conclusions about the exact variation of the optical gap with various factors such as the size of the nanocrystals and the surface passivation. Moreover, the accuracy of the present calculations can be used to pin down the sources of existing discrepancies in the literature.

Keywords: Si nanocrystals, Si nanoclusters, optical properties, quantum dots,

PACS: 36.40-c , 71.15.Mb, 71.24.+q, 71.35.Cc, 73.22.-f, 78.67.Bf, 78.67.-n

1. Introduction

The study of atomic nanoclusters and nanocrystals, which form a link between molecules and crystals, is very active over the last few decades due to its fundamental importance in science and technology. The novel properties of these nanosystems can be found neither in molecules nor in infinite solids. The nanocrystals, contrary to nanoclusters, preserve most of the symmetry properties of the corresponding infinite crystals, of which many times are used as model systems. Nowadays nanocrystals with "diameters" as small as 1 nm can be readily produced by different technological techniques. Among the various types of atomic nanoclusters and nanocrystals the semiconductor and especially the silicon nanocrystals (and nanoclusters) have a special technological importance for nano-electronics and optoelectronics.

The electronic properties of silicon nanocrystals have been already applied in new device designs, such as single-electron transistors and floating-gate memory [1].

The optical properties of silicon nanocrystals are very important for optoelectronics applications. The study of such properties is a very challenging and promising field of research [1-14, 23-39], since the optical properties of the corresponding bulk crystal are rather poor because of the small band gap and the resulting indirect phonon-assisted light emission. The culmination of the silicon nanocrystal research occurred with the observation of visible photoluminescence (PL) in porous silicon and silicon nanocrystals [39, 2-14]. A large number of experimental and theoretical approaches have been carried out in order to explore the properties and resolve the origins of the observed visible photoluminescence [1-14, 23-29], which was attributed to quantum confinement. However, even today there are still issues, which are considered by several researchers as unsettled. A major point of dispute, besides the origin of PL, is the variation of the optical gap as a function the nanocrystal size. This variation is very important for optoelectronics and nano-electronics applications. The majority of the earlier experimental (and theoretical) work gives diverse results as for the size of the Si nanoparticles capable of emitting in the visible. The results of Wolkin et al [3] revealed optical gaps as small as 2.2 eV, for nanoclusters with a diameter of 18 Å (1.8 nm). For nanoclusters of about the same size Wilcoxon et al [2] obtained a similar result (2.5 eV) together with a much larger gap of about 3.2 eV for highly purified samples of the same diameter. Schupler et al [5] have estimated the critical diameter for visible PL to be less than 15 Å.

As it will be demonstrated bellow, most of the existing discrepancies in the experimental results are due to either oxygen contamination (or more generally to the preparation conditions), or experimental uncertainties in the determination of the nanocrystal diameter. The discrepancies in the *ab initio* theoretical calculations, especially in the region of diameters between 1 and 2 nm, are due to poor treatment of exchange and (to a lesser degree) of correlation [7,40]. Another source of error in phenomenological calculations (i.e empirical and semi-empirical approaches with adjustable parameters) for small-size nanocrystals is the choice of fitting key parameters of the method (for instance matrix elements in the tight-binding method) to bulk values. This could lead to significant errors for the gap in such small sizes of nanocrystals, due to the effects of quantum confinement. For the oxygen "contaminated" nanocrystals the role of bonding environment and the possible surface reconstruction are also controversial issues. Recent studies about the role of surface oxygen on the optical properties of silicon nanoclusters report conflicting levels of importance, ranging from minimal to crucial. In most of the realistic calculations the role of surface oxygen, has been ignored or underestimated, despite the evidence given by experiments [2-3, 5, 10-11, 14]. In the last two years, some high level model calculations for oxygenated nanocrystals have appeared [25, 26, 35, 37-38, 40].

The literature for small silicon (nano)clusters is extremely rich and the number of publications ranges to several hundredths or even thousandths [40-46]. Nevertheless here too, controversial results and interpretations do exist despite the widespread believe that most of the major problems are well understood and resolved, especially for the small clusters [40, 46]. Surprisingly enough even Si_6, which is considered as one of the best understood and extensively studied clusters, is full of puzzles and paradoxes about its molecular, electronic and optical structure [40, 46].

The bulk of the calculations presented here have been performed within the framework of static and time-dependent density functional theory (TDDFT) [15], employing the hybrid nonlocal exchange-correlation functional of Becke, Lee, Yang and Parr (B3LYP)[21]. In several cases the non-hybrid functional of Becke and Perdew BP86 [22] has been used for comparison. In addition, several other high-level methods have been used for the structural and electronic properties of Si nanoclusters and the electronic and optical properties of Si nanocrystals. Such methods include high order (2nd, 3rd, 4th) perturbation theory, quadratic configuration interaction (QCI) and coupled clusters (CC) methods mainly for structural and vibrational properties. For the optical properties (absorption spectrum, optical gap) the multi-referenced second order many-body perturbation theory (MR-MP2) [7, 38] has been used in addition to TTDFT/B3LYP. The accuracy of the optical gap obtained by MR-MP2 or TTDFT/B3LYP is better than 0.3 eV [7, 40]. The B3LYP functional, which can efficiently reproduce the band structure of crystalline Si, without the need for ad hoc numerical adjustments [23], is absolutely essential for this type of accuracy. The high accuracy of the results together with the true understanding of what the "agreement" between different results (experimental and theoretical) really means, allows us to resolve controversies and misconceptions in the literature. Many times the different results are in reality dealing with "different" systems and usually the real and the idealized system are not identical. Sometimes, however, there is interesting physics underlying the discrepancies. This is indeed the case of the Si_6 cluster as will be shown below.

The present work is organized as follows: In the next section an overview of the various theoretical approaches used for silicon nanocrystals (and nanoclusters) is presented. In section 3 some representative results for Si nanocrystals are given. The results for silicon nanoclusters, which are restricted for obvious reasons of space and time economy only to the very interesting and representative case of Si_6, are discussed in section 4.

2. Overview of theoretical methods and calculations

The full description of the electronic and optical properties of silicon nanocrystals and nanoclusters, even within the framework of adiabatic and harmonic approximations which uncouple the electronic and nuclear motions, requires the solution of a many-body Schrödinger equation of the form:

$$\left\{ -\frac{1}{2}\sum_{i=1}^{N}\nabla_i^2 + \sum_{i=1}^{N}\upsilon(\mathbf{r}_i) + \frac{1}{2}\sum_{i=1}^{N}\sum_{\substack{j=1\\(j\neq i)}}^{N}\frac{1}{|\mathbf{r}_i - \mathbf{r}_j|} \right\}\Psi(\mathbf{r}_1,\mathbf{r}_2,\cdots,\mathbf{r}_N) = E\Psi(\mathbf{r}_1,\mathbf{r}_2,\cdots,\mathbf{r}_N) \qquad (1),$$

where N the number of electrons ranges from a few hundreds (10^2) up to 10^{24}.
This looks like a formidable task since the N (interacting)-body problem cannot be solved exactly for N larger than 2 (not even for N=1). Equation (1) is equivalent to N coupled equations of the form:

$$\left\{ -\frac{1}{2}\nabla_i^2 + \upsilon(\mathbf{r}_i) + \frac{1}{2}\sum_{\substack{j=1\\(j\neq i)}}^{N}\frac{1}{|\mathbf{r}_i - \mathbf{r}_j|} \right\}\psi_i(\mathbf{r}_i) = \varepsilon_i\psi_i(\mathbf{r}_i) \quad , \quad i = 1,2,3,\cdots,N \qquad (2)$$

The main difficulty is the presence of the interaction term, $\sum_{\substack{j=1\\(j\neq i)}}^{N}\frac{1}{|\mathbf{r}_i - \mathbf{r}_j|}$ in (2) or $\frac{1}{2}\sum_{i=1}^{N}\sum_{\substack{j=1\\(j\neq i)}}^{N}\frac{1}{|\mathbf{r}_i - \mathbf{r}_j|}$ in (1),

which couples the coordinates of the i-th and j-th electron. However, for small molecular systems (small N) as well as infinite crystalline solids of high (translational) symmetry there are very accurate and efficient high-level methods working very well. These methods involve fundamental and well valid approximations with "controllable errors" which can become much smaller than the desired accuracy. The traditional *ab initio* approach to this many-body problem is based on the "one-electron approximation". That is, instead of considering all the electrons together as in eq. (1), according to the one electron (one-body) approximation we look at the electrons one by one as in eq. (2), replacing the complicating fluctuating forces due to the others (the interaction term) by an average force (field) known as the "mean field". This trick reduces the N coupled three-dimensional Schrödinger's equations (2) to one (N similar) much simpler (three-dimensional) equation for each electron. Corrections to this mean field approach which are needed to take into account correlation between interacting electrons are taken care either by perturbation theory or by configuration interaction (CI) method, or by one of the existing variants and combinations of the two [40]. Such a method is the multi-reference second order perturbation theory MR-MP2 [19, 38] used here for the calculation of the optical gap of small nanocrystals. However, these high-level methods require high computational cost which scales with a high power of the system size (N). Therefore these methods are usually restricted to relatively small molecules and clusters. The silicon nanocrystals suitable for technological applications in nanoelectronics are usually too big to be described by the real-space high-level *ab initio* quantum chemistry techniques, used for small systems. At the same time they are too small to be described by k-space theoretical techniques, used for infinite crystals. Therefore both of these two categories of theoretical approaches, the "atomistic" (real space) and the "crystalline" (k-space), require special attention with specific strategies and further modifications or approximations in order to be applicable to "small nanocrystals", which with atomistic standards are already much too large.

A good alternative to high-level fully *ab initio* methods is the density functional theory (DFT), which in its simplest approximation is known as the local density approximation (LDA). The DFT , which offers an accurate description of medium size systems with current computational power, is a mean-field theory which in principle is exact. In DFT the total energy (Hamiltonian) of the system is expressed in terms of the "exchange and correlation energy functional", which is a unique functional of the electronic density. This functional is not known. Instead, several approximated well known functionals have been proposed and used in the literature. One of the functionals which is used extensively and with great successes lately [7, 38, 40] is the hybrid B3LYP functional [21] which combines the advantages of DFT (accurate correlation) and HF (exact exchange).

For large size systems, where DFT calculations are not possible, the common compromise is the use of non-*ab initio* empirical or phenomenological techniques. These techniques sooner or later become indispensable for the description of larger nanocrystals. Therefore, there is nothing wrong with their use, provided that care has been taken for small nanocrystal sizes, for which the approximations and/or the values of the empirical parameters usually brake down. One way to achieve consistency at small and large sizes of nanocrystals is the use of transferable bond (length and angle)-dependent tight-binding parameters. This has been attempted by the author and coworkers with very promising preliminary results [47].

It should be emphasized at this point that both DFT and empirical methods (in addition to the high-level *ab initio*) work satisfactorily for structural and energetical properties which are obtained by the calculation of the ground state energy and wave function. However, for the description of the optical gap and more generally of the absorption spectrum correlation in both the ground and the excited states are essential. The time-dependent DFT (TDDFT) is the most economical method for the efficient description of the optical absorption and other time-dependent and excited states properties. The accuracy of the TDDFT method is greatly enhanced by the use of the hybrid B3LYP functional. The accuracy of the optical gap calculated by the TDDFT/B3LYP method is better than 0.3 eV, comparable to the accuracy of the MR-MP2 results [7, 38, 40]. A crude approximation of the optical gap is obtained by the energy difference of the highest occupied (HOMO) orbital from the lowest unoccupied (LUMO) orbital. The HOMO-LUMO gap for an infinite crystalline solid corresponds to the usual band gap, calculated by usual band structure calculations. Sometimes a slightly better approximation, known as ΔSCF method, is used as a computational compromise between the high-level (TD) and the zero level (HOMO-LUMO gap). Instead of a TD calculation, two ground-state calculations are performed (one for the ground and one for the "pseudo-excited" state and their energy difference is considered as the corresponding excitation energy. According to the ΔSCF method an electron is removed from the HOMO and placed in the LUMO orbital and the energy is recalculated. However, this is not a correct or accurate procedure, even when the ground-state calculations are of relatively high level.At least for the case of DFT calculations the ΔSCF procedure implies that the resulting excited state is a spin triplet. As a result, if total energy differences between the excited and ground states are considered as optical absorption energies, then they correspond to singlet-triplet transitions which are forbidden unless there is a strong spin-orbit coupling. Moreover, removing an orbital from HOMO and placing it in LUMO does not necessarily mean that we have constructed a real excited state.

Thus, we can categorize the existing calculations for silicon nanoparticles (especially for the electronic and optical properties) in i) Empirical or semi-empirical methods. ii) Single ground state calculations (HOMO-LUMO gap), mainly based on DFT or LDA. iii) ΔSCF calculations employing dual ground state calculations. Such calculations have been performed in the framework of LDA and, more recently [31.b], of quantum Monte Carlo (QMC) [40]. iv) Excited state calculations usually based on time-dependent density functional theory (TDDFT or TDLDA) or CIS [38, 40]. The present calculations for the optical gap are based on TDDFT/B3LYP and MR-MP2 [7, 38, 40] methods. In addition CIS and ΔSCF/B3LYP calculations [38, 40], as well as TDDFT calculations based on the BP86 [13] functional have been also performed for comparison.

3. Results and discussion for Si nanocrystals

Some representative structures of the Si nanocrystals are shown in fig. 1 in a "ball-and-stick" diagram. The nanocrystals are distinguished in oxygen-free which are terminated by hydrogen atoms and oxygen

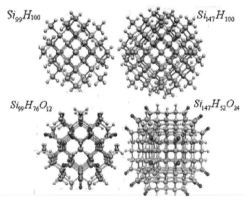

"contaminated" nanocrystals, which are terminated by hydrogen and double-bonded oxygen atoms. Both types of nanocrystals are fully optimized within the T_d symmetry constrain. The largest diameter of the Si nanocrystals is around 2.5 nm. The structures of the nanocrystals, especially of the oxygen contaminated ones were obtained by a special substitution procedure designed to preserve the T_d symmetry [40, 48]. In some cases of oxygen "contaminated" or reconstructed nanocrystals the symmetry was lowered. The DFT and the TDDFT calculations were performed with the TURBOMOLE [17] suite of programs using Gaussian atomic orbital basis sets of split valence [SV(P)]: [4s3p1d]/2s [18] quality which involves 5400 basis functions for the largest systems studied. In table 1 the results for the optical gap of the smaller oxygen-free nanocrystals

Figure 1: Representative structures of the Si_{99} and Si_{147} nanocrystals with hydrogen (small grey spheres) and oxygen (red or dark spheres in black and white version) on their surface.

are presented and compared with four different methods.

Table 1: Comparison of the optical gap of the smaller oxygen free nanocrystals with 5-35 Si atoms (#Si=5-35, #total=17-71) according to ground state (ΔSCF/B3LYP, ΔSCF/QMC) and excited state (TDDFT/B3LYP,BP86, MR-MP2) theoretical techniques.

#Si	ΔSCF B3LYP (eV)	ΔSCF QMC (eV)	TDDFT/ B3LYP (eV)	MR-MP2 (eV)	TTDFT/ BP86 (eV)
5	6.20*	6.9*	6 .66*	6.56*	6.10*
17	4.68	5.7	5.03	5.01	4.44
29	4.40	5.2	4.53	4.45	3.85
35	4.36	5.0	4.42	4.33	3.73

*Experimental value: 6.5 eV

As we can see in table 1 our MR-MP2 and TDDFT/B3LYP results are in excellent agreement with each other and with experiment, wherever reliable experimental data exist (Si$_5$H$_{12}$ in table 1). The TTDFT/BP86 results consistently underestimate the optical gap by about 0.6 eV. This illustrates the importance of the partially exact exchange, included only in B3LYP and not in the BP86 (and similar) functionals. Apparently for much larger nanocrystals the role of exchange should not be so crucial. Considering the ΔSCF results, independent of the formal objections for the validity of the method, we see that the ΔSCF/B3LYP method consistently underestimates the optical gap (it can be considered as a lower limit of the real gap [40]) whereas the ΔSCF/QMC consistently overestimates the gap. In any case, the ΔSCF/B3LYP is certainly in better agreement with experiment (at least for Si$_5$H$_{12}$) and the MR-MP2 method, the accuracy of which is well established [7, 38, 40]. In figure 2 the optical gap obtained by TDDFT with the B3LYP and BP86 functionals is plotted as a function of the size of the nanocrystals. In this figure the present results are compared with other theoretical and experimental results. As we have mentioned in the introduction and elsewhere [7, 38, 40], the differences are due to either experimental preparation (such as oxygen contamination) and characterization (size of the nanocrystal, etc) conditions, or poor treatment of exchange (and correlation) or, finally, poor fitting of the phenomenological parameters [40].

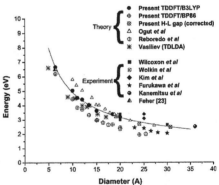

Figure 2: The optical gap as a function of the nanocrystal diameter

Table 2: The optical gap (E_g) obtained with TDDFT/B3LYP for larger nanocrystals. The mark (*) indicates results with only bridging Si-O bonds. No double bonds.

# Si	# H	E_g(TDDFT) B3LYP(eV)	#H	#O	E_g(TDDFT) B3LYP(eV)
47	60	4.04	48	6	2.70
47	-	-	36	12*	3.7*
71	84	3.64	72	6	2.21
71	-	-	60	12*	3.1*
99	100	3.39	76	12	2.45
147	100	3.19	52	24	1.85
281	172	2.61	-	-	-

The results for oxygen containing nanocrystals, especially when oxygen is present in very small amounts as a "contaminant" are discussed elsewhere [48] in this symposium. In table 2 the optical gaps for nanocrystals of both types are compared. As we can see in table 2, although both types of silicon-oxygen bonds, the bridging bonds and the double bonds, lower the gap relative to the gap of the corresponding oxygen-free nanocrystals, only the Si=O bonds can bring the gap down to the visible region. Hydroxyl bonds, O-H, can produce a similar lowering but only if they are present in large amounts [48]. The possibility of adjusting the gap with other controllable ways using Ge or SiGe nanocrystals is discussed in two different papers in this symposium [49, 50].

4. Results and discussion for Si nanoclusters: The Si_6 cluster

The field of small atomic clusters and particularly of semiconductor and especially silicon clusters has been active and very fruitful for a long time (longer than the related field of nanocrystals due to its profound scientific and technological importance [41-46]. And yet, even today several questions and puzzles still remain open, although many of them were believed as settled. This is not surprising for

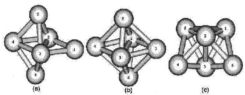

Figure 3: The lowest structures of Si_6 cluster

the larger clusters, for which high level *ab initio* calculations are difficult. However, for small clusters, such as Si_6 which is extensively studied both theoretically and experimentally, and which for long time has been considered as fully understood [43-45], the existing discrepancies and puzzles about its basic properties such as geometrical, electronic and dynamical is a real surprise. Fortunately enough, the puzzles about Si_6 lead to a new and profound understanding not only of what is really happening but also of why and where previous calculations (especially by respected pioneers of the cluster research, such as K. Ragavachari) have apparently failed [40, 44-45]. In figure 3, the three lowest (energetically) structures of Si_6 are shown in a "ball-and-stick-diagram".

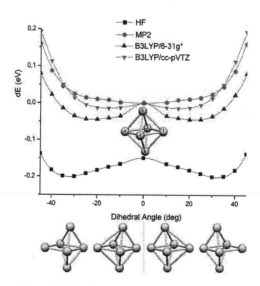

Figure 4: Variation of the relative (to the D_{4h} structure) energy (in eV) with distortion of the dihedral angle from zero to +/- 45° for various models. The HF curve has been shifted vertically for clarity.

The structure in 3(a) is an edge-capped trigonal bipyramid of C_{2v} symmetry. The structure in 3(b) is distorted octahedron of D_{4h} symmetry and the structure in 3(c) is a bicapped tetrahedron. This last structure 3(c) can be also viewed as a face-capped trigonal bipyranmid of C_S or near C_{2v} symmetry. Structures 3(a) and 3(b) were initially considered by Ragavachari [44], on the basis of HF geometry optimization, as the two lowest structures very close one another, whereas structure 3(b) was unstable. Later, on the basis of MP2 geometry optimization Ragavachari discovered exactly the opposite. The structures 3(a) and 3(c) were considered as unstable and the D_{4h} structure was accepted as the ground state, supported also by experimental evidence from Raman spectroscopy [45]. Unfortunately it turns out that this final suggestion is wrong in spite of the experimental support [40, 46]. Figure 4 shows why. Distorting the D_{4h} as in the figure, we can see that the energy changes are extremely small. All methods except MP2 reveal shallow minima for the C_{2v} (shown in figure 4) and the C_S structures (not shown in the figure) and a small bump (maximum in the figure) for the D_{4h}. Higher order perturbation theory and other high-level methods [40, 46] agree with the B3LYP results. This is also verified by table 3. Apparently the simple perturbation theory breaks down due to the flatness of the energy "hyper-surface", as is illustrated in figure 4. As a result the MP2 gradients fail to converge to the "right" structure. Which structure of the three is the "right" structure? The (paradoxical) answer is either "all of them" or "neither one". As we can see in table 3, although the C_S is the lowest possible structure, the energy differences between the three structures are marginal and in any case much smaller than the energies of the low frequency "soft" modes. This is highly suggestive that the three structures could be continuously transforming between themselves at no real energy cost through the soft vibrations. We could recognize this effect as a dynamical Jahn-Teller effect, in analogy to the more familiar (static) Jahn-Teller effect, which is responsible for the (static) distortion of the high-symmetry ideal octahedron into the D_{4h} structure.

Table 3 : Absolute and relative energies of Si_6 (using the $D95^*$ basis set).
The energies of the C_S and C_{2v} structures correspond to the B3LYP optimized geometry.
The energies of the D_{4h} structure are given at the MP2 optimized geometry.

Method	Structures (energies in Hartrees)		
	C_S	C_{2v}	D_{4h}
HF	-1733.383962	-1733.384411	-1733.381269
MP2	-1733.989200	-1733.988409	-1733.990781
	----------	(+0.0215 eV)	(-0.0430 eV)
MP3	-1733.972622	-1733.972747	-1733.971587
MP4SDTQ	-1734.063381	-1734.062753	-1734.062809
CCSD	-1733.987462	-1733.987404	-1733.986721
	----------	(+0.0016 eV)	(+0.0202 eV)
CCSD(T)	**-1734.040527**	**-1734.040374**	**-1734.040007**
	----------	(+0.0042 eV)	(+0.0141 eV)
State	$^1A^{\prime}$	$^1A_1$	$^1A_{1g}$

If this is the case, then the Raman measurements reflect time-averages over the three structures [40, 46]. As it is explained elsewhere [40], a closer look at the experimentally measured spectrum and the one calculated (for all three structures) give more support to this interpretation. Independent of this, as we can see in table 4, it is hard to distinguish between the tree structures on the basis of the experimental Raman data.

The last two columns of table 4 show some selective frequency results for the C_S and C_{2v} structures, for modes compatible with D_{4h} Raman active modes and with large intensities. As we can see, these frequencies are comparable in magnitude to the experimental results and to the values predicted by MP3, MP4, B3LYP (and MP2) for the corresponding D_{4h} structure. If we assume that the experiment cannot always detect all active Raman modes below some threshold value, or cannot resolve frequencies which are "nearby" we can have an additional illustration from a slightly different point of view that the experimental "verification" of the D_{4h} ground state is not really founded. The same data could be interpreted as supporting anyone of the three structures or, better yet, all of them together through a time average, illustrating the structural non-rigidity of Si_6.

Table 4: Vibrations of D_{4h} structure at various levels of theoretical treatment along with some "corresponding" B3LYP of the C_{2v} and C_S structures. Frequencies are given in cm^{-1}. The symbol i indicates imaginary frequencies. Experimental measurements are from ref. 45.

D_{4h}		D_{4h}				C_{2v}	C_s
Exp	Sym.	MP2	MP3	MP4	B3LYP	B3LYP	B3LYP
-----	E_u	52	81i	75i	79i	-------	-------
--	B_{2u}	197	117	149	117	-------	-------
-----	B_{2g}	220	252	225	234	276	276
--	A_{1g}	314	331	319	312	295	295
252	A_{2u}	358	344	347	315	-------	-------
300	B_{1g}	396	418	404	386	361	360
-----	E_g	447	424	421	395	395	370
-	A_{1g}	481	480	476	442		{ 439
386						{ 436	{ 448
404	E_u	482	483	480	443	{ 460	{ 448
458							{ 451

There is an intriguing possibility suggested by this author [46], that the structural non-rigidity of Si$_6$ could be related to its magic property. In this case the enhanced stability of this magic cluster would be a dynamic rather than a static property.

References

[1] M. L. Ostraat et al, Appl. Phys. Lett. 2001, 79 (2001)
[2] J.P. Wilcoxon, G. A. Samara and P. N. Provencio, Phys. Rev. B 60, 2704 (1999).
[3] M. V. Wolkin, J. Jorne and P. M. Fauchet G, Allan and C. Delerue, Phys. Rev. Lett. 82, 197 (1999)
[4] F. A.Reboredo A. Franceschetti and A. Zunger, Phys. Rev. B 61 13073 (2000)
[5] Schuppler, S.L. Friedman, M.A. Marcus, D.L. Adler, Y.H. Xie, F.M. Ross, Y.L. Cha-bal, T.D. Harris, L.E. Brus, W.L. Brown, E.E. Chaban, P.F. Szajowski S.B. Christman and P.H. Citrin, Phys Rev. Lett. 72, 2648 (1994) ; Phys. Rev. B 52, 4910 (1995).
[6] A. Franceschetti and A. Zunger Phys. Rev. Lett. 78, 915 (1997).
[7] C. S. Garoufalis, A. D. Zdetsis and S. Grimme, Phys. Rev. Lett. 87 276402 (2001)
[8] M. Rohlfing and S.G.Louie Phys. Rev. Lett. 80, 3320 (1998).
[9] S. Ogut and J. Chelikowsky, S. G. Louie. Phys. Rev. Lett. 79, 1770 (1997); Phys. Rev. Lett. 80, 3162 (1998)
[10] J. L. Gole and D. A. Dixon, Phys. Rev. B 57 12002 (1998)
[11] L.E. Brus, P.F. Szajowski, W.L. Wilson, T. D. Harris, S. Schuppler, and P.H. Citrin, J. Am. Chem. Soc. 117, 2915 (1995)
[12] K. Kim, Phys. Rev B 57, 13072 (1998)
[13] S. Furukawa and T. Miyasato, Phys. Rev. B 38,5726 (1988)
[14] Y. Kanemitsou, Phys.Rev. B 49, 16845 (1994)
[15] M.E.Casida, in : Recent Advances in density functional methods, Vol. 1, ed. D.P.Chong (World Scientific, Singa- pore, (1995)
[16] R. B. Murphy and R. P. Messmer. Chem. Phys. Lett., 183:443, (1991); J. Chem. Phys., 97:4170, (1992).
[17] TURBOMOLE (Vers. 5.3),Universitat Karlsruhe, 2000.
[18] A. Schafer and H. Horn and R. Ahlrichs, J.Chem. Phys. 97, 2571 (1992)
[19] S. Grimme and M. Waletzke. Phys. Chem. Chem. Phys., 2:2075 (2000)
[20] R. Bauernschmitt, R. Ahlrichs, Chem. Phys. Lett., 256, 454 (1996)
[21] P. J. Stephens and F. J. Devlin and C. F. Chabalowski and M. J. Frisch, J. Phys. Chem, 98, 11623 (1994)
[22] A. D. Becke, Phys.Rev.A38 (1988) 3098; J. P. Perdew, Phys.Rev. B 33 8822 (1986)
[23] J. Muscat, A. Wander and N.M. Harrison, Chem. Phys. Lett. 342. 397 (2001)

[24] I. Vasiliev, S. Ogut and J. Chelikowsky, Phys. Rev. Lett. 86, 1813 (2001)
[25] A.Puzder, A.J. Williamson, Jeffrey C. Grossman and Giulia Galli Phys. Rev. Lett. 88, 097401 (2002).
[26] I. Vasiliev, J. R. Chelikowsky and R. M. Martin Phys. Rev. B 65, 121302-1 (2002)
[27] E. Degoli, G. Cantele, E. Luppi, R. Magri, D. Ninno, O. Bisi, and S. Ossicini, Phys. Rev. B 69, 155411 (2004)
[28] H. Weissker, J. Furthmuller, and F. Bechstedt, Phys. Rev. B67, 245304 (2003)
[29] A. Franceschetti and S. T. Pantelides, Phys. Rev. B 68, 033313 (2003)
[30] A. Puzder, A. J. Williamson, J. C. Grossman, and G. Galli, J. Am. Chem. Soc. 125, 2786 (2003)
[31] A. J. W Williamson, J. C. Grossman, R. Q. Hood, A. Puzder, and G. Galli, Phys. Rev. Lett. 89, 196803 (2002) ; A. Puzder, A. J. Williamson, J. C. Grossman, and G. Galli, J. Chem. Phys., 117, 6721 (2002)
[32] O. Akcakir, J. Therrien, G. Belomoin, N. Barry, J. D. Muller, E. Gratton and M.Nayfeh, App. Phys. Lett. 76, 1857 (2000)
[33] L. Mitas, J. Therrien, R. Twesten, G. Belomoin and M.Nayfeh, App. Phys. Lett. 78, 1918, (2001)
[34] S. Rao, J. Sutin, R. Clegg, E. Gratton, M. H. Nayfeh, S. Habbal, A. Tsolakidis and R. Martin Phys. Rev B69,205319-1 (2004)
[35] Zhiyong Zhou, Louis Brus, and Richard Friesner, NanoLetters, 3, 163, (2003)
[36] M. Luppi and S. Ossicini, J. App. Phys. 94, 2130, (2003)
[37] Zhiyong Zhou, Richard A. Friesner and Louis Brus, J.Am. Chem Soc. 125, 15599, (2003)
[38] A.D. Zdetsis, C. S. Garoufalis and Stefan Grimme, NATO Advanced Research Workshop on "Quantum Dots: Fundamentals, Applications, and Frontiers" (Crete 2003), B.A. Joyce et al. (eds.), 317-332, Kluwer-Springer, 2005
[39] L. T. Canham, Appl. Phys. Lett. 57, 1046 (1990)
[40] A. D. Zdetsis, Optical properties of small size semiconductor nanocrystals and nanoclusters, *Reviews on Advanced Materials Science (RAMS)* (2005)
[41] R. P. Andres et al. J. Mat. Res. 4, 704 (1989)
[42] K. Raghavachari , in Phase Transitions, 24-26, 61-90 (1990)
[43] M. F. Jarrold , Science 252 ,1085 (1991)
[44] K. Raghavachari , J. Chem. Phys. 84, 5672 (1986)
[45] E. C. Honea et al., Nature 366, 42 (1993); E. C. Honea et al., J. Chem. Phys. 110, 12161 (1999)
[46] A. D. Zdetsis, Phys. Rev. A 64, 023202 (2000); A.D. Zdetsis to be published
[47] A. D Zdetsis and N. C. Bacalis, *Properties of silicon nanoparticles via a transferable tight binding Hamiltonian based on ab initio results*, in this symposium, ICCMSE 2005
[48] C. S. Garoufalis and A. D. Zdetsis, *Optical properties of oxygen contaminated Si nanocrystals,* in this symposium, ICCMSE 2005
[49] A. D. Zdetsis and C. S. Garoufalis, *Optical properties and excitation energies of small Ge nanocrystals*, in this symposium, ICCMSE 2005
[50] A. D. Zdetsis, C. S. Garoufalis and E. N. Koukaras, *Optical and electronic properties of mixed SiGe:H nanocrystals,,* in this symposium, ICCMSE 2005

Brill Academic Publishers
P.O. Box 9000, 2300 PA Leiden
The Netherlands

*Lecture Series on Computer
and Computational Sciences*
Volume 3, 2005, pp. 298-308

A Taxonomy for Model Characterization and Parameter Identification in Engineering Economics and Physical Chemistry of Industrial Processes

F.A. Batzias[1]

Department of Industrial Management & Technology,
University of Piraeus,
GR-185 34 Piraeus, Greece

Abstract: This work deals with a knowledge based approach of multicriteria taxonomy of models frequently met in Economics and Physical Sciences. The criteria used have been extracted/formed through data mining techniques by examining over 300 cases of modelling and they are presented herein as bipolar categories according to the primary step of dialectics (thesis-antithesis). A simple alphanumeric coding system is proposed for ease model characterization and implementation is presented in four cases: (i) extractive fermentation in ethanol production, (ii) adsorption by activated Carbon for wastewater treatment with metal recovery, (iii) energy saving within an industrial complex, (iv) determination of compensation for lignite field exploitation. In all these cases, parameter identification is performed and the results are discussed.

Keywords: Taxonomy, modelling, model characterization, parameter identification, engineering economics, fermentation, adsorption, subsidy, compensation.

1. Introduction

Modelling is a way for representing reality and is used to analyze/understand/ simplify a situation, enrich Theory, predict an outcome, or solve a problem. In Science and Engineering, modelling is mainly performed via deduction and induction. Deduction is the logical process of reaching a specific conclusion by reasoning from general premises (specificization). With deductive reasoning we know that if the premises are true, and that we have not committed any reasoning mistake, then our conclusion will be true too. Induction is the logical process of reaching a generalized conclusion by reasoning from specific premises, which share something in common (generalization). Inductive reasoning maintains that if a situation holds in all observed cases so far (including reasoning by analogy to similar cases), then the situation is expected to hold in all cases in the future. It is obvious that, whereas deduction allows one to formulate a specific truth from a general truth, induction simply allows one to formulate a probability of truth from a series of specific observations.

Reactive or proactive abduction frequently serves as an initiative for performing an induction/deduction interplay leading to knowledge improvement under the form of a dynamic (ie., continually enriched) inductive model. Reactive abduction is the logical process of generating a hypothetical explanation for an event, taken either from the main corpus of the corresponding Theory by deduction or from data by induction; the plausibility of reactive abduction depends heavily on the likelihood of alternative explanations. Proactive abduction is the logical process of incorporating a specific/particular/ local pattern in the correct context of a general pattern or Theory by making the least assumption, according to the Principle of Simplicity; the plausibility of proactive abduction depends heavily on the likelihood of propitious embedments in other theoretical contexts.

---

[1] Corresponding author. E-mail: fbatzi@unipi.gr

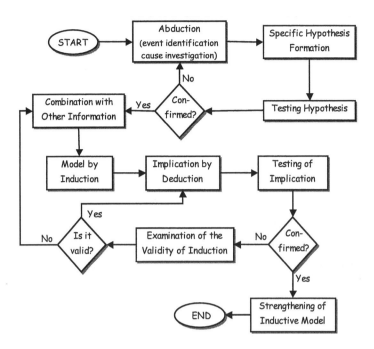

Figure 1. Hypothesis creation and testing through abduction/induction/deduction for building/ improving an inductive model.

Unlike deduction and induction, abduction is a type of critical thinking (usually leading to a hypothesis formation and testing, if possible) rather than formal logic:

The unexpected event Y is observed
The alternative hypotheses A, B, C are selected/synthesized to explain Y.
Hypothesis C is chosen as more plausible and is properly prepared/specificized for testing.

The relation between abduction, induction, and deduction is shown in the logic diagram of Figure 1, where three decision nodes that control the whole procedure of modelling are also quoted.

A taxonomy describes a classification structure based on common semantic content or format similarity, according to a predetermined system or an ontology; the latter provides the full specification of conceptualization of the corresponding knowledge domain, usually under the form of an agreed/controlled vocabulary that describes concepts/objects and the relations between them in a formal way. A computational taxonomy is a hierarchical classification which is automatically derived by computer software using data mining, textual analysis, and statistical inference; the same term is also loosely used to refer to any classification represented in the form of a computer relational database (RDB), provided it contains adequate information, i.e., an empty shell with its computer program do not constitute a taxonomy.

According to Jones and Paton [1], there are five types of taxonomy problems, which may be encountered when moving from an informal description of a domain to a formal representation of hierarchical knowledge:

- Atypical instances: an instance is not a typical example of a class to which it belongs; this may lead to difficulties when identification of the instance as a member of the class and with inheritance of properties from the class to the instance.
- Overlapping siblings: a member of a class is also a member of a sibling of that class.
- Context-sensitive subclasses: whether a hierarchical relationship is said to exist between two classes or not varies between contexts.
- Excluding instances: an instance of a class cannot be included as an instance of any of the immediate sub-classes of that class.

- Non-instance similarity: an entity satisfies the criteria for membership of a class which is not a member of.

The aim of the present work is to synthesize a multicriteria taxonomy for model characterization and parameter identification and apply it in the domains of engineering economics and physical chemistry of industrial processes. The concept of parameter identification covers a wide range from simple value estimation of given model parameters to reconstruction of functions or objects appearing as parameters; the latter case is a so-called 'inverse problem', in the sense that it somehow inverts the process of solving the differential equation(s) under the form of which the model appears. In the present work, the concept of parameter identification is specified as revealing the parameters of the parameters and (possibly) their relations, i.e. penetrating into deeper phenomenological levels (i.e. more analytic/scientific ones) of the original parameters without distorting their initial appearance built on a surface phenomenological level as a first approach.

2. Methodological Framework

To overcome the difficulties emerging when model taxonomy is based on hierarchical schemes, we decided to design a knowledge system based on criteria not necessarily within a strict treelike branching structure where each intermediate component has only one higher-level component. These criteria have been extracted/formed through data mining techniques by examining over 300 cases of modelling and they are presented herein as bipolar categories according to the primary step of dialectics (thesis-antithesis, see [2]) not followed by synthesis as the aim of this work is the multi-dimensional taxonomy and coding. According to this method, models can be distinguished to:

b. qualitative and quantitative, according to the way their terms/variables/parameters are expressed. It is worthwhile noting that predicate logic uses quantifiers like *some* (\$) and *all* ("), in sentences with variables ($x, y, ...$), predicates (F, G, ...), relations (R, ...), identity and logical constants/operators (AND, OR, NOT, IF...THEN). For example, (\$$x$) (\$$y$) (IF F$x$ AND Gy, THEN $x=y$) reads 'For some x and for some y, IF x is F AND y is G, THEN x is identical with y'. Consequently, by using quantifiers, a quasi-quantitative expression is obtained.

c. descriptive and operational; the former serve for describing/informing about a structure or a domain, while the latter are used as a working tool to produce results from some kind of type within which a description might be included.

d. deterministic and probabilistic or stochastic; in the former, for any given input set, the corresponding output is exactly determined, through either mathematical expressions or logical relations; in the latter, for any given input set, the corresponding output is estimated only in terms of probability distributions.

e. static and dynamic; the former indicate lack of change, while the latter indicate some kind of change, usually over time/space; a dynamic model in equilibrium is characterized as steady state model.

f. positive (or analytic) and normative; the former explain what/how happens and (possibly) why, making also predictions as a response to a change of certain initial values, while the latter deal with target setting/achievement, value judgments and policy determination/ evaluation, including multi-criteria optimization and strategic planning.

g. mechanismic[2] (or theoretical) and empirical; the former offer explanations in terms of driving forces and causal relations within a scientific Theory, while the latter offer relations based solely on observations/measurements but without a corresponding theoretical background.

h. linguistic and symbolic[3]; in the former, the terms are words, usually with contextual or even subjective meaning; in the latter, the terms are standardized and defined *a priori*, usually under the form of mathematical or logical symbols.

i. causal and teleological; the former (push-models) include or refer to something that occasions or effects a result while the latter (pull-models) have to do with purposes or ends [teleo-goal + logical] as the main driving forces; the corresponding key words that characterize these

---

[2] Usually met in scientific literature as 'mechanistic', a term adopted herein as the opposite for evolutionary – see bellow.
[3] From another point of view, symbolic models are opposite to physical models which, in their turn, may include analogue models whose behavior in certain respects represents the behavior of economic subsystem under consideration, although the model and the subsystem may be quite different entities. There are various forms of analogue, which are possible, e.g. the use of biological analogues to study organizational behavior, or the use of entropy analogues in thermodynamics to study disorder within a market or a firm.

         models are 'caused by' an event and 'in order to' achieve a goal (and their equivalent expressions)[4].

j.   Proactive and reactive; the former are used for anticipating needs/problems/changes (i.e. controlling a situation by causing something to happen rather than waiting to respond to it after it happens), while the latter are used to eliminate/control/avoid an undesirable event that has already occurred.

k.   Explicit and tacit (or implicit); the former deal with knowledge expressed/codified in word/numbers and shared in the form of data, scientific formulae, guidelines, specifications; the latter deal with personal, not easily visible or expressible knowledge, which usually requires joint, shared activities in order to extract/transfer it.

l.   Individualistic and aggregate; the former deal with beings and single/distinct/particular and further indivisible objects/elements, while the latter deal with sums of individuals/objects, either simply added to form a cluster or united/associated to form an agglomerate.

m.   Partial and total; the former deal with parts, while the latter deal with wholes; when the whole is organized, the terms holistic and systemic are preferable.

n.   Evolutionary and mechanistic; in economics, the former denote adoption of a mechanism similar to that of biological evolution implying that competitive selection is the only driving force that matters, while the latter denote a mechanical view of the system under consideration implying adoption of the reductionist method developed in physics[5], consisted in breaking up a phenomenon into its constituent parts and analyzing them according to simple fundamental laws that have become widely accepted.

o.   Subjective and objective; the former deal with judgment based on personal opinion and knowledge elucidation usually extracted by means of questionnaire and/or interviews; the latter deal with value estimations based on census data representing an initial population or a sample designed/taken according to widely accepted and possibly standardized specifications.

p.   Partonomic and taxonomic; the former include logical operators of the type 'is part of' and/or 'consists of' while the latter include logical operators of the type 'belongs to' a set of elements or entities with common characteristics; the operators or semantic links function within a network of concepts/relations/mechanisms representing a system under the form of an ontology[6].

q.   Rational and intuitive; The former are based on (usually deductive) logic, for setting up an explicit expression to solve the real problem under consideration while the latter are based on tacit knowledge and holistic subjective reasoning.

r.   Parametric and non-parametric; the former refer to functions applicable to relatively wide value-rage of the parameters involved within the function while the latter refer to functions applicable to a relatively narrow value-range of these parameters (in the extreme, applicable for a unique value of each parameter, in which case the parameter is replaced by a numeric coefficient)[7].

s.   Dimensional and dimensionless; the former include terms determined within a space of pre-defined dimensions (e.g. in the space [M, L, T, N, U], where these symbols stand for mass, length, time, money, utility, respectively) while the latter include only dimensionless terms; usually, a model can be expressed in both ways.

t.   Forecasting/predicting and backcasting; the former deal with making estimates of future performance or events based on present knowledge, facts, theory and judgment while the latter

---

[4] All important variables should be included within a causal model, because omission of such a variable may result in invalid inference about the causality structure of the model. Within the same model or family of models causality may change direction (i.e. the cause may become effect and *vice versa*) when the semantic content of at least one variable changes.

[5] In economics, according to Cohen [3], individuals were initially seen as the counterparts to Newtonial particles in classical mechanics, and hence subject to external forces; for example, Adam Smith analyzed the economic system as a collection of self-interested individuals on whom the law of supply and demand acts. Later, with the revolution of marginal economic analysis, we have the first complete and systematic attempt to create an economic science based on the Newtonian paradigm; economists such as Augustin Cournot, Stanley Jevons, Karl Menger, and Leon Walras systematically applied to economics methods and mathematical techniques derived from classical physics in order to transform economics in an exact science (e.g. see [4]).

[6] In knowledge management, 'ontology' is an explicit formal specification of how to represent the objects, concepts and other entities that are assumed to exist in some area of interest and the relationships that hold among them. Ontologies resemble faceted taxonomies but use richer semantic relationships among terms and attributes, as well as strict rules about how to identify / specify / analyze / synthesize terms and relationships. Since ontologies do more than just control a vocabulary, they are thought of as knowledge representation, suitable to provide information for (a) decision making in complex interdisciplinary domains (like energy markets and environmental systems) and (b) problem solving within these domains (see [5] for the basic concepts and [6] for an application concerning the energy sector).

[7] By extension, a full parameterized computer program is one that can be applied in many cases, because it is based on parameters of wide-range applicability, whereas a non-parameterized computer program is especially designed for solving only a specific problem regarding the corresponding situation.

deal with how desirable futures can be attained, involving working backwards from a particular desirable future end-point to the present in order to determine the physical feasibility of that future and what policy measures would be required to reach that point; the former are positive/analytic while the latter are normative model (based on pure causality and teleology/causality, or push- and pull-modelling, respectively)[8].

u. Simple and complex; the former are usually relations between dependent and a set of really independent variables, while the latter consist usually of a dense network of various relations among interdependent variables/components, including possibly different phenomenological levels and certainly feedback loops leading to non-linearities[9]; a metric of the network density can be used as a measure of complexity, provided that all variables/components belong to same phenomenological level, otherwise an ad hoc subjective index should be applied referring to the amount of information needed to make the model simpler or understood.

v. Linear and non-linear; the former refer to some kind of simple proportionality between dependent and one or more independent variables, while the latter refer to disproportionality introduced by structure and enhanced by function; in the latter case, chaotic[10] behavior is possible under conditions of instability.

w. Black box and white box, according to existence and transparency of the internal explanatory mechanism, which is responsible for transforming input to output; in the former, the internal mechanism either does not exist or is not described for any reason, while in the latter the internal mechanism is complete and transparent; gray box models are intermediates: in them, the internal mechanism either is known to some depth which allows for satisfactory functioning, at least in common situations, or is completely known but not fully represented on purpose (in which case, the part not described is usually replaced by simple rules of a surface phenomenological level); e.g. a computer program for DSS in exe form may have a subroutines as black boxes either because of proprietary reasons or because there no need for the user to know (to avoid confusion and save learning time/effort – in such a case, the black-boxing process is called *encapsulation*[11]).

x. General and specific, according to breadth of applicability; the former refer to first principles/concepts/relations and describe essential structures/features/functions at universal level while the latter refer to implications and describe characteristics in detail at local level; the former provide a platform for the deduction of the latter (specification) which, in their turn, may infer inductively to form the former (generalization/conceptualization); the knowledge representation of the whole process forms an ontology (including also the corresponding partonomic/ taxonomic models and the respective linguistic variables).

y. Numerical and functional; the former are constituted of representative data, set in certain order to reveal a structure and give the necessary information (referring to the format and the content of each model, respectively), while the latter are constituted of one or more functions, including continuous or discrete independent variables; a functional model can be derived either from theory by deduction or from *ad hoc* selected data / information by induction or

---

[8] Occasionally, scientists working on special disciplines use certain narrower definitions; e.g., in data mining, forecasting, in the form of statistical regression, is considered as one of the two modelling methods by which predictions are made, the other being classification; the distinction between the two is that forecasting predicts values within a series, whereas classification predicts membership in sets.

[9] The former obey primarily to the *Principle of Simplicity* while the later obey primarily to the *Principle of Consistency*. In methodology of science, the Principle of Simplicity is often quoted as Occam's (or Ockham's) razor, attributed to the 14th century logician and Franciscan friar William of Occam (born at Occham, located in Surrey, UK). After him, '*pluralitas non est ponenda sine neccesitate*', i.e. entities should not be multiplied unnecessarily. Ernst Mach advocated the following version of Occam's razor, which he named *Principle of Economy* ; 'Scientist must use the simplest means of arriving at their results and exclude everything not perceived by the senses'. In modelling, we can state that 'if you have more than one equally likely functioning models pick the simplest'. Although modern in its scientific application, forming the basis of methodological reductionism (known also as *Principle of Parsimony*), this principle goes back to Aristotle, who wrote 'Nature operates in the shortest way possible'.

[10] Within the present text, *chaos* is defined as a cryptic form of order, produced by the aid of a random number generator, mainly due to sensitive dependence on initial conditions; given a non-linear model, the existence or non-existence of (possible) chaotic behavior can be proved within a predetermined time / space domain but predictability of specified occurrence of such a form decreases as complexity increases.

[11] Most common Artificial Neural Networks (ANN) are representative examples of black box modelling as they make satisfactory prediction of output provided that they 'learned' their lesson adequately from input / output data ranging within the same interval where the need for prediction; their function is completely empirical, based on internal weighting of input / output combinations of variables using data in the form of matrices (MATLAB is a common high level language / environment for such applications). Batzias and Sidiras [7] have applied a black box model under the form of a neural network in agricultural waste management and compared its output with the results obtained by using a mechanismic / explanatory model.

from both (usually the format is deduced from theory and the parameter values estimation is induced from data / information)[12].

z. Primary and secondary; the former give the latter by deduction, synthesis or some kind of merging/decomposition/evolution, i.e. the former are the 'parents' of the latter and consequently they may have also grandchildren (tertiary models) etc, but the main unit-characteristic is the parental relation which can be multiplied at *infinitum*[13].

Therefore, a model can be characterized by a vector with alphanumeric elements of the type $1 or $2, where $ is a letter, according to the bipolar categories quoted above (the category under the letter *a* refers to the 'deductive and inductive' category described in Introduction), while the digits 1 and 2 refer to the order of the corresponding pole. For example, a model which is functional and deterministic is coded [b2, c1]. Although this taxonomy does not obey to any treelike structure, we can use the following categorical/prepositional constraints to build up a kind of ontology at a deeper phenomenological level, without distorting the surface level, i.e. the vector of ordered elements:

Universal affirmative (UA), such as 'All A are B'; e.g., all secondary models are deductive.

Universal negative (UN), such as 'No A are B'; e.g., no complex model are linear.

Particular affirmative (PA), such as 'Some A are B'; e.g., some forecasting model are subjective.

Particular negative (PN), such as 'Some A are not B'; e.g., some subjective model are not implicit.

These representative examples are given in Figure 2 with the corresponding Venn Diagrams. Last, for reasons of completeness and consistency, the symbols $0 and $3 are introduced to characterize the model under examination as irrelevant or 'between', respectively; e.g., a quasi-quantitative or a gray box model are *b3* or *w3*, respectively; on the other hand, a deductive simple linear model (i.e. [a1, u1, v1]) is expected to be *p0*.

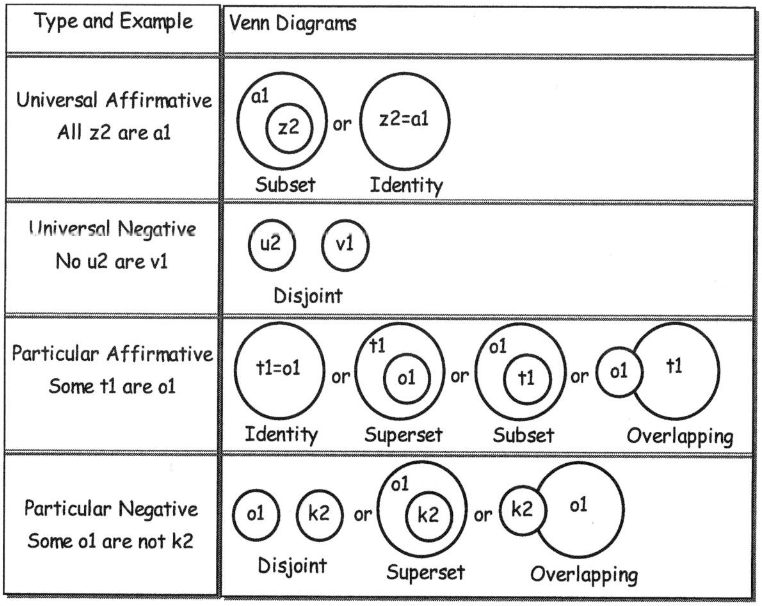

Type and Example	Venn Diagrams
Universal Affirmative All z2 are a1	Subset — Identity
Universal Negative No u2 are v1	Disjoint
Particular Affirmative Some t1 are o1	Identity — Superset — Subset — Overlapping
Particular Negative Some o1 are not k2	Disjoint — Superset — Overlapping

Figure 2. Illustration of the representative examples given as Venn Diagrams.

---

[12] Sometimes the continuous dependent variables given by the corresponding functions may either pass through points of discontinuity or enter regions (strictly speaking, 'spaces') where the meaning of the function, and possibly its structure, changes / transforms (as a necessity to keep going on through the space) giving rise to a new ontology (known as *morphogenesis* in Catastrophe Theory).

[13] A primary model without at least one parent is an axiom constituting the very start of a Theory or a System with mathematical structure, actually coming from an other phenomenological (meta-)level. Within an axiomatic system built on logic, at least one preposition is unproved and, generally, provability is a weaker notion than truth in such systems (including mathematical economics), according to the famous Gödel's Incompleteness Theorem (1931). This means that the very first deduction within an axiomatic economic system comes unavoidably from an unproved proposition established by induction, which, in its tern,

3. Implementation

As an implementation of the methodology described above, for representative examples are presented herein, extracted from over 300 cases examined in total. The criteria used for the selection of these four case examples were (i) satisfactory representativeness, (ii) adequate coverage of the two domains under consideration, i.e. Engineering Economics and physical chemistry (extended properly to include some neighboring fields of applied Biochemistry of Industrial Processes, and (iii) personal involvement of the present author in relevant research (as proved by citing the corresponding published works) to provide evidence that the model investigation is coming from 'inside', not from an external observer; needless to say that such an involvement implies a personal look, not always unbiased (or at least sometime subjective); evidently, this handicap is counterbalanced (i) at the general level of the Knowledge Base (developed for model characterization), by the plethora of the cases examined from 'outside', and (ii) at the specific/local level of the four cases, by the in depth knowledge acquired, enabling penetration into deeper phenomenological levels to minimize misunderstanding and clarifying concepts which can be proved useful in further examination of the rest models in the same knowledge base.

3.1. Extractive Fermentation in Ethanol Production

Extractive fermentation is a composite process that can reduce the effect of end product inhibition through the use of a water-immiscible phase which removes fermentation products *in situ*. Consequently, we have two beneficial actions: (i) removal of inhibitory products as they are formed thus keeping bioreaction rates high, and (ii) reduction of product recovery cost, due to decrease of both, capital and operating cost. An original model developed (by synthesizing / modifying expressions found in previous studies [8,9]) for the needs of the present work is the following:

$$LP^2 + MP + N = 0$$

where

P = aqueous concentration of ethanol produced

L, M, N = model parameters identified/specified as follows:

$$L = \frac{\beta}{S_o^2}\left[\frac{\mu_{max}K_p}{\rho_p}\frac{a_2 S}{K_s + S + bS^2} + \frac{a_0 + S_o(a_1 - y_{c/s})}{Y_{p/s}S_o}\left(1 - \frac{\mu_{max}K_p}{\rho_p D_s}\frac{S}{K_s S + bS^2}\right) - 1\right]$$

$$M = \frac{K_p}{S_o^2}\left[\frac{\mu_{max}}{D_s}\frac{S}{K_s + S + bS^2}\left[\frac{\beta}{\rho_p}(a_0 + S(a_1 - Y_{c/s})) - a_2\right] + \frac{a_0 + (a_1 - y_{c/s})S_o}{Y_{p/s}S_o}\left[\beta - \frac{\mu_{max}}{D_s}\frac{S}{K_s + S + bS^2}\left(\frac{\beta Y_{p/s}S}{\rho_p} - 1\right)\right]\right]$$

$$N = \frac{\mu_{max}K_p}{D_s S_o^2}\frac{S}{K_s + S + bS^2}\left[\frac{S}{S_o}(a_0 + (a_1 - Y_{c/s})S_o) - (a_0 + (a_1 - Y_{c/s})S)\right]$$

where

β = actual distribution coefficient for ethanol between the solvent and the aqueous phases

S, S_o = effluent and influent aqueous substrate concentration (g/L)

μ_{max} = maximum specific growth rate (h^{-1})

K_p = product inhibition constant (g / L)

ρ_p = density of absolute ethanol

D_s = dilution rate based on influent solvent flowrate (h^{-1})

a_0, a_1, a_2 = multilinear regression coefficients

K_s = substrate saturation constant (g / L)

$Y_{p/s}$ = ethanol yield coefficient (g ethanol produced /g substrate consumed)

$Y_{c/s}$ = carbon dioxide yield coefficient (g CO_2 produced /g substrate consumed)

b = inhibition factor

Model Characterization:

[a1, b2, c2, d1, e3, f1, g1, h2, i1, j0, k1, l0, m2, n2, o2, p0, q1, r1, s2, t1, u1, v2, w2, x2, y2, z2]

---

might have not been expressed if we had not an *a priori* sense of order about the world / cosmos, that is the starting point of any deductive system, according to John Stewart Mill.

3.2. Adsorption by Activated Carbon for Wastewater Treatment with Metal Recovery

Adsorption is a separation process based on the adhesion of molecules of liquids, gases and dissolved substances to the surfaces of solids. We have proved that the widely used first order kinetics model can be derived according to the following deductive procedure in the case of adsorption by activated carbon for wastewater treatment with metal recovery [10].

Let z be the driving force for adsorption, as a function of concentration, temperature and surface nature / topology. Under the assumption that the adsorption rate is proportional to this force, which in its turn, decreases steadily when the adsorbed amount L increases, we have the following simple relations:

$$dL/dt = k_1 z \ , \quad -dz = k_2 dL \tag{1}$$

A combination of these equations gives $-dz/dt = -(dL/dt)(dz/dL) = k_1 k_2 t$ or $-dz/z = k_1 k_2 dt$, which is easily integrated to give $\ln(z/z_0) = -k_1 k_2 t$, or $z = z_0 \exp(-k_1 k_2 t)$, where z_o is the value of z at $t = 0$. By substituting the last expression for z into the first equation, we obtain

$$dL/dt = k_1 z_0 \exp(-k_1 k_2 t) \text{ or } \int_0^L dL = k_1 z_0 \int_0^t \exp(-k_1 k_2 t) dt \text{ or } L = (z_0/k_2)\left[1 - \exp(-k_1 k_2 t)\right] \tag{2}$$

For $t = 0$, $L = 0$. We define $L_{inf} = \lim L_{t \to \infty}$. Obviously, $L_{inf} = z_0/k_2$. By introducing this asymptotic value L_{inf} into the last equation, we obtain

$$L = L_{inf}\left[1 - \exp(-k_1 k_2 t)\right] \tag{3}$$

Let C_b, C_{inf}, C_0 be the concentrations of adsorbate in the bulk solution at time t, 0, ∞, respectively. Obviously, $(C_0 - C_b)V = L$ and $(C_0 - C_{inf})V = L_{inf}$ where V is the volume of the solution. Therefore, eqn (3) is transformed in concentration terms as follows: $(C_0 - C_b)V = (C_0 - C_{inf})V\left[1 - \exp(-k_1 k_2 t)\right]$ or, for $k = k_1 k_2$,

$$C_b = C_0 - (C_0 - C_{inf})\left[1 - \exp(-kt)\right] \tag{4}$$

Instead of using this surface phenomenological model, we can derive by deduction a deeper and more explanatory one, so that parameter identification can be achieved. Starting with the mass balances of the metal species at the liquid-solid interface and in the bulk solution, we obtain the following relations:

$$\frac{dC_b}{dt} = -\frac{Sk_{mA}}{V}\left[(C_b - C_1) - (C_s - C_2)\right] \tag{5}$$

$$\frac{dC_s}{dt} = n\frac{Sk_{mA}}{V}\left[(C_b - C_1) - (C_s - C_2)\right] - n\frac{SD_e k_s}{V}(C_s - C_2) \tag{6}$$

where
  V = the volume of the bulk solution, $V = nv$, $n \in \mathrm{R}^+$
  v = the volume of the layer adjacent to the solid
  S = the total outside surface area of the solid
  C_b = the concentration of the metal species in the bulk solution at time t
  C_s = the concentration of the metal species at the liquid solid interface
  k_{mA} = the mass transfer coefficient
  D_e = the effective diffusivity of gold species in the microscope
  k_s = the kinetic constant
  C_1, C_2 = asymptotic values of C_b, C_s for $t \to \infty$

$$C_b = C_1 - \left[\frac{1}{2}(C_{b0} - C_1)k_{mA} + (C_2 - C_{s0})k_{mA} + (C_1 - C_{b0})\right]\exp\left[-\frac{D_e k_s}{V}St\right] \tag{7}$$

where C_{b0}, C_{s0} are the initial values of C_b, C_s at $t = 0$.

This kinetic model is quite general and the values of any three parameters can be estimated through LSNR, provided that the values of the rest of them are determined exogenously either by measuring the corresponding magnitudes experimentally or by calculating them independently. On the other hand, some kinetic models cited in literature can be deduced as special cases of our model. E.g., for $C_1 = C_2 = 0$, eqn (7) is reduced to a form similar to the kinetic model derived by Meng and Han [11].

Model Characterization:
[a1, b2, c2, d1, e2, f1, g1, h2, i1, j0, k1, l0, m2, n2, o2, p0, q1, r1, s1, t1, u1, v2, w3, x3, y2, z2]

3.3. Energy Saving within an Industrial Complex

There are several models, which indicate / propose optimal ways of reducing energy consumption through (i) using suitable materials (for both construction and insulation) (ii) increasing energy efficiency of old equipment and (iii) reengineering that may lead to installing new equipment in an industrial plant, usually with the financial support of the State or Local Authorities. Subsequently, we present such a model, developed for the determination of the optimal subsidy I_{opt}, expressed as a fraction of capital invested in energy saving equipment within an industrial complex (for complete techno-economic modelling, see [12]). Assuming that (i) at least a fraction K of the energy saving (in monetary units) is deducted annually by the State from its welfare budget and (ii) re-investment of annual 'gains' is realized with the common interest rate i, we have the following periodical (herein, annual) gains V_i ($i = 1, 2, ..., t$) for the State, estimated as present value:

$$V_1 = K \cdot F \cdot (1 + i)^{-1}$$

$$V_2 = K \cdot F \cdot (1 + i)^{-2} \cdot (1 + f)$$

$$V_3 = K \cdot F \cdot (1 + i)^{-3} \cdot (1 + f)^2$$

$$: : : : : : : : : : : : : :$$

$$V_t = K \cdot F \cdot (1 + i)^{-t} \cdot (1 + f)^{t-1}$$

$$V = \sum_{i=1}^{t} V_i = K \cdot F \cdot (1 + i)^{-1} \cdot \left[1 + \frac{1+f}{1+i} + \left(\frac{1+f}{1+i}\right)^2 + ... + \left(\frac{1+f}{1+i}\right)^{t-1} \right] \tag{8}$$

The expression within the brackets is a geometric series of the type $1 + x + x^2 + ... + x^{t-1}$ (indicating that energy savings follow an exponential pattern in the time course) the sum of which is given by $(x^{t-1} \cdot x - a)/(x - 1)$, where $a = 1$ and $x = (1 + f) / (1 + i)$, i.e., the first term and the constant multiplier (or ratio), respectively.

$$V = \frac{K \cdot F}{1+i} \cdot \frac{\left(\frac{1+f}{1+i}\right)^{t-1} \cdot \frac{1+f}{1+i} - 1}{\frac{1+f}{1+i} - 1}, \text{ or } V = \frac{K \cdot F}{1+i} \cdot \frac{\left(\frac{1+f}{1+i}\right)^t - 1}{\frac{1+f}{1+i} - 1} \tag{9}$$

where F is the first year energy cost savings (increased each period by a fraction f or $100f\%$, implying exponential growth) and t the time periods (dimensionless[14]) taken into account. On the other hand, we can estimate the opportunity cost as the potential loss L that is the value of the alternatives or other opportunities which have to be foregone in order to subsidize an energy saving investment of initial capital S, with an amount of money $I \cdot S$. If r is the return on the best alternative investment (called 'the second best' in comparison with the first best for the State, which is the subsidized fraction I), then L, expressed similarly in present value, is given by the following relation:

$$L = IS(1 + r)^t/(1 + i)^t \tag{10}$$

Obviously, the optimal value of I is obtained for $V = L$, so neither the State nor the investor make a surplus profit causing a corresponding loss to the other part (condition of equilibrium). From the equations (9) and (10), we obtain

$$\frac{K \cdot F}{1+i} \cdot \frac{\left(\frac{1+f}{1+i}\right)^t - 1}{\frac{1+f}{1+i} - 1} = I_{opt} S \frac{(1+r)^t}{(1+i)^t} \text{ or } I_{opt} = \frac{K \cdot F \cdot (1+i)^{t-1}}{S \cdot (1+r)^t} \cdot \frac{\left(\frac{1+f}{1+i}\right)^t - 1}{\frac{1+f}{1+i} - 1} \tag{11}$$

Model Characterization:
[a1, b2, c2, d1, e3, f2, g3, h2, i2, j1, k1, l3, m3, n0, o2, p0, q1, r1, s2, t2, u1, v2, w2, x3, y2, z2]

---

14 Exponents are always dimensionless; this means that whenever time appears as an exponent, it is expressed a time periods, i.e. as t/t0, where t0 is unit time corresponding to one period.

3.4. Determination of Compensation for Lignite Field Exploitation

The models used for decision making are usually normative. Nevertheless, positive models can also be used when they include a term connected with (at least one) policymaking variable. By rearranging the model to declare this variable as dependent, we obtain a configuration suitable for decision making. Such an example is the modification/application of the multiplier-accelerator by Batzias and Sidiras [13] for estimating the impact of investing the Compensation of Lignite Fields Exploitation paid by the Public Power Corporation (PPC) of Greece to local Authorities of three counties (Arkadia, Florina, Kozani), as a contribution to regional development:

$$Y_t = \alpha(1+\beta)Y_{t-1} - \alpha\beta Y_{t-2} + G_t + H_t \tag{12}$$

This positive model has been derived by deduction from the following three equations:

$$I_t = \beta[C_t - C_{t-1}] + H_t \qquad Y_t = C_t + I_t + G_t \qquad C_t = \alpha Y_{t-1} \tag{13}$$

The first of these three equations represents the Samuelson's acceleration principle, according to which the induced private investment I_t in any period t is proportional to the increase in consumption C_t of that period over the preceding C_{t-1}. In this relation, we have introduced an amount of exogenous investment H_t equal to CLFE. The second equation is actually an equality by definition, connecting the regional income Y_t of the district where CLFE is applied with consumption C_t investment I_t and governmental expenditure G_t (exogenously determined). The coefficient α of proportionality in the third equation is the marginal propensity to consume. By substituting the last two equations in the first, we obtain the positive model (1) for G, H = constants, under the form of a second order difference equation. The form of the complete solution of this equation, under steady state conditions, depends on the type of roots of its homogenous part, but at any case it can be set in the general form $Y_t = f(G,H,...)$, which by inversion gives the amount of H required for supplementary regional (due to CLFE) income to follow a predetermined time-path. Therefore, this positive model can be used as decision aid by local authorities (acting through their daughter company), when negotiating with PPC the CLFE level, by taking also into account other economic indices, like the net regional product and the marginal propensity to consume corresponding to the inhabitants of each lignite producing/extracting region. Model Characterization:
[a1, b2, c2, d1, e2, f1, g1, h2, i1, j1, k1, l2, m2, n2, o2, p0, q1, r1, s1, t3, u2, v2, w2, x1, y2, z2]

4. Discussion and Concluding Remarks

In the case of extractive fermentation in ethanol production, the model can be characterized as only indirectly dynamic [e3], because time enters the parental relations via S and the model is valid under steady state conditions; generally, such a model can be characterized as *e2* if it appears under the form of a differential equation including the derivative of the dependent variable with respect to time. It is worthwhile noting that we have derived the model presented herein by deduction from the following equation based on substrate inhibition kinetics (in absence of such inhibition parameter identification might be approached through [8]).

$$D' = \mu_{max} \frac{S}{K_s + S + bS^2} \frac{K_p}{K_p + P}$$

In the case of adsorption by activated carbon for wastewater treatment with metal recovery, the model is dynamic [e2] because time exists as an independent variable, obtained by solving the system of parental differential equations. Parameter identification is approached by (7), under certain conditions which usually prevail in these systems.

In the case of energy saving within an industrial complex, the model (11) is characterized as normative [f2], although it has been derived by deduction from its partial positive parental relations, since the optimization criteria introduce the normative aspect within the final policy making model. Generally, optimization can be induced into the economic modelling system under the form of an operational research method within which certain equilibrium (if and when achieved) implies a corresponding normative model, suitable for decision and/or policy making; a similar procedure can be applied also in macro economics by inducing economic models into a games empty shell (operational researcher's view) or policy games into an economic modelling system (economist's view, e.g. see [14]).

In the case of determination of compensation for lignite field exploitation, the model (12) is between forecasting and backcasting, although it is positive [f1], in the sense it can be used as a normative one by scanning all possible solutions and choosing the best for achieving a future target, i.e.

policy making according to preset criteria; this is an indirect way to induce optimization techniques, thus changing a model from positive to normative, at least to a certain degree.

In conclusion, the proposed knowledge based approach of multicriteria taxonomy of models performs satisfactorily, thus serving as a methodological framework for model comparison/evaluation/ discrimination; some partial characterizations of certain models may be though rebuttable, depending on the model's final usage. In trying to form a more objective system we might add presumptions to each criterion on the basis of the model's usage; such an addition would make the system complex, without offering any real advantage, because presumptions of this kind are rebuttable themselves and consequently redundant.

References

[1] D.M. Jones and R.C. Paton, Toward principles for the representation of hierarchical knowledge in formal ontologies, *Data & Knowledge Engineering* **31** 99-113(1999).
[2] F. Batzias and Z. Res: *Decision Making*. H.O.U. Press, Patras, Greece 2005.
[3] I.B. Cohen, *Interactions. Some Contacts Between the Natural Sciences and the Social Sciences*. MIT Press, Cambridge, 1994.
[4] B. Ingrao and G. Israel: *The Invisible Hand. Economic Equilibrium in the History of Science*. MIT Press, Cambridge, 1990.
[5] T.R. Gruber, A translation approach to portable ontologies, *Knowledge Acquisition* **5** 199-220(1993).
[6] F.A. Batzias and F.M.P. Spanidis, An ontological approach to knowledge management in engineering companies of the energy sector, *Intern. Series Energy and Development*, WIT Press Computational Mechanics Inc., Boston Massachusetts, USA, 11(2003), 349-359.
[7] F.A. Batzias and D.K. Sidiras, Agricultural waste management: neural network prediction of fermentable sugars yield from crop residues, *Proc. 5th International Conference on Environmental Pollution*, Thessaloniki, Greece, 2000, 712-719.
[8] F. Kollerup and A.J. Dragulis, A mathematical model for ethanol production by extractive fermentation in a continuous stirred tank fermentor, *Biotechnology and Bioengineering* **27** 1335-1346(1985).
[9] E.C. Marcoulaki and F.A. Batzias, Extractant design for enhanced biofuel production through fermentation of cellulogic waste, *Computer Aided Chem. Eng.* **14** 1121-1126(2003).
[10] F.A. Batzias and D.K. Sidiras, Wastewater treatment with gold recovery through adsorption by activated carbon, *Water Pollution* **6** 143-152(2001).
[11] X. Meng and K.N. Han, Adsorption of gold from iodide solution by activated carbon, *Minerals and Metallurgical Processing* **2** 31-36(1994).
[12] A.P. Oliveira Francisco, F.A. Batzias and H.A. Matos, Computer aided determination of optimal subsidy for insalling energy saving equipments within an industrial complex, *Computer Aided Chem. Eng.* **18** 409-414(2004).
[13] F.A. Batzias and D.K. Sidiras, Local sustainable development in districts with current exploitation of lignite fields, *Proc of the 17th Intern. Mining Congress*, Ankara, 2001, 575-582.
[14] G.M. Agiomirgianakis, Monetary policy ganes and international migration of labor in interdependent economies, *Journal of Macroeconomics* **20**(2), 243-266(1998).

Brill Academic Publishers
P.O. Box 9000, 2300 PA Leiden,
The Netherlands

*Lecture Series on Computer
and Computational Sciences*
Volume 3, 2005, pp. 309-320

On the numerical solution of the Sine-Gordon equation in 2+1 dimensions

A. G. Bratsos[1]

Department of Mathematics,
Technological Educational Institution (T.E.I.) of Athens,
122 10 Egaleo, Athens, Greece

1 Introduction

Solitons represent essentially special wave-like solutions to nonlinear dynamic equations. These waves have been found to a variety of nonlinear differential equations such as the Korteweg & de Vries equation, the Schrödinger equation, the Sine-Gordon equation etc. Physical applications of solitons have been found among others to shallow-water waves, optical fibres, Josephson-junction oscillators etc. The development of analytical solutions of soliton type equations, especially the inverse scattering transform (see Ablowitz and Segur [1] etc.) are known long time ago.

In higher dimensions models which possess soliton-like solutions have attracted great interest for numerical investigation. The two-dimensional Sine-Gordon (SG) equation in two space variables or as it is also known in $2 + 1$ dimensions, which belongs in this category, is given by

$$\frac{\partial^2 u}{\partial t^2} + \rho \frac{\partial u}{\partial t} = \frac{\partial^2 u}{\partial x^2} + \frac{\partial^2 u}{\partial y^2} - \phi(x, y) \sin u \qquad (1.1)$$

with $u = u(x, y, t)$ in the region $\Omega = \left\{ (x, y), \ L_x^0 \leq x \leq L_x^1, \ L_y^0 \leq y \leq L_y^1 \right\}$ for $t > t_0$, where the parameter ρ is the so-called *dissipative* term, which is assumed to be a real number with $\rho \geq 0$.

The two-dimensional SG equation arises in extended rectangular Josephson junctions, which consists of two layers of super conducting materials separating by an isolating barrier. A typical arrangement is a layer of lead and a layer of niobium separated by a layer of niobium oxide. A quantum particle has a nonzero significant probability of being able to penetrate to the other side of a potential barrier that would be in penetrable to the corresponding classical particle. This phenomenon is usually referred to as quantum tunneling. When $\rho = 0$, Eq. (1.1) reduces to the undamped SG equation in two space variables, while, when $\rho > 0$ to the damped one.

For the undamped SG equation in higher dimensions exact solutions have been obtained in [11], [9] and [13] using Hirota's method, in [10], [5] and [16] using Lamb's method, in [6] and [14] by Bläcklund transformation and in [12] by Painlevè transcendents. A numerical solution for the undamped SG equation has been given by Christiansen and Lomdahl [4], who used a generalized leapfrog method and by Argyris et *al* [2] with a finite-element technique. Both methods using appropriate initial conditions have been applied successfully with the latter one giving slightly better results. A numerical approach to the undamped SG equation appears in Djidjeli et *al* [7], where the method arises from a two-step, one-parameter leapfrog scheme, which is a generalization to that used by Christiansen and Lomdahl [4] and by Bratsos [3] who used a predictor-corrector

---

[1] Corresponding author. E-mail: bratsos@teiath.gr

scheme. In the case of the damped SG equation, the presence of the dissipative term corresponds to a physically relevant effect in real Josephson junctions (see Nakajima et *al* [15]). This problem has been studied numerically by Djidjeli et *al* [7] and Bratsos [3].

Initial conditions associated with Eq. (1.1) will be assumed to be of the form

$$u\left(x,y,t_0\right) = g\left(x,y\right) \; ; \; L_x^0 \le x \le L_x^1 \, , \; L_y^0 \le y \le L_y^1 \tag{1.2}$$

with initial velocity

$$\left.\frac{\partial u\left(x,y,t\right)}{\partial t}\right|_{t=t_0} = \hat{g}\left(x,y\right) \; ; \; L_x^0 \le x \le L_x^1 \, , \; L_y^0 \le y \le L_y^1 \, . \tag{1.3}$$

In Eq. (1.1) the function $\phi\left(x,y\right)$ may be interpreted as the Josephson current density, while in Eqs. (1.2)-(1.3) the functions $g\left(x,y\right)$ and $\hat{g}\left(x,y\right)$ represent wave modes or kinks and velocity respectively.

Boundary conditions will be assumed to be of the form

$$\frac{\partial u\left(x,y,t\right)}{\partial x} = 0 \text{ for } x = L_x^0 \text{ and } x = L_x^1 \, , \; L_y^0 < y < L_y^1 \tag{1.4}$$

and

$$\frac{\partial u\left(x,y,t\right)}{\partial y} = 0 \text{ for } y = L_y^0 \text{ and } y = L_y^1 \, , \; L_x^0 < x < L_x^1 \, , \tag{1.5}$$

when $t > t_0$.

2   The finite-difference method

2.1   Grid and solution vector

To obtain a numerical solution the region $R = \Omega \times [t > t_0]$ with its boundary ∂R consisting of the lines $x = L_x^0$, L_x^1, $y = L_y^0$, L_y^1 and $t = t_0$ is covered with a rectangular mesh, G, of points with coordinates $(x,y,t) = (x_k, y_m, t_n) = \left(L_x^0 + kh_x, L_y^0 + mh_y, t_0 + n\ell\right)$ with $k,m = 0,1,...,N+1$ and $n = 0,1,...$, in which $h_x = \left(L_x^1 - L_x^0\right)/(N+1)$ and $h_y = \left(L_y^1 - L_y^0\right)/(N+1)$ represent the discretization into $N+1$ subintervals of the space variables, while ℓ represents the discretization of the time variable. The solution of an approximating finite-difference scheme at the same point will be denoted by $u_{k,m}^n$: for the purpose of analyzing stability, the numerical value of actually obtained (subject, for instance, to computer round-off errors) will be denoted by $\tilde{u}_{k,m}^n$.

Let the solution vector be

$$\mathbf{u^n} = \mathbf{u}\left(t_n\right) = \mathbf{u}\left(t\right) = \left[u_{0,0}^n, u_{1,0}^n, ..., u_{N+1,0}^n; \; u_{0,1}^n, u_{1,1}^n, ..., u_{N+1,1}^n; \right.$$

$$\left. ...; \; u_{0,N+1}^n, u_{1,N+1}^n, ..., u_{N+1,N+1}^n \right]^T \tag{2.1}$$

T denoting transpose. Then there are $(N+2)^2$ values to be determined at each time step.

2.2   Development of the method

Replacing the space derivative in Eq. (1.1) by the familiar second-order central-difference approximant and applying Eq. (1.1) to all $(N+2)^2$ mesh points of the grid G at time level $t = n\ell$ with $n = 0,1,...$ subject to the boundary conditions given by Eq. (1.4)-(1.5) it leads to an initial-value problem of the form

$$D^2\mathbf{u}\left(t\right) + \rho D\mathbf{u}\left(t\right) = A\mathbf{u}\left(t\right) - \mathbf{G}\left(\mathbf{u}\left(t\right)\right)$$

$$\mathbf{u}\left(t_0\right) = \mathbf{g} \;, \; D\mathbf{u}\left(t_0\right) = \hat{\mathbf{g}} \;, \; t > 0, \tag{2.2}$$

where $D = \mathrm{diag}\left\{d/dt\right\}$ and $D^2 = \mathrm{diag}\left\{d^2/dt^2\right\}$ are diagonal matrices of order $\left(N+2\right)^2$, $\mathbf{G}\left(\mathbf{U}\left(t\right)\right)$ is a vector of order $\left(N+2\right)^2$ arising from the nonlinear term $\sin u$ of Eq. (1.1) and $A = h_x^{-2}B + h_y^{-2}C$ is a matrix of order $\left(N+2\right)^2$ with B a block diagonal matrix with tridiagonal blocks B_1 of order $N+2$ given by

$$B = \begin{bmatrix} B_1 & & & \\ & B_1 & & \\ & & \cdots & \\ & & & B_1 \end{bmatrix} \quad \text{with } B_1 = \begin{bmatrix} -2 & 2 & & & \\ 1 & -2 & 1 & & \\ & \cdot & \cdot & \cdot & \\ & & 1 & -2 & 1 \\ & & & 2 & -2 \end{bmatrix} \tag{2.3}$$

and C a block tridiagonal matrix with diagonal blocks given by

$$C = \begin{bmatrix} -2I & 2I & & & \\ I & -2I & I & & \\ & \cdot & \cdot & \cdot & \\ & & I & -2I & I \\ & & & 2I & -2I \end{bmatrix} \tag{2.4}$$

with I the identity matrix of order $N+2$.

Using the relations

$$\mathbf{u}\left(t \pm \ell\right) = \exp\left(\pm \ell D\right)\mathbf{u}\left(t\right) \tag{2.5}$$

it leads to the following three-time level recurrence relation for solving Eq. (1.1)

$$\mathbf{u}\left(t + \ell\right) = \left[\exp\left(\ell D\right) + \exp\left(-\ell D\right)\right]\mathbf{u}\left(t\right) - \mathbf{u}\left(t - \ell\right) \;; \; t = \ell, 2\ell, \ldots \tag{2.6}$$

The numerical methods will be developed by replacing the matrix-exponential term in the recurrence relation (2.6) by rational replacements, which are also known as the $\left(\mu, \nu\right)$ Padé approximants, of the form

$$\exp\left(\ell D\right) \approx \left(I + a_1 \ell D + b_1 \ell^2 D^2\right)^{-1}\left(I + c_1 \ell D + d_1 \ell^2 D^2\right) \tag{2.7}$$

with a_1, b_1, c_1 and d_1 parameters, which are real numbers having appropriate values for each type of approximant.

2.3 The (1,1) Padé approximant

From the family of methods arising from Eq. (2.6) using the expression (2.7) only the following *implicit* one for $a_1 = -1/2$, $b_1 = 0$, $c_1 = 1/2$ and $d_1 = 0$, which is also known as the Crank-Nickolson method, will be examined. Then

$$\left(\tilde{I} - \frac{1}{4}\ell^2 D^2\right)\left[\mathbf{u}\left(t + \ell\right) + \mathbf{u}\left(t - \ell\right)\right] = \left(2\tilde{I} + \frac{1}{2}\ell^2 D^2\right)\mathbf{u}\left(t\right)$$

or using Eq. (2.2) as

$$\left(\tilde{I} + \frac{1}{4}\ell^2 \rho D\right)\mathbf{u}\left(t + \ell\right) - \frac{1}{4}\ell^2\left(D^2 + \rho D\right)\mathbf{u}\left(t + \ell\right) = 2\left(\tilde{I} - \frac{1}{4}\ell^2 \rho D\right)\mathbf{u}\left(t\right)$$

$$+ \frac{1}{2}\ell^2\left(D^2 + \rho D\right)\mathbf{u}\left(t\right) - \left(\tilde{I} + \frac{1}{4}\ell^2 \rho D\right)\mathbf{u}\left(t - \ell\right) + \frac{1}{4}\ell^2\left(D^2 + \rho D\right)\mathbf{u}\left(t - \ell\right). \tag{2.8}$$

Then Eq. (2.8) using Eq. (2.2) and the approximations

$$D\mathbf{u}\,(t+\ell) = \frac{\mathbf{u}\,(t+\ell) - \mathbf{u}\,(t)}{\ell} + O\,(\ell) \text{ as } \ell \to 0, \tag{2.9}$$

$$D\mathbf{u}\,(t) = \frac{\mathbf{u}\,(t+\ell) - \mathbf{u}\,(t-\ell)}{2\ell} + O\,(\ell^2) \text{ as } \ell \to 0, \tag{2.10}$$

$$D\mathbf{u}\,(t-\ell) = \frac{\mathbf{u}\,(t) - \mathbf{u}\,(t-\ell)}{\ell} + O\,(\ell) \text{ as } \ell \to 0 \tag{2.11}$$

leads to the following three-time level *implicit* finite-difference scheme

$$\left(1 + \frac{1}{2}\ell\rho\right)\mathbf{u}\,(t+\ell) - \frac{1}{4}\ell^2 A\mathbf{u}\,(t+\ell) + \frac{1}{4}\ell^2 \mathbf{G}\,(\mathbf{u}\,(t+\ell))$$

$$= 2\mathbf{u}\,(t) + \frac{1}{2}\ell^2 A\mathbf{u}\,(t) - \frac{1}{2}\ell^2 \mathbf{G}\,(\mathbf{u}\,(t))$$

$$- \left(1 - \frac{1}{2}\ell\rho\right)\mathbf{u}\,(t-\ell) + \frac{1}{4}\ell^2 A\mathbf{u}\,(t-\ell) - \frac{1}{4}\ell^2 \mathbf{G}\,(\mathbf{u}\,(t-\ell)), \tag{2.12}$$

which forms the following nonlinear system

$$\mathbf{F}\,(\mathbf{u}\,(t+\ell)) = \mathbf{0} \tag{2.13}$$

for the unknown vector $\mathbf{u}\,(t+\ell) = \mathbf{u}^{n+1}$.

Let $r_x = \ell/h_x$, $r_y = \ell/h_y$, $p_x = \ell^2/h_x$ and $p_y = \ell^2/h_y$. Eq. (2.12), when applied to the general mesh point (x_k, y_m, t_n) of the grid G, gives the following three-time level implicit scheme for the unknown vector $\mathbf{u}^{n+1}$

$$\left(1 + \frac{1}{2}\ell\rho\right)u_{k,m}^{n+1} - \frac{1}{4}r_x^2\left(u_{k-1,m}^{n+1} - 2u_{k,m}^{n+1} + u_{k+1,m}^{n+1}\right) - \frac{1}{4}r_y^2\left(u_{k,m-1}^{n+1} - 2u_{k,m}^{n+1} + u_{k,m+1}^{n+1}\right)$$

$$+ \frac{1}{4}\ell^2\phi_{k,m}\sin u_{k,m}^{n+1} = 2u_{k,m}^n + \frac{1}{2}r_x^2\left(u_{k-1,m}^n - 2u_{k,m}^n + u_{k+1,m}^n\right)$$

$$+ \frac{1}{2}r_y^2\left(u_{k,m-1}^n - 2u_{k,m}^n + u_{k,m+1}^n\right) - \frac{1}{2}\ell^2\phi_{k,m}\sin u_{k,m}^n - \left(1 - \frac{1}{2}\ell\rho\right)u_{k,m}^{n-1}$$

$$+ \frac{1}{4}r_x^2\left(u_{k-1,m}^{n-1} - 2u_{k,m}^{n-1} + u_{k+1,m}^{n-1}\right) + \frac{1}{4}r_y^2\left(u_{k,m-1}^{n-1} - 2u_{k,m}^{n-1} + u_{k,m+1}^{n-1}\right) - \frac{1}{4}\ell^2\phi_{k,m}\sin u_{k,m}^{n-1}. \tag{2.14}$$

2.3.1 Local truncation error

The local truncation error of the method is

$$L\,(x,t) = -\frac{1}{12}\left(2\ell^2\frac{\partial^4 u}{\partial t^4} + h^2\frac{\partial^4 u}{\partial x^4}\right) - \frac{1}{720}\left(58\ell^2\frac{\partial^6 u}{\partial t^6} + h^4\frac{\partial^6 u}{\partial x^6}\right) + O\,(\ell^6 + h^6), \tag{2.15}$$

where the first term on the right-hand side in Eq. (2.15) is the principal part, which tends to zero as h, ℓ tend to zero simultaneously, so the method is also consistent with Eq. (1.1).

2.3.2 Stability analysis

The Fourier method of analyzing stability will be used. This method entails considering a small error

$$Z_m^n = u_{k,m}^n - \tilde{u}_{k,m}^n \tag{2.16}$$

of the form

$$Z_{k,m}^n = e^{\alpha n \ell} e^{i\beta k h_x} e^{i\gamma m h_y} \; ; \; i = \sqrt{-1}, \tag{2.17}$$

where α is complex number and β, γ are real and finding the criteria under which the von Neumann necessary criterion for stability

$$\left| e^{\alpha \ell} \right| \leq 1 + K\ell, \tag{2.18}$$

where K is a non-negative constant independent of h_x, h_y and ℓ is satisfied. Obviously Eq. (2.18) makes no allowance for growing solution if $K = 0$.

Then Eqs. (2.16)-(2.18), when applied to Eq. (2.14) after the Maclaurin's expansion

$$\sin u_{k,m}^n = \sum_{j=0}^{+\infty} (-1)^j \frac{\left(u_{k,m}^n \right)^{2j+1}}{(2j+1)!} \tag{2.19}$$

of the term $\sin u$, lead to

$$\left(1 + \frac{1}{2}\ell\rho\right) Z_{k,m}^{n+1} - \frac{1}{4}r_x^2 \left(Z_{k-1,m}^{n+1} - 2Z_{k,m}^{n+1} + Z_{k+1,m}^{n+1} \right) - \frac{1}{4}r_y^2 \left(Z_{k,m-1}^{n+1} - 2Z_{k,m}^{n+1} + Z_{k,m+1}^{n+1} \right)$$

$$+ \frac{1}{4}\ell^2 \, \phi_{k,m} \, S_{k,m}^{n+1} \, Z_{k,m}^{n+1} - 2Z_{k,m}^n - \frac{1}{2}r_x^2 \left(Z_{k-1,m}^n - 2Z_{k,m}^n + Z_{k+1,m}^n \right)$$

$$- \frac{1}{2}r_y^2 \left(Z_{k,m-1}^n - 2Z_{k,m}^n + Z_{k,m+1}^n \right) + \frac{1}{2}\ell^2 \phi_{k,m} \, S_{k,m}^{n+1} \, Z_{k,m}^n + \left(1 - \frac{1}{2}\ell\rho\right) Z_{k,m}^{n-1}$$

$$- \frac{1}{4}r_x^2 \left(Z_{k-1,m}^{n-1} - 2Z_{k,m}^{n-1} + Z_{k+1,m}^{n-1} \right) - \frac{1}{4}r_y^2 \left(Z_{k,m-1}^{n-1} - 2Z_{k,m}^{n-1} + Z_{k,m+1}^{n-1} \right)$$

$$+ \frac{1}{4}\ell^2 \phi_{k,m} \, S_{k,m}^{n+1} \, Z_{k,m}^{n-1} = 0, \tag{2.20}$$

where

$$S_{k,m}^\mu = \sum_{j=0}^{+\infty} \frac{(-1)^j}{(2j+1)!} \left[\left(u_{k,m}^\mu \right)^{2j} + \left(u_{k,m}^\mu \right)^{2j-1} \tilde{u}_{k,m}^\mu + \dots + \left(\tilde{u}_{k,m}^\mu \right)^{2j} \right]$$

$$\approx \sum_{j=0}^{\infty} \frac{(-1)^j}{(2j)!} (u_s)^{2j} = \cos u_s \tag{2.21}$$

with $\mu = n+1, n, n-1$ after linearization of the term in square brackets and $u_s = \max_{k,m=0,1,\dots,N+1} \left\{ u_{k,m}^0 \right\}$.
Eq. (2.20) using Eq. (2.17), after canceling both sides by $e^{\alpha n\ell} e^{i\beta k h} e^{i\gamma mh}$ and Euler's identity $e^{ix} = \cos x + i \sin x$, leads to the following stability equation

$$\left(1 + r_x^2 \sin^2 \frac{\beta h_x}{2} + r_y^2 \sin^2 \frac{\gamma h_y}{2} + \frac{\ell^2}{4}\phi_{k,m} \cos u_s + \frac{\ell\rho}{2} \right) \xi^2$$

$$- 2 \left(1 - r_x^2 \sin^2 \frac{\beta h_x}{2} - r_y^2 \sin^2 \frac{\gamma h_y}{2} - \frac{\ell^2}{4}\phi_{k,m} \cos u_s \right) \xi$$

$$+1 + r_x^2 \sin^2 \frac{\beta h_x}{2} + r_y^2 \sin^2 \frac{\gamma h_y}{2} + \frac{\ell^2}{4} \phi_{k,m} \cos u_s - \frac{\ell \rho}{2} = 0, \tag{2.22}$$

where $\xi = e^{\alpha \ell}$ is the amplification factor.

Eq. (2.22) is of the form

$$\check{A}\xi^2 - 2\check{B}\xi_2 + \check{C} = 0 \tag{2.23}$$

with $\check{A}, \check{B}, \check{C} \in \Re$. It can be assumed that $\cos u_s > 0$ and consequently that $\check{A} > 0$. Let ξ_1, ξ_2 be the roots of Eq. (2.23). Then $\xi_1 + \xi_2 = \check{B}/\check{A}$, so the von Neuman's necessary criterion for stability given by Ineq. (2.18) will be always satisfied, when

$$\left| \check{B} \right| \leq \check{A}. \tag{2.24}$$

From Ineq. (2.24) using known properties it follows that if $\rho = 0$, then

$$\ell \leq \left\{ 2 \left[\frac{4}{3} \left(\frac{1}{h_x^2} + \frac{1}{h_y^2} \right) + \frac{\tilde{\phi}}{3} \right]^{-1} \right\}^{1/2}, \tag{2.25}$$

where

$$\tilde{\phi} = \sup_{(x,y) \in [L_x^0, L_x^1] \times [L_y^0, L_y^1]} |\phi(x,y)| \tag{2.26}$$

with $\tilde{\phi} < +\infty$, while, when $\rho > 0$

$$M\ell^2 + \frac{\ell \rho}{2} - 2 \leq 0 \tag{2.27}$$

with $M = \frac{4}{3} \left(\frac{1}{h_x^2} + \frac{1}{h_y^2} \right) + \frac{\tilde{\phi}}{3}$ and $M > 0$. Let ℓ_1, ℓ_2 be the roots of the relevant to Ineq. (2.27) equation. Obviously $\ell_1 \ell_2 < 0$, so both the roots are real and distinct. If $\Delta = \frac{1}{4}\rho^2 + 8M$ is the discriminant of Eq. (2.27) and $\ell_1 = \frac{-\rho/2 - \sqrt{\Delta}}{2M}, \ell_2 = \frac{-\rho/2 + \sqrt{\Delta}}{2M}$ are the roots, then Ineq. (2.27) leads to the following restriction for the time step

$$\ell \leq \ell_2. \tag{2.28}$$

2.4 The Predictor-Corrector scheme

To avoid solving the nonlinear system arising from Eq. (2.13) the following *Predictor-Corrector* scheme was used.

2.4.1 Predictor

Following Bratsos [3] the Predictor scheme is

$$\hat{u}_{k,m}^{n+1} = 2u_{k,m}^n - u_{k,m}^{n-1} + r_x^2 \left(u_{k+1,m}^n - 2u_{k,m}^n + u_{k-1,m}^n \right)$$

$$+r_y^2 \left(u_{k,m+1}^n - 2u_{k,m}^n + u_{k,m-1}^n \right) - \ell^2 \phi_{k,m} \sin u_{k,m}^n, \tag{2.29}$$

which is explicit for the unknown vector $\hat{u}^{n+1}$.

From the stability analysis of the predictor scheme it follows the following restriction for the time step

$$\ell \leq \left(\frac{1}{4} \tilde{\phi} + \frac{1}{h_x^2} + \frac{1}{h_y^2} \right)^{-1/2}. \tag{2.30}$$

2.4.2 Corrector

It is proposed the following scheme

$$\left(1+\frac{1}{2}\ell\rho\right)\mathbf{u}\left(t+\ell\right)=\frac{1}{4}\ell^{2}A\hat{\mathbf{u}}\left(t+\ell\right)-\frac{1}{4}\ell^{2}\mathbf{G}\left(\hat{\mathbf{u}}\left(t+\ell\right)\right)$$

$$+2\mathbf{u}\left(t\right)+\frac{1}{2}\ell^{2}A\mathbf{u}\left(t\right)-\frac{1}{2}\ell^{2}\mathbf{G}\left(\mathbf{u}\left(t\right)\right)$$

$$-\left(1-\frac{1}{2}\ell\rho\right)\mathbf{u}\left(t-\ell\right)+\frac{1}{4}\ell^{2}A\mathbf{u}\left(t-\ell\right)-\frac{1}{4}\ell^{2}\mathbf{G}\left(\mathbf{u}\left(t-\ell\right)\right). \tag{2.31}$$

3   Numerical results

In this section it is examined the behavior of the finite-difference scheme arising from Eq. (2.14) to selected examples used by Christiansen and Lombahl [4], Argyris et *al* [2], Djidjeli et *al* [7] and Bratsos [3]. In all experiments it was used $t_0=0$, $h_x=h_y=h=0.2$ for the space and $\ell=0.01$ for the time step.

3.1   Superposition of two orthogonal line solitons

Following Christiansen and Lombahl [4], if one line soliton operates along the x-axis and the other on the y-axis, an exact solution to the sine-Gordon equation can be obtained. Numerical solutions for the case of $\phi\left(x,y\right)=1$ with initial conditions

$$g\left(x,y\right)=4\left[\tan^{-1}\left(\exp x\right)+\tan^{-1}\left(\exp y\right)\right];-6\leq x,y\leq6 \tag{3.1}$$

and

$$\hat{g}\left(x,y\right)=0;\quad-6\leq x,y\leq6 \tag{3.2}$$

have been calculated and are presented in Figs. 1-3 at $t=0$, 1 and 3, 4 when $\rho=0$ and in Fig. 4 at $t=3$ when $\rho=1$ and 0.3.

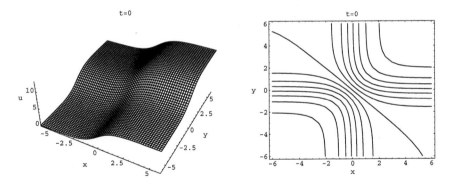

Figure 1: Line solitons. Superposition of two orthogonal line solitons at $t=0$ when $\rho=0$.

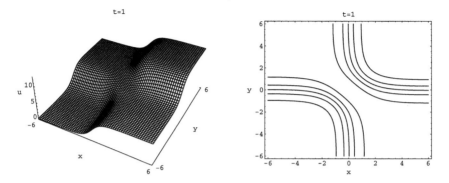

Figure 2: Line solitons. Superposition of two orthogonal line solitons at $t = 1$ when $\rho = 0$.

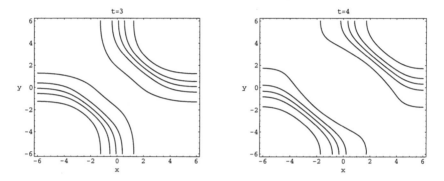

Figure 3: Line solitons. Superposition of two orthogonal line solitons at $t = 3$ and $t = 4$ when $\rho = 0$.

3.2 Ring solitons

3.2.1 Circular ring solitons

Circular ring solitons are obtained for $\phi(x, y) = 1$ with initial conditions

$$g(x, y) = 4 \tan^{-1} \exp\left(3 - \sqrt{x^2 + y^2}\right) \; ; -7 \le x, y \le 7 \tag{3.3}$$

and

$$\hat{g}(x, y) = 0 \; ; \; -7 \le x, y \le 7. \tag{3.4}$$

Figs. 5-8 show circular ring solitons at $t = 0$, 2.8, 5.6 and 8.4 when $\rho = 0$ in terms of $z = \sin(u/2)$.

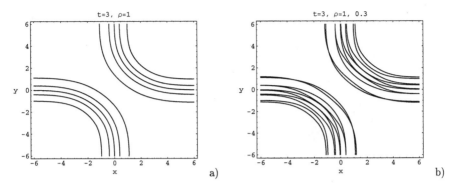

Figure 4: Line solitons. In Fig. 4a the contour shows the superposition of two orthogonal line solitons at $t = 3$ when $\rho = 1$, while in Fig. 4b the joined contours at $t = 3$ when $\rho = 1, 0.3$.

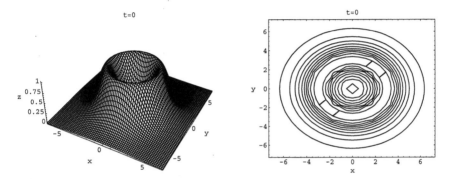

Figure 5: Ring solitons. Circular ring soliton at $t = 0$ when $\rho = 0$.

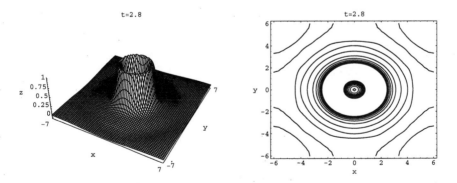

Figure 6: Ring solitons. Circular ring soliton at $t = 2.8$ when $\rho = 0$.

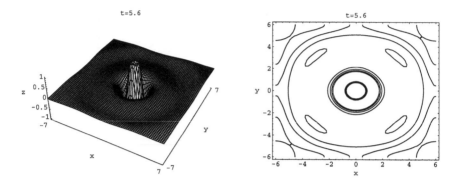

Figure 7: Ring solitons. Circular ring soliton at $t = 5.6$ when $\rho = 0$.

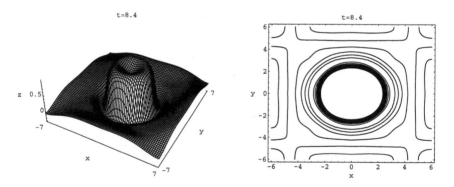

Figure 8: Ring solitons. Circular ring soliton at $t = 8.4$ when $\rho = 0$.

4   Conclusions

An implicit finite-difference method arising from the use of the method of lines based on rational approximant to the matrix-exponential term it has been proposed for the numerical solution of the two-dimensional Sine-Gordon equation. Even if the method leads to a nonlinear systeme, a new method introduced by Bratsos [3], which uses a predictor-corrector procedure with two explicit schemes was successfully applied. The proposed method was tested to a variety of known from the bibliography analogous problems and the results have shown a very good performance.

Acknowledgements

This research was co-funded by 75% from E.E. and 25% from the Greek Government under the framework of the Education and Initial Vocational Training Program - Archimedes, Technological Educational Institution (T.E.I.) Athens project *"Computational Methods for Applied Technological Problems"*.

References

[1] M.J. Ablowitz and H. Segur, *Solitons and the Inverse Scattering Transform*, SIAM Studies in Applied Mathematics 4, Society for Industrial and Applied Mathematics, Philadelphia, 1981.

[2] J. Argyris, M. Haase and J.C. Heinrich, Finite element approximation to two-dimensional Sine-Gordon solitons, *Computer Methods in Applied Mechanics and Engineering*, **86** 1-26(1991).

[3] A.G. Bratsos, An explicit numerical scheme for the Sine-Gordon equation in 2+1 dimensions, *Appl. Num. Anal. Comp. Math.* **2** No. 2, 189-211(2005).

[4] P.L. Christiansen and P.S. Lomdahl, Numerical solutions of 2+1 dimensional Sine-Gordon solitons, *Physica 2D* 482-494(1981).

[5] P.L. Christiansen and O.H. Olsen, On Dynamical Two-Dimensional Solutions to the Sine-Gordon Equation, *Z. Angew. Math. Mech.* **59** T30(1979).

[6] P.L. Christiansen and O.H. Olsen, Ring-shaped quasi-soliton solutions to the two and three-dimensional Sine-Gordon equations, *Physica Scripta 20* 531-538(1979).

[7] K. Djidjeli, W.G. Price and E.H. Twizell, Numerical solutions of a damped Sine-Gordon equation in two space variables, *Journal of Engineering Mathematics* **29** 347-369(1995).

[8] R.K. Dodd, J.C. Eilbeck, J.D. Gibbons and H.C. Morris, *Solitons and Nonlinear Wave Equations*, London Academic Press, 1982.

[9] J.D. Gibbon and G. Zambotti, The Interaction of *n*-Dimensional Soliton Wave Fronts, *Nuovo Cimento* **25**B, 1(1976).

[10] G. Grella and M. Marinaro, Special Solution of the Sine-Gordon Equation in 2+1 Dimensions, *Lett. Nuovo Cimento* **23** 459(1978).

[11] R. Hirota, Exact three-soliton solution of the two-dimensional Sine-Gordon equation, *J. Phys. Soc. Japan* **35** 1566(1973).

[12] P. Kaliappan and M. Lakshmanan, Kadomtsev-Petviashvili and two-dimensional sine-Gordon equations: reduction to Painlevè transcendents, *J. Phys. A: Math. Gen.* **12** L249(1979).

[13] K.K. Kobayashi and M. Izutsu, Exact Solution of the n-Dimensional Sine-Gordon Equation, *J. Phys. Soc. Japan* **41** 1091(1976).

[14] G. Leibbrandt, New exact solutions of the classical Sine-Gordon equation in $2+1$ and $3+1$ dimensions, *Phys. Rev. Lett.* **41** 435-438(1978).

[15] K. Nakajima, Y. Onodera, T. Nakamura and R. Sato, Numerical analysis of vortex motion on Josephson structures, *Journal of Applied Physics* **45**(9) 4095-4099(1974).

[16] J. Zagrodzinsky, Particular solutions of the Sine-Gordon equation in $2+1$ dimensions, *Phys. Lett.* **72**A 284-286(1979).

Brill Academic Publishers
P.O. Box 9000, 2300 PA Leiden
The Netherlands

Lecture Series on Computer
and Computational Sciences
Volume 3, 2005, pp. 321-332

Comparison of Fuzzy-based MCDM and Non-fuzzy MCDM in Setting a New Fee Schedule for Orthopedic Procedures in Taiwan's National Health Insurance program

Chih-Young Hung[1][1] Yuan-Huei Huang[1,2] Pei-Yeh Chang[3]

[1]Institute of Management of Technology, National Chiao Tung University
[2]Department of Surgery, Hsin-Chu General Hospital, Department of Health
[3]Department of Pediatric Surgery, Chang Gung Children's Hospital

Received 10 August, 2005; accepted 12 August 2005

Abstract: In this paper, we for the first time propose a fuzzy-based multiple criteria decision-making (FMCDM) in setting a new fee schedule for orthopedic procedures in a national health insurance program. Among numerous approaches and techniques documented in the literature for setting fee schedule, the most frequently adopted approach is experts' opinions. Alternative methods within this approach include: in-depth interviews, focus group, nominal group method, and the Delphi method. Regrettably, these various methods reach qualitative conclusions mainly. Such a deficiency brings great difficulties in quantifying and summarizing practical for setting a fee schedule. To mitigate these weaknesses, this research proposes a FMCDM framework that allows experts to express their opinions more precisely than with traditional methods. The results of the illustrative case show that the new fee schedule as obtained from applying the FMCDM model is more convincing than previous schedule and more persuasive to the references for the policy setting. In addition, it has one major advantage over the traditional non-fuzzy based counterpart. The method eliminated the undesirable occurrences of argument among experts in the research process that hinder the progress of research.

Keywords: fee schedule, orthopedic procedures, national health insurance, fuzzy multiple criteria decision making

Mathematics Subject Classification: AMS-MOS

AMS-MOS: 03E72 Fuzzy set theory

1. Introduction

Over the past two decades, the Health Care Financing Administration of the United States (US) has periodically sponsored comprehensive national surveys of physician payment systems in the Medicare and Medicaid programs (Schieber and Langenbrunner, 1989). Not only the US, but also many other countries with health insurance programs review payment systems periodically. The main reasons are to address continually rising health care costs (Huber and Orosz, 2003) and to ensure an equitable allocation of resources (Hsiao, Braun, Dunn, et al., 1988a, 1988b). Taiwan's National Health Insurance (NHI) program was established on 1995(Chiang, 1997). Following the implementation of NHI, the ratio of overall payments to surgical specialties, including orthopedic, was quite low due to many innate restrictions of the fee schedule in the payment systems. In 2001, the ratio of annual surgical expenditures was only 6.18 percent of the total health spending by the Bureau of National Health Insurance (BNHI)(Huang et al, 2004). In contrast to the Medicare system implemented in the United States, the surgical expenditure ratio accounted for 18.3 percent in 1991, and 17.8 percent in 1995 (Folland, 2001).

Although the payment systems of BNHI experienced adjustment many times after its initial implementation, some serious problems from previous programs (such as Labor Insurance and Government Employee Insurance) are still unavoidable. The inherent structure causing irrational fee schedule biases has not changed from past adjustments (Cheng and Chiang, 1998). BNHI's fee

---

[1] Corresponding author. Director of Institute of Technology Management in the Faculty of Management, National Chiao Tung University, Taiwan. E-mail: cyhung@cc.nctu.edu.tw

schedules are established jointly by the insurer and the contracted medical care institutions. However, there are apparent conflicts between these two negotiating parties due to their opposing interests. Because of the lack of objective data and suitable arbitration points, it is very difficult to achieve mutual agreement. Only with support by objective evidence or fair arbitration can the two parties come to mutual satisfaction. Therefore, the core issue is to design a fair method for setting fee schedule that can effectively integrate opinions from different standpoints.

Among numerous methods and techniques documented in the literature for setting fee schedules, the most frequently adopted approach is experts' opinions. Related techniques include: in-depth interviews (Giacomini and Cook, 2000), focus group (Powell and Single, 1996), nominal group method (Pope and Mays, 1995), and the Delphi method (Gupta and Clarke, 1996). Regrettably, these research methods, which range from personal interviews to extensive negotiations, reach qualitative conclusions mainly (Chapple and Rogers, 1998; Patricia, 1996; Pope and Mays, 1995). Such a deficiency brings great difficulties in quantifying and summarizing practical for setting a fee schedule.

To improve previous weaknesses, this research aims to design a fuzzy-based multiple criteria decision-making (FMCDM) framework to precisely capture experts' opinions. The FMCDM framework includes: the Saaty's analytical hierarchy process (AHP) (Saaty, 1977, 1980; Belton and Gear, 1983), fuzzy approach (Zadeh, 1965, 1975; Dubis and Prade, 1978), and multiple criteria decision-making (MCDM) techniques (Bellman and Zadeh, 1970; Tzeng and Teng, 1994). FMCDM analysis has been used in different areas as it succeeds in quantifying expert opinion systematically (Bohanec et al, 2000; Huang and Wu, 2005; Sylla and Wen, 2002; Tang and Tzeng, 1999; Teng and Tzeng, 1996). Through application, this research coordinates the opinions of orthopedic surgeons in the Taiwan Surgical Association to establish a new fee schedule for orthopedic procedures by both fuzzy-based and non-fuzzy MCDM approaches.

2. Methods

In a simple environment or using a single measurement index, the traditional minimum cost, maximum profit or the cost efficiency methods can be employed to conduct alternative evaluation. However, in an increasingly complex and diversified decision making environment, there is much correlated information that needs to be analyzed and traditional analysis is not suitable for problem solving (Bellman and Zadeh, 1970; Tzeng and Teng, 1994). Therefore, this research uses the MCDM to evaluate different relative values of orthopedic procedures in the new fee schedule. In 1989, Hsiao's Resource Based Relative Value Scale (RBRVS) changed the Medicare fee schedule for physician payment, making reimbursement dependent on the time, skill, and resources needed to deliver service. (Hsiao, Braun, Dunn, et al., 1988a, 1988b). RBRVS could be said a fee schedule which was setting based on multiple criteria decision making, however, it was expensive to building up (Schieber and Langenbrunner, 1989).

Since evaluators may have different perceptions on different objectives and criteria, in terms of their importance and the possible adverse consequence, the evaluation is conducted in an uncertain and fuzzy environment. This fuzzy evaluation design allows evaluators to express their opinions in fuzzy expression manners. Based on the above reasons, the FMCDM is selected to conduct this evaluation ((Bellman and Zadeh, 1970; Buckley, 1985).

We examined the fee schedule for orthopedic procedures in the draft announced by the BNHI authority on March, 21, 2003, and designed a study to obtain a new fee schedule. The study was conducted in three steps: first, to classify all existing orthopedic procedures into 12 groups through questionnaire survey of experts' opinions; second, to construct the main FMCDM framework; and finally, to perform an empirical study of the FMCDM and non-fuzzy models and obtain a new fee schedule.

2-1 Questionnaire survey of orthopedic experts' opinions

In establishing a fee schedule, the opinions of experts with special expertise are a valuable and often important source of qualitative data analysis. Nine senior surgeons from the Taiwan Surgical Association were requested to express their opinions about classifying existing orthopedic procedures promulgated by the BNHI into groups. As a result, all 194 orthopedic procedures were classified into 12 groups according to the similarity of characteristics of the organ systems, operation sites, and the difficulty of techniques of each procedure. Finally, one procedure was chosen from each group as a representative procedure. The content of classified groups is shown in Table 1.

Table 1 The classified groups of orthopedic procedures and the representative procedure.

Classified Group	Items	Representative procedure
1.General orthopedic procedures	24	Bone or osteochondral graft
2. Procedures of tendon.	21	Rupture of Achilles tendon primary suture
3. Open reductions.	33	Open reduction for fracture of tibia
4. Close reductions.	17	Close reduction for fracture of tibia, humerus
5. Amputations	15	Amputation of limbs--leg, arm, forearm
6.Procedures of anthroplastic	13	Cord decompression for ANFH (trephing)
7. Procedures of artificial joint.	17	Total hip replacement
8.Procedures of nerve and spine	21	Diskectomy-lumbar
9.Procedures of tumors	6	Wide excision-bone tumor, malignant
10. Procedures of sport medicine	9	Rotator cuff tear repair—massive
11.Procedures of hand, foot, and ankle	9	Hallux valgus (Chevron)
12.Pediatric orthopedic procedures	9	Congenital dislocation of hips-open reduction

2-2 Constructing the main FMCDM framework

In considering the proportion of different orthopedic procedures, surgeons of different sub-specialties diverge greatly in converting practice costs, risks, workload and specialty training costs into fee schedule (Hsiao, Braun, Dunn, et al., 1988a, 1988b). In order to integrate these hetero-perceptions, we adopted a fuzzy multiple criteria decision-making (FMCDM) model, which was introduced by Bellman and Zadeh in 1970. The FMCDM model for establishing the new fee schedule for orthopedic procedures, as illustrated in Figure 1, contains four evaluation aspects, sixteen criteria, and eleven representative procedures. The resource-based relative value scale (RBRVS) used in US Medicare is the blueprint for the evaluation aspects and criteria.

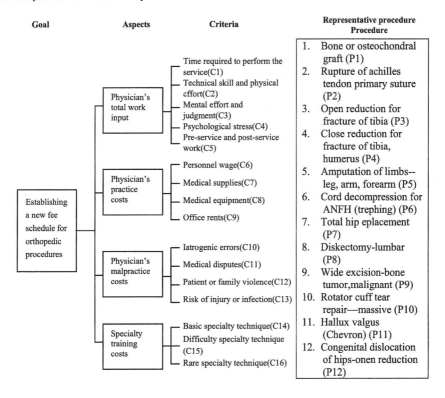

Figure 1. The fuzzy multiple criteria decision making framework for establishing the new fee schedule for orthopedic procedures.

The evaluation process includes two steps:

2-2-1 Using AHP to Evaluate the relative weightings

The AHP weighting is determined by the evaluators who conduct pairwise comparisons, so as to reveal the comparative importance of two criteria. If there are evaluation aspects /criteria, then the decision-makers have to conduct pairwise comparison. Moreover, the relative importance derived from these pairwise comparisons allows a certain degree of inconsistency within a domain. Saaty used the principle eigenvector of the pairwise comparison matrix derived from the scaling ration to find the comparative weight among the criteria of the hierarchy system (Saaty, 1977, 1980).

Suppose we wish to compare a set of n aspects/criteria in pairs according to their relative importance (weights). Denote the aspects /criteria by $c_1, c_2,...,c_n$ and their weights by $w_1, w_2,..., w_n$. If $W = (w_1, w_2,..., w_n)^t$ is given, the pairwise comparison may be represented by a matrix A of the following formula:

$$(A - \lambda_{max} I) W = 0 \tag{1}$$

Equation (1) denotes that A is a matrix of pairwise comparison values derived from intuitive judgment (perception) for ranking order. In order to find the principle eigenvector, we must find the eigenvector w_i with respective λ_{max} which satisfies $AW = \lambda_{max} W$. The comparative importance derived from the pairwise comparisons allows a certain degree of inconsistency within a domain. Saaty used the principle eigenvector of the pairwise comparison matrix contrived by scaling ration to find the comparative weight among the criteria (Saaty, 1977, 1980).

2-2-2 Obtain the performance value

The evaluators choose a score for each surgical procedure based on their subjective judgment. By doing this, we can use the methods of fuzzy theory to estimate the payment level of each procedure in a fuzzy environment. Since Zadeh introduced fuzzy set theory (Zadeh, 1965), and Bellman and Zadeh (1970) described the decision-making method in fuzzy environments, an increasing number of studies have dealt with uncertain fuzzy problems by applying fuzzy set theory. The application of fuzzy theory to get the performance values can be described as follows:

(i) fuzzy numbers:

According to the definition made by Dubis and Prade (1978), the fuzzy number $\tilde{A}$ is a fuzzy set, and its membership function is $\mu\tilde{A}(x)$: $R \rightarrow [0,1]$, where x represents the policy tools. It is common to use triangular fuzzy numbers (TFNs) $\mu\tilde{A}(x) = (L, M, U)$, for fuzzy operations, as shown in equation (2) and Figure 2.

$$\mu_{\tilde{A}}(x) = \begin{cases} (x-L)/(M-L) & \text{if } L \leq x \leq M \\ (U-x)/(U-M) & \text{if } M \leq x \leq U \\ 0 & \text{otherwise} \end{cases} \tag{2}$$

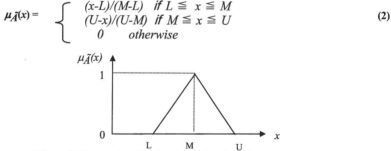

Figure 2. The membership function of the triangular fuzzy number

According to the nature of triangular fuzzy numbers and the extension principle put forward by Zadeh (1965), the algebraic calculation of the triangular fuzzy number can be displayed as follows:

- Addition of a fuzzy number ⊕

$$(L_1, M_1, U_1) \oplus (L_2, M_2, U_2) = (L_1+L_2, M_1+M_2, U_1+U_2) \tag{3}$$

-Multiplication of a fuzzy number ⊡

$$(L_1, M_1, U_1) \square (L_2, M_2, U_2) = (L_1 L_2, M_1 M_2, U_1 U_2) \tag{4}$$

-Any real number k:

$$k \square \mu_{\tilde{A}}(x) = k \square (L, M, U) = (kL, kM, kU) \tag{5}$$

-Subtraction of a fuzzy number ⊖

$$(L_1, M_1, U_1)\ominus(L_2, M_2, U_2) = (L_1-L_2, M_1-M_2, U_1-U_2) \qquad (6)$$

-Division of a fuzzy number $\oslash$

$$(L_1, M_1, U_1)\oslash (L_2, M_2, U_2) = (L_1/L_2, M_1/M_2, U_1/U_2) \qquad (7)$$

(ii) Linguistic Variable:

A linguistic variable is a variable whose values are words or sentences in a natural or artificial language. For example, the expression "maximize technical skill and physical effort" and "mental effort and judgment" represents a linguistic variable in the context in these problems. Linguistic variables may take on effect-values such as "very high, "high", "fair", "low", and "very low". The use of linguistic variables is rather widespread at present and the linguistic effect values of orthopedic procedures found in this study are primarily used to assess the linguistic ratings given by the evaluators. Furthermore, linguistic variables are used as a way to measure the performance value for each aspect/criterion (Zadeh, 1965, 1975).

(iii) Fuzzy multiple criteria decision-making:

Bellman and Zadeh (1970) were the first to probe into the decision-making problem under a fuzzy environment, and they heralded the initiation of FMCDM. This study uses this method to evaluate various orthopedic procedures and ranks each procedure according to its score. The following will be the method and procedures of the FMCDM theory. One is management of evaluation criteria. Under the measurement of linguistic variables to demonstrate the criteria performance by expressions such as "very high", "high", "fair", "low" and "very low", the evaluators were asked to conduct their judgments and each linguistic variable can be indicated by a triangular fuzzy number (TFN) within the scale range of 0-100. Also, the evaluators can subjectively assume their personal range of the linguistic variable (Buckley 1985).

Let E_{ij}^{k} indicate the fuzzy performance value of evaluator k toward procedure i under aspect /criterion j, and let the performance of the aspect /criterion be indicated by the set S, then,

$$E_{ii}^{k} = (LE_{ij}^{k}, ME_{ij}^{k}, UE_{ij}^{k}), j \in S \qquad (8)$$

Since the perception of each evaluator varies according to the evaluator's experience and knowledge, and the definitions of the linguistic variables vary as well, the study uses the notion of average value so as to integrate the fuzzy judgment value of m evaluators, that is,

$$E_{ij} = (1/m)\square(E_{ij}^{1} \oplus E_{ij}^{2} ...\oplus E_{ij}^{k}) \qquad (9)$$

The sign $\square$ indicates fuzzy multiplication, the sign $\oplus$ denotes fuzzy addition, E_{ij} is the average fuzzy number of the judgment of the decision-maker, and it can be displayed by a triangular fuzzy number as follows:

$$E_{ij} = (LE_{ij}, ME_{ij}, UE_{ij}) \qquad (10)$$

The proceeding end-point values

$$LE_{ij} = (1/m)\square(\sum_{k=1}^{m} LE_{ij}^{k}), \quad ME_{ij} = (1/m)\square(\sum_{k=1}^{m} ME_{ij}^{k}), \quad UE_{ij} = (1/m)\square(\sum_{k=1}^{m} UE_{ij}^{k})$$

can be solved by the method introduced by Buckley (1985).

(iv) Fuzzy synthetic decision:

The weights of each aspect /criterion of orthopedic procedures as well as fuzzy performance values has to be integrated by the calculation of fuzzy numbers so as to be located at the fuzzy performance value of the integral evaluation, which is of the procedures of fuzzy synthetic decision. According to the weight w_i derived by AHP, the weight vector can be obtained while the fuzzy performance matrix E of each of the alternatives can also be obtained from the fuzzy performance value of each alternative under n criteria, that is,

$$W = (w_1,..., w_j,..., w_n)^t \qquad (11)$$
$$E = (E_{ij}) \qquad (12)$$

$$R = E \Leftrightarrow \mathbf{W} \tag{13}$$

and the sign "$\Leftrightarrow$" indicates the operation of the fuzzy numbers, including addition and multiplication. Since the operation of fuzzy multiplication is rather complex, it is usually denoted by the approximate multiplied result of the fuzzy multiplication and the approximate fuzzy number R, of the synthetic decision of each procedure. The expression then becomes,

$$R_i = (LR_i, MR_i, UR_i), \forall i \tag{14}$$

Where

$$LR_i = \sum_{k=1}^{m} LE_{ij} \square w_j \tag{15}$$

$$MR_i = \sum_{k=1}^{m} ME_{ij} \square w_i \tag{16}$$

$$UR_i = \sum_{k=1}^{m} UE_{ij} \square w_i \tag{17}$$

(v) Ranking the procedures (fuzzy number):

The result of fuzzy synthetic decision of each procedure is a fuzzy number. Therefore, it is necessary that the non-fuzzy ranking method for fuzzy numbers be employed during the comparison of the strategies. In other words, the procedure of defuzzification is to locate the best non-fuzzy performance value (BNP). Methods of such defuzzified fuzzy ranking generally include mean of maximal (MOM), center of area (COA), and α-cut, three kinds of method (Zhau & Govind, 1991; Teng and Tzeng, 1996; Tang and Tzeng, 1999). To utilize the COA method to find out the BNP is a simple and practical method and there is no need to bring in the preference of any evaluators. For those reasons, the COA method is used in this study. The BNP value of the fuzzy number Ri can be found by the following equation:

$$BNP_i = [(UR_i - LR_i) + (MR_i - LR_i)] / 3 + LR_i, \forall i \tag{18}$$

According to the value of the derived BNP, the evaluation of each orthopedic procedure can then proceed.

2-3 Empirical study of FMCDM model

A temporary Board meeting of the Taiwan Surgical Association was held on June 18, 2003. Thirty-five orthopedic surgeons participated in answering our FMCDM questionnaire. Based on the FMCDM framework, every expert could express his subjective preference in pairs of comparisons among the hierarchy system. These subjective and qualitative judgments were then transformed through mathematical programming into objective and quantitative results.

Before answering the questionnaire, we fully explained the purpose and background of this research to every participant, as well as the steps and sources of the FMCDM model. In addition, all other information was provided, such as the Medicare's RBRVS schedule, the current payment scheme, the proposal draft by the BHI authority, and the official declaration of surgical procedures from previous years.

3. Results

100% (35/35) of the surveys were returned by respondents, and 88% (31/35) of those returned were valid. The invalid surveys were those that could not pass the AHP consistency verification tests or weren't completely answered.

3-1 Evaluating the criteria weights

See Table 2 for the analysis of relative weightings in valid survey responses. In the first-tier evaluation aspects, the rankings of factors among the 31 surgery respondents were: (1) physician's total work input (0.475); (2) specialty training costs (0.190); (3) physician's practice costs (0.185); and (4) physician's malpractice costs (0.150). In the second-tier valuation criteria, of the 16 factors, the top five ranked as follows: (1) Time required to perform the service (0.138); (2) mental effort and judgment

(0.106); (3) technical skill and physical effort (0.091); (4) personnel wage (0.083); and (5) psychological stress (0.076).

Table 2 The relative weightings in 31 valid survey responses in orthopedics.

Aspects and Criteria	First-tier evaluation (Aspects)	Second-tier evaluation (Criteria)
Physician's total work input	**0.475(1)**	
Time required to perform the service		**0.138(1)**
Technical skill and physical effort		**0.091(3)**
Mental effort and judgment		**0.106(2)**
Psychological stress		**0.076(5)**
Pre-service and post-service work		0.063
Physician's practice costs	**0.185(3)**	
Personnel wage		**0.083(4)**
Medical supplies		0.045
Medical equipments		0.041
Office rents		0.016
Physician's malpractice costs	**0.150(4)**	
Iatrogenic errors		0.036
Medical disputes		0.046
Patient or family violence		0.027
Risk of injury or infection		0.040
Specialty training costs	**0.190(2)**	
Basic specialty technique		0.062
Difficulty specialty technique		0.066
Rare specialty technique		0.061

3-2 Estimating the performance matrix

The relative weighting of a total of 16 criteria can also be interpreted as a vector and ranked as follow: $\bar{W}$ = (0.138, 0.091, 0.106, 0.076, 0.063, 0.083, 0.045, 0.041, 0.016, 0.036, 0.046, 0.027, 0.04, 0.062, 0.066, 0.061). The subjective value on the 12 representative procedures of the respondent i can be expressed as another vector $U_{\text{respondent }i}$ (i = 1~31). $\bar{W}$ is then multiplied with $U_{\text{respondent }i}$, and a synthetic value matrix on the new fee schedule is obtained as the equation:

New fee schedule = $\bar{W} \times U_{\text{respondent }i}$

$$= \begin{bmatrix} \text{Criterion 1} \\ \text{Criterion 2} \\ .... \\ .... \\ \text{Criterion 16} \end{bmatrix} \begin{bmatrix} \text{Procedure 1, Procedure 2,..........., Procedure 12} \end{bmatrix}$$

$$= \begin{bmatrix} u_{i,1}^1 & u_{i,1}^2 & \cdot \cdot \cdot & u_{i,1}^{12} \\ u_{i,2}^1 & u_{i,2}^2 & \cdot \cdot \cdot & u_{i,2}^{12} \\ \cdot \cdot \cdot & \cdot \cdot \cdot & \cdot \cdot \cdot \cdot & \cdot \cdot \cdot \\ \cdot \cdot \cdot & \cdot \cdot \cdot & \cdot \cdot \cdot \cdot & \cdot \cdot \cdot \\ u_{i,16}^1 & u_{i,16}^2 & \cdot \cdot & u_{i,16}^{12} \end{bmatrix}$$

$$= [u_{i1}, u_{i2}, u_{i3}, ... u_{i12}] \tag{19}$$

Where u_{i1} represents the new fee schedule of respondent i regarding procedure 1; u_{i2} represents the new fee schedule of respondent i regarding procedure 2; ...and u_{i12} represents the new fee schedule of respondent i regarding procedure 12. The results composed by their average of 31 experts are shown in Table 3 to Table 6.

Two sets of vector $U_{respondent\ i}$ $(i = 1\sim31)$ are proposed, one is fuzzy-based, which experts can express their own subject preference value freely according the rule of fuzzy set theory; the other is non-fuzzy MCDM, which the value of utility is assigned prospectively.

3-3 Comparison of fuzzy MCDM and non-fuzzy MCDM

From the results of Table 3 to Table 6, we can see the synthetic value matrix obtained form fuzzy-based MCDM are no significant difference to that of non-fuzzy MCDM. However, in the fuzzy-based matrix, the utility value can express freely by each experts, and can got a crisp value, or a range of value according to the different defuzzification methods. This feature gives more elastic decision-making space to the policy makers when setting a fee schedule, and increases the successful rate of negotiation.

Table 3. Summary of the fuzzy-based synthetic value matrix composed by the average of 31 experts.

Procedure Criteria	P1	P2	P3	P4	P5	P6	P7	P8	P9	P10	P11	P12
C1	3.68	5.56	7.01	4.04	6.08	5.81	7.83	7.04	10.27	7.71	4.96	6.73
C2	3.04	3.86	4.74	3.54	4.16	4.57	5.84	5.22	7.15	5.58	4.37	5.78
C3	4.49	5.37	6.02	4.26	5.51	6.08	8.31	7.54	8.95	7.58	6.24	8.30
C4	2.31	3.03	3.61	2.39	3.23	3.42	4.69	4.75	5.87	4.03	3.59	4.68
C5	1.93	2.05	2.50	2.08	2.85	2.89	3.62	3.47	4.32	2.91	2.11	3.61
C6	3.18	3.66	4.37	2.75	4.16	4.41	5.46	4.71	6.16	4.35	3.78	5.09
C7	1.74	1.84	3.02	1.59	2.07	2.17	3.61	2.56	3.15	2.36	1.98	2.34
C8	1.15	1.58	1.89	1.09	1.60	1.54	2.25	2.05	3.01	1.93	1.60	2.18
C9	0.43	0.62	0.76	0.41	0.63	0.61	0.89	0.81	1.18	0.73	0.58	0.81
C10	1.42	1.37	2.23	1.61	1.73	1.87	2.64	2.45	2.24	1.78	1.47	1.84
C11	1.38	1.36	2.05	1.52	1.62	1.60	2.47	2.57	2.04	1.60	1.44	1.89
C12	0.49	0.65	0.94	0.67	0.75	0.72	0.99	1.11	0.94	0.80	0.65	0.84
C13	1.38	1.29	1.55	1.10	1.51	1.46	2.07	1.98	1.74	1.45	1.15	1.28
C14	2.94	3.67	3.70	2.89	3.55	4.28	5.02	5.07	5.46	5.09	4.90	5.41
C15	1.68	2.02	1.99	1.95	1.98	2.24	3.10	3.22	4.96	3.55	2.59	4.16
C16	1.34	2.10	1.42	1.46	1.95	2.45	2.17	2.58	4.76	3.06	2.52	3.92
Total	32.58	40.02	47.80	33.36	43.38	46.12	60.95	57.15	72.19	54.51	43.93	58.91

Table 4. Summary of the non-fuzzy synthetic value matrix composed by the average of 31 experts.

Procedure Criteria	P1	P2	P3	P4	P5	P6	P7	P8	P9	P10	P11	P12
C1	2.98	4.81	6.43	3.35	5.32	4.91	7.41	6.28	10.20	7.06	4.43	6.29
C2	2.10	3.13	4.02	2.72	3.31	3.96	5.40	4.69	7.08	4.92	3.79	5.39
C3	3.76	4.77	5.35	3.42	4.82	5.62	8.05	7.16	9.04	7.30	5.77	8.16
C4	1.82	2.55	3.15	2.05	2.80	3.02	4.36	4.49	5.88	3.79	3.09	4.49
C5	1.75	1.91	2.36	1.88	2.74	2.78	3.53	3.38	4.39	2.75	1.85	3.55
C6	2.75	3.30	4.05	2.51	3.84	4.10	5.30	4.58	6.42	4.14	3.53	4.97
C7	1.52	1.73	3.13	1.50	1.93	2.11	3.89	2.52	3.21	3.12	1.88	2.32
C8	0.85	1.31	1.71	0.74	1.32	1.24	2.21	1.84	3.06	1.72	1.38	2.02
C9	0.30	0.55	0.71	0.29	0.55	0.51	0.88	0.73	1.19	0.67	0.50	0.75
C10	1.30	1.32	2.13	1.56	1.67	1.75	2.58	2.41	2.20	1.70	1.32	1.79
C11	1.08	1.15	1.89	1.29	1.46	1.46	2.41	2.48	1.93	1.43	1.27	1.83
C12	0.48	0.66	0.94	0.70	0.82	0.75	1.04	1.12	0.96	0.78	0.66	0.87
C13	1.35	1.23	1.54	1.12	1.54	1.47	2.13	1.92	1.78	1.45	1.14	1.35
C14	2.26	3.20	3.18	2.24	2.97	3.92	4.85	5.04	5.47	5.01	4.58	5.19
C15	0.87	1.35	1.30	1.41	1.38	1.66	2.76	2.79	4.97	3.08	2.01	3.78
C16	0.95	1.73	1.08	1.13	1.51	2.18	1.83	2.29	4.73	2.78	2.33	3.82
Total	26.14	34.71	42.98	27.92	37.99	41.44	58.63	53.73	72.52	51.70	39.53	56.57

Table 5. Summary of the fuzzy-based synthetic value by each expert.

Procedures Respondent	P1	P2	P3	P4	P5	P6	P7	P8	P9	P10	P11	P12
1	26.18	34.60	39.11	14.01	46.26	48.80	67.88	44.08	78.82	56.42	40.78	52.57
2	38.37	51.04	59.39	60.29	54.53	52.90	69.62	62.57	87.12	65.27	56.57	62.45
3	49.35	41.25	46.01	44.46	43.50	40.03	51.78	48.64	67.79	44.46	50.32	62.60
4	25.88	35.25	37.09	17.83	36.75	43.45	79.67	78.07	62.66	58.70	29.06	40.48
5	18.93	35.38	54.04	21.00	36.58	31.17	67.48	51.36	85.96	44.94	40.54	45.78
6	36.55	43.20	61.26	32.23	56.44	60.91	71.88	70.72	84.20	61.98	54.47	64.67
7	19.32	23.78	41.38	16.46	33.53	33.53	56.59	31.83	61.69	49.54	28.51	52.49
8	24.21	40.27	59.90	48.81	68.09	74.72	85.62	85.62	87.69	73.66	50.18	80.91
9	35.96	43.15	47.36	34.96	42.73	48.70	54.11	59.46	75.36	60.08	53.25	66.86
10	27.38	29.15	35.27	19.83	29.80	31.00	49.10	55.83	67.62	51.17	41.49	54.68
11	47.63	54.47	54.71	42.65	46.20	56.72	59.30	64.26	66.92	60.72	58.90	58.76
12	23.19	32.43	36.49	29.55	26.29	33.29	40.05	59.09	55.97	35.79	28.70	38.36
13	34.76	41.50	46.96	40.04	40.95	51.11	53.12	49.78	53.12	48.60	33.51	48.68
14	35.06	47.40	50.79	35.36	49.82	50.63	60.22	58.89	71.12	52.14	37.47	61.04
15	53.83	57.13	65.27	46.11	60.68	62.21	74.28	66.80	85.67	74.84	62.21	81.81
16	36.51	49.27	52.46	39.06	45.49	49.29	66.21	49.36	66.75	64.75	54.39	66.54
17	38.36	33.89	36.86	22.03	32.33	34.77	55.26	52.38	73.45	53.58	39.39	63.89
18	38.59	43.93	50.04	45.09	40.68	46.93	62.41	56.89	70.32	54.27	47.28	56.80
19	11.41	18.51	28.69	15.37	31.37	27.64	62.51	62.51	74.22	34.65	22.92	61.20
20	48.78	48.55	51.38	50.15	50.63	49.27	55.38	59.39	61.96	53.17	52.61	56.50
21	27.46	37.66	42.09	36.87	48.58	53.86	57.57	57.42	69.66	52.37	41.09	68.62
22	37.77	45.46	45.21	36.59	43.51	48.83	59.07	57.42	60.56	52.42	51.21	52.19
23	18.62	23.26	28.98	16.87	28.18	23.62	45.21	39.87	61.31	43.09	27.33	44.26
24	20.32	23.99	31.64	28.35	21.61	44.91	52.31	46.50	75.98	41.84	33.57	52.26
25	68.03	63.81	73.35	48.32	65.07	61.59	79.27	77.84	87.61	70.48	71.71	81.80
26	23.33	41.39	55.84	35.94	55.05	55.20	59.73	57.95	69.84	59.10	39.89	67.04
27	42.50	41.50	57.61	30.86	45.66	54.22	71.83	69.04	84.38	52.66	31.96	56.18
28	16.81	29.25	31.52	20.13	36.41	31.32	36.20	35.70	54.29	33.07	26.70	50.43
29	40.34	47.90	49.65	46.21	48.60	51.04	58.81	48.40	69.37	50.03	50.39	58.45
30	28.07	41.61	63.52	37.91	37.66	49.10	65.36	46.54	77.89	62.46	41.56	52.02
31	16.66	40.53	47.93	20.92	41.88	48.95	61.79	67.33	88.48	73.50	63.80	65.90
Total	1010	1240	1482	1034	1345	1430	1890	1772	2238	1690	1362	1826
Average	32.58	40.02	47.80	33.36	43.38	46.12	60.95	57.15	72.19	54.51	43.93	58.91
SD	12.76	10.35	11.39	12.40	11.13	11.74	11.16	12.17	10.59	11.02	12.43	10.86

Table 6. Summary of the non-fuzzy synthetic value by each expert.

Procedures Respondent	P1	P2	P3	P4	P5	P6	P7	P8	P9	P10	P11	P12
1	23.98	33.28	32.00	10.00	44.80	47.52	65.40	43.22	81.71	54.48	39.70	51.38
2	15.48	28.54	38.32	42.90	33.12	30.26	61.24	47.14	87.10	50.95	34.56	46.84
3	47.20	35.96	41.66	39.30	37.38	35.29	50.62	42.30	73.14	37.36	47.40	64.70
4	29.35	38.21	38.73	22.83	39.41	44.77	74.81	73.21	60.55	57.25	32.89	42.39
5	15.13	34.73	53.71	17.83	36.01	29.95	67.15	51.03	85.63	45.61	40.21	45.45
6	26.78	32.10	48.76	21.88	44.54	49.40	64.14	59.42	79.96	51.36	42.40	54.10
7	22.82	28.23	42.99	21.31	36.13	36.13	56.23	34.77	60.43	50.51	31.85	52.97
8	16.44	31.18	53.24	42.56	66.17	73.94	87.42	87.42	90.18	73.02	45.48	83.06
9	31.91	39.00	43.39	30.85	39.25	46.59	52.99	57.24	75.95	58.55	49.93	66.71
10	30.81	32.49	37.21	24.13	33.01	33.97	49.75	55.45	66.03	51.31	43.57	54.39
11	27.34	37.86	38.80	18.02	26.30	45.10	48.04	61.30	67.98	54.60	48.94	48.38
12	21.11	31.01	33.75	28.57	24.87	32.41	39.57	54.75	56.95	33.49	26.89	36.87
13	20.70	33.44	42.22	31.26	33.08	47.48	50.60	45.50	49.88	44.78	20.54	44.40
14	31.46	44.44	47.42	31.04	46.22	47.22	58.22	57.02	70.04	49.40	34.02	59.30
15	53.07	55.71	62.89	46.87	58.55	60.67	70.61	64.45	80.87	71.23	60.67	77.01
16	16.83	39.19	47.03	20.81	33.37	39.73	72.27	40.93	72.21	68.93	50.27	72.55
17	36.12	29.32	34.60	17.30	26.82	30.62	53.86	50.98	74.48	54.89	37.46	65.10
18	26.34	34.60	42.22	36.34	29.96	46.60	59.78	51.48	70.22	50.26	38.40	53.46
19	16.23	23.43	31.99	19.93	34.13	31.15	60.53	60.53	70.51	37.33	26.85	59.31
20	22.41	21.49	32.81	27.89	29.79	24.37	48.79	64.83	77.33	39.95	37.71	55.49
21	32.66	40.96	44.38	40.18	49.22	37.92	57.16	57.50	68.00	53.14	43.68	66.76
22	18.91	29.61	29.11	16.81	25.91	31.57	52.05	49.49	55.77	38.75	37.07	40.15
23	23.06	31.76	38.26	23.98	36.44	33.34	54.36	50.50	75.14	55.78	35.76	53.00
24	16.77	23.65	32.17	27.51	22.99	45.95	52.57	46.67	74.15	44.31	35.95	52.81
25	52.17	48.05	60.85	35.97	50.59	45.29	63.89	60.31	78.41	52.05	53.95	66.51
26	9.97	33.11	45.61	27.39	43.57	42.99	50.78	46.01	67.23	48.01	32.33	61.31
27	40.55	39.55	57.81	28.65	43.71	52.15	74.43	69.49	87.25	50.55	29.75	54.23
28	20.71	31.57	33.83	23.93	38.17	33.41	37.75	37.43	53.61	34.89	29.66	50.83
29	17.91	34.05	37.09	29.11	35.97	39.99	59.33	35.99	79.91	37.93	39.11	60.15
30	24.45	37.07	61.21	34.67	34.65	47.71	63.43	45.29	73.69	58.33	37.53	51.31

Table 6 (continued)

31	21.65	42.29	48.33	25.71	43.49	49.15	59.75	63.85	83.77	68.79	61.02	62.70
Total	810.3	1076	1332.	865.5	1178	1285	1818	1665.	2248	1578	1226	1754
Average	26.14	34.71	42.98	27.92	37.99	41.44	58.63	53.73	72.52	50.90	39.53	56.57
SD	10.68	7.13	9.38	8.79	9.71	10.26	10.52	11.56	10.32	10.42	9.48	10.57

4. Discussion and Conclusion

In the past, expert opinions from various groups were usually gathered to determine fee schedule for each orthopedic or surgical procedure. If expert opinions were not in consensus then there was no way to come to a conclusion and it would pose a great obstacle toward rationally adjusting a fee schedule. The FMCDM model developed in this study and verified by results show precise principles, easy operation, and effective gathering of the opinions of surgical experts. The complicated determination of relative values for fee schedule for orthopedic procedures can be shown by using a simplified hierarchy structure. After expert evaluation and mathematical calculations, each influencing factor can be concretely shown as a ranked value and thus, a new fee schedule values can be obtained.

The RBRVS system has been implemented in the US for more than one decade (Harris-Shapiro, 1998; Marc, et al, 2002; Rotarius, 2001). Grimaldi's review of the results concluded that physician workloads accounted for 50% of Medicare payments, while practice costs accounted for 46%, and malpractice insurance accounted for 4% (Grimaldi, 2002). Compared with the AHP results, orthopedic surgeons in Taiwan list physician workloads (47.5 %) as the major consideration in their decision-making, not far from results in the US. However, the following items are quite different: training costs of sub-specialties (19 %); practice cost (18.5 %); and malpractice cost (15 %).

Table 3 to Table 6 lists the synthetic value matrix of the fuzzy-based MCDM approach and that of non-fuzzy approach. It can be seen that there are few differences between the data of the two groups. However, the results of non-fuzzy MCDM present each expert view of the new orthopedic fee schedule that the other members will not necessarily accept. The relative values derived from the fuzzy-based MCDM model, combines strict model computation and expert opinions, the results are very convincing and the possibility of accepting by the other experts is greater.

The biggest difference between the FMCDM model and traditional methods is that our model has more precise calculations and the results are both qualitative and quantitative. It can tackle problems that formerly quantitative data and research methods could not solve easily. Examples of these problems are differences in expert opinions, difficulty in coming to a consensus, analytical results that are qualitative data, tests and conclusions not easily verified, and experts' bias sampling that would severely skew reference points. The new method effectively consolidates expert opinions, is suitable for integrating opinions from different groups, and can solve the complex problem of many evaluating factors. The values derived from the FMCDM model are convincing and can serve as an important reference for setting or revising the fee schedule.

In summary, this study consolidated analytic hierarchy process, fuzzy theory and multiple criteria decision-making techniques to form the FMCDM model. In the model, the synthetic value matrix for 12 representative procedures was derived using strict mathematical calculations. The source data combined with relative weightings from the AHP results and experts' opinions of each orthopedic procedure. Thus, the final data are both qualitative and quantitative. The result is more convincing than previous programs and more persuasive to be the references for the NHI policy setting. Furthermore, research in compliance with this method of determining a fee schedule can be expected in the future. Beyond the procedures of orthopedic surgery proposed in this study, we suggest further research in cross-boards, cross-specialties, and overall fee schedules of BNHI.

Acknowledgements

This study is deeply indebted to the cooperation provided by the board of the Taiwan Surgical Association, and the Bureau of National Health Insurance, Department of Health, Executive Yuan for providing research funding (BNHI Research and Development Project, Doc. No.:DOH91–NH-1040).

References

1. J.J. Buckley: *Ranking alternatives using fuzzy numbers*. Fuzzy Sets and Systems **15(1)** 21-31(1985).

2. R.E. Bellman and L.A. Zadeh, Decision-making in a fuzzy environment, *Management Science* **17** 141-146(1970).

3.   V. Belton and A.E. Gear, On a shortcoming of Saaty's method of analytic hierarchies, *Omega* **11(3)** 227-230(1983).

4.   M. Bohanec, B. Zupan and V. Rajkovic, Applications of qualitative multi-attribute decision models in health care, *International Journal of Medical Informatics* **59** 191-205(2000).

5.   A. Chapple and A. Rogers, Explicit guidelines for qualitative research: a step in the right direction, a defense of the soft option, or a form of sociological imperialism? *Fam Pract* **15** 556–561(1998).

6.   S.H. Cheng and T.L. Chiang, Disparity of medical care utilization among different health insurance schemes in Taiwan, *Social Science & Medicine* **47(5)** 613-620(1998).

7.   T.L. Chiang, Taiwan's 1995 health care refor,. *Health policy* **39** 225-239(1997).

8.   D. Dubis and H. Prade, Operations on fuzzy numbers, *International Journal of Systems Science* **9** 613-626(1978).

9.   S. Folland, Government regulation - principal regulatory mechanisms, in *The economics of health and health care*, 3rd edn. (New Jersey, Prentice-Hall, Inc, Upper Saddle River,), pp. 445-469(2001).

10.   M.K. Giacomini, and D.J.Cook, Users' guides to the medical literature, XXIII: qualitative research in health care—A, are the results of the study valid? *JAMA* **284**:357–362(2000).

11.   P.L. Grimaldi, Medicare fees for physician services are resource-based, *Journal of Health Care Finance* **28(3)**:88-104(2002).

12.   U.G. Gupta and R.E. Clarke, Theory and applications of the Delphi technique: a bibliography (1975-1994), *Technological Forecasting and Social Change* **53** 185-211(1996).

13.   J. Harris-Shapiro, RBRVS revisited, *Journal of Health Care Finance New York* **25** 49-54(1998).

14.   W.C. Hsiao, P. Braun and D.Dunn, et al, Resource-based relative values, an overview, *JAMA* **260(16)** 2347-2352(1988a).

15.   W.C. Hsiao, P. Braun and D. Dunn, et al, Special report: results and policy implications of the resource-based value study, *New England Journal of Medicine* **319(13)** 881-888(1988b).

16.   L.C. Huang and R. Y. Wu, Applying fuzzy analytic hierarchy process in the managerial talent assessment model – an empirical study in Taiwan's semiconductor industry, *International Journal of Technology Management* **30** 105–130 (2005).

17.   M. Huber and E. Orosz, Health Expenditure Trends in OECD Countries, 1990-2001, *Health Care Financing Review* **25(1)** 1-22(2003).

18.   Y.H. Huang, K.I. Wang and C.H. Hung, et al, Possible reasons for the decline of surgical specialty's manpower after implantation of the National Health Insurance program, *Taiwan Medical Journal* **47** 40-44(2004).(in Chinese)

19.   J. Marc K. Kesteloot and D. De Graeve,. A typology for provider payment systems in health care, *Health Policy* **60** 255-273(2002).

20.   Q.K. Malterud, Qualitative research: standards, challenges, and guidelines, *The Lancet* **358(9280)** 483-488(2001).

21.   L. Patricia, Limitations of quantitative research in the study of structural adjustment, *Social Science & Medicine* **42** 313-324(1996).

22.   C. Pope and N. Mays, Qualitative research: reaching the parts other methods cannot reach: an introduction to qualitative methods in health and health care services research, *BMJ* **311** 42-5 (1995).

23.   R.A. Powell and H.M. Single, Methodology matters--V focus group, *International Journal for Quality in Health Care* **8** 499-504 (1996).

24.   T. Rotarius, An RBRVS approach to financial analysis in health care organizations, *The Health Care Manager* **19(3)** 17-23(2001).

25.   T. L. Saaty, A scaling method for priorities in hierarchical structures, *Journal of Mathematical Psychology* **15(2)** 234-281(1977).

26.   Saaty, T. L. (1980). The analytic hierarchy process, New York, McGraw-Hill.

27.   G.J. Schieber and J.C. Langenbrunner, Physician payment research efforts at HCFA, *Health Affairs* **8(1)** 214-218(1989).

28.   C. Syllaand H. J. Wen, A conceptual framework for evaluation of information technology

investments, *International Journal of Technology Management* **24** 236–261(2002).

29. M.T. Tang and G.H. Tzeng, A hierarchy fuzzy MCDM method for studying electronic marketing strategies in the information service industry, *Journal of International Information Management* **8** 1-22(1999).

30. J.Y. Teng and G.H. Tzeng, Fuzzy multicriteria ranking of urban transportation investment alternative, *Transportation Planning and Technology* **20** 15-31(1996).

31. G. H. Tzeng, and J. Y. Teng, Multicriteria evaluation for strategies of improving and controlling air-quality in the super city: A case of Taipei city, *Journal of Environmental Management* **40(3)** 213-229(1994).

32. L.A. Zadeh, Information and control, *Fuzzy Sets* **8** 338-353 (1965).

33. L.A. Zadeh, The concept of a linguistic variable and its application to approximate reasoning, Parts 1, 2, and 3, *Information Science* **8** 199-249, **8** 301-357, **9** 43-80 (1975).

34. R. Zhau and R. Goving, Algebraic characteristics of extended fuzzy numbers, *Information Science* **54(1)** 103-130(1991).

Brill Academic Publishers
P.O. Box 9000, 2300 PA Leiden,
The Netherlands

*Lecture Series on Computer
and Computational Sciences*
Volume 3, 2005, pp. 333-342

Monotone Iterative Method for Numerical Solution of Nonlinear ODEs in MOSFET RF Circuit Simulation

Yiming Li[1],[a],[b], Shao-Ming Yu[b],[c] and Chuan-Sheng Wang[b],[d]

[a]Department of Communication Engineering, National Chaio Tung University, Hsinchu 300,
Taiwan
[b]Microelectronics and Information Systems Research Center, National Chaio Tung University,
Hsinchu 300, Taiwan
[c]Department of Computer and Information Science, National Chaio Tung University, Hsinchu
300, Taiwan
[d]Department of Mathematics, National Tsing Hua University, Hsinchu city, Hsinchu 300, Taiwan

Received 10 July, 2005; accepted in revised form 1 August 2005

Abstract: In this paper, we model the metal-oxide-semiconductor field effect transistor (MOSFET) radio frequency (RF) circuit as a system of nonlinear ordinary differential equations. Then we solve them with the waveform relaxation method, the monotone iterative method, and Runge-Kutta method directly in time domain. With the monotone iterative method, we prove each decoupled and transformed circuit equation converges monotonically. In comparison with the HSPICE outputs, results calculated with our method are stable and robust in both the time and frequency domains. Convergence properties for the monotone iterative and outer iterative loops are also presented and discussed. This method provides an alternative in the time domain numerical solution of MOSFET RF circuit equations.

Keywords: Nonlinear Circuit Model, MOSFET Device, Ordinary Differential Equation, Monotone Iterative Method

Mathematics Subject Classification: Initial value problems, Multistep Runge-Kutta and extrapolation methods, Stability and convergence of numerical methods.

PACS: 02.30.Hq, 02.60.Cb, 02.60.Jh, 02.60.Lj, 02.60.Nm.

1 Introduction

Metal-oxide-semiconductor field effect transistor (MOSFET) devices used for designing radio frequency (RF) integrated circuit becomes a tendency because of the successful experience in digital circuit design [1, 2]. Numerical methods for the RF circuit provide an efficient alternative in the development of integrated RF components, such as filter, low-noise amplifier, mixer, and power amplifier. Due to the unusually high linearity of MOSFETs at high frequencies, these active devices have been of great interests for RF and wireless applications in the recent years. General approach to analyze the inter-modulation distortion and two-tone characteristics for an MOSFETs is to solve a set of equivalent circuit ordinary differential equations (ODEs) in the frequency domain. The harmonic balanced method is a standard approach for solving such RF problems [3, 4]. This frequency domain approach has its merits and limitations in studying the physical properties

---

[1]Corresponding author. E-mail: ymli@faculty.nctu.edu.tw

of MOSFET with time variations. Another approach to the analysis of electrical characteristics for a MOSFET RF circuit is to solve a set of equivalent circuit ODEs in time domain. The time domain results are then further calculated with fast Fourier transformation (FFT) for obtaining its spectrum. However, the decoupled and discretized ODEs in circuit simulation are often solved with conventional Newton's iterative (NI) method [5]. The well-known HSPICE circuit simulator is right adopted the NI method and its variants in its numerical kernel. It is known that the NI method is a local method; in general, it has a quadratic convergence property in a sufficiently small neighborhood of the exact solution, and hence it encounters convergence problem for practical engineering applications [6, 7].

In this work, we apply the monotone iterative (MI) method [11, 12, 13, 14, 15] to simulate MOSFET RF characteristics with exploiting the basic nonlinear property in the equivalent circuit model. The MI method was successfully applied to the semiconductor device simulation in our recent work [11, 12]. By considering the Kirchhoff's current law for each node, the circuit governing equations are formulated in terms of the nodal voltages $(V_G, V_D, V_S, V_{DX}, V_{SX}, V_{GX})$. The circuit model is decoupled into several independent ODEs with the waveform relaxation decoupling scheme. The basic idea of the decoupled method for circuit simulation is similar to the well-known Gummel's decoupling method for device simulation [10, 8, 9, 11, 12]; it is that the circuit equations are solved sequentially. In the circuit model, the first equation is solved for V_G^{j+1} given the previous states $V_X^j, X = D, S, GX, SX$, and DX, respectively. For the second equation is solved for V_D^{j+1} given $V_X^j, X = G, S, GX, SX$, and DX, respectively. We have the same procedure for other ODEs. Each decoupled ODE is transformed and solved with Runge-Kutta (RK) method and the monotone iterative method. We prove the MI method converges monotonically for all decoupled circuit equations. It means that we can solve the circuit ODEs with arbitrary initial guesses in the time domain. Numerical results including convergence properties for a deep-submicron MOSFET circuit operated with a two-tone input signal have been reported to demonstrate the robustness and accuracy of the method.

This paper is organized as follows. In Sec. 2, we state the MOSFET RF circuit to formulate the ODE system. Sec. 3 shows the numerical algorithms for the system of ODEs and proves the convergence properties of the monotone iterative method for each decoupled ODEs. A computational procedure is also introduced in this section. Sec. 4 shows numerical results for a n-type MOSFET circuit operated under two-tone RF range. Different convergence behaviors are also discussed in this section. Finally, Sec. 5 draws the conclusions and suggests future works.

2 A MOSFET Radio-Frequency Circuit Model

As shown in Fig. 1(a) , based on the node current flow conservation and utilize the EPFL-EKV (EKV) large signal equivalent circuit model, shown in Fig. 1(b), for the MOSFET device [16, 17], the mathematical model is formulated. At nodes B, D, S, and GX we have the following equations.

$$(C_{gs} + C_{gs0})(\frac{dV_S}{dt} - \frac{dV_G}{dt}) + (C_{gd} + C_{gd0})(\frac{dV_D}{dt} - \frac{dV_G}{dt}) +$$
$$(C_{gb} + C_{gb0})(\frac{dV_B}{dt} - \frac{dV_G}{dt}) + \frac{V_{GX} - V_G}{R_G} = 0, \tag{1}$$

$$-I_{DS} - I_{DB} + (C_{gd} + C_{gd0})(\frac{dV_G}{dt} - \frac{dV_D}{dt}) + C_{bd}(\frac{dV_B}{dt} - \frac{dV_D}{dt}) + \frac{V_{DX} - V_D}{R_D} = 0, \tag{2}$$

$$(C_{gs} + C_{gs0})(\frac{dV_G}{dt} - \frac{dV_S}{dt}) + C_{bs}(\frac{dV_B}{dt} - \frac{dV_S}{dt}) + \frac{V_{SX} - V_S}{R_S} + I_{DS} = 0, \tag{3}$$

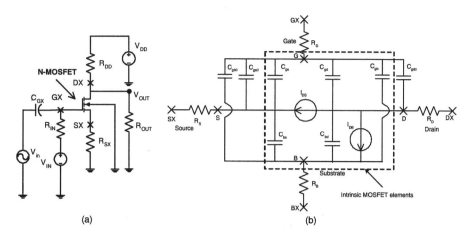

Figure 1: (a) A MOSFET RF circuit (b) and the MOSFET EKV equivalent circuit.

$$C_{GX}\left(\frac{dV_{in}}{dt} - \frac{dV_{GX}}{dt}\right) + \frac{V_G - V_{GX}}{R_G} + \frac{V_{IN} - V_{GX}}{R_{IN}} = 0 \tag{4}$$

Similarly, at nodes SX and DX, we formulate the equations as follows:

$$\frac{V_S - V_{SX}}{R_S} + \frac{-V_{SX}}{R_{SX}} = 0 \tag{5}$$

$$\frac{V_D - V_{DX}}{R_D} + \frac{V_{DD} - V_{DX}}{R_{DD}} + \frac{-V_{DX}}{R_{OUT}} = 0 \tag{6}$$

The Eqs. (1)-(4) are the ODEs, and the Eqs. (5) and (6) are the algebraic equations. These equations are subject to proper initial values at time $t = 0$ for all unknowns to be solved. All currents I and capacitances C above are nonlinear functions of unknown variables. These nonlinear terms are modelled by the EKV. For finding the characteristic of the RF circuit behavior shown in Fig. 1 , there are 4 coupled ODEs with the nonlinear current and capacitance models have to be solved and the unknowns to be calculated in the system of ODEs are V_G, V_D, V_S, and V_{GX}, respectively. Due to the exponential dependence of current and capacitance models, we note that the system consists of strongly coupled nonlinear ODEs.

3   Decoupled Algorithm and Monotone Iterative Method

We propose here a decoupled and globally convergent simulation technique to solve the system ODEs in the large-scale time domain directly. Firstly, under the steady state condition, we find the DC solution as the starting point to compute other time dependent solutions. For a specified time period T, to solve these nonlinear ODEs in the time domain, the computational scheme consists of following steps:

1. Let an initial time t and time step Δt be given.

2. Use the decoupling method to decouple all Eqs. (1)-(6).

3. Each decoupled ODE is solved sequentially with the MI and RK methods .

4. Convergence test for each MI loop.

5. Convergence test for overall outer loop.

6. If the specified stopping criterion is reached for the outer loop, then go to step (7), else update the newer results and back to step (3).

7. If $t < T$, $t = t + \Delta t$ and repeat the steps (3)-(6) until the time step meets the specified time period T .

3.1   The Decoupled Algorithm for the System ODEs

All coupled ODEs must decouple by the well-known waveform relaxation (WR) method [8, 9]. We state here the WR method in details. Consider a general nonlinear system of $\mathcal{N}$ ODEs with associated initial conditions,

$$\frac{dV}{dt} = f(V, t), V(0) = V_0, \tag{7}$$

where $t \in [0, T]$, $T > 0$, $f : \mathbb{R}^{\mathcal{N}} \times [0, T] \longrightarrow \mathbb{R}^{\mathcal{N}}$. $V_0 = [V_{1,0}, V_{2,0}, \cdots, V_{\mathcal{N},0}] \in \mathbb{R}^{\mathcal{N}}$ is the initial vector of V, and $V(t) = [V_1(t), V_2(t), \cdots, V_{\mathcal{N}}(t)] \in \mathbb{R}^{\mathcal{N}}$ is the solution vector at time t. The system can be written as follows,

$$\begin{cases} \frac{d}{dt} V_1 = f_1(V_1, V_2, \cdots, V_{\mathcal{N}}, t), & V_1(0) = V_{1,0} \\ \frac{d}{dt} V_2 = f_2(V_1, V_2, \cdots, V_{\mathcal{N}}, t), & V_2(0) = V_{2,0} \\ \vdots \\ \frac{d}{dt} V_{\mathcal{N}} = f_{\mathcal{N}}(V_1, V_2, \cdots, V_{\mathcal{N}}, t), & V_{\mathcal{N}}(0) = V_{\mathcal{N},0} \end{cases} \tag{8}$$

The WR method for solving (7) is a continuous-time iterative method. Therefore, given a function which approximates the solution, it calculates a new approximation along the whole time-interval of interest. Clearly, it differs from most standard iterative techniques in that its iterates are functions in time instead of scalar value. The iteration formula is chosen in such a way that one avoids having to solve a large system ODEs. A particularly simple, but often very effective iteration scheme is written below. It maps the old iterate V^{j-1}.

$$\begin{cases} \frac{d}{dt} V_1^j = f_1(V_1^j, V_2^{j-1}, \cdots, V_{\mathcal{N}}^{j-1}, t), V_1^j(0) = V_{1,0} \\ \frac{d}{dt} V_2^j = f_2(., V_2^j, V_3^{j-1}, \cdots, V_{\mathcal{N}}^{j-1}, t), V_2^j(0) = V_{2,0} \\ \vdots \\ \frac{d}{dt} V_{\mathcal{N}}^j = f_{\mathcal{N}}(., V_2^j, \cdots, V_{\mathcal{N}-1}^j, V_{\mathcal{N}}^j, t), V_{\mathcal{N}}^j(0) = V_{\mathcal{N},0} \end{cases} \tag{9}$$

It is obvious similar to the Gauss-Seidel method for iteratively solving linear and nonlinear systems of algebraic equations, so-call the Gauss-Seidel waveform relaxation scheme. It converts the task of solving a differential equation in $\mathcal{N}$ variables into the task of solving a sequence of differential equations in a single variable. A closely related iteration is the Jacobi waveform relation scheme, the iteration formula is given by,

$$\begin{cases} \frac{d}{dt} V_1^j = f_1(V_1^j, V_2^{j-1}, \cdots, V_{\mathcal{N}}^{j-1}, t), V_1^j(0) = V_{1,0} \\ \frac{d}{dt} V_2^j = f_2(V_1^{j-1}, V_2^j, V_3^{j-1}, \cdots, V_{\mathcal{N}}^{j-1}, t), V_2^j(0) = V_{2,0} \\ \vdots \\ \frac{d}{dt} V_{\mathcal{N}}^j = f_{\mathcal{N}}(V_1^{j-1}, \cdots, V_{\mathcal{N}-1}^{j-1}, V_{\mathcal{N}}^j, t), V_{\mathcal{N}}^j(0) = V_{\mathcal{N},0} \end{cases} \tag{10}$$

To solve the described Jacobi waveform relation scheme shown above, the algorithms is used for solving such system ODEs. For a given specified time step t and the previous calculated results, the decoupling algorithm solves the circuit equations sequentially, for instance, the V_G in Eq. (1) is solved for given the previous results $(V_D, V_S, V_{GX}, V_{DX}, V_{SX})$. The V_D in Eq. (2) is solved for newer given V_G and $(V_S, V_{GX}, V_{DX}, V_{SX})$. The V_S in Eq. (3) is solved for newer given (V_G, V_D) and (V_{GX}, V_{DX}, V_{SX}). We have similar procedure for other unknowns. A computational procedure is summarized in the following **Algorithm**.

Algorithm:

Given $j = 0$;
Choose $V_x^0(t)$ for $t \in [0, T]$, $x = 1, 2, \cdots, \mathcal{N}$
do
    $j = j + 1$
    for $x = 1, 2, \cdots, \mathcal{N}$
      solve $\frac{d}{dt} V_x^j = f_x(V_1^j, V_2^j, \cdots$
        $\cdots V_{x-1}^j, V_x^j, V_{x+1}^{j-1}, \cdots, V_{\mathcal{N}}^{j-1}, t)$
      with $V_x^j(0) = V_{x,0}$
    end for
while $|V_x^j - V_x^{j-1}| < $ **TOL**, for $x = 1, 2, \cdots, \mathcal{N}$

Each decoupled ODE is solved with the MI method [11, 12, 13, 14, 15]. In the following subsection, we state the MI method and show the convergence property.

3.2 The Monotone Iterative Method for Decoupled ODE

To clarify the MI method for the numerical solution of the decoupled nonlinear ODEs, we write the above decoupled ODEs as the following form

$$\frac{dV_X^j}{dt} = f(V_X^j, t),$$
$$V_X^j(0) - V_{X,0}^j \tag{11}$$

where V_X^j is the unknowns to be solved, j is the decoupling index $j = 0, 1, 2, \cdots$. We note that the f is the collection of the nonlinear functions and $f \in C[\mathbf{R} \times I, \mathbf{R}]$ and $I = [0, T]$. We may assume the upper and lower solutions are $\overline{V}_X^j$ and $\underline{V}_X^j$ exist in the circuit for a fixed index j and X. and $\overline{V}_X^j \geq \underline{V}_X^j$, we can prove the solution existence in the set Ω.

$$\Omega = \{(V_X^j, t) \mid \overline{V}_X^j \geq V_X^j \geq \underline{V}_X^j, \forall t \in I\} \tag{12}$$

for each decoupled circuit ODE.

We define the MI parameter $\lambda = \frac{\partial f}{\partial V_X}$ and insert the λ into Eq. (11), then we have the MI equation

$$\frac{dV_X^j}{dt} = f(\eta, t) - \lambda(V_X^j - \eta) \tag{13}$$

where $\overline{V}_X^j \leq \eta \leq \underline{V}_X^j$ is a value in $[0, T]$.

In the following, we use Theorem 1 to provide the existence of the solution of the system ODEs. We also describe a monotone constructive method for the simulation methodology of the circuit

ODEs. The constructed sequences will converge to the solution of Eq. (11) for all decoupled ODEs in the circuit simulation. In this condition, instead of original nonlinear ODE to be solved, a transformed ODE is solved with such as the RK method with the MI scheme formulated as Eq. (13). Next, we state the main result for the solution of each decoupled MOSFET circuit ODEs by proving Theorem 2 and 3.

Theorem 1 *Let $\overline{V}_X^j$ and $\underline{V}_X^j$ are the upper and lower solutions of Eq. (11) in $C^1[\mathbf{R} \times I, \mathbf{R}]$ such that $\overline{V}_X^j \geq \underline{V}_X^j$ in the time interval I and $f \in C[\mathbf{R} \times I, \mathbf{R}]$. Then there exists a solution V_X^j of Eq. (11) such that $\overline{V}_X^j \geq \underline{V}_X^j$ in the time interval I.*

Proof It is a direct result with the continuous property of f, here the comparison theorem is applied [13, 14, 15, 18].

We note that the nonlinear function f is nonincreasing function of the unknown V_X^j and the upper and lower solutions $\overline{V}_X^j(0)$ and $\underline{V}_X^j(0)$ of Eq. (11) in I can be found for each decoupled ODE. Further more, we can prove there exists a unique solution V_X^j of Eq. (11) in I and $\overline{V}_X^j(0) \geq V_X^j \geq \underline{V}_X^j(0)$.

Theorem 2 *Let the $f \in C[\mathbf{R} \times I, \mathbf{R}]$, $\overline{V}_X^j(0)$ and $\underline{V}_X^j(0)$ are the upper and lower solutions of Eq. (11) in I. Sice $f(V_X^j, t) - f(\widetilde{V}_X^j, t) \geq -\lambda(V_X^j - \widetilde{V}_X^j)$, $\overline{V}_X^j(0) \geq V_X^j \geq \widetilde{V}_X^j \geq \underline{V}_X^j(0)$ and $\lambda \geq 0$. There exist sequences $\{\overline{V}_{X_n}^j\}_{n=1}^{\infty}$ converge to $\overline{V}_X^j$ uniformly and $\{\underline{V}_{X_n}^j\}_{n=1}^{\infty}$ converge to $\underline{V}_X^j$ uniformly as $n \longrightarrow \infty$ monotonically in I.*

Proof For $\mathcal{V} \in C[\mathbf{R} \times I, \mathbf{R}]$, such that $\overline{V}_X^j \geq \mathcal{V} \geq \underline{V}_X^j$ we consider the following transformed ODE equation for the fixed j and X

$$\frac{dV_X}{dt} = f(\mathcal{V}, t) - \lambda(V_X^j - \mathcal{V})$$
$$V_X^j(0) = V_{X_0}^j, \tag{14}$$

then $\forall \mathcal{V}$, $\exists!$ V_X^j of Eq. (11) in I.
Define $\Psi \mathcal{V} = V_X^j$ then we can verify:

(i) $\Psi \overline{V}_X^j(0) \geq \overline{V}_X^j(0)$ and $\Psi \underline{V}_X^j(0) \leq \underline{V}_X^j(0)$

(ii) Ψ is a monotone operator in

$$\left[\overline{V}_X^j(0), \underline{V}_X^j(0) \right] \equiv \left[V_X^j \in C[I, \mathbf{R}] \mid \right.$$
$$\left. \overline{V}_X^j(0) \geq V_X^j \geq \underline{V}_X^j(0) \right] \tag{15}$$

Now we construct two sequences by using the mapping Ψ: $\Psi \overline{V}_{X_n}^j = \overline{V}_{X_{n+1}}^j$ and $\Psi \underline{V}_{X_n}^j = \underline{V}_{X_{n+1}}^j$ and by above observations, the following relation holds

$$\overline{V}_{X_0}^j \geq \cdots \geq \overline{V}_{X_n}^j \geq \underline{V}_{X_n}^j \geq \cdots \geq \underline{V}_{X_0}^j \tag{16}$$

in I. Hence $\overline{V}_{X_n}^j \overset{unif.}{\longrightarrow} \overline{V}_X^j$ and $\underline{V}_{X_n}^j \overset{unif.}{\longrightarrow} \underline{V}_X^j$ as $n \longrightarrow \infty$ monotonically in I. Furthermore the $\overline{V}_{X_n}^j$ and $\underline{V}_{X_n}^j$ satisfy

$$\frac{d\underline{V}_{X_{n+1}}^j}{dt} = f(\underline{V}_{X_n}^j, t) - \lambda(\underline{V}_{X_{n+1}}^j - \underline{V}_{X_n}^j)$$
$$\underline{V}_{X_n}^j(0) = V_{X_0}^j, \tag{17}$$

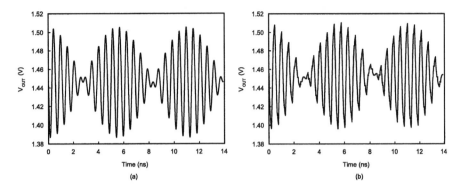

Figure 2: (a) Our (b) and HSPICE time domain outputs.

and

$$\frac{d\overline{V}^j_{X_{n+1}}}{dt} = f(\overline{V}^j_{X_n}, t) - \lambda(\overline{V}^j_{X_{n+1}} - \overline{V}^j_{X_n})$$

$$\overline{V}^j_{X_n}(0) = V^j_{X_0},\tag{18}$$

respectively. Thus $\overline{V}^j_X$ and $\underline{V}^j_X$ are the solutions of Eq. (11).

Theorem 3 *For decoupled ODEs, the nonlinear function f is nonincreasing in V^j_X and $f(V^j_{X_1}, t) - f(V^j_{X_2}, t) \geq -\lambda(V^j_{X_1} - V^j_{X_2})$, $\forall V^j_{X_1} \geq V^j_{X_2}$. Thus $\left\{\overline{V}^j_{X_n}\right\}^{\infty}_{n=1}$ and $\left\{\underline{V}^j_{X_n}\right\}^{\infty}_{n=1}$ converge uniformly and monotonically to the unique solution V^j_X of Eq. (11).*

Proof By using Theorem 2 and note the nonincreasing property of f, the result is followed directly.

4 Results and Discussion

As shown in Fig. 1 , the input signal V_{IN} is the DC bias and the expression of the two-tone input signal V_{in} is as follows:

$$V_{in} = V_m \sin(2\pi f_1 t) + V_m \sin(2\pi f_2 t),\tag{19}$$

The input two-tone signal has an amplitude $V_m = 0.005$ V, and fundamental frequencies f_1 and f_2 are 1.71 and 1.89 GHz, respectively. Figure 2 shows our simulation results and the HSPICE results in time domain, respectively. We also can find the difference in the Fig. 2 at the time between 2.5 ns and 3.5 ns, the HSPICE results show some non-smooth calculation, but our calculation does not have this phenomenon [11]. With the time domain results, we calculate the spectra of the output power by the FFT. Figure 3a shows the corresponding spectra with our time domain results. We can find the 3rd-order intermodulation (IM3) products at $2f_1 - f_2$ and $2f_2 - f_1$ clearly. Figure 3b shows the corresponding spectra with HSPCIE time domain result. However, it is difficult to identify the two IM3 products.

The stopping criteria for the monotone iterative loop is the maximum norm error $< 10^{-10}$ and for the outer iterative loop is the maximum norm error $< 10^{-9}$, respectively. Figures 4 and 5 show

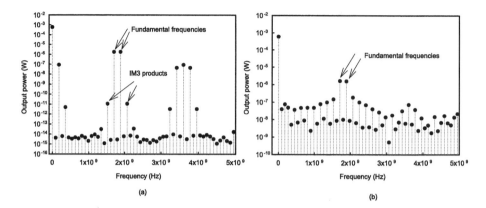

Figure 3: (a) Our (b) and HSPICE output power in frequency domain.

the convergence property of our simulation kernel. Figure 4a shows the inner loop iteration versus the maximum norm error at the beginning ($t = 0s$) of the simulation. Figure 4b also shows the inner loop iteration versus the maximum norm error in time point at $t = 13.8\ fs$. This two figures show the inner loop iteration has strictly convergent behavior and have $n\log(n)$ convergence where n is the number of iterations. Figure 5 shows the outer loop iteration convergence property in different time periods. As we can see the speed of convergence is almost around n^2.

5   Conclusions

A novel numerical method for circuit simulation that based on waveform relaxation and monotone iterative methods has been proposed. First of all, the waveform relaxation algorithm is applied to decouple a set of coupled nonlinear ODEs. Each ODE is numerically solved with the monotone iterative method. Mathematically, we have proved each decoupled nonlinear ODE solved by the monotone iterative method converges monotonically. By studying a MOSFET RF circuit with two-tone input signals, computational results have been reported in this work to demonstrate the robustness and accuracy of the method. Good convergence properties have been obtained in several explored examples. The proposed numerical method for solving nonlinear ODE provides an alternative in the computer simulation of integrated circuits. This numerical method for the time domain solution of circuit ODEs can be generalized to simulate more complicated integrated circuit with more transistors. Furthermore, it is ready for parallelization and can be systematically extended to simulate integrated circuits in frequency domain.

Acknowledgment

This work is supported in part by the National Science Council (NSC) of Taiwan under Contract NSC-93-2215-E-429-008 and Contract NSC-94-2752-E-009-003-PAE, by the Ministry of Economic Affairs, Taiwan under Contract 93-EC-17-A-07-S1-0011, by the Taiwan Semiconductor Manufacturing Company under a 2004-2005 grant, and by the Toppoly Optoelectronics Corp. under a 2003-2005 grant.

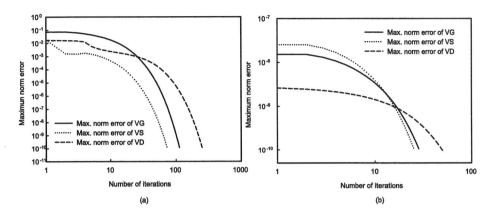

Figure 4: The maximum norm error versus the number of inner iterations at the (a) $t = 0s$ and (b) $t = 13.8\ fs$.

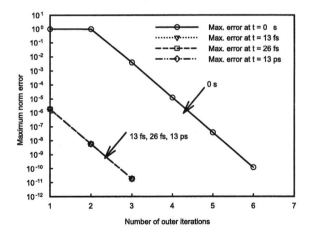

Figure 5: The maximum norm error versus the number of outer iterations.

References

[1] W. Alan Davis, K. Agarwal, *Radio Frequency Circuit Design*, New York: John Wiley & Sons, Inc, 2001.

[2] K. Chang, *RF and Microwave Wireless Systems*, New York: John Wiley & Sons, Inc, 2000.

[3] B. Trpyanovsky, Z. Yu , R. W. Dutton, Physics-based simulation of nonlinear distortion in semiconductor devices using the harmonic balance method, *Computer Methods in Applied Mechanics and Engineering* **181** 467-482 (2000).

[4] K. S. Kundert, J. K. White, and A. S. Vinventelli, *Steady-State Methods for Simulating Analog and Microwave Circuits*, Kluwer Academic, 1990.

[5] G. F. Carey, et al., Circuit, *Device and Process Simulation - Mathematical and Numerical Aspects*, Wiley, New York, 1996.

[6] Y. Inoue, S. Kusanobu, Theorems on the unique initial solution for globally convergent homotopy methods, *IEICE Trans. Fundametals* **E86-A** 2184-2191 (2003).

[7] Y. Inoue, S. Kusanobu, A partical approach for the fixed-point homotopy method using a solution-tracing circuit, *IEICE Trans. Fundametals* **E85-A** 287-298 (2002).

[8] V. Stefan, *Parallel multigrid waveform relaxation for parabolic problems*, Teubner, 1993.

[9] E. Lelarasmee, et al., The waveform relaxation method for time-domain analysis of large scale integrated circuits, *IEEE Trans. on CAD of IC and Systems* **1** 131-145 (1982).

[10] H. K. Gummel, A self-consistent iterative scheme for one-dimensional steady state transport calculations, *IEEE Trans. Electron Devices* **ED-11** 455-465 (1964).

[11] Y. Li, K.-Y. Huang, A novel numerical approach to heterojunction bipolar transistors circuit simulation, *Computer Physics Communications* **152** 307-316 (2003).

[12] Y. Li, A parallel monotone iterative method for the numerical solution of multi-dimensional semiconductor Poisson equation, *Computer Physics Communications* **153** 359-372 (2003).

[13] L. Colatz, *Functional Analysis and Numerical Mathematics*, Academic Press, London, 1966.

[14] C. V. Pao, *Nonlinear parabolic and elliptic equations*, Plenum Press, New York, 1992.

[15] J. M. Ortega and W. C. Rheinboldt, *Iterative solution of nonlinear equations in several variables*, Academic Press, New York, 1970.

[16] C. Enz, F. Krummenacher, E. Vittoz, An analytical MOS transistor model valid in all regions of Operation and dedicated to low-voltage and low-current applications, *Journal on Analog Integrated Circuits and Signal Processsing, Kluwer Academic Publishers* 83-114 (1995).

[17] J.-M. Sallese, Advancements in DC and RF MOSFET Modeling with the EPFL-EKV Charge Based Model, Special session: MOS Transistor: Compact Modeling and Standardization Aspects, 8th International Conference MIXDES, Zakopane, Poland, 2001.

[18] V. Lakshmikantham and S. Leela, *Differential and integral inequalities, Vol. I, II*, Academic Press, New York, 1968.

Brill Academic Publishers
P.O. Box 9000, 2300 PA Leiden,
The Netherlands

Lecture Series on Computer
and Computational Sciences
Volume 3, 2005, pp. 343-359

Semiparametric Smooth Transition Modeling for Nonlinear Time Series

Dimitrios D. Thomakos[1]

Department of Economics,
School of Management and Economics,
University of Peloponnese,
GR-221 00 Tripolis, Greece

Received 20 July, 2005; accepted August 5, 2005

Abstract: In this paper I propose an extension to the class of semiparametric time series models by incorporating a weight function to allow for smooth transitions between a linear autoregressive and a nonparametric component. The weight function and nonparametric component may depend on auxiliary variables. The inclusion of the weight function broadens the flexibility of the semiparametric model and allows for testing against smooth transition parametric models and against linearity. If no auxiliary variables are used the model is a special case of the functional coefficient autoregressive model, while if auxiliary variables are used the model's solution is a distributed lag in nonlinear functions. The nonparametric component is approximated using a Fourier series expansion and the model is estimated using the MINPIN approach proposed in Andrews (1994). An empirical illustration is provided using the U.S. quarterly unemployment rate data, previously analyzed in Montgomery *et al.* (1998).

Keywords: semiparametric, nonparametric, smooth transition, autoregression, nonlinearity

Mathematics Subject Classification: 60G10, 60G12, 62F10, 62G05, 62M05, 62M10, 62M20, 62P20

1 Introduction

Over the last twenty years there has been a growing interest in the time series literature for nonlinear models. Non-linearities have been identified in many areas of application and a number of models have been proposed to deal with phenomena such as non-normality, asymmetric responses or multi-modality. Most of the earlier models have been parametric in nature while later a number of semi and nonparametric models have been proposed and applied. Examples of parametric nonlinear models include the threshold autoregressive model, see [26] and references therein, various forms of the smooth transition autoregressive model, see [16] and references therein, the bilinear model in [14] and the exponential autoregressive model in [20]. Examples of nonparametric nonlinear models include the nonlinear additive autoregressive model (possibly with exogenous variables) (NAARX) in [7], the functional coefficient autoregressive (FAR) in [8] and the functional coefficient time series (FCTS) model in [5]. See also [23] and [24] for model selection procedures in nonparametric time series models. An examples of semiparametric time series model is [13], which is a time series adaptation of the semiparametric model in [21].

---

[1] Email for correspondence: thomakos@uop.gr

In this paper I propose a semiparametric model that extends the model in [13] or [21] by incorporating a weight function that allows for smooth transitions between the linear and nonparametric component of the model. The weight function and the nonparametric component may depend on auxiliary variables. The inclusion of the weight function considerably broadens the flexibility of the semiparametric model: under suitable restrictions, the proposed model contains within it the semiparametric model of [13] or [21], the smooth transition autoregressive model, the linear autoregressive model, the nonparametric (auto)regression model and the dynamic regression model. As the proposed model combines parametric and nonparametric information it also has some similarity to models that combine a parametric and nonparametric regression using a scalar parameter. In the absence of any auxiliary variables the model belongs to the class of NAARX and FAR models and, under some conditions, it inherits their properties. If auxiliary variables are present then the model's causal solution is a distributed lag in nonlinear functions. The nonparametric component of the model can be approximated using a series expansion, as in [13], and I explicitly consider Fourier series. Approximation results based on series expansions relevant for this study can be found in [32]. Estimation is then based on the MINPIN approach[2] in [1]. Alternatively, the results in [15] could be used (with appropriate modifications to a time series context) but I do not pursue their approach here.

The rest of the paper is organized as follows. In the next section I outline the model and its assumptions, find its causal solution, and define the form of the minimum mean squared error predictor. In section 3 I discuss the Fourier series approximation of the unknown nonparametric component. In section 4 I discuss the estimation and specification of the model. Section 5 has a brief empirical illustration using the unemployment dataset previously analyzed in [19]. Section 6 offers some concluding remarks. Technical conditions and the computation of point and interval forecasts from the proposed model are discussed in the appendix.

2   The model

Consider the real-valued time series $\zeta_t \stackrel{\text{def}}{=} \{y_t, x_t, z_t, \epsilon_t\}_{t \in \mathbb{Z}}$, defined on a complete probability space $(\Omega, \mathcal{A}, \mathcal{P})$, and call y_t the dependent variable, x_t the transfer variable, z_t the transition (or threshold) variable and ϵ_t the noise variable. Let $\delta \geq 0$ be a non-negative integer, define $\boldsymbol{x}_{t-\delta} \stackrel{\text{def}}{=} (x_{t-\delta}, x_{t-\delta-1}, \ldots, x_{t-\delta-q})^\top$ for some $q \geq 0$, and let $\{\mathcal{T}(\boldsymbol{x}_{t-\delta}) : \mathbb{R}^{q+1} \to \mathbb{R}\}$ be a measurable, smooth function of lags of the transfer variable; call $\mathcal{T}(\cdot)$ the transfer function and δ the transfer delay. The form that the transfer function takes is assumed unknown. Finally, let $d > 0$ be a positive integer, let $\boldsymbol{\lambda} \stackrel{\text{def}}{=} (\lambda_1, \lambda_2, \ldots, \lambda_l)^\top \in \Lambda \subset \mathbb{R}^l$ be a parameter vector and consider a measurable, smooth and bounded function $\{w(z_{t-d}; \boldsymbol{\lambda}) : \mathbb{R} \times \Lambda \to [0,1]\}$; call $w(z_{t-d}; \boldsymbol{\lambda})$ the weight function, d the transition delay and define $\bar{w}(z_{t-d}; \boldsymbol{\lambda}) = 1 - w(z_{t-d}; \boldsymbol{\lambda})$. Finally, define $\boldsymbol{y}_{t-1} \stackrel{\text{def}}{=} (y_{t-1}, y_{t-2}, \ldots, y_{t-p})^\top$ and $\boldsymbol{\rho} \stackrel{\text{def}}{=} (\rho_1, \rho_2, \ldots, \rho_p)^\top$ for some $p > 0$ with $\rho_p \neq 0$. The dependent variable is then assumed to follow the model:

$$y_t = w(z_{t-d}; \boldsymbol{\lambda})\boldsymbol{y}_{t-1}^\top \boldsymbol{\rho} + \bar{w}(z_{t-d}; \boldsymbol{\lambda})\mathcal{T}(\boldsymbol{x}_{t-\delta}) + \epsilon_t \qquad (1)$$

This is a semiparametric smooth transition autoregressive model with 'exogenous' variables (S-STARX) model, with the transitions taking place between a linear autoregression and a nonparametric regression. An important aspect of smooth transition models like this one, is their adaptability, something that can be potentially useful in the model's forecasting performance. One of the necessary conditions for superior forecasting performance of a nonlinear model is that the forecasting period contains some 'nonlinear' features, see [16]. If it does not, a nonlinear model

---

[2]The acronym stands for estimators that "...MINimize a criterion function that may depend on a Preliminary Infinite dimensional Nuisance parameter estimator", see [1].

may not outperform a linear one. By an appropriate choice of the transition variable z_{t-d}, the model should place more weight in the nonlinear specification on such time periods that the data exhibit nonlinearities. The adaptability of the model is further increased by allowing for an arbitrary transfer function rather than imposing a particular parametric form, as in standard smooth transition models. Taking $z_{t-d} = y_{t-d}$ and $\boldsymbol{x}_{t-\delta} = \boldsymbol{y}_{t-\delta} \stackrel{\text{def}}{=} (y_{t-\delta}, y_{t-\delta-1}, \ldots, y_{t-\delta-q})^{\top}$ one obtains a S-STAR model like:

$$y_t = w(y_{t-d}; \boldsymbol{\lambda}) \boldsymbol{y}_{t-1}^{\top} \boldsymbol{\rho} + \bar{w}(y_{t-d}; \boldsymbol{\lambda}) \mathcal{T}(\boldsymbol{y}_{t-\delta}) + \epsilon_t \tag{2}$$

The weight function can take a variety of forms but, as in the rest of the literature, the most frequently applied ones will be logistic function $w(z_{t-\delta}; \boldsymbol{\lambda}) = \{1 + \exp\left[-\lambda_1 (z_{t-d} - \lambda_2)\right]\}^{-1}$ and exponential function $w(z_{t-d}; \boldsymbol{\lambda}) = 1 - \exp\left[-\lambda_1 (z_{t-d} - \lambda_2)^2\right]$, both with the restriction $0 \leq \lambda_1$. For the logistic and exponential weight functions the transition variable acts as a threshold: as $z_{t-d} \to \lambda_2$ then (for fixed λ_1) $w(z_{t-d}; \boldsymbol{\lambda}) \to c$, where c is a constant, and the weight function becomes a step function. The same is also true as $\lambda_1 \to \infty$. Unlike parametric smooth transition models, λ_1 can take the value of zero, as the model is identified even when $\lambda_1 = 0$. This allows for various forms of nesting (and testing) other models within equation (1). Take, for example, the logistic function as the weight function and impose the restriction $\lambda_1 = 0$. Then, the model becomes a standard semiparametric autoregressive, distributed lag model:

$$y_t = \boldsymbol{y}_{t-1}^{\top} \boldsymbol{\rho}^* + \mathcal{T}^*(\boldsymbol{x}_{t-\delta}) + \epsilon_t \tag{3}$$

where the * indicates a scaling by $1/2$. A practically relevant form for the model in (3) above takes $x_t = t/n$, i.e. $\mathcal{T}^*(t/n)$ is the nonparametric mean function (n denotes the sample size). Parametrizing the transfer function as $\mathcal{T}^*(\boldsymbol{x}_{t-\delta}; \boldsymbol{\beta}) = \boldsymbol{x}_{t-\delta}^{\top} \boldsymbol{\beta}^*$ I obtain a dynamic regression model:

$$y_t = \boldsymbol{y}_{t-1}^{\top} \boldsymbol{\rho}^* + \boldsymbol{x}_{t-\delta}^{\top} \boldsymbol{\beta}^* + \epsilon_t \tag{4}$$

Therefore, the parametric linear specification is immediately testable against a smooth transition alternative or a semiparametric smooth transition alternative. Similarly, if the exponential function is chosen as the weight function then the restriction $\lambda_1 = 0$ implies a purely nonparametric (possibly auto) regression or a linear autoregression:

$$\begin{aligned} y_t &= \mathcal{T}(\boldsymbol{x}_{t-\delta}) + \epsilon_t \\ y_t &= \boldsymbol{y}_{t-1}^{\top} \boldsymbol{\rho} + \epsilon_t \end{aligned} \tag{5}$$

depending on the position of $w(z_{t-d}; \boldsymbol{\lambda})$ and $\bar{w}(z_{t-d}; \boldsymbol{\lambda})$, which can of course be switched. This immediately provides for a simple test of the null hypothesis of linearity against a S-STARX. This is especially relevant for the case of a smooth transition autoregression. Taking $\delta = 1$, $q = p - 1$, parametrizing the transfer function as $\mathcal{T}(\boldsymbol{y}_{t-1}; \boldsymbol{\varphi}) = \boldsymbol{y}_{t-1}^{\top} \boldsymbol{\varphi}$ and re-arranging, I obtain the smooth transition autoregressive model:

$$y_t = \boldsymbol{y}_{t-1}^{\top} \boldsymbol{\varphi} + w(z_{t-d}; \boldsymbol{\lambda}) \boldsymbol{y}_{t-1}^{\top} \boldsymbol{\varphi}^* + \epsilon_t \tag{6}$$

where $\boldsymbol{\varphi}^* \stackrel{\text{def}}{=} \boldsymbol{\rho} - \boldsymbol{\varphi}$. Therefore, with a suitable placement of the exponential weights, the parametric smooth transition autoregressive can be tested against the S-STAR model of equation (2).

To simplify the exposition, it is convenient to re-express equation (1) using the companion form of the autoregressive component. Define the $(p \times 1)$ vector $\boldsymbol{y}_t \stackrel{\text{def}}{=} (y_t, y_{t-1}, \ldots, y_{t-p+1})^{\top}$, the $(p \times 1)$ vector $\mathcal{T}(\boldsymbol{x}_{t-\delta}) \stackrel{\text{def}}{=} (\mathcal{T}(\boldsymbol{x}_{t-\delta}), 0, \ldots, 0)^{\top}$, the $(p \times 1)$ vector $\boldsymbol{\epsilon}_t \stackrel{\text{def}}{=} (\epsilon_t, 0, \ldots, 0)^{\top}$, the $(p \times p)$ weight matrix $w(z_{t-\delta}; \lambda)$ and the $(p \times p)$ companion matrix α as:

$$w(z_{t-\delta}; \lambda) \stackrel{\text{def}}{=} \begin{bmatrix} w(z_{t-\delta}; \lambda) & 0 & \cdots & 0 & 0 \\ 0 & 1 & \cdots & 0 & 0 \\ 0 & 0 & \cdots & 0 & 0 \\ \vdots & \vdots & \vdots & \vdots & \vdots \\ 0 & 0 & \cdots & 0 & 1 \end{bmatrix}, \alpha \stackrel{\text{def}}{=} \begin{bmatrix} \rho_1 & \rho_2 & \cdots & \rho_{p-1} & \rho_p \\ 1 & 0 & \cdots & 0 & 0 \\ 0 & 1 & \cdots & 0 & 0 \\ \vdots & \vdots & \vdots & \vdots & \vdots \\ 0 & 0 & \cdots & 1 & 0 \end{bmatrix} \quad (7)$$

and then re-write equation (1) as:

$$y_t = w(z_{t-\delta}; \lambda)\alpha y_{t-1} + \overline{w}(z_{t-d}; \lambda)\mathcal{T}(x_{t-\delta}) + \epsilon_t \quad (8)$$

where $\overline{w}(z_{t-d}; \lambda) \stackrel{\text{def}}{=} I_p - w(z_{t-d}; \lambda)$. If the model contains only lags of the dependent variable, as in equation (2), and $p < d^* \stackrel{\text{def}}{=} \max\{\delta + q, d\}$, then augment y_t to be of dimension $(d^* \times 1)$ and modify the rest of the elements of equation (8) appropriately. In both cases (with or without $\{x_t, z_t\}$ in the right-hand side), equation (8) takes the generic form of a Markov chain as:

$$y_t = \Phi(y_{t-1}) + T(x_{t-\delta})I(x_t, z_t) + \epsilon_t \quad (9)$$

where $I(x_t, z_t)$ is the indicator function for the inclusion of $\{x_t, z_t\}$.

There is extensive literature dealing with the probabilistic properties of the such Markov chains, mainly arising from nonlinear autoregressive models, i.e. when $I(x_t, z_t) = 0$. Ergodicity and geometric ergodicity play a key role in this type of models. Geometric ergodicity is particularly useful as it implies stationarity and strong mixing, with a geometric mixing rate, see [3]. For conditions of ergodicity and geometric ergodicity on nonlinear autoregressive models see, among others, [29], [22] and [26]. I give one such set of conditions in the appendix. When these conditions hold then theorem 1.1 of [7] applies and y_t is geometrically ergodic.

Alternatively, when $I(x_t, z_t) = 1$, and therefore the model includes exogenous variables, one can make direct assumptions about the variables in ζ_t, the transfer function $\mathcal{T}(\cdot)$ and obtain a 'closed form' solution for the model. In this case there is a variety of assumptions that one can make, inlcuding ergodicity or strong mixing for ζ_t or strong mixing for $\{x_t, z_t, \epsilon_t\}$ and near-epoch dependence for y_t. One plausible set of such assumptions, along with some auxiliary assumptions, are also presented in the appendix. Using these assumptions I can obtain the (causal) solution of the model use equation (9) and assumption [A7]. By recursive substitution m periods back, we obtain the representation:

$$y_t = H_m(z_{t-d}; \lambda)y_{t-m} + \sum_{j=0}^{m-1} H_j(z_{t-d}; \lambda)\left[\overline{w}(z_{t-d-j}; \lambda)\mathcal{T}(x_{t-\delta-j}) + \epsilon_{t-j}\right] \quad (10)$$

and letting $m \to \infty$ we get:

$$y_t = \sum_{j=0}^{\infty} H_j(z_{t-d}; \lambda)\left[\overline{w}(z_{t-d-j}; \lambda)\mathcal{T}(x_{t-\delta-j}) + \epsilon_{t-j}\right] \quad (11)$$

or, in terms of the original model in equation (1), we get:

$$y_t = \sum_{j=0}^{\infty} h_{11,j}(z_{t-d}; \lambda)\left[\overline{w}(z_{t-d-j}; \lambda)\mathcal{T}(x_{t-\delta-j}) + \epsilon_{t-j}\right] \quad (12)$$

with $h_{11,j}(z_{t-d}; \lambda)$ denoting the $(1,1)$ element of $H_j(z_{t-d}; \lambda)$. Therefore, the model is expressible as a nonlinear distributed lag in the transfer variable, the transition variable, the noise variable, the weight function and the parameter vector λ with absolutely summable, time-varying coefficients.

The response pattern of the dependent variable at time t from an impulse in the transfer variable at $t - \delta - j$ is easily obtained from the final form of equation (12) as:

$$\omega_j(\lambda) \stackrel{\text{def}}{=} \frac{\partial y_t}{\partial x_{t-\delta-j}} = h_{11,j}(z_{t-d}; \lambda)\bar{w}(z_{t-d-j}; \lambda) \cdot \frac{\partial \mathcal{T}(x_{t-\delta-j})}{\partial x_{t-\delta-j}} \tag{13}$$

and it follows from the assumptions that $\lim_{j\to\infty} \omega_j(\lambda) = 0$ and $\omega_\infty(\lambda) \stackrel{\text{def}}{=} \sum_{j=0}^{\infty} \omega_j(\lambda) < \infty$, with $\omega_\infty(\lambda)$ being the 'long-run' response coefficient.

Let $D \stackrel{\text{def}}{=} \min\{\delta, d\}$ and assume for the moment that $D > 0$. Consider forecasting the dependent variable at some horizon $D \geq h \geq 1$, so that actual values of x_t and z_t are used in forming the forecast.[3] The minimum mean squared error predictor of y_{t+h} made at time t is given by $y_{t+h}^t \stackrel{\text{def}}{=} \mathsf{E}\left[y_{t+h}|\mathcal{F}_t\right]$ and we therefore have:

$$y_{t+h}^t = w(z_{t-d+h}; \lambda)y_{t+h-1}^{t\top}\rho + \bar{w}(z_{t-d+h}; \lambda)\mathcal{T}(x_{t-\delta+h}) \tag{14}$$

and using equation (8) and recursive substitution we can obtain explicit expressions for the predictor as:

$$y_{t+h}^t = \boldsymbol{F}_0(z_{t-d}; \lambda)y_t + \sum_{j=1}^{h} \boldsymbol{F}_j(z_{t-d}; \lambda)\overline{\boldsymbol{w}}(z_{t-d+j}; \lambda)\mathcal{T}(x_{t-\delta+j}) \tag{15}$$

or, in terms of the original series:

$$y_{t+h}^t = f_{11,0}(z_{t-d}; \lambda)y_t + \sum_{j=1}^{h} f_{11,j}(z_{t-d}; \lambda)\bar{w}(z_{t-d+j}; \lambda)\mathcal{T}(x_{t-\delta+j}) \tag{16}$$

where $\boldsymbol{F}_j(z_{t-d}; \lambda)$ is the partial product sequence $\boldsymbol{F}_j(z_{t-d}; \lambda) = \prod_{i=j+1}^{h} C(z_{t-d+i}; \lambda)$, satisfying the same conditions as the $\boldsymbol{H}_j(z_{t-d}; \lambda)$ sequence, and where $f_{11,j}$ denotes the corresponding $(1, 1)$ elements for $j = 0, 1, \ldots, h$. The predictor, therefore, takes the form of a finite, nonlinear distributed lag. To actually compute the forecast we require values for the autoregressive coefficient vector ρ, the weight function parameter vector λ and a suitable approximation to the unknown transfer function $\mathcal{T}(x_{t-\delta})$.

3 Fourier Series Approximation of $\mathcal{T}(\cdot)$

For the proposed model to be easily implementable we need such an approximation for $\mathcal{T}(x_{t-\delta}) = \sum_{j=0}^{q} \mathcal{T}_j(x_{t-\delta-j})$ that it would allow us to apply a standard method for estimating the model's parameters. Series approximations are well suited to the problem at hand. In addition, they convert a semiparametric problem into a parametric one for which regression techniques can be applied. We therefore consider approximating the jth transfer function $\mathcal{T}_j(x_{t-\delta-j})$ using a truncated, tapered Fourier series expansion.[4] For an available sample of size n, and writing $m_j = m_{jn} = o(n)$ for the

---

[3]For $h > D$ the predictor cannot be computed using actual values of x_t and z_t and one would need to use predicted values obtained from univariate (parametric or nonparametric) models for x_t and z_t respectively.

[4]It is implicitly assumed in the rest of the discussion that the transfer variable is scaled to be in the unit interval before taking its Fourier series. Such a scaling is easily achieved as follows. Let (κ_1, κ_2) be two constants such that $\kappa_1 = \min_t x_t - c$, for small $c > 0$, and $\kappa_2 > \max_t x_t$ and make the transformation $(x_t - \kappa_1)/\kappa_2$.

truncation point, we have:

$$T_j(x_{t-\delta-j}) \approx T_{m_j}(x_{t-\delta-j}; \phi_j) \stackrel{\text{def}}{=} \phi_{j0} + 2 \sum_{s=1}^{m_j} \psi_s \phi_{js} \cos(\pi s x_{t-\delta-j}) \tag{17}$$

where $\{\phi_{js}\}_{s=0}^{m_j}$ are the Fourier coefficients $\phi_{js} \stackrel{\text{def}}{=} \int_0^1 T_j(x) \cos(\pi s x) dx$, and where $\{\psi_s\}_{s=1}^{m_j}$ are the taper weights. We also use $\phi_j = (\phi_{j0}, \phi_{j1}, \ldots, \phi_{jm_j})^\top$ to denote the $(m_j + 1 \times 1)$ vector of the Fourier coefficients.[5] The approximation error $r_{m_j}(x; \phi_j) \stackrel{\text{def}}{=} T_j(x) - T_{m_j}(x; \phi_j)$ obeys $\lim_{m_j \to \infty} r_{m_j}(x; \phi_j) = 0$ uniformly in x and all $j = 0, 1, \ldots, q$.

As always, there is a trade-off by considering a parametric approximation to a nonparametric function. First, there is now a total of $k \stackrel{\text{def}}{=} \sum_{j=0}^{q}(m_j + 1) + l + p$ parameters to be estimated from the data: the p autoregressive coefficients, the l coefficients of the transition vector λ and the $\sum_{j=0}^{q}(m_j+1)$ parameters of the Fourier expansions. Second, the transfer delay δ and the transition delay d also need to be infered or estimated from the data. Note that, as the truncation parameters depends on the sample size, the number of parameters to be estimated will, in principle, vary with the sample size as well.

A variety of different weights can be used for tapering, for example:

T1. The constant weights $\psi_s = 1$.

T2. the Fejér weights $\psi_s = 1 - (s-1)/m_j$.

T3. the modified Fejér weights

$$\psi_s = \left\{ \begin{array}{ll} 1, & 1 \le s \le m_j/2 \\ 2(1 - s/m_j), & m_j/2 \le s \le m_j \\ 0, & s > m_j/2 \end{array} \right\} \tag{18}$$

T4. the cosine weights $\psi_s = \cos\left[(\pi s)/(2m_j + 1)\right]$.

4 Estimation and Model Selection

4.1 Estimation

The model to be estimated can now be written as:

$$y_t = w(z_{t-d}; \lambda) y_{t-1}^\top \rho + \bar{w}(z_{t-d}; \lambda) \sum_{j=0}^{q} T_{m_j}(x_{t-\delta-j}; \phi_j) + e_t \tag{19}$$

where $e_t \stackrel{\text{def}}{=} \epsilon_t + \bar{w}(z_{t-d}; \lambda) \sum_{j=0}^{q} r_{m_j}(x_{t-\delta-j}; \phi_j)$. If the selection parameters $(p, q, \delta, d, m_0, \ldots, m_q)$ and the form of the weight function $w(z_{t-d}; \lambda)$ were known, estimation of the model's coefficients

---

[5] Augmenting the approximation of equation (17) with the level and square terms of the jth lag of the transfer variable we obtain the Fourier flexible form $T_{f,m_j}(x_{t-\delta-j}; \phi_j, \beta_j) \stackrel{\text{def}}{=} \beta_0 x_{t-\delta-j} + \beta_1 x_{t-\delta-j}^2 + 2 \sum_{s=0}^{m_j} \psi_s \phi_{js} \cos(\pi s x_{t-\delta-j})$. For simplicity, we consider the approximation given in equation (17) in what follows, although the flexible form is used in the empirical application.

$\boldsymbol{\theta} \stackrel{\text{def}}{=} (\boldsymbol{\rho}^\top, \boldsymbol{\lambda}^\top, \boldsymbol{\phi}_0^\top, \dots, \boldsymbol{\phi}_q^\top)^\top$ could be accomplished using nonlinear least squares (NLLS). Assume that a sample of n observations was available and let $v \stackrel{\text{def}}{=} \max\{p, d^*\}$. Then, the NLLS estimator $\widehat{\boldsymbol{\theta}}_n$ would be given by:

$$Q(\widehat{\boldsymbol{\theta}}_n) \stackrel{\text{def}}{=} \inf_{\boldsymbol{\theta}} \; n^{-1} \sum_{t=v+1}^{n} \left[y_t - w(z_{t-d}; \boldsymbol{\lambda}) \boldsymbol{y}_{t-1}^\top \boldsymbol{\rho} - \bar{w}(z_{t-d}; \boldsymbol{\lambda}) \sum_{j=0}^{q} \mathcal{T}_{m_j}(x_{t-\delta-j}; \boldsymbol{\phi}_j) \right]^2 \tag{20}$$

If each $\mathcal{T}_j(\cdot)$ has a finite Fourier series expansion of order m_j then the model in these equations is a standard nonlinear regression with dependent observations. Some additional regularity conditions, as for example in [31], will ensure that the NLLS estimator is consistent and asymptotically normal.

However, if the transfer functions do not have a finite Fourier series expansion then, even for known (p, q, δ, d), we need to use a method that provides consistent estimators of $\mathcal{T}_j(\cdot)$ as $m_{jn} \to \infty$. To do so the results in [1] can be used to obtain a consistent and asymptotically normal MINPIN estimator. The Fourier coefficients are now treated as (possibly infinite dimensional) nuisance parameters and the vector of interest becomes $\boldsymbol{\vartheta} \stackrel{\text{def}}{=} (\boldsymbol{\rho}^\top, \boldsymbol{\lambda}^\top)^\top$. Define the weighted obsrvations for the dependent variable as $g_{t1} \equiv g_1(y_t, z_{t-d}; \boldsymbol{\lambda}) \stackrel{\text{def}}{=} y_t / \bar{w}(z_{t-d}; \boldsymbol{\lambda})$ and the 'doubly' weighted observations for the lags of the dependent variable as $g_{t2} \equiv g_2(y_{t-1}, \dots, y_{t-p}, z_{t-d}; \boldsymbol{\vartheta}) \stackrel{\text{def}}{=} [w(z_{t-d}; \boldsymbol{\lambda}) / \bar{w}(z_{t-d}; \boldsymbol{\lambda})] \boldsymbol{y}_{t-1}^\top \boldsymbol{\rho}$. Then, let $\mathcal{T}_1(\boldsymbol{x}_{t-\delta}, \boldsymbol{\lambda}) \stackrel{\text{def}}{=} \mathsf{E}[g_{t1} | \boldsymbol{x}_{t-\delta}]$ and $\mathcal{T}_2(\boldsymbol{x}_{t-\delta}; \boldsymbol{\vartheta}) \stackrel{\text{def}}{=} \mathsf{E}[g_{t2} | \boldsymbol{x}_{t-\delta}]$ and, using the independence of ϵ_t with $\{x_s, z_s\}$ for all (t, s), note that:

$$\mathcal{T}(\boldsymbol{x}_{t-\delta}) \equiv \mathcal{T}_1(\boldsymbol{x}_{t-\delta}; \boldsymbol{\lambda}) - \mathcal{T}_2(\boldsymbol{x}_{t-\delta}; \boldsymbol{\vartheta}) \tag{21}$$

For any admissible value of $\boldsymbol{\vartheta}$, $\{g_{t1}\}_{t \in \mathbb{Z}}$ and $\{g_{t2}\}_{t \in \mathbb{Z}}$ are strictly stationary and strong mixing sequences, with the same mixing coefficients as in [A4]. Therefore, the results in [32] apply[6] and consistent estimators for these two conditional means can be obtained using Fourier series approximations. To this end, let the truncation point m_j obey the conditions in [32] and minimize, for each $i = 1, 2$ and $j = 0, 1, \dots, q$:

$$\sum_{t=v+1}^{n} \left[g_{ti} - \phi_{ij0} - 2 \sum_{s=1}^{m_j} \phi_{ijs} \cos(\pi s x_{t-\delta-j}) \right]^2 \tag{22}$$

where for simplicity we ommitted the taper weights. The Fourier coefficient estimates can be estimated using ordinary LS or can be approximated by:

$$\widehat{\phi}_{ijs} \approx \frac{1}{n-v} \sum_{t=v+1}^{n} (g_{ti} - \widehat{\phi}_{j0}) \cos(\pi s x_{t-\delta-j}) \quad s \geq 1 \tag{23}$$

where $\widehat{\phi}_{ij0} \approx (n-v)^{-1} \sum_{t=v+1}^{n} g_{ti}$. Therefore, consistent estimators for $\mathcal{T}_1(\boldsymbol{x}_{t-\delta}; \boldsymbol{\lambda})$ and $\mathcal{T}_2(\boldsymbol{x}_{t-\delta}; \boldsymbol{\vartheta})$ are given by:

$$\begin{aligned} \widehat{\mathcal{T}}_1(\boldsymbol{x}_{t-\delta}; \boldsymbol{\lambda}) &\stackrel{\text{def}}{=} \sum_{j=0}^{q} \left[\widehat{\phi}_{1j0} + 2 \sum_{s=1}^{m_j} \widehat{\phi}_{1js} \cos(\pi s x_{t-\delta-j}) \right] \\ \widehat{\mathcal{T}}_2(\boldsymbol{x}_{t-\delta}; \boldsymbol{\vartheta}) &\stackrel{\text{def}}{=} \sum_{j=0}^{q} \left[\widehat{\phi}_{2j0} + 2 \sum_{s=1}^{m_j} \widehat{\phi}_{2js} \cos(\pi s x_{t-\delta-j}) \right] \end{aligned} \tag{24}$$

---

[6]See theorem 4.2 (p. 474) in [32].

Using the estimators from equation (27) and defining:

$$q_t(\boldsymbol{\vartheta}) \stackrel{\text{def}}{=} y_t - w(z_{t-d}; \boldsymbol{\lambda})\boldsymbol{y}_{t-1}^{\top}\boldsymbol{\rho} - \bar{w}(z_{t-d}; \boldsymbol{\lambda})\left[\widehat{\mathcal{T}}_1(\boldsymbol{x}_{t-\delta}; \boldsymbol{\lambda}) - \widehat{\mathcal{T}}_2(\boldsymbol{x}_{t-\delta}; \boldsymbol{\vartheta})\right] \tag{25}$$

the MINPIN estimator for $\widehat{\boldsymbol{\vartheta}}_n$ is then obtained from:

$$Q(\widehat{\boldsymbol{\vartheta}}_n) \stackrel{\text{def}}{=} \inf_{\boldsymbol{\vartheta}} \ n^{-1} \sum_{t=v+1}^{n} q_t(\boldsymbol{\vartheta})^2 \tag{26}$$

Under certain conditions[7] the estimator obtained from the above equation is consistent and asymptotically normal. Additional details about the MINPIN estimators, including their asymptotic covariance matrices, can be found in [1]. Using $\widehat{\boldsymbol{\vartheta}}_n$ we can then obtain a consistent estimator for the transfer function $\mathcal{T}(\boldsymbol{x}_{t-\delta})$ as:

$$\widehat{\mathcal{T}}(\boldsymbol{x}_{t-\delta}) = \widehat{\mathcal{T}}_1(\boldsymbol{x}_{t-\delta}; \widehat{\boldsymbol{\lambda}}_n) - \widehat{\mathcal{T}}_2(\boldsymbol{x}_{t-\delta}; \widehat{\boldsymbol{\vartheta}}_n) \tag{27}$$

4.2 Model Selection

Irrespective of the approach used in estimation, the practical problems that need to be addressed are (a) the choice of the autoregressive order p, the transfer and transition delays δ and d, the transfer order q and the truncation points $m_0, \ldots, m_q$, (b) the choice of the weight function $w(z_{t-d}; \boldsymbol{\lambda})$, and (c) the (possible) choice of taper weights $\{\psi_s\}_{s=1}^{m_j}$. Probably the most critical among these problems is the choice of the truncation points, as they determine the consistency of the series estimators. The results in [32] give some guidelines for the choice of the truncation points based on asymptotic theory, but even these authors argue in favor of a data-based method for selecting appropriate values for m_j's in any particular application. Therefore, we consider the generalized cross-validation (GCV) procedure, see [9], which is also used in [13]. An alternative GCV approach is given in [33], which we approriately adapt to the present context. We describe both of these GCV methods next. To simplify the exposition lets consider the, practically useful, case where all $q+1$ truncation points are the same .

Suppose that (p, q, δ, d) are known and let $\mathcal{M}_n \stackrel{\text{def}}{=} \left\{\left[an^{1/6-\gamma}\right], \ldots, \left[bn^{1/2-\gamma}\right]\right\}$ denote a set of possible values for the truncation point m_n, where $0 < a < b < \infty$ are given constants and $\gamma > 0$ and small. The powers of n are taken from the suggested values in [32]. The constants (a, b) allow for some flexibility in the choice of the initial range in $\mathcal{M}_n$ and are used in both [13] and [33]. For example, for $n = 1000$ and $a = 0.5$, $b = 1.5$ and $\gamma = 10^{-3}$ the range of $\mathcal{M}_n$ is $\{1, \ldots, 47\}$. In practice, a smaller/larger set $\mathcal{M}_n$ can always be selected by some appropriate choice of (a, b, γ). The GCV procedure used in [13] selects $\widehat{m}_n$ that minimizes the penalized sum of squares function $\mathcal{G}(m)$:

$$\mathcal{G}(\widehat{m}_n) \stackrel{\text{def}}{=} \inf_{m \in \mathcal{M}_n} \left[1 - \frac{m+1}{n}\right]^{-2} \cdot Q(\widehat{\boldsymbol{\vartheta}}_n) \tag{28}$$

and therefore this approach selects a value for the truncation point based on 'in-sample' fit. The GCV procedure used in [33] splits the sample into two subsets of $n_1 < n$ training observations and $n_2 \stackrel{\text{def}}{=} n - n_1$ validation observations. Using the n_1 training observations estimate the model over all values in $\mathcal{M}_n$, compute the forecast error variances corresponding to the one-step-ahead (static) forecasts, say $\widetilde{y}_{t+1}^t(m)$, for the n_2 validation observations and select that value $\widehat{m}_n$ that minimizes the forecast error variance function $\mathcal{F}(m)$:

$$\mathcal{F}(\widehat{m}_n) \stackrel{\text{def}}{=} \inf_{m \in \mathcal{M}_n} n_2^{-1} \sum_{t=n_1+1}^{n} \left[y_t - \widetilde{y}_{t+1}^t(m)\right]^2 \tag{29}$$

---

[7]See assumptions SE (p. 50), N (p. 55) and C (p. 65) in [1].

and therefore this approach selects a value for the truncation point based on 'out-of-sample' predictive ability. A GCV approach related to this one can be found in [5].

In practice the other model selection parameters will not be known and will have to be determined, along with the truncation points, from the data as well. A feasible modeling procedure would be to combine the GCV selection of the truncation point with a model selection criterion, like AIC or BIC.[8] The total number of models that need to be considered are $G \stackrel{\text{def}}{=} p \times (q+1) \times \delta \times d$, which can be a large number even for small values of (p, q, δ, d). For each of the $g = 1, 2, \ldots, G$ models select $\widehat{m}_{n,g}$, using one of the GCV procedures discussed previously, use $\widehat{m}_{n,g}$ to compute the value of AIC or BIC and select those values for (p, q, δ, d) that minimize the model selection criterion. Consider, for example, the BIC criterion and let $\boldsymbol{\mu}_g \stackrel{\text{def}}{=} (p, q, \delta, d)^\top$ to be the vector of model selection parameters in the gth model. We then have:

$$\widehat{\mu} \stackrel{\text{def}}{=} \operatorname*{argmin}_{\boldsymbol{\mu}_g \in G} \ \log Q(\widehat{\boldsymbol{\vartheta}}_n) + [(q+1)(\widehat{m}_{n,g}+1) + l + p] \frac{\log n}{n} \tag{30}$$

The use of a model selection criterion can be refined by tests of significance for the estimated coefficients and specification tests for model reduction. Such tests are being considered in greater detail in [25].

5 Empirical Illustration

In this section I analyze the U.S. quarterly civilian unemployment rate and initial unemployment claims dataset that was previously used in [19]. In that paper the authors examined the forecasting performance of a number of linear and nonlinear models for the quarterly civilian unemployment rate. Their results indicate that a TAR model significantly improved the forecasting performance when the forecasts are made in periods of rapidly rising unemployment. Their TAR model suggests a different behavior for the series, depending on whether the economy is at a stable or expanding period or is at a contracting period. I choose to analyze the same dataset for a number of reasons: first, the change in the unemployment rate can be taken to be a nonlinear time series (see the tests in [28] (p. 165) and below); second, the unemployment claims series can be used as the transfer variable with transfer delay $\delta = 1$ (see the results in [19] and below); third, a number nonlinear features of the unemployment series, claimed for by the TAR model in [19], become visible through the estimation of the weight function $w(z_{t-d}; \boldsymbol{\lambda})$; fourth, the countercyclical nature of the unemployment series suggests a function of real output as the transition variable. Since this analysis is for illustrative purspuses, I do not attempt to examine the relative forecasting performance of the model presented here or test against parametric alternative; rather I attempt to see if a number of the salient features of the unemployment series can be captured and interpreted using the methodology proposed in the previous sections.

Table 1. Descriptive statistics for (y_t, x_t, z_t)

Series	Min	Mean	SD	Max
y_t	-0.9667	0.0083	0.3911	1.667
x_t	-1.0310	0.0044	0.3213	1.7126
z_t	-3.9220	-0.0017	1.2144	2.8491

The sample span used is different from that in [19], with the series beginning on 1952:Q1 and ending in 1998:Q3. In the following analysis I use all available observations. Below, y_t denotes the change in the quarterly civilian unemployment rate, x_t denotes as the change in the initial

---

[8]A similar approach is taken in [10] for model selection in self-exciting TAR models.

unemployment claims and z_t denotes the transition variable. As noted in [16], the transition variable in this type of model should be slowly varying without a deterministic trend. In the analysis that follows, I used the residual from a quadratic trend of real GDP as the transition variable which appers to satisfy these conditions. Descriptive statistics are given in table one. The cross-correlation functions between $P_{yx} \stackrel{\text{def}}{=} (y_{t\pm k}, x_t)$ and between $P_{yz} \stackrel{\text{def}}{=} (y_{t\pm k}, z_t)$ are given in table two.[9] The signs of the contemporaneous correlations, $k = 0$, are as expected, positive for the first pair and negative for the second pair. For P_{yx} the highest cross-correlation occurs at $k = 0$ and $k = 1$, with the next highest occuring at $k = 2$. This suggests that a potential transfer delay is $\delta = 1$ with transfer lag $q = 1$. For P_{yz} the highest cross-correlation occurs at $k = 4$, $k = 5$ and $k = 6$. This suggests that a potential transition delay may between $d = 4$ and $d = 6$.

Table 2. Cross-correlations between $P_{yx} \stackrel{\text{def}}{=} (y_{t\pm k}, x_t)$ and $P_{yz} \stackrel{\text{def}}{=} (y_{t\pm k}, z_t)$

k	-6	-5	-4	-3	-2	-1	0	1	2	3	4	5	6
P_{yx}	-0.06	-0.15	-0.32	-0.26	-0.19	0.07	0.62	0.62	0.37	0.19	-0.02	-0.13	-0.03
P_{yz}	-0.19	-0.23	-0.26	-0.31	-0.32	-0.31	-0.19	0.01	0.18	-0.30	0.36	0.36	0.33

Note: the 95% standard error bound is ±0.1441.

Finally, in table three I present some F-tests for linearity, see [27], for both y_t and x_t. The results agree with those in [28] and linearity is rejected for both series at less than the 5% level.

Table 3. Linearity F-tests for y_t and x_t

	Tests for y_t			
Lag	2	3	4	5
F-test	4.3754	2.7249	2.8476	2.8739
p-value	0.0053	0.0148	0.0027	0.0005
	Tests for x_t			
Lag	2	3	4	5
F-test	5.0066	3.8714	2.4052	1.8031
p-value	0.0023	0.0012	0.0107	0.0379

In specifying an appropriate model, I used the following maximum values for the model's selection parameters: $p = 2$, $q = 1$, $\delta = 1$ and $d = 6$, for a total of $G = 24$ potential models. The choice of the autoregressive value $p = 2$ is in agreement with the TAR model order in [19]. For each model the GCV criterion of equation (28) was used to select an appropriate value $\hat{m}_n$, with a grid $\mathcal{M}_n = \{1, 2, \ldots, 15\}$. The flexible Fourier form with no tapering was used in approximating the nonparametric component of the model (see footnote 5) and the model's parameters were estimated using the Nelder-Mead simplex method.[10] Following the suggestion in [16] (pp. 123-124), the weight function's argument was scaled by the variance/standard deviation of the transition variable, depending on whether the exponential or logistic weight functions were used. The two models reported were selected based on their BIC criterion value and their residual properties. Tables four and five have the estimation results for the exponential and logistic weight functions respectively.[11] The reported z statistics were obtained using the appropriate covariance matrix of equation given in [25].

---

[9]Note that the CCF are used here to tentatively identify the delay parameters. The final selection of the model's parameters was done following section 4.2. Besides, both pairs exhibit considerable lagged feedback.

[10]Starting values for the iterations were obtained by a linear autoregressive model for ρ and by assuming $\lambda_1 = 1$ and $\lambda_2 = 0.5$.

[11]When using the exponential weight function the weights were switched so that $\bar{w}(z_{t-d}; \lambda)$ was in front of the linear autoregressive component.

For the model using the exponential weight function the chosen values for the model's selection parameters were $(p = 1, q = 1, \delta = 1, d = 6)$ with $\widehat{m}_n = 3$. Therefore, the approximation of the nonparametric component used three cosine term plus a constant, the level and the square of x_t. All coefficients in $\vartheta \stackrel{\text{def}}{=} (\rho_1, \lambda_1, \lambda_2)^\top$ are estimated to be significant at the 5% level, with the autoregressive and threshold parameters λ_2 being significant at the 1% level. The hypothesis that $\lambda_1 = 0$, corresponding to a linear AR(1) specification, is rejected at the 5% level in favor of the smooth transition semiparametric alternative.[12] The estimated value of λ_1 is 15.5390 while the estimated threshold value of λ_2 is 0.5861, and was exceeded sixty times in the data. The large value of λ_1 indicates that the transitions are almost instantaneous around the threshold value. The estimated residual variance is 0.0714 and the BIC criterion value is -2.2167. The estimated residuals appear to be uncorrelated, as the Ljung-Box statistic with $h = 6$ lags[13] is equal to 3.4073 and is less than its corresponding 5% critical value of 12.59.

Table 4. Estimation results using the exponential weight function
$(p = 1, q = 1, \delta = 1, d = 6, \widehat{m}_n = 3)$

Parameter	ρ_1	λ_1	λ_2
Estimate	0.9484	15.5390	0.5861
z ratio	3.8436	1.7850	10.8466

$Q(\widehat{\vartheta}_n) = 0.0714$, BIC = -2.2167, $r^2 = 0.5177$, LB(6) = 3.4073

Note: $Q(\widehat{\vartheta}_n)$ is the residual variance, BIC is the Schwartz model selection criterion, r^2 is the squared correlation coefficient between actual and fitted values and LB(h) is the Ljung-Box test for residual autocorrelation of order h.

For the model using the logistic weight function the chosen values for the model's selection parameters were similar $(p = 1, q = 1, \delta = 1, d = 5)$ with $\widehat{m}_n = 3$. That is, only the transfer delay was different from the previous model. The estimated value of λ_1 is 2.4549 while the estimated threshold value of λ_2 is 1.7105, both being significant at the 1% level. However, the estimated autoregressive coefficient is only significant at the 10% level (with a one-sided test), with a z ratio of only 1.3373. The threshold value for this model is higher than before and was exceeded only 10 times in the data. The large value of the threshold (and the smaller value of λ_1) implies a much smoother transition in this case. The estimated residual variance is equal to 0.0701 and is slightly lower than the corresponding estimate for the case of the exponential weight function. The BIC criterion value is also slightly lower and equal to -2.2352. The estimated residuals appear uncorrelated, with the Ljung-Box statistic with $h = 6$ lags to be equal to 10.5368. Both models appear to fit the data well. The squared correlation coefficients between the actual and fitted values are over 50% for both models.

---

[12]Note that $\lambda_1 \geq 0$ and therefore a one-sided test can be used.
[13]The number of lags was chose using the rule $h = [\log(n)]$. Higher values of h did not reveal any remaining autocorrelation.

Table 5. Estimation results using the logistic weight function
$(p = 1, q = 1, \delta = 1, d = 5, \widehat{m}_n = 3)$

Parameter	ρ_1	λ_1	λ_2
Estimate	2.0225	2.4549	1.7105
z ratio	1.3373	3.7210	2.4027

$Q(\widehat{\vartheta}_n) = 0.0701$, BIC = -2.2352, $r^2 = 0.5571$, LB(6) = 10.5368

Overall, the results from this analysis are in broad agreement with the results in [19] about the behavior of the change in unemployment series. Of additional interest would be (a) to test the semiparametric smooth transition model applied here agaist a parametric smooth transition autoregression or smooth transition regression with the change in unemployment claims as an input variable, and (b) to examine if additional forecasting gains can be obtained using the semiparametric model against the TAR model in [19] or other suitable nonlinear models.

6 Concluding Remarks

The model proposed in this paper paper extends the class of smooth transition models by allowing the transitions to take place between a linear parametric and a nonparametric component. The parametric component is taken to be a linear autoregression while the nonparametric component is approximated using a Fourier series expansion. Under suitable conditions, the model's parameters can be estimated with the MINPIN approach in [1]. The structure of the model allows for a number of useful specification tests to be performed, including nested testing against parametric alternatives and testing for linearity. An application to the U.S. civilian unemployment rate illustrates the use of the proposed model.

The practical usefulness of the proposed model needs to be further evaluated by (a) testing the additivity assumption on the transfer function, (b) by exploring in greater detail the specification test proposed in [25] and (c) by applying the model to previously analyzed datasets to see whether fit and forecasting improvements can be obtained from the inclusion of the less restrictive nonparametric transfer function, as opposed to parametric ones.

Acknowledgment

An earlier version of this paper was presented in the 2002 Joint Statistical Meetings of the American Statistical Assocation in New York. I would like to thank the participants of my JSM session for the useful comments and suggestions. I would also like to thank Thanasis Stengos for his careful reading of the manuscript and his constructive criticisms that considerably enhanced the presentation of the paper.

References

[1] D. W. K. Andrews, 1994. "Asymptotics for semiparametric econometric models via stochastic equicontinuity", *Econometrica*, 62, pp. 43-72.

[2] D. W. K. Andrews and Y. J. Whang, 1990. "Additive interactive regression models: circumvention of the curse of dimensionality", *Econometric Theory*, 4, pp. 466-479.

[3] R. C. Bradley, 1986. "Basic properties of strong mixing conditions", in *Dependence in Probability and Statistics*, E. Eberlein and S. Taqqu eds., pp. 165-192. Boston, Birkhäuser.

[4] A. Buja, T. Hastie and R. Tibshirani, 1989. "Linear smoothers and additive models", *Annals of Statistics*, 17, pp. 453-510.

[5] Z. Cai, J. Fan and Q, Yao, 2000. "Functional coefficient models for nonlinear time series", *Journal of the American Statistical Association*, 95, pp. 941-956.

[6] R. Chen, J. Liu and R. S. Tsay, 1995. "Additivity tests for nonlinear autoregressive models", *Biometrika*, 82, pp. 369-383.

[7] R. Chen and R. S. Tsay, 1993a. "Functional coefficient autoregressive models", *Journal of the American Statistical Association*, 88, pp. 298-308.

[8] R. Chen and R. S. Tsay, 1993b. "Nonlinear additive ARX models", *Journal of the American Statistical Association*, 88, pp. 955-967.

[9] P. Craven and G. Wabha, 1979. "Smoothing noisy data with spline functions: estimating the correct degree of smoothing by the method of generalized cross-validation", *Numerical Mathematics*, 31, pp. 377-403.

[10] J. G. DeGooijer, 2001. "Cross-validation criteria for SETAR model selection", *Journal of Time Series Analysis*, 22, pp. 267-281.

[11] R. L. Eubank, J. D. Hart, D. G. Simpson and L. A. Steffanski, 1995. "Testing for additivity in nonparametric regression", *Annals of Statistics*, 23, pp. 1896-1920.

[12] J. Fan, C. Zhang and J. Zhang, 2001. "Generalized likelihood ratio statistics and Wilks phenomenon", *Annals of Statistics*, pp. 153-193.

[13] J. Gao, 1998. "Semiparametric regression smoothing of nonlinear time series", *Scandinavian Journal of Statistics*, 25, pp. 521-39.

[14] C. W. J. Granger and A. P. Andersen, 1978. *An Introduction to Bilinear Time Series Models*, Göttingen, Vandenhoeck and Ruprecht.

[15] P. Gozalo and O. Linton, 2000. "Local nonlinear least squares: using parametric information in nonparametric regression", *Journal of Econometrics*, 99, pp. 63-106.

[16] C. W. J. Granger and T. Teräsvirta, 1993. *modeling Nonlinear Economic Relationships*, Cambridge, Oxford University Press.

[17] T. Hastie and R. Tibshirani, 1986. "Generalized additive models", *Statistical Science*, 1, pp. 297-310.

[18] T. Hastie and R. Tibshirani, 1987. "Generalized additive models: some applications", *Journal of the American Statistical Association*, 82, pp. 371-386.

[19] A. L. Montgomery, V. Zarnowitz, R. S. Tsay and G. C. Tiao, 1998. "Forecasting the U.S. unemployment rate", *Journal of the American Statistical Association*, 93, pp. 478-493.

[20] V. Haggan and T. Ozaki, 1981. "Modeling nonlinear vibrations using an amplitude dependent autoregressive time series model", *Biometrika*, 68, pp. 189-196.

[21] P. M. Robinson, 1988. "Root-N Consistent Semiparametric Regression", *Econometrica*, 56, pp. 931-954.

[22] D. Tjøstheim, 1990. "Nonlinear time series and Markov chains", *Advances in Applied Probability*, 22, pp. 587-611.

[23] D. Tjøstheim and B. H. Auestad, 1994a. "Nonlinear identification of time series: projections", *Journal of the American Statistical Association*, 89, pp. 1398-1409.

[24] D. Tjøstheim and B. H. Auestad, 1994b. "Nonparametric identification of nonlinear time series: selecting significant lags", *Journal of the American Statistical Association*, 89, pp. 1410-1419.

[25] D. D. Thomakos, 2003. "Model Specification Tests for Semiparametric Smooth Transition Models", mimeo.

[26] H. Tong, 1990. *Non-Linear Time Series: A Dynamical System Approach*, Oxford University Press.

[27] R. S. Tsay, 1986. "Non-linearity tests for time series", *Biometrika*, 73, pp. 461-466.

[28] R. S. Tsay, 2002. *Analysis of Financial Time Series*, New York, John Wiley.

[29] R. L. Tweedie, 1975. "Sufficient conditions for ergodicity and recurrence of Markov chains on a general state space", *Stochastic Processes and their Applications*, 3, pp. 385-403.

[30] Y. J. Whang and D. W. K. Andrews, 1993. "Tests of specification for parametric and semiparametric models", *Journal of Econometrics*, 57, pp. 277-318.

[31] H. White and I. Domowitz, 1984. "Nonlinear regression with dependent observations", *Econometrica*, 52, pp. 142-162.

[32] H. White and J. M. Wooldridge, 1991. "Some results on sieve estimation with dependent observations", in *Nonparametric and Semiparametric Methods in Econometrics and Statistics*, W. A. Barnett, J. Powel and G. Tauchen eds., Cambridge University Press.

[33] Q. Yao and H. Tong, 1998. "Cross-validatory bandwidth selections for regression estimation based on dependent data", *Journal of Statistical Planning and Inference*, 68, pp. 387-415.

Appendix

Obtaining geometric ergodicity in the context of the model in equation (2) requires certain assumptions about the nature of ϵ_t, the transfer function $\mathcal{T}(\cdot)$ and the autoregressive coefficients $\{\rho_i\}_{i=1}^p$. Since the model falls into the class of FAR and NAAR models of [7] and [8], one can use the results of these papers to obtain geometric ergodicity as follows. Let $\boldsymbol{Y}_{t-1}^* \stackrel{\text{def}}{=} (y_{t-\delta}, y_{t-\delta-1}, \dots, y_{t-\delta-q} \cap y_{t-d})^\top$ and re-write the model of equation (2) as:

$$y_t = \sum_{i=1}^p \varphi_i(\boldsymbol{Y}_{t-1}^*; \lambda) y_{t-i} + \varphi_0(\boldsymbol{Y}_{t-1}^*; \lambda) + \epsilon_t \qquad (31)$$

where $\varphi_i(\boldsymbol{Y}_{t-1}^*; \lambda) \stackrel{\text{def}}{=} w(y_{t-d}; \lambda)\rho_i$ and $\varphi_0(\boldsymbol{Y}_{t-1}^*; \lambda) \stackrel{\text{def}}{=} \bar{w}(y_{t-d}; \lambda)\mathcal{T}(y_{t-\delta})$, which is in the form of a FAR model.[14] Since $w(y_{t-d}; \lambda)$ and $\bar{w}(y_{t-d}; \lambda)$ are bounded we have that $|\varphi_i(\boldsymbol{Y}_{t-1}^*; \lambda)| \leq c_w |\rho_i|$ and $|\varphi_0(\boldsymbol{Y}_{t-1}^*; \lambda)| \leq \bar{c}_w |\mathcal{T}(y_{t-\delta})|$ for some constants $(c_w, \bar{c}_w)$. Suppose now that the following (sufficient) conditions are satisfied:

GE0. $\{\epsilon_t\}_{t\in\mathbb{Z}}$ is a sequence of independent and identically distributed random variables with mean zero and finite variance σ_ϵ^2.

GE1. ϵ_t is independent of y_{t-i} for all $i > 0$.

GE2. The density function of ϵ_t is continuous and positive everywhere on $\mathbb{E}$, where $\mathbb{E}$ is the compact support of ϵ_t on $\mathbb{R}$.

GE3. All the roots of $\mu^p - c_w \sum_{i=1}^p |\rho_i| \mu^{p-i} = 0$ are inside the unit circle.

GE4. The transfer function is $m \geq 1$ times continuously differentiable on $\mathbb{Y}^{q+1} \subset \mathbb{R}^{q+1}$, where $\mathbb{Y}$ is the compact suppport of y_t.

Assumptions [GE0] to [GE3] are the same as in [7] while assumption [GE4] implies that the transfer function is uniformly bounded with m bounded derivatives[15] for all $y_{t-\delta} \in \mathbb{Y}^{q+1}$.

If one considers the full model of equation (1) with auxiliary variables, a plausible extended set of assumptions, that covers the behavior of all the variables is given below.

A0. The density functions of $\{y_t, x_t\}$ are continuous and positive everywhere on $\{\mathbb{Y}, \mathbb{X}\}$, where $\{\mathbb{Y}, \mathbb{X}\}$ are the corresponding compact supports of $\{y_t, x_t\}$ on $\mathbb{R}$.

A1. $\{y_t, x_t, z_t\}$ are adapted to $\mathcal{F}_t \stackrel{\text{def}}{=} \sigma(y_s, x_s, z_s)_{s\leq t} \subset \mathcal{A}$, the σ-field generated by events occuring up to time t.

A2. $\{\epsilon_t\}_{t\in\mathbb{Z}}$ is a sequence of independent and identically distributed random variables with mean zero and finite moments of order $\nu_\epsilon \stackrel{\text{def}}{=} 2(r_\epsilon + 1)$ for some $r_\epsilon \geq 1$.

A3. ϵ_t is independent of $\{y_{t-i}, x_s, z_s\}$ for $i > 0$.

A4. $\{y_t, x_t, z_t\}_{t\in\mathbb{Z}}$ are jointly strictly stationary and strong mixing sequences with mixing coefficients $\left\{\alpha(m) \stackrel{\text{def}}{=} \alpha_0 \xi^m \; ; 0 < \xi < 1, \alpha_0 > 0, m > 0\right\}$ and finite moments of order $\nu \stackrel{\text{def}}{=} 2(r+1)$ for some $r > 0$.

---

[14] Defining $f_i(\boldsymbol{Y}_{t-1}^*; \lambda) \stackrel{\text{def}}{=} \varphi_i(\boldsymbol{Y}_{t-1}^*; \lambda) y_{t-i}$ and $f_0(\boldsymbol{Y}_{t-1}^*; \lambda) \stackrel{\text{def}}{=} \varphi_0(\boldsymbol{Y}_{t-1}^*; \lambda)$ we obtain the NAAR form of the model.

[15] This assumption also implies square integrability of the transfer function and its derivatives.

A5. The transfer function $\mathcal{T}(\boldsymbol{x}_{t-\delta})$ is additive to each of the transfer lags, that is

$$\mathcal{T}(\boldsymbol{x}_{t-\delta}) \stackrel{\text{def}}{=} \sum_{j=0}^{q} \mathcal{T}_j(x_{t-\delta-j}) \tag{32}$$

where each $\mathcal{T}_j(\cdot)$ is $m \geq 1$ times continuously differentiable on $\mathbb{X}$.

A6. For $p \geq 1$, define the matrix $\boldsymbol{C}(z_{t-d}; \boldsymbol{\lambda}) \stackrel{\text{def}}{=} \boldsymbol{w}(z_{t-d}; \boldsymbol{\lambda})\boldsymbol{\alpha}$ and the partial product sequence $\boldsymbol{H}_m(z_{t-d}; \boldsymbol{\lambda}) \stackrel{\text{def}}{=} \prod_{j=0}^{m-1} \boldsymbol{C}(z_{t-d-j}; \boldsymbol{\lambda})$, for $m = 1, 2, \ldots$ with the convention that $\boldsymbol{H}_j(z_{t-d}; \boldsymbol{\lambda}) = \boldsymbol{I}_p$ for $j \leq 0$. For all values of z_{t-d} and all admissible parameter vectors $\boldsymbol{\lambda}$, all the characteristic roots of $\boldsymbol{C}(z_{t-d}; \boldsymbol{\lambda})$ are less than one in absolute value and the sequence satisfies $\lim_{m \to \infty} \boldsymbol{H}_m(z_{t-d}; \boldsymbol{\lambda}) = \boldsymbol{0}$ element-wise.

The boundedness assumption [A0], used in combination with [A5], is needed for the approximation of the unknown transfer function. Assumptions [A2] and [A3] define the nature of the noise variable (exogenous to both auxiliary variables x_t and z_t) while assumption [A4] specifies the nature of the observable time series in the model. Joint stationarity and strong mixing with mixing coefficients that decay geometrically is also used in [13] and [32].[16] The geometric decay of the mixing coefficients also allows for the imposition of less stringent moment restrictions. Assuming that the unknown transfer function is additive, as in [A5] and [13], is restrictive but not considerably so since: (a) making it avoids the 'curse of dimensionality' that affects all nonparametric approximation problems and (b) is readily testable (as in the case of the NAARX model in [8]; see also [6] and [11]). Additive regression models, that impose a similar condition with $\mathcal{T}(\boldsymbol{x})$ being a nonparametric regression function, have been considered, among others, by [4] and [17], [18]. Note that interactions between different lags of the transfer variable can be incorporated in this additive context by writing the transfer function as:

$$\mathcal{T}(\boldsymbol{x}_{t-\delta}) \stackrel{\text{def}}{=} \sum_{j=0}^{q} \mathcal{T}_j(x_{t-\delta-j}) + \sum_{j=0}^{q} \sum_{s=j+1}^{q} \mathcal{T}_{js}(x_{t-\delta-j}, x_{t-\delta-s}) \tag{33}$$

which is of the form of an additive interactive regression of order two of [2]. However, we only consider the form of the transfer function given in [A5]. The rest of this assumption imposes a boundedness condition, similar to [GE4], on the transfer functions. Finally, assumption [A6] is used in obtaining the solution of the model with absolutely summable coefficients.

Finally, estimates of out-of-sample point forecasts for y_{t+h}, with $D \geq h \geq 1$, can be obtained using the recursive formula in equation (14) and by substituting the estimators $w(z_{t-d+h}; \widehat{\boldsymbol{\lambda}}_n)$ and $\widehat{\mathcal{T}}(\boldsymbol{x}_{t-\delta+h})$ in place of their corresponding unknown counterparts. The forecasts are functions of the estimated truncation point $\widehat{m}_n$ and we therefore have:

$$y_{n+h}^n(\widehat{m}_n) = w(z_{n-d+h}; \widehat{\boldsymbol{\lambda}}_n)\boldsymbol{y}_{n-1+h}^{n\top}(\widehat{m}_n)\widehat{\boldsymbol{\rho}}_n + \bar{w}(z_{n-d+h}; \widehat{\boldsymbol{\lambda}}_n)\widehat{\mathcal{T}}(\boldsymbol{x}_{n-\delta+h}) \tag{34}$$

If the forecasting horizon h is greater than D then the forecasts become functions of estimates of point forecasts of $x_{n-\delta+h}$ and z_{n-d+h}, obtained separately from appropriate models for x_t and z_t. Let $\widetilde{\boldsymbol{\xi}}_{n+h}^n \stackrel{\text{def}}{=} (\widetilde{x}_{n+h}^n, \widetilde{z}_{n+h}^n)^\top$ be the forecasts for the transfer and transition variable respectively. In this case, there are different methods for computing the forecast, see [16]. The easiest (but not necessarily best) methods are based on (a) the naïve forecast:

$$y_{n+h}^n(\widehat{m}_n, \widetilde{\boldsymbol{\xi}}_{n+h}^n) = w(\widetilde{z}_{n-d+h}; \widehat{\boldsymbol{\lambda}}_n)\boldsymbol{y}_{n-1+h}^{n\top}(\widehat{m}_n, \widetilde{\boldsymbol{\xi}}_{n-1+h}^n)\widehat{\boldsymbol{\rho}}_n + \bar{w}(\widetilde{z}_{n-d+h}; \widehat{\boldsymbol{\lambda}}_n)\widehat{\mathcal{T}}(\widetilde{\boldsymbol{x}}_{n-\delta+h}) \tag{35}$$

---

[16]Note that, as mentioned before, the same rate of decay for the mixing coefficients obtains, under geometric ergodicity, in the case no exogenous variables.

which directly substitutes the forecasts, $\tilde{x}^n_{n+h}$ and $\tilde{z}^n_{n+h}$, in the forecast function, and (b) the bootstrap forecast:

$$y^n_{n+h}(\widehat{m}_n, B) = \frac{1}{n}\sum_{j=1}^{n} w(\tilde{z}_{n-d+h,j}; \widehat{\lambda}_n)y^{n\top}_{n-1+h}(\widehat{m}_n, B) + \bar{w}(\tilde{z}_{n-d+h,j}; \widehat{\lambda}_n)\widehat{\mathcal{T}}(\tilde{x}_{n-\delta+h,j}) \qquad (36)$$

where $\tilde{x}^n_{n+h,j} \stackrel{\text{def}}{=} \tilde{x}^n_{n+h} + e_{xj}$ and $\tilde{z}^n_{n+h,j} \stackrel{\text{def}}{=} \tilde{z}^n_{n+h} + e_{zj}$, with e_{xj} and e_{zj} being the estimated residuals from the forecasting models for x_t and z_t.

The bootstrap can also be used to estimate forecast intervals as follows. Use the centered residuals from the model $q^*_t(\widehat{\vartheta}_n)$ and generate bootstraped residuals from their. Generate N bootstrap sample observations $Y^* \stackrel{\text{def}}{=} \{y^*_{tj}\}^{n,N}_{t=1,j=1}$ and compute the corresponding forecasts using one of the forecast functions in equations (34)-(36), say $Y^{*n}_{n+h} \stackrel{\text{def}}{=} \{y^{*n}_{n+i,j}\}^{h,N}_{i=1,j=1}$, where we suppress the dependence of the forecasts on $\widehat{m}_n$, the auxiliary variables and the estimated parameters. The empirical distributions of the bootstraped forecasts, for $i = 1, 2, \ldots h$, can now be used to obtain forecast intervals for any desired confidence level $(1 - \alpha) \times 100\%$. The arithmetic averages

$$\bar{y}^{*n}_{n+i} \stackrel{\text{def}}{=} N^{-1}\sum_{j=1}^{N} y^{*n}_{n+i,j}$$

and medians $\tilde{y}^{*n}_{n+i} \stackrel{\text{def}}{=} F^{-1}(Y^{*n}_{n+i}; 0.5)$ of the empirical distributions can also serve as additional estimates of point forecasts.

Brill Academic Publishers
P.O. Box 9000, 2300 PA Leiden
The Netherlands

Lecture Series on Computer
and Computational Sciences
Volume 3, 2005, pp. 360-383

Estimation of Diffusion Coefficients in Polymer Solutions

George D. Verros[1]
Department of Electrical Engineering,
Technological Educational Institute of Lamia,
GR-351 00 Lamia, Greece

Nikolaos A. Malamataris
Department of Mechanical Engineering,
Technological Educational Institute of W. Macedonia,
GR-501 00 Kila, Greece

Received 26 June, 2005; accepted 12 August, 2005

Abstract: In the present work a comprehensive methodology is proposed for the estimation of the Fickian diffusion coefficients in multi-component polymer systems. The idea is to study the evaporation process from multi-component polymer systems as a one-dimensional numerical experiment and to use polymer solution weight vs. time data to estimate the unknown parameters of the diffusion coefficient correlations based on the free volume theory. For this purpose, the evaporation process is modeled as a coupled heat and mass transfer problem with a moving boundary and the Galerkin finite element method is used to solve simultaneously the non-linear governing equations. The Fickian diffusion coefficients in a ternary (non)solvent(1)-solvent(2)-polymer system were calculated as a function of the solvents self-diffusion coefficients, the constituent chemical potentials and the process conditions by applying well established principles such as the Gibbs-Duhem theorem for diffusing systems. . This method is successfully applied to the estimation of the Fickian ternary diffusion coefficients in the water-acetone-cellulose acetate system and is valid over the whole range of temperature and concentration for practical applications in membrane technology. Additionally, how water affects the morphology of the final cellulosic membrane is discussed in detail by studying the concentration profiles of the constituents of the casting solution.

Keywords: Multi-component Diffusion, Solvent Evaporation, Free Volume Theory, Asymmetric Membrane Morphology, Mathematical Modeling,

1. Introduction

The diffusion coefficients in solvent(s)-polymer systems are of major importance for a number of industrial processes, including membrane manufacture[1-6], foam and coating formation[7], devolatilization[8] and the effectiveness of polymerization reactors at high conversion[9-14]. In addition the diffusion of small molecules in a polymer matrix is essential for miscellaneous polymer products such as controlled drug delivery systems, barrier materials and membranes for separation[5-7]. The industrial importance of diffusion coefficients in polymer systems has led to extensive work for their measurement and estimation. Measurement techniques include sorption and desorption, radiotracer methods, chromatography, nuclear magnetic resonance (NMR) experiments as reviewed by Crank and Park[15], Cussler[16-17], Tyrrell and Harris[18]. Predictive models based on the free volume concepts as developed by Vrentas - Duda and coworkers[19-21] may be used for the estimation of polymer-solvent diffusion coefficients. The existing experimental techniques and theoretical models have been developed before the successful implementation of powerful numerical tools like the finite

---

[1] Corresponding author.. E-mail: gdverros@otenet.gr, verros@vergina.eng.auth.gr

element method in the study of transport phenomena. The wide use of computer-aided analysis enables simpler methods such as the solvent evaporation method to be used for the estimation of diffusion coefficients in polymer solutions. The aim of the present work is to review the recent advance in the field of the solvent evaporation method.

This review is organized as follows: In the first section the fundamental aspects of multi-component diffusion are reviewed; In the next section, the solvent evaporation method is presented; The fundamental aspects of the method such as the process description, the governing equations along with the appropriate initial and boundary conditions and the methodology for solving the governing equations are described in detail. Finally, in the last sections the solvent evaporation method is applied to the ternary water-acetone-CA system, to estimate the ternary diffusion coefficients, results are presented and conclusions are drawn.

2. Fundamental Aspects of Multi-component Diffusion

The multi-component diffusion is based on the friction factor formalism. The origin of the friction coefficients concept can be found in the works of Einstein and Sutherland for binary diffusion[18]. Onsager[22], based on non-equilibrium thermodynamics, applied the friction factor concepts in multi-component systems for the fist time. In his view, the dissipation function $\mathbf{F}$, representing the free energy variation rate as a function of the local molar flux densities ($\mathbf{J_i}$, $\mathbf{J_k}$) and the friction factors (ζ_{ik}) is given as:

$$\mathbf{F} = \frac{1}{2}\sum_{i=1}^{N}\sum_{k=1}^{N}\zeta_{ik}(\mathbf{J_i J_k}) = \frac{1}{2}\sum_{i=1}^{N}\sum_{k=1}^{N}\zeta_{ik}c_i c_k(\mathbf{v_i v_k})$$

(1)

Where the local velocity vector $\mathbf{v}_i$ is defined as $\mathbf{J}_i/c_i$, where c_i is the molar concentration of the i-th substance.

When all the components in the mixture have the same velocity $\mathbf{v}$, the free energy density remains constant and the dissipation function $\mathbf{F}$ is zero. Consequently,

$$\frac{1}{2}\mathbf{v}^2\sum_{i=1}^{N}\sum_{k=1}^{N}\zeta_{ik}c_i c_k = 0 \quad \sum_{k=1}^{N}\zeta_{ik}c_i = 0$$

$$k = 1,2, ..N \; ; \, i = 1,2, ..N$$

(2)

He also showed that the rate of free energy variation reaches a maximum when the local gradients of chemical potential have the following form:

$$\frac{d\mu_i}{dz} = -\sum_{k=1}^{N}c_k\zeta_{ik}(\mathbf{v_i} - \mathbf{v_k})$$

$$i = 1,2, ..N$$

(3)

The friction coefficients defined in equations (1)-(3) are related to the usual Fickian diffusion equations using the equations for the diffusion flux ($\mathbf{j_i^{\#}}$) relative to the volume average velocity $\mathbf{v}^{\#23}$:

$$\mathbf{j_i^{\neq}} = -\sum_{j=1}^{N-1}D_{ij}\nabla\rho_j = \rho_i(\mathbf{v_i} - \mathbf{v}^{\neq}) \quad \mathbf{v}^{\neq} = \sum_{i=1}^{N}u_i\mathbf{v_i} \quad \sum_{i=1}^{N}\mathbf{j_i^{\neq}}V_i = 0$$

(4)

Where V_i represents the specific volume and u_i stands for the volume fraction of the i-th substance respectively. D_{ij} are the Fickian diffusion coefficients.

By using equations (1)-(4), one can directly write the Fickian diffusion coefficients D_{ij} in terms of the friction coefficients ζ_{ij}. Unfortunately, the concentration and temperature dependence of the friction coefficients is still unknown. Therefore, one has to resort to additional equations describing the friction coefficients variation in terms of well defined quantities such as self-diffusion coefficients.

According to Bearman[24] the self diffusion coefficient of the i-th substance is given as a function of the friction coefficients and the molar concentrations as:

$$D_i = \frac{RT}{c_i\zeta_{i^*i} + \sum_{\substack{j=1 \\ j \neq i}}^{N} c_j\zeta_{ij}}$$

(5)

ζ_{i^*i} represents the friction factor of i-th substance isotopes. In the above notation, a distinction was made between the isotopes friction coefficients as a result of the self-diffusion experimental measurement by the radiotracer technique. An excellent treatment of the subject is given by Tirrell and Harris[18].

In the above equations, there are more friction coefficients to define than available equations (high degree of freedom). This was the starting point for several workers[25-29] to develop physical theories. Most theories relate friction coefficients to standard properties such as molar volume, molecular weight etc. These theories have been recently reviewed by Price and Romdhane[30]. According to these workers the developed physical theories can be classified in terms of the following equation:

$$\frac{\zeta_{ij}}{\zeta_{ik}} = \frac{\alpha_j \ V_j \ M_j}{\alpha_k \ V_k \ M_k}$$

(6)

Where α_i are physical constants or functions used to scale the friction coefficients, V_i represents specific volume and M_i stands for the molecular weight of the i-th substance, respectively. The resulting theories for various α_i values are summarized in Table 1.

Table 1: Models for Multi-component Diffusion

Model	References
$\alpha_i = 1$, i=1,2..N	Alsoy & Duda [26]
$\alpha_i = V_i^{-1}$, i=1,2..N	Zielinski & Hanley [27]
$\alpha_i = 0$, i≠N, $\alpha_N \neq 0$	Dabral et al. [28]

As it can be shown, most theories assume constant friction coefficients ratio. A deeper question arises from equation (6): Is the ratio of the friction coefficients constant in diffusing polymer solutions?

As it was shown by Bearman[31] the mutual diffusion coefficient in the case of binary solutions and the constant friction factors ratio is given as a function of solvent mass concentration ρ_1, chemical potential μ_1 and the self-diffusion coefficient D^*_1 by the following equation[32]:

$$D = \frac{D^*_1}{RT} \frac{\partial\mu_1}{\partial \ln\rho_1}$$

(7)

This equation can be used to express the binary Fickian diffusion coefficient in terms of the solvent self-diffusion coefficient. The authors are aware that more rigorous approaches correlating the self- and the binary diffusion coefficients over a wide solvent concentration range including the dilute region appeared in the open literature[33]. However, one could use equation (7) correlating the self-diffusion coefficient with the binary diffusion coefficient to check the validity of the assumptions made in its derivation such as the constant friction coefficients ratio.

This task is achieved by using the free-volume theory[33-34] to describe the dependence of the solvent self-diffusion coefficient on temperature and concentration as well as Flory-Huggins thermodynamics[35] for chemical potential. As it was shown by several workers, the binary and self-diffusion coefficients for binary solvent-polymer systems are satisfactorily correlated by the above equation thus validating the assumption of the constant friction coefficients ratio for the binary systems[36-38

However, Zielinski and Alsoy[39] in a subsequent work have raised doubts about the models based on the assumption of constant friction factor ratios in multi-component solutions. Moreover, there is no clear reason to apply one theory (see Table 1) instead of another.

The aim of this work is to overcome these difficulties and suggest a unique theory for diffusion in multi-component polymer solutions. The starting point in our analysis is the definition of Onsager L_{ij} mobility coefficients[22]. Following Onsager we shall assume that the molar fluxes J_i in a n-component system can be expressed as a linear combination of the vector of chemical potential gradients X_i and a $N x N$ matrix of Onsager L_{ij} mobility coefficients:

$$J_i = \|L_{ij}\| X_i \; ; \; X_i = -\text{grad}\mu_i \; ; \; J_i = c_i \left(v_i - v^R \right) \tag{8}$$

Where v^R is refers to the velocity of an arbitrary reference framework. By inverting the above equations one gets:

$$X_i = \|L_{ij}\|^{-1} J_i = \|\zeta_{ij}\| J_i \; ; \; i = 1,2..N \; ; \; j=1,2..N \tag{9}$$

By applying the Gibbs-Duhem theorem for isothermal and isobaric systems equation (9) is written as [18]:

$$\sum_{k=1}^{N} c_k \frac{d\mu_k}{dz} = 0$$

or

$$\sum_{k=1}^{N}\sum_{i=1}^{N} \zeta_{ik} c_k J_i = \sum_{i=1}^{N} J_i \sum_{k=1}^{N} c_k \zeta_{ik} = 0$$

or

$$\sum_{k=1}^{N} c_k \zeta_{ik} = 0 \qquad i=1,2..N \tag{10}$$

By multiplying the above equation by $(v_i - v^R)$ and subtracting the result from equation (9) equation (3) is derived [18,22]:

$$\frac{d\mu_i}{dz} = -\sum_{k=1}^{N} c_k \left[\zeta_{ik}(v_i - v^R) - \zeta_{ik}(v_k - v^R) \right] = -\sum_{k=1}^{N} c_k \left[\zeta_{ik}(v_i - v_k) \right] \tag{11}$$

Finally, Miller[40] using equations (10) and (11) derived the Onsager reciprocal conditions[22]:

$$\zeta_{ik} = \zeta_{ki} \quad i = 1,2..N \; ; \; k=1,2..N \tag{12}$$

By eliminating the concentrations in equation (10) and applying the Onsager reciprocal rule the following equation is derived [41]:

$$\zeta_{ij}^2 = \zeta_{ii}\zeta_{jj} \tag{13}$$

This equation was also proposed by Price and Romdhane[30]. The importance of the above equation is shown in subsequent paragraphs
It is suggested by most workers in the field (see Table 1) that the ratio of friction coefficients is constant. Equation (10) is re-written as

$$\zeta_{ii} + \sum_{\substack{j=1 \\ j \neq i}}^{n} \zeta_{ji}(c_j / c_i) = 0 \qquad i=1,2..N \tag{14}$$

The following equation also holds [18]:

$$\sum_{i=1}^{n} c_i \upsilon_i = 1 \tag{15}$$

By using equations (13) - (15), taking into account the constant ratio of friction coefficients (arbitrary physical constants) and taking the derivatives with respect to molar concentration, one directly derives the following equation for the ratio of friction coefficients:

$$(\zeta_{ii}/\zeta_{jj})^{0.5} = \upsilon_i/\upsilon_j \quad \underset{or}{\zeta_{ik}/\zeta_{jk}} = \upsilon_i/\upsilon_j = \frac{V_j}{V_k}\frac{M_j}{M_k} \tag{16}$$

This equation, also proposed by Alsoy and Duda (Table 1), simultaneously satisfies the Onsager reciprocal condition as well as the Gibbs-Duhem relation along with the concept of constant friction coefficients ratio.

Regarding the multi-component system consisting of n-1 low molecular weight substances and a polymer, the application of the Gibbs-Duhem theorem leads to the calculation of the friction coefficients if the low molecular weight substances self diffusion coefficients are known. For example in the ternary system (non)-solvent(1)-solvent(2)-polymer(3) system the ternary diffusion coefficients, D_{ij}, are related to the friction coefficients and the thermodynamic properties by the following equations directly derived from eq. (4) and (11):[42]

$$D_{11} = -\frac{V_1}{N^2 E}\left(E_{22}\frac{\partial\mu_1}{\partial u_1} - E_{12}\frac{\partial\mu_2}{\partial u_1}\right) \quad D_{12} = -\frac{V_2}{N^2 E}\left(E_{22}\frac{\partial\mu_1}{\partial u_2} - E_{12}\frac{\partial\mu_2}{\partial u_2}\right),$$

$$D_{21} = -\frac{V_1}{N^2 E}\left(E_{11}\frac{\partial\mu_2}{\partial u_1} - E_{21}\frac{\partial\mu_1}{\partial u_1}\right) \quad D_{22} = -\frac{V_2}{N^2 E}\left(E_{11}\frac{\partial\mu_2}{\partial u_2} - E_{21}\frac{\partial\mu_1}{\partial u_2}\right) \tag{17}$$

where N is Avogadro's number; E_{ij} and E are defined as

$$E_{11} = \frac{V_1 u_2 \zeta_{12}}{\upsilon_2 u_3} - \frac{RTV_1(1-u_2)}{N^2 D_{T1}u_1 u_3} \quad E_{12} = \frac{(1-u_1)\zeta_{12}}{M_2 u_3} - \frac{RTV_2}{N^2 D_{T1}u_3}$$

$$E_{21} = \frac{(1-u_2)\zeta_{12}}{M_1 u_3} - \frac{RTV_1}{N^2 D_{T2}u_3} \quad E_{22} = \frac{V_2 u_1 \zeta_{12}}{\upsilon_1 u_3} - \frac{RTV_2(1-u_1)}{N^2 D_{T2}u_2 u_3}$$

$$E = -\frac{\zeta_{12}^2}{M_1 M_2 u_3} + \frac{R^2 T^2 V_1 V_2}{N^4 D_{T1} D_{T2} u_1 u_2 u_3} \quad D_{Ti} = \frac{D_i^*}{1-(D_i^*/D_i^0)} \approx D_i^* \quad D_i^0 = \frac{RTM_i \upsilon_i}{u_i \zeta_{ii} N^2} \tag{18}$$

Here, M_i and υ_i are the molecular weight and pure molar volume of the i-th component, and ζ_{ij} represents the friction coefficient between components i and j, respectively. R represents the universal gas constant, T stands for temperature in Kelvin and V_i is the specific volume of the i-th component. In the above equations u_i represents the volume fraction of the i-th substance respectively.

The chemical potentials μ_i are directly calculated as a function of polymer solution temperature and volume fractions in the ternary system using Flory-Huggins theory[35].

D_i^* represent the self diffusion coefficient of the i-th component in the ternary solution and are given as follows[43]:

$$D_1^* = D_{01} \exp\left(\frac{-\left(\omega_1 V_1^* + \omega_2 V_2^* \xi_{13}/\xi_{23} + \omega_3 V_3^* \xi_{13}\right)}{V_{FH}/\gamma} \right)$$

(19)

$$D_2^* = D_{02} \exp\left(\frac{-\left(\omega_1 V_1^* \xi_{23}/\xi_{13} + \omega_2 V_2^* + \omega_3 V_3^* \xi_{23}\right)}{V_{FH}/\gamma} \right)$$

(20)

$$V_{FH}/\gamma = \sum_{i=1}^{3} \frac{K_{1i}}{\gamma}(K_{2i} - T_{gi} + T)\omega_i$$

(21)

D_{0i} is a pro-exponential factor, V_{FH} is the average hole free volume per kg of the solution and γ is an overlap factor, which is introduced, because the same free volume is available to more than one molecule. V_i^* is the specific critical hole free volume of the i-th component required for a diffusion jump and ξ_{i3} represent the ratio of the critical molar volume of the jumping unit of ith-solvent to that of the polymer. K_{1i} and K_{2i} are free volume parameters for the i-th component and T_{gi} is the glass transition temperature. In the above equations ω_i represents the weight fraction of the i-th substance respectively.

In the equations (17)-(21) the Fickain diffusion coefficients D_{ij} are related not only to self-diffusion coefficients but also to the ζ_{12} friction coefficient which must be determined. This could be achieved by estimating ζ_{12} from binary diffusion coefficient data for the solvent(1)-solvent(2) system in the limit of zero polymer concentration[44-49] or by physical theories (Table 1). An alternative way is presented in this work; The Gibbs-Duhem equations (Eq. 10) for a diffusing ternary system are written as:

$$\zeta_{12} = \sqrt{\zeta_{11}\zeta_{22}} \, , \, c_1\zeta_{11} + c_2\zeta_{12} + c_3\zeta_{13} = 0 \, ,$$
$$c_2\zeta_{22} + c_1\zeta_{12} + c_3\zeta_{23} = 0 \, , \, c_3\zeta_{33} + c_2\zeta_{23} + c_3\zeta_{13} = 0$$

(22)

The friction coefficients ζ_{13} and ζ_{23} are related to ζ_{12} in terms of the respective Bearman equations for the self-diffusion coefficients (eq. 15) as follows[42]:

$$\zeta_{13} = \frac{\upsilon_3}{u_3}\left(\frac{RT}{N^2 D_{T1}} - \frac{u_2}{\upsilon_2}\right) \, ; \quad \zeta_{23} = \frac{\upsilon_3}{u_3}\left(\frac{RT}{N^2 D_{T2}} - \frac{u_1}{\upsilon_1}\right) \, ; \quad D_{Ti} = \frac{D_i^*}{1-(D_i^*/D_i^0)} \approx D_i^* \, ;$$

$$D_i^0 = \frac{RTM_i\upsilon_i}{u_i\zeta_{i*i}N^2} \, ; i=1,2$$

(23)

According to Vrentas and Duda[33-34] one could assume $(D_i^*/D_i^0) = 0$ in the case of moderate concentrated solution ($\omega_3 > 0.1$) as the error introduced by the above simplification is quite small. Equations (22)-(23) result for a ternary system in a simple quadratic equation which is directly solved to calculate ζ_{12} and consequently D_{ij} in terms of the self-diffusion coefficients and the process conditions. In the following section, the previously described methodology is applied to estimate the ternary diffusion coefficients by the solvent evaporation method.

3. The Solvent Evaporation Method

3.1 Method Description
This method is based on a very simple experimental set-up. More specifically, the evaporation rate from polymer solutions cast in the form of thin films on a glass substrate is measured gravimetrically (Figure 1). To prevent heat exchange between the glass slides and the microbalance pan, the glass

slides were placed on an insulating block resting upon the microbalance pan. The microbalance is exposed to the ambient air, so that accumulation of solvent vapor in the vicinity of the equipment is avoided and zero solvent concentration in the bulk

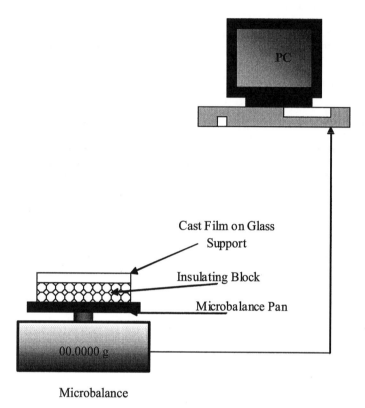

Figure 1: The solvent evaporation method

The idea of the solvent evaporation method is to study the evaporation process from appropriate polymer solutions cast on a glass plate as a one-dimensional numerical experiment and to estimate by comparison with gravimetric data the unknown diffusion parameters of the Vrentas-Duda self-diffusion equations (Eq. 19-21). In this method gravimetric data of the solvent evaporation rate is compared with predictions utilizing non-linear regression analysis. The objective function requires the sum of squares of differences between predicted and measured solvent evaporation rate to be minimal. The primary unknown in the parameter estimation procedure is ξ (see eq. 19-21). If the resulting fitting between model predictions and experimental data was not satisfactory, then, quantities D_0 and E are added to the estimation procedure, thus leading to the complete estimation of diffusion coefficients.

This a not a new idea; Ataka and Sasaki[50] first proposed that gravimetric measurements of the solvent evaporation rate could be used for estimation of the diffusion coefficients in polymer solutions. However, this idea was only realized recently by the implementation of powerful numerical methods such as Galerkin finite elements in the study of computational transport phenomena. Price et al.[51] were the first who got estimates of the diffusion coefficients by fitting drying rate in order to optimize the performance of industrial dryers. However, Verros and Malamataris[37,49,52] further developed this method. The evaporation process was studied as a one-dimensional numerical experiment utilizing the Galerkin finite element method. The numerical technique provides simultaneous solution of the model equations and yields in comparison with gravimetric data for the evaporation rate, the diffusion coefficients of the solvent(s) in polymer over a wide range of temperature and composition.

The systems studied include the acetone-CA system[37], the acetone-water[49], the poly(vinyl acetate) solution with acetone, methyl acetate or chloroform[52]. The estimated diffusion coefficients based on the free volume theory were in satisfactory agreement with experimental data measured by other techniques[53-62].

The aim of the present work is to calculate the diffusion coefficients in ternary solution by the solvent evaporation method. This method was successfully applied in our previous work for the determination of water-CA system.

However, recent advances in the field of multi-component diffusion, as is developed in the section 2, leads to re-examination of the method for the multi-component systems. The task of the present work is to apply the solvent evaporation method in the system water-acetone-CA by taking into account the recent developments in multi-component diffusion. In what follows the main issues of this method such as the model equations and the Galerkin Finite Element Method are discussed in detail.

3. 2 Model Equations

Figure 2 depicts the solvent evaporation process of a ternary system. The solution is cast on a solid support with thickness L_{sup}, which rests upon an insulating block. Heat is exchanged between the solution and the support as well as between the solution and the environment. Initially, the whole system (solution, ambient air, support) is at temperature T_0 and the polymer solution thickness is L_0. At time t=0 the ternary polymer solution is exposed to the environment; the solvent(s) begin to evaporate. At time **t** the polymer solution extends from coordinate **z=0** at the upper glass plate to **z=L(t)** at the gas-liquid moving interface, while glass support has a constant thickness extending from **z=0** to **z=-L$_{sup}$**. From a modeling point of view this is a coupled heat and mass transfer process with a moving boundary. Since diffusion is much slower than the relaxation mechanisms of the polymer chains in the solution, we assume pure Fickian diffusion[44-49,63-66]. Additionally, due to the relative small initial thickness of the polymer solution (order of μm) compared to the width and length (order of cm), the process is considered as 1-dim model.

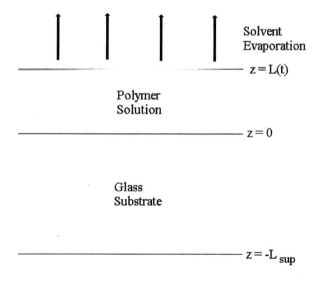

Figure 2: Schematic representation of the solvent evaporation process

The dimensionless governing equations, that describe the conservation of mass and energy in an one phase ternary system and the support are:

$$\frac{\partial u_1}{\partial \tau} = \frac{\partial}{\partial \eta}\left(C_1 \frac{\partial u_1}{\partial \eta}\right) + \frac{\partial}{\partial \eta}\left(C_2 \frac{\partial u_2}{\partial \eta}\right) ; \quad \eta = z/L_0 ; \quad \tau = D_0 t / L_0 ; C_1 = D_{11}/D_0 ;$$
$$C_2 = (V_1/V_2)D_{12}/D_0 ; \quad 0 < \eta < s = L(t)/L_0 \quad \quad (24)$$

$$\frac{\partial u_2}{\partial \tau} = \frac{\partial}{\partial \eta}\left(C_3 \frac{\partial u_1}{\partial \eta}\right) + \frac{\partial}{\partial \eta}\left(C_4 \frac{\partial u_2}{\partial \eta}\right) ; C_3 = (V_2/V_1)D_{21}/D_0 ; C_4 = D_{22}/D_0 ; \quad 0 < \eta < s \quad (25)$$

$$C_5 \frac{\partial \Theta}{\partial \tau} = \frac{\partial}{\partial \eta}\left(C_6 \frac{\partial \Theta}{\partial \eta}\right); \quad \Theta = T/T_0 ; \quad C_5 = \rho C_p / \rho_0 C_{p0} ; \quad C_6 = k/D_0 \rho_0 C_p ; \quad 0 < \eta < s \quad (26)$$

$$C_7 \frac{\partial \Theta_{sup}}{\partial \tau} = \frac{\partial}{\partial \eta}\left(C_8 \frac{\partial \Theta_{sup}}{\partial \eta}\right) ; \quad C_7 = \rho_{sup}C_{psup}/ \rho_0 C_{p0} ; \quad C_8 = k_{sup}/\rho_0 C_{p0} D_0$$
$$-L_{sup}/L_0 < \eta < 0 \quad \quad (27)$$

u_i and V_i are the volume fractions and the partial specific volume of i-th substance, respectively. In this work subscript 1 refers to solvent(1), 2 denotes solvent(2) and 3 represents polymer . $\eta = z/L_0$ is the dimensionless space coordinate; $\tau = D_0 t / L_0^2$ is dimensionless time; $\Theta = T/T_0$ is the dimensionless temperature; $s = L(t)/L_0$ is the dimensionless position of the moving boundary. D_{ij} are the appropriate Fickian diffusion coefficients for the ternary system. Their estimation is the major task of this work, as discussed in the introduction. Equations for D_{ij} are presented in the previous section. T represents the temperature, t denotes time and D_0 is a scaling factor having the units of diffusion coefficient. ρ, C_p and k represent density, specific heat capacity and the thermal conductivity of the polymer solution, respectively. C_{p0}, ρ_0 are scaling factors having units of specific heat capacity and density, respectively. Subscript 'sup' denotes properties and variables of the support.

Initial and boundary conditions for the diffusion equation:

$$u_1 = u_{10} ; u_2 = u_{20} ; \quad\quad\quad\quad\quad\quad\quad\quad\quad\quad \tau = 0 \quad\quad (28)$$

$$C_1 \partial u_1/\partial \eta + C_2 \partial u_2/\partial \eta = C_9 ; \quad C_9 = -L_0 N_W V_1/ D_0 \quad\quad \eta = s \quad\quad (29)$$

$$C_3 \partial u_1/\partial \eta + C_4 \partial u_2/\partial \eta = C_{10} ; C_{10} = -L_0 N_A V_2/ D_0 \quad\quad \eta = s \quad\quad (30)$$

$$\partial u_1/\partial \eta = 0 ; \partial u_2/\partial \eta = 0 \quad\quad\quad\quad\quad\quad\quad\quad \eta = 0 \quad\quad (31)$$

Equation (28) gives the initial concentration for the solvent(s) . Equations (29)-(30) are mass balances at the moving interface and equation (31) specifies zero mass flux at the glass plate. N_w and N_A are the mass flux at the gas-liquid interface of the solvent (1) and (2), respectively
Initial and boundary conditions for the energy equations:

$$\Theta = \Theta_{sup} = 1 \quad\quad\quad\quad\quad\quad\quad\quad\quad\quad\quad\quad \tau = 0 \quad\quad (32)$$

$$C_6 \partial \Theta/\partial \eta = C_{11} (\Theta - 1) + C_{12} (1 - \Theta_s^4) - C_{13} ; \quad C_{11} = L_0 h/ D_0 \rho_0 C_{p0} ;$$
$$C_{12} = L_0 \varepsilon \sigma T_0^3 / D_0 \rho_0 C_{p0} ; \quad C_{13} = ((\Delta H_1) N_A + (\Delta H_2) N_W)L_0/(T_0 D_0 \rho_0 C_{p0}) ; \quad \eta = s \quad\quad (33)$$

$$C_8 \partial \Theta_{sup}/\partial \eta = C_6 \partial \Theta/\partial \eta \quad\quad\quad\quad\quad\quad\quad\quad \eta = 0 \quad\quad (34)$$

$$\partial \Theta_{sup}/\partial \eta = 0 \quad\quad\quad\quad\quad\quad\quad\quad\quad\quad \eta = -L_{sup}/L_0 \quad\quad (35)$$

Equation (32) gives the initial temperature in the support and the solution. Equation (33) is an energy balance at the polymer solution-gas interface, taking into account heat transfer to the polymer solution due to free convection and radiation as well as latent heat loss due to acetone and water evaporation. Equation (34) accounts for continuity of heat flux at the glass plate-polymer solution interface, while equation (35) stands for perfect insulation of the glass support lower surface. Θ_s denotes the dimensionless temperature of the liquid-gas interface, h is the heat transfer coefficient, ε is the emissivity of the polymer solution, σ denotes the Stefan-Boltzman constant and ΔH_i is the i-th substance latent heat of vaporization.

Finally, the instantaneous dimensionless solution thickness $s = L(t)/L_0$ is obtained from the following differential equation :

$$u_3 ds/d\tau = C_9 + C_{10}; \qquad\qquad \tau = 0, s = 1 \qquad (36)$$

An alternative equation is obtained from the conservation of the polymer mass:

$$\int_0^s (1 - u_1 - u_2)d\eta = \int_0^1 (1 - u_{10} - u_{20})d\eta \qquad (37)$$

Eq. (36),(37) define the instantaneous position of the moving boundary in terms of the polymer volume fraction $u_3 = (1 - u_1 - u_2)$. Eq. (36) is a differential mass balance and Eq. (37) is the integral mass balance for the polymer, since only acetone and water evaporate from the polymer solution. A comparison between the computational efficiency and the predictions of Eq. (36) and (37) was given in the finite element formulation section. These equations are solved along with the governing equations of the model (eq. (24)-(27)) to give the concentration profiles of acetone and water along with the temperature of the solution and the support and the position of the moving boundary as a function of time.

3.3 Finite Element Formulation

The computational domain is discretized in 70 finite elements. The unknown volume fractions, u_j and the temperature Θ are expanded in terms of quadratic Galerkin basis functions, φ^i as:

$$u_j = \sum_{i=1}^{3} u_j^i \varphi^i \qquad j = 1,2 \ ; \qquad \Theta = \sum_{i=1}^{3} \Theta^i \varphi^i \qquad (38)$$

In the nodes of the support, the energy equation (Eq. 27) is evaluated, the only unknown is the temperature and this part of the domain is fixed. In the polymer solution both the energy and mass equations are evaluated (Eq. 24-26), each node has the temperature and the volume fractions as unknowns and this part of the domain deforms according to the motion of the moving boundary. Finally, the boundary position, s, (Eq. 36-37) is added as unknown at the last node of the computational domain. The governing equations weighted integrally with the basis functions resulting in the following mass R_M^i, energy R_E^i and kinematic R_K^N residuals:

$$R_{M1}^i = \int_0^s \left[\frac{\partial u_1}{\partial \tau} - \frac{\partial}{\partial \eta}\left(C_1 \frac{\partial u_1}{\partial \eta} \right) - \frac{\partial}{\partial \eta}\left(C_2 \frac{\partial u_2}{\partial \eta} \right) \right] \varphi^i d\eta \qquad (39)$$

$$R_{M2}^i = \int_0^s \left[\frac{\partial u_2}{\partial \tau} - \frac{\partial}{\partial \eta}\left(C_3 \frac{\partial u_1}{\partial \eta} \right) + \frac{\partial}{\partial \eta}\left(C_4 \frac{\partial u_2}{\partial \eta} \right) \right] \varphi^i d\eta \qquad (40)$$

$$R_{E,sol}^{i} = \int_{0}^{s}\left[C_5\frac{\partial\Theta}{\partial\tau} - \frac{\partial}{\partial\eta}\left(C_6\frac{\partial\Theta}{\partial\eta}\right)\right]\varphi^i d\eta \qquad (41)$$

$$R_{E,sup}^{i} = \int_{-L/L_{sup}}^{0}\left[C_7\frac{\partial\Theta_{sup}}{\partial\tau} - \frac{\partial}{\partial\eta}\left(C_8\frac{\partial\Theta_{sup}}{\partial\eta}\right)\right]\varphi^i d\eta \qquad (42)$$

$$R_K^N = u_3 ds/d\tau - C_9 - C_{10} \text{ or } R_K^N = \int_0^s(1-u_1-u_2)d\eta - \int_0^1(1-u_{10}-u_{20})d\eta \qquad (43)$$

In order to account for the moving boundary, the total time derivatives for the volume fraction and the temperature were calculated, introducing convective terms in the governing equations[67]:

$$du_i/d\tau = \partial\partial u_i/\partial\tau + (d\eta/d\tau)(\partial u_i/\partial\eta) \quad ; \quad i=1,2 \qquad (44)$$

$$d\Theta/d\tau = \partial\Theta/\partial\tau + (d\eta/d\tau)(\partial\Theta/\partial\eta) \qquad (45)$$

By substituting the above equations into the weighted residuals (39)-(41) we obtain the following equations:

$$R_{M1}^{i} = \int_{0}^{s}\left[\frac{\partial u_1}{\partial\tau} - \frac{d\eta}{d\tau}\frac{\partial u_1}{\partial\eta} - \frac{\partial}{\partial\eta}\left(C_1\frac{\partial u_1}{\partial\eta}\right) - \frac{\partial}{\partial\eta}\left(C_2\frac{\partial u_2}{\partial\eta}\right)\right]\varphi^i d\eta \qquad (46)$$

$$R_{M2}^{i} = \int_{0}^{s}\left[\frac{\partial u_2}{\partial\tau} - \frac{d\eta}{d\tau}\frac{\partial u_2}{\partial\eta} - \frac{\partial}{\partial\eta}\left(C_3\frac{\partial u_1}{\partial\eta}\right) - \frac{\partial}{\partial\eta}\left(C_4\frac{\partial u_2}{\partial\eta}\right)\right]\varphi^i d\eta \qquad (47)$$

$$R_{E,sol}^{i} = \int_{0}^{s}\left[C_5\frac{\partial\Theta}{\partial\tau} - \frac{d\eta}{d\tau}C_5\frac{\partial\Theta}{\partial\eta} - \frac{\partial}{\partial\eta}\left(C_6\frac{\partial\Theta}{\partial\eta}\right)\right]\varphi^i d\eta \qquad (48)$$

In order to decrease the order of differentiation and project the Neuman (natural) boundary conditions of the problem, integration by parts (divergence theorem in one dimension) is applied and the weighted residuals become:

$$R_{M1}^{i} = \int_{0}^{s}\left[\frac{\partial u_1}{\partial\tau}\varphi^i - \frac{d\eta}{d\tau}\frac{\partial u_1}{\partial\eta}\varphi^i + \frac{\partial\varphi^i}{\partial\eta}\left(C_1\frac{\partial u_1}{\partial\eta}\right) + \frac{\partial\varphi^i}{\partial\eta}\left(C_2\frac{\partial u_2}{\partial\eta}\right)\right]d\eta$$
$$- \left(C_1\frac{\partial u_1}{\partial\eta} + C_2\frac{\partial u_2}{\partial\eta}\right)\varphi^i\Bigg|_{\eta=0}^{\eta=s} \qquad (49)$$

$$R_{M2}^i = \int_0^s \left[\frac{\partial u_2}{\partial \tau} \varphi^i - \frac{d\eta}{d\tau} \frac{\partial u_2}{\partial \eta} \varphi^i + \frac{\partial \varphi^i}{\partial \eta} \left(C_3 \frac{\partial u_1}{\partial \eta} \right) + \frac{\partial \varphi^i}{\partial \eta} \left(C_4 \frac{\partial u_2}{\partial \eta} \right) \right] d\eta$$

$$- \left(C_3 \frac{\partial u_1}{\partial \eta} + C_4 \frac{\partial u_2}{\partial \eta} \right) \varphi^i \Bigg|_{\eta=0}^{\eta=s}$$

(50)

$$R_{E,sol}^i = \int_0^s \left[C_5 \frac{\partial \Theta}{\partial \tau} \varphi^i - \frac{d\eta}{d\tau} \frac{\partial \Theta}{\partial \eta} C_5 \varphi^i + \frac{\partial \varphi^i}{\partial \eta} \left(C_6 \frac{\partial \Theta}{\partial \eta} \right) \right] d\eta - C_6 \frac{\partial \Theta}{\partial \eta} \varphi^i \Bigg|_{\eta=0}^{\eta=s}$$

(51)

$$R_{E,sup}^i = \int_{-L/L_{sup}}^{0} \left[C_7 \frac{\partial \Theta_{sup}}{\partial \tau} \varphi^i + \frac{\partial \varphi^i}{\partial \eta} \left(C_8 \frac{\partial \Theta_{sup}}{\partial \eta} \right) \right] d\eta - C_8 \frac{\partial \Theta_{sup}}{\partial \eta} \varphi^i \Bigg|_{\eta=-L_{sup}/L_0}^{\eta=0}$$

(52)

In equations (49)-(52) the Neuman boundary conditions at the ends of the computational domain are substituted by the corresponding equations (29)-(31) and (33)-(35). It is worth noting that due to the integration by parts, the continuity of the heat fluxes at the upper surface of the glass plate (Eq. 34) is satisfied automatically.

The residuals are evaluated numerically using three point Gaussian integration. The time integration follows the Euler backward scheme. A system of non-linear algebraic equations results that is solved with the Newton-Raphson iterative method according to scheme $q^{(n+1)}=q^{(n)}-J^{-1}R(q^{(n)})$, where $q^{(n)}$ is the vector of unknowns of the n-th iteration and J is the Jacobian matrix of residuals R with respect to the

nodal unknowns $q^{(n)}$. The banded matrix of the resulting linear equations is solved with a frontal solver[68] at each iteration in the case of differential moving boundary equation (Eq. 36) and with full Gaussian elimination in the case of the integral moving boundary equation (Eq. 37). The time step was equal to 10^{-4}. The computer program exhibits quadratic convergence in 4-6 iterations at each time step. Any additional mesh refinement or time step decrease has an improvement of less than 10^{-6} in the accuracy of the solution.

A detailed presentation of the finite element technique that enables the simultaneous solution of the primary unknowns of the problem (volume fractions and temperature) with the moving boundary can be found elsewhere[69-70]. Finally, the two moving boundary equations give identical results for arbitrary initial conditions. However, the differential moving boundary equation is superior to the corresponding integral equation since it requires considerably smaller computer memory and CPU time for execution, due to the application of frontal methods.

It should be noted, that apart from the complexities of the model due to the presence of a moving boundary and the non-linear boundary conditions, the solution of the problem is further complicated due to the dependence of the model parameters on temperature and concentration, which are introduced in the next section.

3.4 The Model Parameters

Cellulose acetate membranes are widely used in a number of industrial processes such as separations, solution concentration, water desalination, waste purification, etc. These membranes are manufactured by two major processes[5]: the dry cast and the wet-cast phase-inversion process. In the dry cast process a solution of CA/solvent/non-solvent is allowed to solidify by solvent evaporation. In the wet-cast phase-inversion process the casting solution is partially concentrated by solvent evaporation and then is solidified by immersion in a low temperature non-solvent bath. Several recipes and modifications based on practical experience have appeared in the literature. Most recipes utilize acetone as a solvent.

The industrial importance of the acetone evaporation as a major step in the dry cast and as a precursor step in the wet-cast phase-inversion process has led to extensive experimental[1-7,50,71-74], and modeling[44-48,65-66] studies. In both experimental and modeling studies it is evident that the solvent evaporation process is controlled by the diffusion of the solvent in the polymer solution. Therefore, the magnitude of the ternary diffusion coefficients is of fundamental importance for the cellulose acetate membrane formation. Similar conclusions have been made in the modeling of the polymeric coating drying process[75-77] which exhibits analogous coupled heat and mass transfer phenomena.

In spite of its industrial importance little is known for the ternary diffusion coefficients. The aim of the following sub-sections is to identify the model parameters for the water-acetone-CA system.

Thermophysical properties of the polymer solution and the glass substrate. The density, the specific heat capacity and the thermal conductivity of the polymer solution can be calculated by a simple addition rule, assuming constant partial properties:

$$P = P_1\omega_1 + P_2\omega_2 + P_3 (1-\omega_1-\omega_2) \tag{53}$$

where **P** is the property of the solution, P_i and ω_i denote the corresponding property and the weight fraction of the i-th pure substance. The above equation of state was validated in the case of polymer solution density against experimental data for the limiting cases of acetone-CA[12] and acetone-water solutions and the maximum difference for the ternary polymer solution density did not exceed 3%.

The weight fraction ωi is related to the volume fraction, u_i as follows:

$$\omega_i = (u_i/V_i) \Big/ \sum_{i=1}^{3} (u_i/V_i) \tag{54}$$

The thermophysical properties of the polymer solution constituents and the glass support are given as a function of the solution temperature in standard references[78-80].

Diffusion prameters. The parameters K_{1i}/γ and K_{2i}-T_{gi} (eq. 19-21) can be obtained by fitting viscosity-temperature data of the pure substances. These parameters are for the acetone equal to $1.86.10^{-6}$ m^3/kgK, -53.33 K and for water equal to $2.18.10^{-6}$ m^3/kgK, -152.29 K, respectively[81]. The CA free volume parameters K_{13}/γ and K_{23}-T_{g3}, the pro-exponential factor D_{02} and the parameter ξ_{23} are equal to 5.10^{-7} m^3/kgK, -240 K, 3.610^{-8} m^2/s and , respectively. According to our previous work[37] the diffusion coefficients of acetone in CA predicted by using the above set of free volume parameters for acetone and CA are in excellent agreement with experimental data[53-55]. The critical hole volumes, V_i^*, for acetone, water and CA (39.8% acetyl content) are equal to $9.43.10^{-4}$, $1.071.10^{-4}$ and $6.1.10^{-4}$ m^3/kg according to group contribution methods[82]. This leaves parameters ξ_{13} and D_{01} to be estimated, which are the actual unknowns of the free volume theory. The calculation of ξ_{13}, D_{01} is identical to estimating the ternary diffusion coefficients. In the discussion of the results it is explained in detail how ξ_{13} and D_{01} are obtained by solving the model equations of the problem and utilizing the gravimetric data of available laboratory experiments.

Heat and mass transfer coefficients. For the prediction of heat transfer coefficient, **h** (see Eq. 33), an empirical correlation was utilized describing the heat transfer under free convection conditions to a cooled, horizontal square plate, facing upward[78]:

$$\frac{hw}{K^f} = 0.54(Gr\, Pr)^{0.25} \tag{55}$$

w is the width of the plate and K^f is the thermal conductivity of the gas phase. The superscript "f" denotes properties evaluated at the mean gas phase temperature $T_f = (T_s + T_\infty)/2$ and the mean acetone $x_{af} = (x_{as} + x_{a\infty})/2$ or water vapor mole fraction, $x_{wf} = (x_{ws} + x_{w\infty})/2$, with $x_{a\infty} = 0$, $x_{w\infty} = 0$ and $T_\infty = T_0$. x_{as} and x_{ws} are the acetone and water molar fractions at the gas-liquid interface, while $x_{a\infty}$ and $x_{w\infty}$ are the mole fractions of acetone and water far away from the polymer solution interface.

Gr and **Pr** are the Grashof and the Prandtl number of the gas phase, respectively. These numbers have their standard definitions. In order to evaluate them, the thermal conductivity and the specific heat capacity of the gas phase are calculated from pure substance data and weight fractions assuming ideal gas behavior. Data for the thermophysical properties of acetone and water vapor as well as air is available in the literature[78-80]. The gas phase viscosity is evaluated from pure substance viscosity data using the kinetic theory of gases[83].

Acetone and water mass flux at the gas phase (see Eq. 29-30) were calculated from a multi-component film model assuming pseudo-stationary conditions or equivalently zero air flux. The resulting solution of the Stefan-Maxwell equations describing the multi-component diffusion is given by the following equations[84]:

$$\frac{N_A}{M_2 D_{AC}} + \frac{N_W}{M_1 D_{WC}} = \frac{C_T}{\delta} \ln\left(\frac{1 - x_{w\infty} - x_{a\infty}}{1 - x_{ws} - x_{as}}\right) \tag{56}$$

$$\frac{N_A}{M_2} + \frac{N_W}{M_1} = D_{AW} \frac{C_T}{\delta} \ln\left(\frac{ABx_{w\infty} - Bx_{a\infty} + C}{ABx_{ws} - Bx_{as} + C}\right) \tag{57}$$

$$A = \left(\frac{(1/D_{AW}) - (1/D_{AC})}{(1/D_{AW}) - (1/D_{WC})}\right); \ B = (N_A + N_W)/N_A; \ C = \left(\frac{(1/D_{AC}) - (1/D_{WC})}{(1/D_{AW}) - (1/D_{WC})}\right) \tag{58}$$

D_{AC}, D_{WC} and D_{AW} are the diffusion coefficients of acetone in air, water in air and acetone in water vapor, respectively. Values or alternative methods for their estimation are given in the literature[78] The diffusion coefficients in the gas phase were corrected with respect to the mean gas phase temperature using the correlation of Fuller[85]. C_T is the overall molar concentration and δ is the thickness of the free convection layer. The latter was estimated from the binary acetone-air diffusion coefficient and the respective mass transfer coefficient k_G as follows:

$$\delta = k_G / c_T D_{AC}; \ k_G = 0.54(GrSc)^{0.25}\left(\frac{c_T D_{AC}}{w}\right)\frac{M_2(1 - x_{as})}{Px_{as}}\ln\left(\frac{1}{1 - x_{as}}\right) \tag{59}$$

where P denotes the total pressure of the gas phase and Sc is the Schmidt number. The above mass transfer coefficient k_G, is calculated from the heat transfer coefficient, h, by invoking the heat and mass transfer analogy and using a correction term for high fluxes, obtained from film theory[23].
The acetone, x_{as}, and the water, x_{ws}, mole fraction at the interface, can be written in terms of their activities on the polymer solution side, a_1 and a_2 as:

$$x_{as} = a_1 P_1^{sat} / P \quad ; \quad x_{ws} = a_2 P_2^{sat} / P \tag{60}$$

where P_i^{sat} is the pure substance vapor pressure calculated by Antoine's equation[78] The activities are related to the respective chemical potential in the solution by the following equation:

$$\alpha_i = e^{\frac{\Delta\mu_i}{RT}} \quad ; \quad i = 1,2 \tag{61}$$

where the chemical potentials are calculated from Flory-Huggins theory.
Finally, the emissivity of the ternary solution was equal to 0.8 according to the thermographic measurements of Greenberg et al[74].
Polymer Solution Thermodynamics. The solvent(s) chemical potentials and their derivatives with respect to volume fractions are calculated using Flory-Huggins theory[35,86]. According to this theory the following equations hold:

$$\frac{\Delta\mu_1}{RT} = \ln u_1 - 1 - u_1 - (\upsilon_1/\upsilon_2)u_2 - (\upsilon_1/\upsilon_3)u_3 + (u_2\chi_{12} + u_3\chi_{13})(u_2 + u_3)$$

$$- (\upsilon_1/\upsilon_2)u_2u_3\chi_{13} - h_1h_2u_2\frac{d\chi_{12}}{dh_2} - (\upsilon_1/\upsilon_2)u_2u_3^2\frac{d\chi_{23}}{du_3}$$

$$\frac{\Delta\mu_2}{RT} = \ln u_2 - 1 - u_2 - (\upsilon_2/\upsilon_1)u_1 - (\upsilon_2/\upsilon_3)u_3 + ((\upsilon_2/\upsilon_1)u_1\chi_{12} + u_2\chi_{13})(u_1 + u_3)$$

$$- (\upsilon_1/\upsilon_2)u_1u_3\chi_{13} - (\upsilon_2/\upsilon_1)h_1h_2u_1\frac{d\chi_{12}}{dh_2} - u_2u_3^2\frac{d\chi_{23}}{du_3}$$

$$h_1 = u_2/(u_1 + u_2) \; ; \; h_2 = 1 - h_1 \tag{62}$$

Where subscript 1 refers to water, subscripts 2 and 3 to acetone and CA respectively; μ is the chemical potential, u represents the volume fraction and χ_{ij} are the Flory-Huggins interaction parameters. The Flory-Huggins interaction parameter χ_{12} was calculated from fitting excess energy data for the water-acetone system, χ_{13} was calculated from swelling experiments between water and CA while χ_{23} was obtained from osmotic pressure data.

$$\chi_{12} = 0.661 + \frac{0.417}{1 - 0.755h_2} \; ; \; \chi_{13} = 1.4 \; ; \; \chi_{23} = 0.535 + 0.11u_3 \tag{63}$$

A deeper question arises from the above equations: Is Flory-Huggins theory able to predict the thermodynamics of this complex system exhibiting phase separation phenomena? The answer was given by Smolders and coworkers[87] by comparing the experimental phase separation curve (binodal) as determined by structural analysis with the theoretical one obtained by Flory-Huggins thermodynamics. As can be seen in Figure 3 a satisfactory agreement is obtained between the experimental and the theoretical bimodal for the ternary system Water/Acetone/CA at 20 ^{0}C

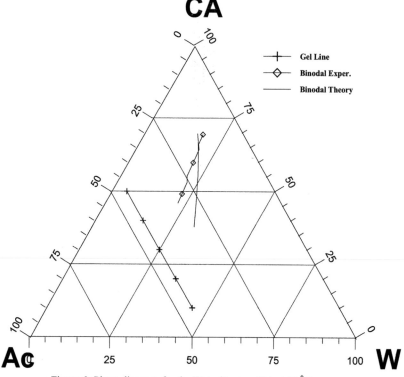

Figure 3: Phase diagram for the Water/Acetone/CA at 20 ^{0}C

Please note that in the above Figure the phase separation curve (binodal) lies inside of the gelation area. This means that the solution if it is left for some time, first exhibits gelation increasing its turbidity, turning to a milky solution as aggregates are formed. After this point crystallites are slowly formed and phase separation occurs due to solid-liquid demixing. The experimental binodal was obtained by rapidly quenching the ternary solution in liquid nitrogen before any aggregation occurs (structural analysis experiments). In this way the experimental binodal occurs due to liquid-liquid demixing and it is in good agreement with the theoretical predictions of Flory-Huggins thermodynamics. In other words, Flory-Huggins thermodynamics is valid until the aggregates formation point as it is determined by the changes in turbidity.

4. Results and Discussion

In order to check the idea for the estimation of diffusion coefficients proposed in this work we chose to compare the numerical results with the experimental work of Shojaie et al[48,72-74]. Shojaie et al. performed, to the best of our knowledge, the most complete experimental study of the evaporation from the water/acetone/CA system, so far. In particular, they measured the polymer solution weight vs. time, they utilized thermographic imaging techniques to measure the polymer solution surface temperature and real time reflectometry to detect the onset of aggregate formation. The on-line turbidity data of Shojaie et al. was also taken into account to detect the one-phase region of the solution by considering their gravimetric & temperature data until the point that turbidity of the solution rapidly increased due to possible crystalline formation causing further phase separation. The experimental set-up is identical to the computational domain in Figure 1. The experimental conditions are given in Table 2.

Table 2: Model Parameters[48,72-74]

Quantity	Value	Units
Initial Temperature,T_0	297 ± 0.5	K
Length of the Glass Support, L_{sup}	1.210^{-3}	m
CA Acetyl Content	39.8%	
CA Molecular Weight	40,000	g/mol
Casting Surface	Calculated from initial conditions	
Experiment I		
Initial Acetone Weight Fraction	0.7	
Initial Water Weight Fraction	0.2	
Initial Length, L_0	266	μm
Experiment II		
Initial Acetone Weight Fraction	0.75	
Initial Water Weight Fraction	0.1	
Initial Length, L_0	195	μm
Experiment III		
Initial Acetone Weight Fraction	0.83	
Initial Water Weight Fraction	0.02	
Initial Length, L_0	246	μm

In the model that we have presented in the previous sections, the unknown parameters of the self-diffusion coefficient correlation are the quantities ξ_{13} and D_{01} (see eq. 19). They are estimated by fitting the model predictions of the polymer solution weight vs. time (evaporation rate) with the corresponding data of Shojaie et al.[72-74] using nonlinear regression analysis.

The calculated values for ξ_{13} and D_{01} are 0.245 and 6.9 10^{-9} m²/s. The estimated value for the water diffusion coefficient at 20 ^{0}C in the limit of pure CA is 1.61 10^{-11} m²/s. The measured diffusion coefficient of water vapor in CA is 1.5 10^{-12}-2.3 10^{-12} m²/s[59-61]. It should be noted that the estimated values of the CA-liquid water in the limit of zero acetone concentration are in the same order of magnitude as the water vapor in cellulose acetate. The observed discrepancy is attributed to the inherent differences between the sorption vapor technique and the present method.

In the estimation procedure for the values of D_{01} and ξ_{13}, only the first two experiments of Table 2 have taken into account, due to the sufficient high water concentration. The maximum evaporation time in each experiment was determined by the onset of phase separation. Figure 4a illustrates the resulting fitting. The satisfactory agreement between experimental data and model predictions completely justifies the assumptions of Fickian diffusion and constant partial volumes.

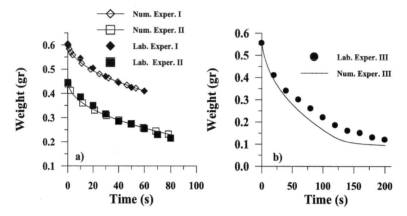

Figure 4: Comparison of model predictions for polymer solution weight with experimental data [48,72-74]. a) Laboratory experiments I-II. b) Laboratory experiment III.

The predictive abilities of the model as well as the accuracy of the model parameters was validated against the polymer solution weight vs. time data of the third experiment of Table 2 and against the variation of the polymer solution surface temperature versus time as shown in Figures 5a and 5b.

In particular, Figure 4b depicts the variation of the of polymer solution weight versus time. In the beginning of the process, where mass transfer is primarily determined by the diffusion in the polymer film, the agreement between numerical and experimental results is excellent. After the first 60s, the evaporation rate decelerates and the process is mainly governed by heat and mass transfer at the polymer solution surface. The discrepancy between numerical predictions and experimental data in that region is mainly due to inaccuracies in the empirical correlation for the determination of heat and mass transfer coefficients. In Figure 5, how the polymer solution surface temperature changes with time is shown. The comparison between numerical versus experimental data shows a discrepancy of the order of 2 ^{0}C in Figures 5a-5b and of the order of 5 ^{0}C in Figure 6 for times greater than 160 s. Again, the explanation for the discrepancy is attributed to the uncertainty of the empirical correlation for heat and mass transfer between polymer solution and ambient air.

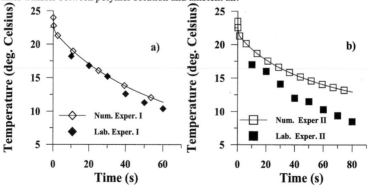

Figure 5: Model predictions for polymer solution surface temperature versus experimental data [48,72-74]. a) Laboratory experiment I. b) Laboratory experiment II.

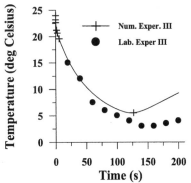

Figure 6: Model predictions for polymer solution surface temperature versus experimental
data[48,72-74]. Laboratory experiment III.

Indeed, by changing the magnitude of heat and mass transfer coefficient (eq 55 and 59) the polymer solution surface temperature changes by 6 ^{0}C as shown in Fig. 7 for the experimental condition of exp.1, which is the observed discrepancy between numerical and experimental data in Fig. 6b. However, the polymer solution weight vs. time data remains almost unaffected from the change in heat and mass transfer coefficients as shown in Figure 8, which explains the better agreement between numerical and experimental results in Fig. 4a. Additionally, the results of Fig. 8, show that the evaporation process is mainly controlled by the diffusion in the polymer solution, which is the limiting step.

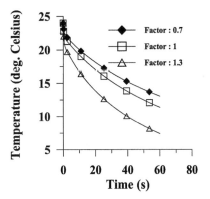

Figure 7: Effect of heat and mass transfer coefficients on polymer solution surface temperature.
Conditions as in experiment I.

Since the major source of error in the model is due to empirical correlations for heat and mass transfer, the fact that their magnitude does not affect the polymer solution weight vs. time data in the selected time period for the estimation procedure, guarantees the reliability of the results of Fig. 4. Possible errors due to the uncertainty in the estimation of the emissivity are negligible as shown elsewhere[37]. Additionally, the computational method employed in this work enables a deep insight into the morphology of the resulting membrane, by studying the concentration profiles of the constituents of the solution. In Fig. 9, how the acetone volume fraction changes with the dimensionless length of the solution at three different time instances is shown. The acetone concentration increases with increasing length up to a certain point very close to the gas liquid interface, where a steep decrease is observed.

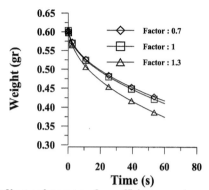

Figure 8: Effect of heat and mass transfer coefficients on polymer solution weight.
Conditions as in experiment I.

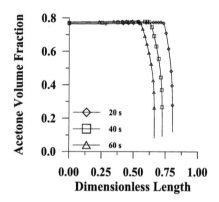

Figure 9: Typical acetone concentration profiles. Conditions as in experiment I.

In Fig. 10, how the water volume fraction varies with dimensionless length of the solution at the same time instances is illustrated. A local minimum and maximum are observed in the water concentration profiles.

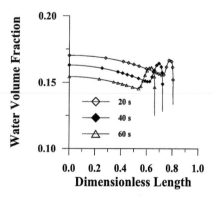

Figure 10: Typical water concentration profiles. Conditions as in experiment I.

The corresponding variation of the CA profiles is given in Fig. 11. Since the fluxes of acetone and water are unidirectional from the bottom up to the interface, the existence of local and global extrema in the concentration profiles are surprising at first glance.

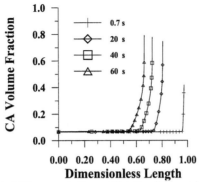

Figure 11: Typical CA concentration profiles. Conditions as in experiment I.

In order to explain this phenomenon, a closer look at the values of the diffusion coefficients should be made. Indeed, Fig. 12 depicts the variation of D_{22} (see eq. 17) with the acetone relative weight fraction defined as $\omega_2/(\omega_2+\omega_3)$ at different water concentrations. It is observed that as the water weight fraction increases, the diffusion coefficient becomes negative and this behavior is enhanced at higher relative acetone weight fractions. Therefore, the increase and the global maximum in the acetone volume fraction in Fig. 9 are attributed to the negative values of ternary diffusion coefficient D_{22}. However, due to its high volatility, acetone evaporates near the surface of the solution and a steep gradient is caused from the global maximum up to the gas liquid interface. As the acetone and water weight fraction decrease, D_{22} becomes positive, as calculated in Fig. 12, and water stops acting as an obstacle in the diffusion of acetone.

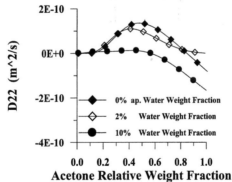

Figure 12: Variation of D_{22} ternary diffusion coefficient with the relative weight fraction of acetone for various water concentrations. Temperature 20 ^{0}C.

Due to the small volatility of water its flux may be considered negligible, so that the gradient of the water concentration is proportional to the negative product of the acetone concentration gradient and the ratio of D_{12}/D_{11} (see eq. 24). For the magnitude of the CA concentration observed in the solution (see Fig 11), the calculated diffusion coefficient D_{11} is positive as shown in Fig 13.

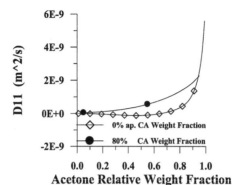

Figure 13: Variation of D_{11} ternary diffusion coefficient with the relative weight fraction of acetone for various CA concentrations. Temperature 20 ^{0}C.

Thus, the existence of the local extrema in the water concentration are caused by the change in sign of diffusion coefficient D_{12} and acetone concentration gradient. The calculated values of D_{12} are shown in Fig 14 for different CA weight fraction.

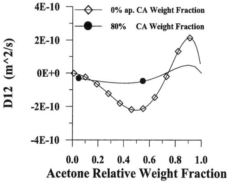

Figure 14: Variation of D_{12} ternary diffusion coefficient with the relative weight fraction of acetone for various CA concentrations. Temperature 20 ^{0}C.

The relative acetone weight fraction is defined in Figures 13 and 14 as $\omega_2/(\omega_1+\omega_2)$. From the bottom of the solution up to the local minimum of water volume fraction both D_{12} and acetone concentration gradient are positive and the concentration decreases up to this point. Then, the concentration gradient of acetone becomes negative and D_{12} is still positive up to an acetone relative weight fraction of 0.75. Thus the water concentration reaches a local maximum and then decreases due to the change in sign of D_{12}. As this change of sign takes place almost at the same concentration of acetone, independently of the CA concentration, the value of the local maximum of water is almost independent of time.

It should b noted here that similar results were shown in our previous work[49]. The only difference is the applied method to calculate ζ_{12}. In this work ζ_{12} is calculated using Gibbs-Duhem equation for diffusing systems, while in our previous work[49] it was estimated from binary diffusion data for the acetone-water system at the limit of zero polymer concentration.

An independent test of the diffusion coefficients that could indicate potential problems would be of great value. Kirkaldy et al.[88] examined the behavior of isothermal, isobaric systems and showed that Onsager consistency and thermodynamic stability require that the matrix of diffusion coefficients have real and positive eigenvalues. For constant diffusion coefficients, this assures that computed concentration profiles will relax nonperiodically. For a ternary system, necessary and sufficient conditions for the diffusion matrix to have real and positive eigenvalues are

$$D_{11}+D_{22}>0,\ D_{11}D_{22}-D_{12}D_{21}\geq 0,\ (D_{11}+D_{22})^2\geq 4(D_{11}D_{22}-D_{12}D_{21}) \qquad (64)$$

These conditions are also satisfied for the diffusion coefficients presented earlier. Regarding the application of isothermal conditions in this system it was proved by several workers[49, 72-74] that the temperature gradients along the solution thickness are very small. Therefore, our system is isothermal regarding space coordinates and the Gibbs-Duhem equation as well as the above equations could be used.

Finally, the concentration profiles of CA shown in Fig. 11 exhibit steep gradients near the gas liquid interface of the solution. This calculation is in accordance with observations of SEM[2]. The region of high CA concentration is known to the membranologists as skin, which is responsible for the separation properties of asymmetric membranes. The presence of water causes this phenomenon, because in the evaporation process of acetone from binary CA-acetone solutions steep profiles have neither been calculated[37] nor observed[71]. Regarding the phase separation phenomena our numerical experiments indicate that the system is outside the liquid-liquid demixing area (see Figure 3) and inside the gelation area. Therefore, phase separation occurs after the increase in turbidity due to possible crystallite formation and liquid-solid demixing.

It should be noted, that the calculated CA concentration profiles justify the ability of the present model to estimate the diffusion coefficients near the pure polymer limit, as the polymer concentration at the solution-air interface becomes very high.

5. Conclusions

We have developed a simple and effective method for estimating ternary diffusion coefficients by studying the non-solvent/solvents evaporation process. The method combines the advantages of both numerical and laboratory experiments. The numerical experiment consists of formulating the mass and energy conservation equations of the evaporation process along with the appropriate boundary and initial conditions and of solving the resulting non-linear system of equations with the finite element method. The laboratory experiment consists of acquiring gravimetric data of the polymer solution weight vs. time. The unknown parameters of the equation based on free volume theory, for the self-diffusion coefficient are determined using non-linear regression analysis. This method is applied for the estimation of diffusion coefficients in the ternary system water/acetone/CA over a wide range of concentration and temperature. The model is validated against polymer solution surface temperature experimental measurements. It was also found that the limiting step of the evaporation process is the diffusion in the solution. Although attempts to solve the model equations of this work have been made before, the employment of the finite element method enables the complete solution of the problem. The calculation of the concentration profiles of the constituents of the polymer solution permit fundamental understanding of the multi-component diffusion mechanisms that occur in membrane formation. Finally, it was shown that the addition of water causes skin formation. The present method may be applied to industrial processes such as asymmetric membrane formation, controlled release and reverse osmosis due to simplicity and accuracy.

Acknowledgments

This work has been sponsored by the Grant Archimedes of the Hellenic Science Foundation. George Verros thanks Ms Kate Somerscales for her help in preparing the manuscript.

References

[1] [S. Loeb, S. Sourirajan, *Adv. Chem. Ser.*, **38**, 117 (1962)
[2] R.L. Riley, J.O. Gardner, V. Merten, *Science*, **143**, 801 (1964)
[3] B. Kunst, S. Sourirajan, *J. Appl. Pol. Sci.* **14**, 1983 (1970)
[4] I. Pinnau,. W.J. Koros, *J. Appl. Pol. Sci.* **43**, 1491 (1991)
[5] R.E. Kesting, Synthetic *Polymeric Membrane A Structural Perspective,* J. Wiley and Sons, NY 1985
[6] D.R. Paul, Y.P. Yampol´skii (Eds), *Polymeric Gas Separation Membranes*, CRC Press, Boca Raton(USA), 1994
[7] D. Stoye, W. Freitag (Eds), *Paints, Coatings and Solvents*, Wiley-VCH, Weinheim (FRG), 1998
[8] R.J. Albalak (Ed.) *Polymer Devolatilization*, Marcel Dekker, New York, 1996
[9] G.T. Russell, D.H Napper, R.G. Gilbert, *Macromolecules* **21(7)**, 2133 (1988)

[10] G.T. Russell, D.H Napper, R.G. Gilbert, *Macromolecules* **21(7)**, 2141 (1988)
[11] D.S Achilias, C. Kiparissides, *Macromolecules* **25(14)**, 3739 (1992)
[12] G.A O'Neil, M.B. Wisnudel, J.M. Torkelson, *Macromolecules* **29(23)**, 7477 (1996).

[13] G.A O'Neil, J.M. Torkelson *Macromolecules* **32(2)**, 411 (1999)
[14] G.D. Verros, T. Latsos, D.S. Achilias, *Polymer* **46**, 539 (2005)
[15] J. Crank, *The Mathematics of Diffusion*, Oxford University Press (1975)
[16] E.L. Cussler, *Multicomponent Diffusion*, Elsevier, (1976)
[17] E.L Cussler, *Diffusion Mass Transfer in Fluid Systems*; Cambridge University Press (1984)
[18] H.J.V. Tyrrell, K.R. Harris, *Diffusion in Liquids. A Theoretical and Experimental Study*,
[19] Butterworths, 1984
[20] S.T. Ju, J.L. Duda, J.S. Vrentas, *Ind. Eng. Chem. Prod. Res. Dev.*, **20**, 330 (1981)
[21] J.M. Zielinski, J.L. Duda, *AIChE J.* **38**, 405 (1992)
[22] J.S. Vrentas, C.M. Vrentas, N. Faridi *Macromolecules*, **29**, 3272 (1996)
[23] L. Onsager, *Proc. Nat. Acad. Sci. USA* **46**, 241 (1945)
[24] R.B. Bird, R. Stewart, E.N. Lightfoot, *Transport Phenomena*, John Wiley & Sons, NY 1960
[25] R.J. Bearman, *J. Phys. Chem.* **65**, 1961 (1961)
[26] C.F. Curtiss, B. Bird, *Proc. Nat. Acad. Sci. USA* **93**, 7440 (1996)
[27] S. Alsoy, J.L. Duda, *AIChE J.* **45**, 896 (1999)
[28] J.M. Zielinski, B.F. Hanley, *AIChE J.* **45(1)**, 1 (1999).
[29] M. Dabral, L.F. Francis, L.E. Scriven *AIChE J.* **48(1)**, 25 (2002)
[30] E.B. Nauman, J. Savoca, *AIChE J.* **47**, 1016 (2001)
[31] P.E. Price, Jr; I.H. Romdhane, , *AIChE J.*, **49**, 313 (2003)
[32] R. J. Bearman, *J. Chem. Phys.* **32**, 1308 (1960)
[33] L. S. Darken, *Trans. Amer. Inst. Min. Met. Eng.* **175**, 184 (1948)
[34] J.S. Vrentas, J.L. Duda, *J. Polym. Sci, Part B: Polym. Phys.* **15**, 403 (1977)
[35] J.S. Vrentas, J.L. Duda, *J. Polym. Sci, Part B: Polym. Phys.* **15**, 417 (1977).
[36] P. J. Flory, *Principles of Polymer Chemistry*, Cornell Univ. Press, Ithaca, NY 1953
[37] S.S. Shojaie, W.B Krantz, A.R. Greenberg, *J. of Mat. Proc & Man. Sci.* **1**, 181 (1992)
[38] G.D. Verros, N.A. Malamataris, *Ind. Eng. Chem. Res.*, **38**, 3572 (1999)
[39] J. M. Zielinski, *J. Poly. Sci.: Part B: Poly. Phys.* **34**, 2759 (1996)
[40] J. M. Zielinski, S. Alsoy, *J. Poly. Sci.: Part B: Pol. Phys.* **39**, 1496 (2001)
[41] D.G. Miller, *J. Phys. Chem, **70**, 2639 (1966)
[42] G.D. Verros, N.A. Malamataris, submitted to *Polymer* (2005).
[43] J.S. Vrentas, J.L. Duda, H.C. Ling, *J. Appl. Pol. Sci.* **30**, 4499 (1985)
[44] J.S.Vrentas, J.L. Duda, H.C. Ling, *J. Polym. Sci. Part B: Polym. Phys.* **22**, 459 (1984)
[45] A.J. Reuvers, J.W.A Van den Berg, C.A. Smolders, *J. Membr. Sci.* **34**, 45 (1987)
[46] C.S. Tsay, A.J. McHugh, *J. Polym. Sci. Part B: Polym. Phys.* **28**, 1327 (1990)
[47] P. Radovanovic, S.W. Thiel, S.T Hwang, *J. Membr. Sci.* **65**, 231 (1992)
[48] L.P. Cheng, Y.S. Soh, A.H. Dwan, C.C. Gryte, *J. Polym. Sci. Part B: Polym. Phys.* **32**, 1413 (1994).
[49] S.S. Shojaie, W.B. Krantz, A.R. Greenberg, *J. Membr. Sci.* **94**, 255 (1994)
[50] G.D. Verros, N.A. Malamataris, *Macromol. Theory Simul.*, **10**, 737 (2001)
[51] M. Ataka, K. Sasaki, *J. Membr. Sci.*, **11**, 11 (1982)
[52] P.E. Price Jr., S. Wang, I.H. Romdhane, *AIChE J.*, **43**, 1925 (1997)
[53] G.D. Verros, N.A. Malamataris, *Chem. Eng. Comm.* **190**, 334 (2003)
[54] G.S. Park, *Trans. Faraday Soc.*, **57**, 2314 (1961)
[55] J.E. Anderson, R. Ullman, *J. Appl. Physics* **44**, 4303 (1973)
[56] M. Sanopoulou, P.P Roussis, J.H. Petropoulos *J. Polym. Sci. Part B: Polym. Phys.* **33**, 993 (1995)
[57] A.J. Reuvers, J.W.A Van den Berg, C.A. Smolders, *J. Membr. Sci.* **34**, 67 (1987)
[58] R.J Kokes., F.A. Long, J.L. Hoard *J. Chem. Phys.*, **20**, 1711 (1952)
[59] S.T. Ju, H.T. Liu, J.L Duda, J.S. Vrentas, *J. Appl. Pol. Sci*, **26**, 3735 (1981)
[60] D. Arnould, J.L. Laurence, *Ind. Eng. Chem. Res.*, **31**, 218 (1992)
[61] A.M. Thomas, *J. Appl. Chem.* **1**, 141 (1951).
[62] P.P. Roussis, *Polymer* **22**, 768 (1981)
[63] P.P. Roussis, *Polymer* **22**, 1058 (1981)
[64] C. Cohen, G.B. Tanny, S. Prager, *J. Polym. Sci, Part B: Polym. Phys.* **17**, 477 (1979)
[65] G.D. Verros, N.A. Malamataris, *Comput. Mech.*, **27**, 332 (2001)

[66] H. Matsuyama, M. Teramoto, T. Uesaka, *J. Membr, Sci.* **154**, 271 (1997)

[67] S.A Alsoy-Altinkaya, B. Ozbas, *J. Membr. Sci.*, *230*, 71 (2004)

[68] P. Murray, G.F. Carey, *J. Comp. Phys.*, **74**, 440 (1988)

[69] P. Hood, *Int. j. Numer. Methods Eng.* **10**, 379 (1974)

[70] S.F. Kistler, L.E. Scriven, In *Computational Analysis of Polymer Processing.* (Eds J.R.A Pearson,. S.M. Richardson), Appl. Sci. Publ.: London, 1983, pp. 243-299

[71] B.A. Finlayson, *Numerical Methods for Problems with Moving Fronts*; Ravenna Park Publishing Inc., Seatle, 1992, p. 257-316

[72] S.B. Tantekin-Ersolmaz, Ph.D. Thesis, University of Colorado, Boulder(USA), 1990

[73] S.S. Shojaie, Ph.D. Thesis, University of Colorado, Boulder (USA), 1992

[74] S.S. Shojaie, W.B. Krantz, A.R. Greenberg, *J. Membr. Sci.* **94**, 281 (1994)

[75] A.R. Greenberg, S.S. Shojaie, W.B. Krantz, S.B. Tantenkin-Ersolmaz, *J. Membr. Sci.* **107**, 249 (1995).

[76] J.S. Vrentas, C.M. Vrentas, *J. Polym. Sci. Part B: Polym. Phys.* **32**, 187 (1994)

[77] R.A. Cairncross, S. Jeyadef, R.F. Dunham, K. Evans, L.F. Francis, L.E. Scriven, *J. Appl. Pol. Sci.* **58**, 1279 (1995)

[78] B. Guerrier, C. Bouchard, C. Allain, C. Benard, *AIChE J.* **44(4)**, 791 (1998)

[79] R.H. Perry, D. Green, *Perry's Chemical Engineers' Handbook*; 6th edn, McGraw-Hill: New York, 1984, Section 3;

[80] J.A. Dean, *Lange's Handbook of Chemistry*; 11th edn, McGraw-Hill: New York, 1973, Sections 9 and 10

[81] Verein Deutscher Ingenieure. *VDI-Warmeatlas;* VDI-Verlag GmbH : Dusseldorf, 1974, p. Db1-Dc27

[82] S. Hong, *Ind. Eng. Chem. Res.*, **34**, 2536 (1995)

[83] R.N. Haward, *J. Macromol. Sci. Rev. Macromol. Chem.*, **C4**, 191 (1970)

[84] C.R. Wilke, *J. Chem. Phys.* **18**, 517 (1950)

[85] R. Taylor, R. Krishna, *Multicomponent Mass Transfer*; Wiley & Sons, 1992, p. 162-165

[86] E.N. Fuller, P.D. Schettler, J.C. Giddings, *Ind. Eng. Chem.* **58(5)**, 18 (1966)

[87] L. Yilmaz, A.J. McHugh, *J. Appl. Pol. Sci.*, **31**, 997 (1986)

[88] A.J. Reuvers, F.W. Altena, C.A. Smolders, *J. Pol. Sci Part B : Pol. Phys.*, **24**, 793 (1986)

[89] J. S. Kirkaldy, D. Weichert, Zia-Ul-Haq, *Can. J. Phys.*, **41**, 2166 (1963)

Brill Academic Publishers
P.O. Box 9000, 2300 PA Leiden,
The Netherlands

Lecture Series on Computer
and Computational Sciences
Volume 3, 2005, pp. 384-398

Parameter Optimization Algorithm with Improved Convergence Properties for Adaptive Learning

G.D. Magoulas[†1], M.N. Vrahatis[‡2]

†School of Computer Science and Information Systems,
Birkbeck University of London, London WC1E 7HX, UK

‡Computational Intelligence Laboratory, Department of Mathematics,
University of Patras, GR-26110 Patras, Greece

Received 10 August, 2004; accepted 12 August, 2004

Abstract: The error in an artificial neural network is a function of adaptive parameters (weights and biases) that needs to be minimized. Research on adaptive learning usually focuses on gradient algorithms that employ problem–dependent heuristic learning parameters. This fact usually results in a trade–off between the convergence speed and the stability of the learning algorithm. The paper investigates gradient–based adaptive algorithms and discusses their limitations. It then describes a new algorithm that does not need user–defined learning parameters. The convergence properties of this method are discussed from both theoretical and practical perspective. The algorithm has been implemented and tested on real life applications exhibiting improved stability and high performance.

Keywords: Feedforward neural networks, Supervised training, Back-propagation algorithm, Heuristic learning parameters, Non-linear Jacobi process, Globally convergent algorithms, Local convergence analysis.

Mathematics Subject Classification: 65K10, 49D10, 68T05, 68G05

1 Introduction

Let us first define the notation we will use in the paper. We use a unified notation for the weights. Thus, for a feedforward neural network (FNN) with a total of n weights, $\mathbb{R}^n$ is the n–dimensional real space of column weight vectors w with components $w_1, w_2, \ldots, w_n$ and w^* is the optimal weight vector with components $w_1^*, w_2^*, \ldots, w_n^*$; E is the batch error measure defined as the sum–of–squared–differences error function over the entire training set; $\partial_i E(w)$ denotes the partial derivative of $E(w)$ with respect to the ith variable w_i; $g(w) = \Big(g_1(w), \ldots, g_n(w)\Big)$ defines the gradient $\nabla E(w)$ of the sum–of–squared-differences error function E at w, which is computed by applying the chain rule on the layers of an FNN (see [48]), while $H = [H_{ij}]$ defines the Hessian $\nabla^2 E(w)$ of E at w. Also, throughout this paper $\mathrm{diag}\{e_1, \ldots, e_n\}$ defines the $n \times n$ diagonal matrix with elements $e_1, \ldots, e_n$, $\Theta^n = (0, 0, \ldots, 0)$ denotes the origin of $\mathbb{R}^n$ and $\rho(A)$ is the spectral radius of matrix A.

The Back-Propagation (BP) algorithm [48] is widely recognized as a powerful tool for training FNNs. It minimizes the error function using the Steepest Descent (SD) method [15] with constant

---

[1]E-mail: gmagoulas@dcs.bbk.ac.uk
[2]E-mail: vrahatis@math.upatras.gr

learning rate η:

$$w^{k+1} = w^k - \eta g(w^k), \qquad k = 0, 1, 2, \ldots \tag{1}$$

The SD method requires the assumption that E is twice continuously differentiable on an open neighborhood $\mathcal{S}(w^0)$, where $\mathcal{S}(w^0) = \{w : E(w) \leqslant E(w^0)\}$ is bounded, for some initial weight vector w^0. It also requires that η is chosen to satisfy the relation $\sup \|H(w)\| \leqslant \eta^{-1} < \infty$ in the level set $\mathcal{S}(w^0)$ [16, 17]. The approach adopted in practice is to apply a small constant learning rate value $(0 < \eta < 1)$ in order to secure the convergence of the BP training algorithm and avoid oscillations in a direction where the error function is steep. However, this approach considerably slows down training since, in general, a small learning rate may not be appropriate for all the portions of the error surface. Furthermore, it affects the convergence properties of training algorithms (see [25, 29]). Nevertheless, there are theoretical results that guarantee convergence when the learning rate is constant. This happens when the learning rate is proportional to the inverse of the Lipschitz constant which, in practice, is not easily available [2, 31].

Attempts to adaptive learning are usually based on the following approaches: (i) start with a small learning rate and increase it exponentially, if successive iterations reduce the error, or rapidly decrease it, if a significant error increase occurs [4, 56], (ii) start with a small learning rate and increase it, if successive iterations keep gradient direction fairly constant, or rapidly decrease it, if the direction of the gradient varies greatly at each iteration [8] and (iii) for each weight an individual learning rate is given, which increases if the successive changes in the weights are in the same direction and decreases otherwise. The well known *delta-bar-delta* method [18] and Silva and Almeida's method [49] follow this approach. Another method, named *quickprop*, has been presented in [11]. Quickprop is based on independent secant steps in the direction of each weight. Riedmiller and Braun in 1993 proposed the *Rprop* algorithm. The algorithm updates the weights using the learning rate and the sign of the partial derivative of the error function with respect to each weight. Note that all these adaptation methods employ heuristic learning parameters in an attempt to secure converge of the BP algorithm to a minimizer of E and avoid oscillations.

A different approach is to exploit the local shape of the error surface as described by the direction cosines or the Lipschitz constant. In the first case the learning rate is a weighted average of the direction cosines of the weight changes at the current and several previous successive iterations [23], while in the second case the learning rate is an approximation of the Lipschitz constant [31].

A variety of approaches adapted from numerical analysis have also been applied, in an attempt to use not only the gradient of the error function but also the second derivative in constructing efficient supervised training algorithms to accelerate the learning process. However, training algorithms that apply nonlinear conjugate gradient methods, such as the Fletcher–Reeves or the Polak–Ribiere methods [34, 53], or variable metric methods, such as the Broyden–Fletcher–Goldfarb–Shanno method [5, 58], or even Newton's method [6, 39], are computationally intensive for FNNs with several hundred weights: derivative calculations as well as subminimization procedures (for the case of nonlinear conjugate gradient methods) and approximations of various matrices (for the case of variable metric and quasi-Newton methods) are required. Furthermore, it is not certain that the extra computational cost speeds up the minimization process for nonconvex functions when far from a minimizer, as is usually the case with the neural network training problem [5, 9, 35]. Thus, the development of improved gradient-based BP algorithms receives significant attention of neural network researchers and practitioners.

The training algorithm introduced in this paper does not use a user–defined initial learning rate, instead it self–determinates the search direction and the learning rates at each epoch. It provides stable learning and robustness to oscillations. The paper is organized as follows. In Section 2 the class of adaptive learning algorithms that employ a different learning rate for each weight is presented and the advantages as well as the disadvantages of these algorithms are discussed. The new algorithm is introduced in Section 3 and its convergence properties are investigated in Section 4.

Experimental results are presented in Section 5 to evaluate and compare the performance of the new algorithms with several other BP methods. The paper ends, in Section 6, with concluding remarks.

2   Adaptive Learning and the Error Surface

The eigensystem of the Hessian matrix can be used to determine the shape of the error function E in the neighborhood of a local minimizer [1, 18]. Thus, studying the sensitivity of the minimizer to small changes by approximating the error function with a quadratic one, it is known that, in a sufficiently small neighborhood of w^*, the directions of the principal axes of the corresponding elliptical contours (n–dimensional ellipsoids) will be given by the eigenvectors of $H(w^*)$, while the lengths of the axes will be inversely proportional to the square roots of the corresponding eigenvalues. Furthermore, a variation along the eigenvector corresponding to the maximum eigenvalue will cause the largest change in E, while the eigenvector corresponding to the minimum eigenvalue gives the least sensitive direction. Therefore, a value for the learning rate which yields a large variation along the eigenvector corresponding to the maximum eigenvalue may result in oscillations. On the other hand, a value for the learning rate which yields a small variation along the eigenvector corresponding to the minimum eigenvalue may result in small steps along this direction and thus, in a slight reduction of the error function. In general, a learning rate appropriate for any one weight direction is not necessarily appropriate for other directions.

Various adaptive learning algorithms with a different learning rate for each weight have been suggested in the literature [11, 18, 33, 41, 46, 49]. This approach allows us to find the proper learning rate that compensates for the small magnitude of the gradient in a flat direction in order to avoid slow convergence, and dampens a large weight change in a steep direction in order to avoid oscillations. Moreover, it exploits the parallelism inherent in the evaluation of $E(w)$ and $g(w)$ by the BP algorithm.

Following this approach Eq. (1) is reformulated to the following scheme:

$$w^{k+1} = w^k - \text{diag}\{\eta_1^k, \dots, \eta_n^k\} g(w^k), \qquad k = 0, 1, 2, \dots \tag{2}$$

The weight vector in Eq. (2) is not updated in the direction of the negative of the gradient; instead, an alternative adaptive search direction is obtained by taking into consideration the weight change, evaluated by multiplying the length of the search step, i.e. the value of the learning rate, along each weight direction by the partial derivative of $E(w)$ with respect to the corresponding weight, i.e. $-\eta_i \partial_i E(w)$. In other words, these algorithms try to decrease the error in each direction, by searching the local minimum with small weight steps. These steps are usually constraint by problem–dependent heuristic parameters, in order to ensure subminimization of the error function in each weight direction.

A well known difficulty of this approach is that the use of inappropriate heuristic values for a weight direction misguides the resultant search direction. In such cases, the training algorithms with an adaptive learning rate for each weight cannot exploit the global information obtained by taking into consideration all the directions. This is the case of many well known training algorithms that employ heuristic parameters for properly tuning the adaptive learning rates [11, 18, 33, 41, 46, 49] and no guarantee is provided that the weight updates will converge to a minimizer of E. In certain cases the aforementioned methods, although originally developed for batch training, can be used for on-line training by minimizing a pattern–based error measure.

3   A Theoretical Derivation of the Adaptive Learning Process

An adaptive learning rate algorithm (see Eq. (2)) seeks for a minimum w^* of the error function and generates with every training epoch a discretized path in the nth dimensional weight space. The limiting value of this path, $\lim_{k \to \infty} w^k$, corresponds to a stationary point of $E(w)$. This path depends on the values of the learning rates chosen in each epoch. Appropriate learning rates help to avoid convergence to a saddle point or a maximum. In the framework of Eq. (2) the learning process can theoretically be interpreted as follows.

Starting from an arbitrary initial weight vector $w^0 \in \mathcal{D}$ (a specific domain), the training algorithm subminimizes, at the kth epoch, in parallel, the n one–dimensional function:

$$E(w_1^k, \ldots, w_{i-1}^k, w_i, w_{i+1}^k, \ldots, w_n^k). \tag{3}$$

First, each function is minimized along the direction i and the corresponding subminimizer $\hat{w}_i$ is obtained. Obviously for this $\hat{w}_i$

$$\partial_i E(w_1^k, \ldots, w_{i-1}^k, \hat{w}_i, w_{i+1}^k, \ldots, w_n^k) = 0. \tag{4}$$

This is a one–dimensional subminimization because all other components of the weight vector, except the ith, are kept constant. Then each weight is updated according to the relation:

$$w_i^{k+1} = \hat{w}_i. \tag{5}$$

In order to be consistent with Eq. (2) only a single iteration of the one–dimensional method in each weight direction is proposed. It is worth noticing that the number of the iterations of the subminimization method is related to the requested accuracy in obtaining the subminimizer approximations. Thus, significant computational effort is needed in order to find very accurate approximations of the subminimizer in each weight direction at each epoch. Moreover, the computational effort for the subminimization method is increased for FNNs with several hundred weights. On the other hand, it is not certain that this large computational effort speeds up the minimization process for nonconvex functions when the algorithm is away from a minimizer w^* [37]. Thus, we propose to obtain $\hat{w}_i$ by minimizing Eq. (3) with one iteration of a minimization method.

The problem of minimizing the error function E along the ith direction

$$\min_{\eta_i \geqslant 0} E(w + \eta_i \, e_i) \tag{6}$$

is equivalent to the minimization of the one–dimensional function

$$\phi_i(\eta) = E(w + \eta_i \, e_i). \tag{7}$$

Since we have n directions we consider n one–dimensional functions $\phi_i(\eta)$. Note that according to experimental work [61] these functions can be approximated for certain learning tasks with quadratic functions in the neighborhood of $\eta \geqslant 0$. In general, we can also use the formulation $\phi(\eta) = E(w + \eta \, d)$, where d is the search direction vector and $\phi'(\eta) = \nabla E(w + \eta \, d)^\top d$.

In our case, we want at the kth epoch to find the learning rate η_i that minimizes $\phi_i(\eta)$ along the ith direction. Since $E(w) \geqslant 0$, $\forall w \in \mathbb{R}^n$ then the w^*, such that $E(w^*)$, minimizes $E(w)$. Thus, by applying one iteration of the Newton method to the one–dimensional equation $\phi_i(\eta) = 0$ we obtain:

$$\eta_i^1 = \eta_i^0 - \frac{\phi_i(\eta^0)}{\phi_i'(\eta^0)}. \tag{8}$$

But $\eta_i^0 = 0$ and $\phi_i'(\eta^0) = g(w)^\top d_i$. Since, $d_i = e_i$ the Eq. (8) is reformulated as

$$\eta_i = -\frac{E(w^k)}{\partial_i E(w^k)}. \tag{9}$$

This is done at the kth epoch in parallel, for all weight directions to evaluate the corresponding learning rates. Then, Eq. (5) takes the form

$$w_i^{k+1} = w_i^k - \frac{E(w^k)}{\partial_i E(w^k)}. \tag{10}$$

Eq. (10) constitutes the weight update formula of the new BP training algorithm with an adaptive learning rate for each weight.

The iterative scheme (10) takes into consideration information from both the error function and the magnitude of the gradient components. When the gradient magnitude is small, the local shape of E in this direction is flat, otherwise it is steep. The value of the error function indicates how close to the global minimizer this local shape is. The above pieces of information help the iterative scheme (10) to escape from flat regions with high error values, which are located far from a desired minimizer.

4   Convergence Analysis

First, we recall two concepts which will be used in our convergence analysis.

1) The Property A^π: Young [60] has discovered a class of matrices described as having *property A* that can be partitioned into block–tridiagonal form, possibly after a suitable permutation. In Young's original presentation, the elements of a matrix $A = [a_{ij}]$ are partitioned into two groups. In general, any partitioning of an n–dimensional vector $x = (x^{(1)}, \ldots, x^{(m)})$ into block components $x^{(p)}$ of dimensions n_p, $p = 1, \ldots, m$ (with $\sum_{p=1}^m n_p = n$) is uniquely determined by a partitioning $\pi = \{\pi_p\}_{p=1}^m$ of the set of the first n integers, where π_p contains the integers $s_p + 1, \ldots, s_p + n_p$, $s_p = \sum_{j=1}^{k-1} n_j$. The same partitioning π also induces a partitioning of any $n \times n$ matrix A into block matrix components A_{ij} of dimensions $n_i \times n_j$. Note that the matrices A_{ii} are square.

Definition 1 [3]: The matrix A has the property A^π if A can be permuted by $PAP^\top$ into a form that can be partitioned into block–tridiagonal form, that is,

$$PAP^\top = \begin{bmatrix} D_1 & L_1^\top & & & & \mathcal{O} \\ L_1 & D_2 & L_2^\top & & & \\ & \ddots & \ddots & \ddots & \\ & & L_{r-2} & D_{r-1} & L_{r-1}^\top \\ \mathcal{O} & & & L_{r-1} & D_r \end{bmatrix},$$

where the matrices D_i, $i = 1, \ldots, r$ are nonsingular.

2) The Root–convergence factor: It is useful for any iterative procedure to have a measure of the rate of its convergence. In our case, we are interested in how fast the weight update equation (10), denoted $\mathcal{P}$, converge to w^*. A measure of the rate of its convergence is obtained by taking appropriate roots of successive errors. To this end we use the following definition.

Definition 2 [37]: Let $\{w^k\}_{k=0}^\infty$ be any sequence that converges to w^*. Then the number

$$R\{w^k\} = \lim_{k \to \infty} \sup \|w^k - w^*\|^{1/k}, \tag{11}$$

is the *root–convergence factor*, or *R–factor* of the sequence of the weights. If the iterative procedure $\mathcal{P}$ converges to w^* and $C(\mathcal{P}, w^*)$ is the set of all sequences generated by $\mathcal{P}$ which convergence to w^*, then

$$R(\mathcal{P}, w^*) = \sup\{R\{w^k\}; \{w^k\} \in C(\mathcal{P}, w^*)\}, \tag{12}$$

is the *R–factor* of $\mathcal{P}$ at w^*.

Our convergence analysis consists of two parts. In first part results regarding the local convergence properties of the algorithm are presented. In the second part appropriate conditions are proposed to guarantee the global convergence of the algorithm, i.e. the convergence to a minimizer from any starting point.

4.1 Local Convergence

In the first part, which concerns the local convergence of the algorithm the objective is to show that there is a neighborhood of a minimizer of the error function for which convergence to the minimizer can be guaranteed. Furthermore, it is interesting to evaluate the asymptotic rate of convergence of the algorithm, i.e. the speed of the algorithm as it convergence to a minimizer. Of course this is not necessarily related to its convergence speed when it is away from the minimizer.

Theorem 1: Let $E : \mathcal{D} \subset \mathbb{R}^n \to \mathbb{R}$ be twice continuously differentiable in an open neighborhood $\mathcal{S}_0 \subset \mathcal{D}$ of a point $w^* \in \mathcal{D}$ for which $\nabla E(w^*) = \Theta^n$ and the Hessian, $H(w^*)$ is positive definite with the property A^π. Then there exists an open ball $\mathcal{S} = \mathcal{S}(w^*, r)$ in $\mathcal{S}_0$ such that any sequence $\{w^k\}_{k=0}^\infty$ generated by the iterative procedure $\mathcal{P}$ (10) converges to w^* which minimizes E and $R(\mathcal{P}, w^*) < 1$.

Proof: Clearly, the necessary and sufficient conditions for the point w^* to be a local minimizer of the function E are satisfied by the hypothesis $\nabla E(w^*) = \Theta^n$ and the assumption of positive definitiveness of the Hessian at w^* (see for example [37]). Finding such a point is equivalent to obtaining the corresponding solution $w^* \in \mathcal{D}$ of the following system:

$$\nabla E(w) = \Theta^n, \tag{13}$$

by applying the nonlinear Jacobi process [37] and employing any one–dimensional method for the subminimization process [15, 37, 42, 43].

Now, consider the decomposition of $H(w^*)$ into its diagonal, strictly lower–triangular and strictly upper–triangular parts

$$H(w^*) = D(w^*) - L(w^*) - L^\top(w^*). \tag{14}$$

Since, $H(w^*)$ is symmetric and positive definite, then $D(w^*)$ is positive definite [55]. Moreover, since $H(w^*)$ has the property A^π, the eigenvalues of

$$\Phi(w^*) = D(w^*)^{-1} \left[L(w^*) + L^\top(w^*) \right], \tag{15}$$

are real and $\rho(\Phi(w^*)) = \varrho < 1$ [3]; then there exists an open ball $\mathcal{S} = \mathcal{S}(w^*, r)$ in $\mathcal{S}_0$, such that, for any initial weight vector $w^0 \in \mathcal{S}$, there is a sequence $\{w^k\}_{k=0}^\infty \subset \mathcal{S}$ which satisfies the method (10) such that $\lim_{k \to \infty} w^k = w^*$ and $R(\mathcal{P}, w^*) = \varrho < 1$ [37, 44, 57]. Thus the Theorem is proved.

The proof of Theorem 1 shows that the asymptotic rate of convergence of the algorithm is not enhanced if one takes more than one iteration of the one–dimensional minimization method. Instead, the asymptotic rate of convergence only depends on the spectral radius ρ. Note also that this theorem is applicable to any training algorithm that adapts a different learning rate for each weight using one–dimensional subminimization methods and does not employ heuristics.

The local convergence analysis of the new algorithm was developed under appropriate assumptions and provides useful insight into the new method. However, in practice, neural network users want a guarantee that the training algorithm will reduce the error at each epoch and that the error will not fluctuate. Particularly, neural network practitioners are interested in techniques will satisfy the above mentioned requirements when the initial weights are far from the neighborhood of a minimizer. Unfortunately, it is well known that far from the neighborhood of a minimizer, the error function has broad flat regions adjoined with narrow steep ones. It is possible that this kind of morphology of the error function will cause the iterative scheme (10) to create very large learning rates, due to the small values of the denominator, pushing the neurons into saturation and thus the algorithm will exhibit pathological convergence behavior.

Therefore we want that the iterative scheme (10) will generate weight iterates that achieve a sufficient reduction in the error function at each epoch. Only these weight iterates will be considered to be accepted. Searching for an acceptable weight vector rather than a minimizer along the current search direction usually reduces the number of function evaluations per epoch and the same goes for the total number of function evaluations required to successfully train the FNN. This is due to the fact that training starts way from a local minimum of the error function and exact minimization steps along the search direction do not usually help, because of the nonlinearity of the error function. On the other hand, when the current iterate w^k is close to the minimizer a "better" approximator of the minimizer w^{k+1} can be found without much difficulty. This issues are investigated below in the framework of global convergence of the algorithm.

4.2 Globally Convergent Algorithms

In order to ensure global convergence of the adaptive algorithm, i.e. convergence to a local minimizer of the error function from any starting point, the following assumptions are needed [10, 24]:

a) *The error function E is a real–valued function defined and continuous everywhere in $\mathbb{R}^n$, bounded below in $\mathbb{R}^n$,*

b) *for any two points w and $v \in \mathbb{R}^n$, ∇E satisfies the Lipschitz condition*

$$\|\nabla E(w) - \nabla E(v)\| \leqslant L\|w - v\|, \tag{16}$$

where $L > 0$ denotes the Lipschitz constant.

The effect of the above assumptions is to place an upper bound on the degree of the nonlinearity of the error function and to ensure that the first derivatives are continuous in w. If these assumptions are fulfilled the algorithm can be made globally convergent by determining the learning rates in such a way that the error function is exactly minimized along the current search direction at each epoch. To this end an iterative search, which is often expensive in terms of error function evaluations, is required. To alleviate this situation it is preferable to determine the learning rates so that the error function is sufficiently decreased at each epoch, accompanied by a significant change in the value of w.

The following conditions, associated with the names of Armijo, Goldstein and Price [37], are used to formulate the above ideas and to define a criterion of acceptance of any weight iterate:

$$E(w^k + \lambda^k \eta^k) - E(w^k) \leqslant \sigma_1 \lambda^k \nabla E(w^k)^\top \eta^k, \tag{17}$$

$$\nabla E(w^k + \lambda^k \eta^k)^\top \eta^k \geqslant \sigma_2 \nabla E(w^k)^\top \eta^k, \tag{18}$$

where $0 < \sigma_1 < \sigma_2 < 1$ and η^k is the column vector $(\eta_1^k, \ldots, \eta_n^k)$. Thus, by using appropriate values for the learning rates we seek to satisfy Conditions (17)–(18): the first condition ensures that using λ^k, the error function is reduced with every epoch of the algorithm and the second condition prevents λ^k, from becoming too small. Furthermore, these conditions can be used to enhance the new training algorithm with tuning techniques that are able to handle arbitrary large learning rates. Note that the value $\sigma_1 = 0.5$ is usually suggested in the literature [2, 35].

A simple technique to tune the length of the minimization step, so that it satisfies Conditions (17)–(18) at each epoch, is to decrease the learning rates by a reduction factor $1/q$, where $q > 1$ [37].

This has the effect that λ^k is decreased by the largest number in the sequence $\{q^{-m}\}_{m=1}^{\infty}$, so that the Condition (17) is satisfied. We remark here that the selection of q is not critical for successful learning, however it has an influence on the number of error function evaluations required to obtain an acceptable weight vector. Thus, some training problems respond well to one or two reductions in the learning rates by modest amounts (such as $1/2$) and others require many such reductions, but might respond well to a more aggressive learning rate reduction (for example by factors of $1/10$, or even $1/20$). On the other hand, reducing λ^k too much can be costly since the total number of epochs will be increased. Consequently, when seeking to satisfy the Condition (17) it is important to ensure that the learning rates are not reduced unnecessarily so that the Condition (18) is not satisfied. Since, in the algorithm, the gradient vector is known only at the beginning of the iterative search for an acceptable weight vector, the Condition (18) cannot be checked directly (this task requires additional gradient evaluations at each epoch of the training algorithm), but is enforced simply by placing a lower bound on the acceptable values of λ^k. This bound on the learning rates has the same theoretical effect as Condition (18) and ensures global convergence [10, 52].

Another approach to perform learning rates reduction is to estimate the appropriate reduction factor at each iteration. This is achieved by modeling the decrease in the magnitude of the gradient vector as the learning rates are reduced. To this end, quadratic and cubic interpolations are suggested that exploit the available information about the error function. Relative techniques have been proposed by Dennis and Schnabel [10] and Battiti [4].

5   Applications

In this section we give comparative results for five batch training algorithms in 1000 simulation runs: Back-propagation with constant learning rate (BP); Back-propagation with constant learning rate and constant Momentum [48], named BPM ; Adaptive Back-propagation with adaptive momentum (ABP) proposed by Vogl [56]; Back-propagation with an adaptive learning rate for each weight proposed by Silva and Almeida [49], named SA; Back-propagation with a self–determined learning rate for each weight (BPS).

The selection of initial weights is very important in FNN training [59]. A well known initialization heuristic for FNNs is to select the weights with uniform probability from an interval (w_{min}, w_{max}), where usually $w_{min} = -w_{max}$. However, if the initial weights are very small the backpropagated error is so small that practically no change takes place for some weights and more iterations are necessary to decrease the error [47, 48]. In the worst case the error remains constant and the learning stops in an undesired local minimum [27]. On the other hand, very large values of weights speed up learning but they can lead to saturation and to flat regions of the error surface where training is considerably slow [28, 30, 47].

A common choice for the range of the initial weights is the interval $(-1, +1)$ (see [21, 22, 40, 46]). This interval has been used to randomly choose initial weights for our experiments. The values of the heuristic learning parameters used in each problem are shown in Table 1. The initial learning rate has been carefully chosen so that the BP algorithm indicates rapid convergence without oscillating towards a global minimum and has been kept the same for all the algorithms tested. Then all other heuristics have been tuned. For the momentum term m we have tried 9 different values ranging from 0.9 to 0.1. Much effort has been made to properly tune the learning rate increment and decrement factors, η_{inc}, u and η_{dec}, d, respectively. To be more specific, various different values, up to 2, have been tested for the learning rate increment factor, while different values between 0.1 and 0.9 have been tried for the learning rate decrement factor. The error ratio parameter, denoted *ratio* in Table 1, has been set equal to 1.04. This value is generally suggested in the literature [56] and indeed it has been found to work better than others tested. All the combinations of these parameter values have been tested on 3 simulation runs, starting from the

Table 1: Learning parameters used in the experiments

Algorithm	Texture classification	Vowel spotting
BP	$\eta_0 = 0.001$	$\eta_0 = 0.0034$
BPM	$\eta_0 = 0.001$	$\eta_0 = 0.0034$
	$m = 0.9$	$m = 0.7$
ABP	$\eta_0 = 0.001$	$\eta_0 = 0.0034$
	$m = 0.9$	$m = 0.1$
	$\eta_{inc} = 1.05$	$\eta_{inc} = 1.07$
	$\eta_{dec} = 0.5$	$\eta_{dec} = 0.8$
	$ratio = 1.04$	$ratio = 1.04$
SA	$\eta_0 = 0.001$	$\eta_0 = 0.0034$
	$u = 1.05$	$u = 1.3$
	$d = 0.6$	$d = 0.7$
BPS	*	*

* no heuristics required

same initial weights, to find the best available in terms of accelarated training.

A consideration that is worth mentioning is the difference between gradient and error function evaluations at each epoch: for the BP, the BPM, the ABP and the SA one gradient evaluation and one error function evaluation are necessary at each epoch; for BPS there is a number of additional error function evaluations when the Goldstein/Armijo condition (17) is not satisfied. Thus, we compare the algorithms in terms of both gradient and error function evaluations. However, the reader has to consider the fact that a gradient evaluation is more costly than an error function evaluation (for example Møller [34] suggests to count a gradient evaluation three times more than an error function evaluation).

5.1 Texture Classification Problem

The first experiment is a texture classification problem. A total of 12 Brodatz texture images [7]: 3, 5, 9, 12, 15, 20, 51, 68, 77, 78, 79, 93 (see Figure 1 in [31]) of size 512×512 is acquired by a scanner at 150dpi. From each texture image 10 subimages of size 128×128 are randomly selected, and the co–occurence method, introduced by Haralick [20] is applied. In the co–occurence method, the relative frequencies of gray–level pairs of pixels at certain relative displacements are computed and stored in a matrix. As suggested by other researchers [36, 54], the combination of the nearest neighbor pairs at orientations 0°, 45°, 90° and 135° are used in the experiment. 10 sixteenth-dimensional training patterns are created from each image. A 16–8–12 FNN (224 weights, 20 biases) with sigmoid activations is trained to classify the patterns to 12 texture types. The termination condition is a classification error $CE < 3\%$.

Detailed results regarding the training performance of the algorithms are presented in Table 2, where μ denotes the mean number of gradient or error function evaluations required to obtain convergence, σ the corresponding standard deviation, Min/Max the minimum and maximum number of gradient or error function evaluations, and % denotes the percentage of simulations that converge to a global minimum. Obviously, the number of gradient evaluations is equal to the number of error function evaluations for the BP, the BPM, the ABP and the SA.

The results of Table 2 suggest that BPS significantly outperforms BP and BPM in the number of gradient and error function evaluations as well as in the percentage of successful simulations. ABP exhibits the best performance of all methods tested, however it requires fine tuning five

Table 2: Comparative results for the texture classification problem

Algorithm	Gradient Evaluation			Function Evaluation			Success
	μ	σ	Min/Max	μ	σ	Min/Max	%
BP	15839	3723.3	8271/25749	15839	3723.3	8271/25749	96
BPM	12422	2912.1	4182/18756	12422	2912.1	4182/18756	94
ABP	560	270.4	310/2052	560	270.4	310/2052	100
SA	704	2112.6	82/18750	704	2112.6	82/18750	88
BPS	791	512.3	417/5318	2185	1405.9	1116/14006	100

heuristic parameters. SA is also faster than BPS needing only tuning three heuristic parameters, but has smaller percentage of success than BPS.

The successfully trained FNNs are tested for their generalization capability using patterns from 20 subimages of the same size randomly selected from each image. To evaluate the generalization performance of the FNN the *max* rule is used, i.e. a test pattern is considered to be correctly classified if the corresponding output neuron has the greatest value among the output neurons. The average success rate classification for each algorithm was: BP=90%, BPM=90%, ABP=93.5%, SA=93.6%, BPS=93%. These results confirm that increased training speed achieved by ABP, SA and BPS affect by no means their generalization capability.

5.2 Vowel Spotting Problem

In the second experiment a 15–15–1 FNN (240 weights and 16 biases), based on neurons of hyperbolic tangent activations, is used for vowel spotting. Vowel spotting provides a preliminary acoustic labeling of speech, which can be very important for both speech and speaker recognition procedures. The speech signal, coming from a high quality microphone in a very quiet environment, is recorded, sampled at 16KHz and digitized at 16–bit precision. The sampled speech data is then segmented into 30ms frames with a 15ms sliding window in overlapping mode. After applying a Hamming window, each frame is analyzed using the Perceptual Linear Predictive (PLP) speech analysis technique to obtain the characteristic features of the signal. The choice of the proper features is based on a comparative study of several speech parameters for speaker independent speech recognition and speaker recognition purposes [50]. The PLP analysis includes: spectral analysis, critical–band spectral resolution, equal–loudness pre–emphasis, intensity–loudness power law, autoregressive modeling. It results in a 15th dimensional feature vector for each frame.

The FNN is trained as speaker independent using labelled training data from a large number of speakers from the TIMIT database [14] and classifies the feature vectors into $\{-1, +1\}$ for the non–vowel/vowel model. The network is part of a text-independent speaker identification and verification system which is based on using only the vowel part of the signal [13].

The fact that the system uses only the vowel part of the signal makes the cost of mistakenly accepting a non-vowel and considering it as a vowel much more than the cost of rejecting a vowel and considering it as non-vowel. A mistaken decision regarding a non-vowel will produce unpredictable errors to the speaker classification module of the system that uses the response of the FNN and is trained only with vowels [12, 13].

Thus, in order to minimize the false acceptance error rate which is more critical than the false rejection error rate we polarize the training procedure by taking 317 non–vowel patterns and 43 vowel patterns. The training terminates when the classification error is less than 2%. After training, the generalization capability of the successfully trained FNNs is examined with 769 feature vectors taken from different utterances and speakers. In this examination, a small set of rules is used. These rules are based on the principle that the cost of rejecting a vowel is much less

Table 3: Comparative results for the vowel spotting problem

Algorithm	Gradient Evaluation			Function Evaluation			Success
	μ	σ	Min/Max	μ	σ	Min/Max	%
BP	905	1067.5	393/6686	905	1067.5	393/6686	63
BPM	802	1852.2	381/9881	802	1852.2	381/9881	57
ABP	1146	1374.4	302/6559	1146	1374.4	302/6559	73
SA	250	157.5	118/951	250	157.5	118/951	36
BPS	362	358.5	96/1580	1756	1692.1	459/7396	64

than the cost of mistakenly accepting a non-vowel and considering it as a vowel and concern the distance, the duration and the amplitude of the responses of the FNN, [12, 13, 51]. The results of the training phase are shown in Table 3, where the abbreviations are as in Table 2.

The results of Table 3 show that BPS compares favorably on the number of gradient evaluations to BP, BPM and ABP. ABP exhibits the slowest convergence but has the most reliable performance regarding the number of successful simulations. Note that SA provides the fastest training, but also exhibits the worst percentage of success. On the other hand, it has the best performance regarding the generalization performance. BPS also improves the error rate achieved by the BP by 1%. Thus, the average performance of the BPS is satisfactory.

With regards to generalization, the adaptive methods provide comparable performance which on the the average improves of the error rate percentage achieved by BP–trained FNNs by 2%. However, BPM trained FNNs did not provide any improvement.

6 Conclusions

The papers analyzed adaptive learning with a different learning rate for each weight as a nonlinear Jacobi process. Along this line an algorithm with a self-determined learning rate for each weight has been presented. A model for the analysis of the local and global convergence of the algorithm has been proposed. It was shown that the asymptotic rate of convergence of the training algorithm does not depend on the multi–step subminimization methods. However this result is not necessarily related with the convergence speed when the algorithm is away from the minimizer. With regards to global convergence, the Goldstein/Armijo conditions have been appropriately adapted to secure the convergence of the algorithm.

The algorithm compares satisfactory with other popular training algorithms without using highly problem–dependent heuristic parameters. Its performance in the two reported experiments is promising.

In a future contribution we will focus on the effects of property A^π in FNN training as well as on the issue of global efficiency, or global rate of convergence, of the training algorithm, which is related to the estimation of the error function reduction at every epoch.

References

[1] G. S. Androulakis, G. D. Magoulas and M. N. Vrahatis, Geometry of learning: visualizing the performance of neural network supervised training methods, *Nonlinear Analysis,Theory, Methods and Applications*, 30 (1997), 4539–4544.

[2] L. Armijo, Minimization of functions having Lipschitz continuous first partial derivatives, *Pacific Journal of Mathematics*, 16 (1966), 1–3.

[3] O. Axelsson, *Iterative Solution Methods*. Cambridge University Press, New York, 1996.

[4] R. Battiti, Accelerated backpropagation learning: two optimization methods, *Complex Systems*, 3 (1989), 331–342.

[5] R. Battiti, First- and second-order methods for learning: between steepest descent and Newton's method, *Neural Computation*, 4 (1992), 141–166.

[6] S. Becker and Y. Le Cun, Improving the convergence of the back–propagation learning with second order methods. In *Proceedings of the 1988 Connectionist Models Summer School*, D.S. Touretzky, G.E. Hinton, and T.J. Sejnowski eds., 29-37, Morgan Koufmann, San Mateo, CA, 1988.

[7] P. Brodatz, *Textures - a photographic album for artists abd designers*, Dover, NY, 1966.

[8] L. W. Chan and F. Fallside, An adaptive training algorithm for back–propagation networks, *Computers Speech and Language*, 2 (1987), 205–218.

[9] J. E. Dennis and J. J. Moré, Quasi-Newton methods, motivation and theory, *SIAM Review*, 19 (1977), 46–89.

[10] J. E. Dennis and R. B. Schnabel, *Numerical Methods for Unconstrained Optimization and nonlinear equations*, Prentice-Hall, Englewood Cliffs, NJ, 1983.

[11] S. E. Fahlman, Faster-learning variations on back–propagation: an empirical study, *Proceedings of the 1988 Connectionist Models Summer School*, D.S. Touretzky, G.E. Hinton, and T.J. Sejnowski, eds., 38–51, Morgan Kaufmann, San Mateo, CA, 1989.

[12] N. Fakotakis and J. Sirigos, A high-performance text-independent speaker recognition system based on vowel spotting and neural nets, *Proceedings of the IEEE International Conference on Acoustic Speech and Signal Processing*, 2, 661–664, 1996.

[13] N. Fakotakis and J. Sirigos, A high-performance text-independent speaker identification and verification system based on vowel spotting and neural nets. To appear in IEEE Trans. Speech and Audio proccooing.

[14] W. Fisher, V. Zue, J. Bernstein and D. Pallet, An acoustic-phonetic data base, *Journal of Acoustical Society of America*, Suppl. A, 81 (1987), 581–592.

[15] P. E. Gill, W. Murray and M. H. Wright, *Practical Optimization*, Academic Press, NY., 1981.

[16] A. A. Goldstein, Cauchy's method of minimization, *Numerische Mathematik*, 4 (1962), 146–150.

[17] A.A. Goldstein, On steepest descent, *SIAM J. Control*, 3 (1965), 147–151.

[18] R. A. Jacobs, Increased rates of convergence through learning rate adaptation, *Neural Networks*, 1 (1988), 295-307.

[19] E. M. Johansson, F. U. Dowala, and D. M. Goodman, Back–propagation learning for multilayer feed forward neural networks using the conjugate gradient method, *Int. J. of Neural Systems*, 2 (1991), 291–302.

[20] R. Haralick, K. Shanmugan and I. Dinstein, Textural features for image classification, *IEEE Trans. System, Man and Cybernetics*, 3 (1973), 610-621.

[21] Y. Hirose, K. Yamashita and S. Hijiya, Back–propagation algorithm which varies the number of hidden units, *Neural Networks*, 4 (1991), 61–66.

[22] M. Hoehfeld and S. E. Fahlman, Learning with limited numerical precision using the cascade-correlation algorithm, *IEEE Trans. on Neural Networks*, 3 (1992), 602–611.

[23] H.-C. Hsin, C.-C. Li, M. Sun and R. J. Sclabassi, An adaptive training algorithm for back–propagation neural networks, *IEEE Transactions on System, Man and Cybernetics*, 25 (1995), 512–514.

[24] C.T. Kelley, *Iterative methods for linear and nonlinear equations*, SIAM publications, Philadelphia, 1995.

[25] C.M. Kuan and K. Hornik, Convergence of learning algorithms with constant learning rates, *IEEE Trans. Neural Networks*, 2 (1991), 484–488.

[26] S. Y. Kung, K. Diamantaras, W. D. Mao, J. S. Taur, Generalized perceptron networks with nonlinear discriminant functions, *Neural Networks Theory and Applications*, Mammone R.J. and Zeevi Y.Y., eds., 245–279, Academic Press, Boston, 1991.

[27] Y. Lee, S.-H. Oh and M. W. Kim, An analysis of premature saturation in backpropagation learning, *Neural Networks*, 6 (1993), 719–728.

[28] P.J.G. Lisboa and S. J. Perantonis, Complete solution of the local minima in the XOR problem, *Network*, 2 (1991), 119–124.

[29] R. Liu, G. Dong and X. Ling, A convergence analysis for neural networks with constant learning rates and non–stationary inputs, *Proceedings of the 34th Conf. on Decision and Control*, New Orleans, 1278-1283, 1995.

[30] G. D. Magoulas, M. N. Vrahatis, and G. S. Androulakis, A new method in neural network supervised training with imprecision, *Proceedings of the IEEE 3rd International Conference on Electronics, Circuits and Systems*, 287–290, 1996.

[31] G. D. Magoulas, M. N. Vrahatis and G. S. Androulakis, Effective back–propagation with variable stepsize, *Neural Networks*, 10 (1997), 69–82.

[32] G. D. Magoulas, M. N. Vrahatis, T. N. Grapsa, and G. S. Androulakis, Neural network supervised training based on a dimension reducing method, S. W. Ellacot, J. C. Mason and I. J. Anderson, eds., *Mathematics of Neural Networks: Models, Algorithms and Applications*, 245–249, Kluwer, 1997.

[33] A.A. Minai and R. D. Williams, Back-propagation heuristics: a study of the extended delta-bar-delta algorithm, *Proceedings of the IEEE Int. Joint Conf. on Neural Networks*, San Diego, 1990, 595-600.

[34] M. F. Møller, A scaled conjugate gradient algorithm, for fast supervised learning, *Neural Networks*, 6 (1993), 525–533.

[35] J. Nocedal, Theory of algorithms for unconstrained optimization, *Acta Numerica*, 199–242, 1991.

[36] P. P. Ohanian and R. C. Dubes, Performance evaluation for four classes of textural features, *Pattern Recognition*, 25, 8 (1992), 819–833.

[37] J.M. Ortega and W. C. Rheinboldt, *Iterative Solution of Nonlinear Equations in Several Variables*, Academic Press, NY, 1970.

[38] A.G. Parlos, B. Fernandez, A. F. Atiya, J. Muthusami and W. K. Tsai, An accelerated learning algorithm for multilayer perceptron networks, *IEEE Trans. on Neural Networks*, 5 (1994), 493–497.

[39] D. B. Parker, Optimal Algorithms for Adaptive Networks: Second Order Back–Propagation, Second Order Direct Propagation, and Second Order Hebbian Learning, *Proceedings of the IEEE International Conference on Neural Networks*, 2, 593–600, 1987.

[40] B. Pearlmutter, Gradient descent: second–order momentum and saturating error, *Advances in Neural Information Processing Systems 4*, J.E. Moody, S.J. Hanson, and R.P. Lippmann, eds., 887–894, Morgan Kaufmann, San Mateo, CA, 1992.

[41] M. Pfister and R. Rojas, Speeding-up backpropagation -A comparison of orthogonal techniques, *Proceedings of the Joint Conference on Neural Networks*, Nagoya, Japan, 517–523, 1993.

[42] W. H. Press, S. A. Teukolsky, W. T. Vetterling and B. F. Flannery, *Numerical recipes in C*, Cambridge University Press, New York, 1992.

[43] A. Ralston and P. Rabinowitz, *A first course in numerical analysis*, McGraw-Hill, New York, 1978.

[44] W.C. Reinboldt, *Methods for solving systems of nonlinear equations*, SIAM Publications, Philadelphia, Pennsylvania, 1974.

[45] M. Riedmiller and H. Braun, A direct adaptive method for faster backpropagation learning: the Rprop algorithm, *Proceedings of the IEEE International Conference on Neural Networks*, San Francisco, CA, 586–591, 1993.

[46] M. Riedmiller, Advanced supervised learning in multi-layer perceptrons - From backpropagation to adaptive learning algorithms, *International Journal of Computer Standards and Interfaces*, Special issue on Neural Networks, 5, 1994.

[47] A. K. Rigler, J. M. Irvine and T. P. Vogl, Rescaling of variables in backpropagation learning, *Neural Networks*, 4 (1991), 225–229.

[48] D. E. Rumelhart, G. E. Hinton and R. J. Williams, Learning Internal Representations by Error Propagation, D. E. Rumelhart, J. L., and McClelland, eds., *Parallel Distributed Processing: Explorations in the Microstructure of Cognition*, 1, pp. 318–362. MIT Press, Cambridge, Massachusetts, 1986.

[49] F. Silva and L. Almeida, Acceleration techniques for the back–propagation algorithm, *Lecture Notes in Computer Science*, 412, 110–119. Springer-Verlag,Berlin, 1990.

[50] J. Sirigos, N. Fakotakis and G. Kokkinakis, A comparison of several speech parameters for speaker independent speech recognition and speaker recognition, *Proceedings of the 4th European Conference of Speech Communications and Technology*, 1995.

[51] J. Sirigos, V. Darsinos, N. Fakotakis and G. Kokkinakis, Vowel/non-vowel decision using neural networks and rules, *Proceedings of the 3rd IEEE International Conference on Electronics, Circuits, and Systems*, 510–513, 1996.

[52] G. A. Shultz, R. B. Schnabel and R. H. Byrd, A family of trust region based algorithms for unconstrained minimization with strong global convergence properties, University of Colorando, Computer Science TR CU-CS-216-82, 1982.

[53] P. P. Van der Smagt, Minimization Methods for training feedforward neural networks, *Neural Networks*, 7 (1994), 1–11.

[54] J. Strang and T. Taxt, Local frequency features for texture classification, *Pattern Recognition*, 27, 10 (1994), 1397–1406.

[55] R. Varga, *Matrix Iterative Analysis*, Prentice-Hall, Inc., Englewood Cliffs, New Jersey, 1962.

[56] T. P. Vogl, J. K. Mangis, J. K. Rigler, W. T. Zink and D. L. Alkon, Accelerating the convergence of the back-propagation method, *Biological Cybernetics*, 59 (1988), 257–263.

[57] R. G. Voigt, Rates of convergence for a class of iterative procedures, *SIAM J. Numer. Anal.*, 8 (1971), 127–134.

[58] R. L. Watrous, Learning Algorithms for Connectionist Networks: Applied Gradient of Nonlinear Optimization, *Proceedings of the IEEE International Conference on Neural Networks*, 2, 619–627, 1987.

[59] L. F. Wessel and E. Barnard, Avoiding false local minima by proper initialization of connections, *IEEE Trans. Neural Networks*, 3 (1992), 899–905.

[60] D. Young, Iterative methods for solving partial difference equations of elliptic type, *Trans. American Mathematical Society*, 76 (1954), 92–111.

[61] X.-H. Yu, G.-A. Chen and S.-X. Cheng, Dynamic learning rate optimization of the backpropagation algorithm, *IEEE Trans. Neural Networks*, 6 (1995), 669–677.